W9-CXL-616

The Synaptic Organization of the Brain

for Barry Blumberg
from Gert Stephan
May 14, 1963

THE SYNAPTIC ORGANIZATION OF THE BRAIN

THIRD EDITION

Edited by

GORDON M. SHEPHERD

PROFESSOR OF NEUROSCIENCE
YALE UNIVERSITY SCHOOL OF MEDICINE

New York   Oxford
OXFORD UNIVERSITY PRESS
1990

Oxford University Press

Oxford New York Toronto
Delhi Bombay Calcutta Madras Karachi
Petaling Jaya Singapore Hong Kong Tokyo
Nairobi Dar es Salaam Cape Town
Melbourne Auckland

and associated companies in
Berlin Ibadan

Copyright © 1974, 1979, 1990 by Oxford University Press, Inc.

Published by Oxford University Press, Inc.,
200 Madison Avenue, New York, New York 10016

Oxford is a registered trademark of Oxford University Press

All rights reserved. No part of this publication may be reproduced,
stored in a retrieval system, or transmitted, in any form or by any means,
electronic, mechanical, photocopying, recording, or otherwise,
without the prior permission of Oxford University Press.

Library of Congress Cataloging-in-Publication Data
The Synaptic organization of the brain
edited by Gordon M. Shepherd.
— 3rd ed. p. cm.
Rev. ed. of:
The synaptic organization of the brain
Gordon M. Shepherd. 2nd ed. 1979.
Bibliography: p. ISBN 0-19-506255-8 —ISBN 0-19-506256-6 (pbk.)
1. Brain. 2. Synapses. 3. Neural circuitry.
I. Shepherd, Gordon M., 1933– .
II. Shepherd, Gordon M., 1933–
Synaptic organization of the brain.
[DNLM: 1. Brain—physiology. 2. Neurons—physiology.
3. Synapses. WL 300 S992] QP376.S9 1990
612.8'2—dc20 DNLM/DLC for Library of Congress
89-16167 CIP

2 4 6 8 9 7 5 3 1

Printed in the United States of America
on acid-free paper

Preface

The explosion of knowledge in neuroscience has brought with it an increasing need to identify organizing principles. A body of such principles has begun to emerge at the level of synaptic organization, which is the subject of this book.

Synaptic organization may be defined as the study of principles underlying the organization of neurons and synapses into circuits that mediate the functional operations of different brain regions. It is a multidisciplinary subject, requiring the integration of results from experimental studies in molecular neurobiology, neuroanatomy, neurophysiology, neurochemistry, neuropharmacology, development and behavior, as well as theoretical studies of computational neuronal models and neural networks. It is also a multilevel subject, beginning with the properties of the individual synapse and building up through microcircuits and neurons to the local circuits characteristic of a given region.

Previous editions of this book provided introductory accounts to this subject for a number of the best understood brain regions. The aim was to organize the knowledge about a given region in terms of basic principles, whose generality was then tested by making explicit comparisons between regions. This approach has appeared to have some value, both as a guide for workers in the field as well as for teaching undergraduate, graduate and medical school courses in the neurosciences.

In considering the call for a new edition, I felt that the subject had outgrown its origins; introductory accounts are available now in many textbooks. What is most needed now are definitive statements by experts working actively in each region. Happily, a number of colleagues shared this view and indicated they would participate enthusiastically in such an endeavor. The result is the present volume.

As a multiauthor volume, this book is unusual in several ways. Each author is not only a leading expert in a given region, but is also master of the multi-disciplinary approach, having had direct experience with methods in many of the disciplines mentioned above. Each agreed to write a new chapter according to the same sequence of analysis used in previous editions. This starts with *neuronal elements* and *synaptic connections* in order to synthesize a *basic circuit* for that region. It then proceeds to consider *physiological actions*, *neurotransmitters* and *dendritic properties* that underlie function. A final section discusses *functional operations* generated by the synaptic circuits of that region. Readers will therefore recognize the same format of previous editions, the same regions covered in the same sequence, and the same emphasis on comparative principles, now greatly amplified by the expertise of the contributors and refracted through their personal views.

In writing the chapters we placed a premium on including the latest research developments in our respective fields, which meant holding to a very tight publication schedule, not usually feasible with multiple authors. This volume has been a pleasant exception, as all authors performed splendidly in meeting the deadlines, so that less than two years have passed between the initial invitations and the final publication date. During the writing each chapter went through several stages of revision, involving liberal amounts of suggestions and rewriting. I should like to record my thanks to all the authors for the wonderful spirit of cooperative enterprise enlivening our interactions.

A final unusual aspect of this volume is the belief shared by all of the authors in the critical importance of this level of organization for understanding brain function. There is unfortunately a feeling among some neuroscientists that the details of neurons and synapses are too bewildering to admit of simplifying principles, or, alternatively, that animal behavior depends on emergent properties of higher level circuits and systems that render irrelevant the details of neurons and synapses. The present volume should relegate both these arguments to history. The authors have been among the leaders in discovering the roles of different ion channels in neuronal functions, the properties of dendrites and dendritic spines, the long-lasting changes in synaptic properties that may underlie learning and memory, and the specific functions of different types of synaptic circuits. Here they show clearly and persuasively how these properties lie at the heart of the operations characteristic of a given region. This information not only gives insight by itself into the neural basis of brain function, but also provides the means for constructing network simulations based on real neural properties. It is only by incorporating these properties at the synaptic circuit level that we can simulate fully and accurately the mechanisms underlying brain function at the systems level.

Preceding the discussions of individual brain regions, the first chapter provides an introduction to the concept of levels of organization, which is crucial for understanding the ways that synaptic circuits are organized in a hierarchical manner. This is followed by an orientation to the great variety of membrane channels, both voltage- and ligand-gated, which provide the dynamic substrate for much of the information processing that takes place in synaptic circuits. A final appendix presents an introduction to the cable properties of dendrites, which are critical for interpreting physiological recordings from neurons as well as for understanding the integration of membrane activity in dendritic trees. The themes developed in these three sections run as common threads through all the chapters on specific regions and are frequently referred to in those accounts. Together, they constitute the theoretical foundation for much of the subject matter of the book.

It should be noted that all bibliographical references are gathered in a common list at the end of the book. Authors were asked to keep citations to a minimum, in order not to impede the narrative flow. Nonetheless, the irresistible urge for careful documentation, and the wide range of topics covered, has resulted in a reference list that runs to over 1,200 entries. This should be a valuable resource in itself for anyone interested in the broad subject of brain organization.

In a large enterprise such as this there are many debts to acknowledge. For the individual chapters, these are gathered into a list at the front of the book. For the book as a whole, I want especially to thank Jeffrey House of Oxford University Press for his unwavering support; Christof Koch, Charles Greer and David Mc-Cormick for their generous editorial assistance; and Rosalind Corman for superb copy editing. Finally, I want to acknowledge the tremendous amount I have learned from my collaborators in this enterprise.

Hamden, Conn.   G.M.S.
November 12, 1989

Acknowledgments

Chaps. 1, 3, 8 and Appendix. C.K. would like to acknowledge the Air Force Office of Scientific Research, the Office of Naval Research, the National Science Foundation, and the James S. McDonnell Foundation for the generous support of the work upon which these chapters are based.

Chaps. 1, 5 and Appendix. G.M.S. is grateful to Michelle O'Mara for secretarial assistance, and to Sharon Schmiedel, Heather Cameron, Christine Kaliszewski and Sarah Whittaker for preparation of illustrations. The work upon which these chapters is based has been supported by the National Institute of Neurological and Communicative Disorders and Stroke, and the Office of Naval Research.

Chap. 5. C.A.G. is grateful to Heather Cameron and Christine Kaliszewski for preparation of illustrations. His work has been generously supported by the National Institute of Neurological and Communicative Disorders and Stroke.

Chap. 6. P.S. thanks members of his laboratory, past and present, for their contributions to the data and ideas summarized here: Barbara McGuire, Michael Freed, Ethan Cohen, April Firstencel, John Megill, Robert Smith, Noga Vardi, Patricia Masarachia, Ellen Magyarits, and Yoshihiko Tsukamoto. The research was supported by grants from the National Eye Institute. I am grateful to G. Shepherd and P. Masarachia for their editorial suggestions, to Suzanne Leahy for preparing the illustrations, and to K. Gallagher for typing the manuscript.

Chap. 11. T.H.B. and A.M.Z. thank David Amaral, Brenda Claiborne, Anna C. Nobre, and Philip Schwartzkroin for useful comments on this manuscript. We also thank James Vaughn for furnishing an electron micrograph of mossy-fiber synapses, and Claude Keenan for the confocal scanning laser photomicrograph of thorny excrescences. This work was supported by the Office of Naval Research and by the Air Force Office of Scientific Research.

Contents

Contributors, xiii

1. Introduction to Synaptic Circuits, 3
   GORDON M. SHEPHERD AND CHRISTOF KOCH

2. Membrane Properties and Neurotransmitter Actions, 32
   DAVID A. MCCORMICK

3. Peripheral Ganglia, 67
   PAUL R. ADAMS AND CHRISTOF KOCH

4. Spinal Cord: Ventral Horn, 88
   ROBERT E. BURKE

5. Olfactory Bulb, 133
   GORDON M. SHEPHERD AND CHARLES A. GREER

6. Retina, 170
   PETER STERLING

7. Cerebellum, 214
   RODOLFO R. LLINÁS AND KERRY D. WALTON

8. Thalamus, 246
   S. MURRAY SHERMAN AND CHRISTOF KOCH

9. Basal Ganglia, 279
   CHARLES J. WILSON

10. Olfactory Cortex, 317
    LEWIS B. HABERLY

11. Hippocampus, 346
    THOMAS H. BROWN AND ANTHONY M. ZADOR

12. Neocortex, 389
    RODNEY J. DOUGLAS AND KEVAN A. C. MARTIN

13. Appendix: Dendritic Electrotonus and Synaptic Integration, 439
    GORDON M. SHEPHERD AND CHRISTOF KOCH

References, 475   Author Index, 543   Subject Index, 551

Contributors

Paul R. Adams, Ph.D.
Department of Neurobiology
  and Behavior
State University of New York,
  Stony Brook
Stony Brook, New York 11794

Thomas H. Brown, Ph.D.
Department of Psychology
Yale University
New Haven, Connecticut 06520

Robert E. Burke, M.D.
Laboratory of Neural Control
National Institute of Neurological
  Disorders and Stroke
National Institutes of Health
Bethesda, Maryland 20892

Rodney Douglas, M.B., Ch.B.,
  Ph.D.
MRC Anatomical Neuropharmacology
  Unit
Department of Pharmacology
University of Oxford
South Parks Road
Oxford OX1 3UD
England

Charles A. Greer, Ph.D.
Sections of Neurosurgery
  and Neuroanatomy
Yale University School of Medicine
New Haven, Connecticut 06510

Lewis B. Haberly, Ph.D.
Department of Anatomy
University of Wisconsin
Madison, Wisconsin 53706

Christof Koch, Ph.D.
Computation and Neural Systems
  Program
California Institute of Technology
Pasadena, California 91125

Rodolfo R. Llinás, M.D., Ph.D.
Department of Physiology
  and Biophysics
New York University Medical Center
New York, New York 10016

Kevan A. C. Martin, D. Phil.
MRC Anatomical Neuropharmacology
  Unit
Department of Pharmacology
University of Oxford
South Parks Road
Oxford OX1 3UD
England

David A. McCormick, Ph.D.
Section of Neuroanatomy
Yale University School of Medicine
New Haven, Connecticut 06510

Gordon M. Shepherd, M.D.,
  D. Phil.
Section of Neuroanatomy
Yale University School of Medicine
New Haven, Connecticut 06510

S. Murray Sherman, Ph.D.
Department of Neurobiology
  and Behavior
State University of New York,
  Stony Brook
Stony Brook, New York 11794

Contributors

Peter Sterling, Ph.D.
Department of Anatomy
University of Pennsylvania
Philadelphia, Pennsylvania 19104

Kerry D. Walton, Ph.D.
Department of Physiology
   and Biophysics
New York University Medical Center
New York, New York 10016

Charles J. Wilson, Ph.D.
Department of Anatomy
   and Neurobiology
University of Tennessee, Memphis
Memphis, Tennessee 38163

Anthony M. Zador
Department of Psychology
Yale University
New Haven, Connecticut 06520

The Synaptic Organization of the Brain

1

INTRODUCTION TO SYNAPTIC CIRCUITS

GORDON M. SHEPHERD AND CHRISTOF KOCH

In this chapter we introduce some of the basic principles underlying the organization of neurons into circuits and networks within different regions of the nervous system. These circuits form the elementary functional units (cf. Shepherd, 1972; Koch and Poggio, 1987) of which the nervous system is constructed and with which it computes (or "processes information"). Our aim is to describe some of the basic operations that all regions must perform, as well as provide an introduction to the adaptations for specific types of information processing that are unique for each of the regions considered in subsequent chapters.

THE TRIAD OF NEURONAL ELEMENTS

Figure 1.1 illustrates that the brain consists of many local regions, or centers, and many pathways between them. At each center, the *input fibers* make synapses onto the cell body (*soma*), and/or the branched processes (*dendrites*) emanating from the cell body of the nerve cells contained therein. Some of these neurons send out a long axon that, in turn, carries the signals to other centers. These are termed *principal, relay,* or *projection* neurons. Other cells are concerned only with local processing within the center. These are termed *intrinsic* neurons, *local* neurons, or *inter*neurons. An example of this latter type is shown in the cerebral cortex in Fig. 1.1. The distinction between a principal and an intrinsic neuron cannot be rigid, since principal neurons also take part in local interactions. It is, nonetheless, a useful way of characterizing nerve cells, which we will use throughout this book.

The principal and intrinsic neurons, together with the incoming input fibers, are the three types of neuronal constituents common to most regions of the brain. We will refer to them as a *triad* of neuronal elements. The relations between the three elements vary in different regions of the brain, and these variations underlie the specific functional operations of each region.

THE SYNAPSE AS THE BASIC UNIT OF CIRCUIT ORGANIZATION

Interactions between the triad of neuronal elements are mediated by numerous junctions termed *synapses*. It follows that the synapse is the elementary structural

3

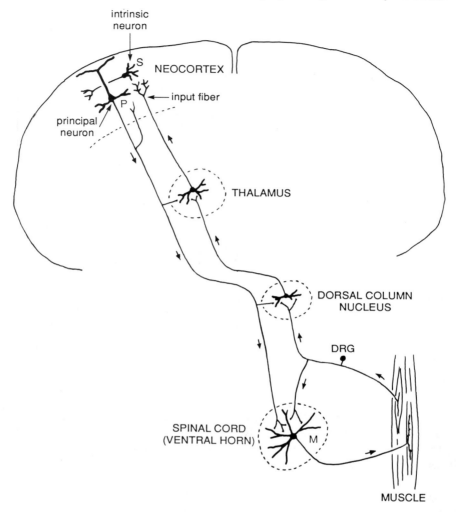

FIG. 1.1. Examples of local regions and some interregional pathways formed through the long axons of principal neurons. M, motoneuron; DRG, dorsal root ganglion cell; P, pyramidal (principal) neuron; S, stellate (intrinsic) neuron.

and functional unit for the construction of neural circuits. Traditionally, most concepts of neural organization have assumed that a synapse is a simple connection that can impose either excitation or inhibition on a receptive neuron. Current experimental evidence indicates that this assumption, like that of the simple neuron, needs to be enlarged.

Figure 1.2 summarizes the current view of the synapse. Most synapses involve the apposition of the plasma membranes of two neurons to form a punctate junction, also termed an active zone. The *punctate* character is an important property; the width of a synaptic contact is commonly 1 μm or less, permitting a

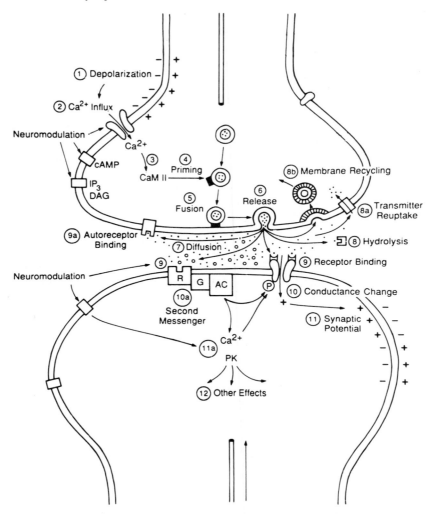

FIG. 1.2. A summary of some of the main mechanisms involved in immediate signaling at the synapse. Steps 1–12 are described in the text. Abbreviations: IP$_3$, inositol triphosphate; CaM II, Ca/calmodulin-dependent protein kinase II; DAG, diacylglycerol; PK, protein kinase; R, receptor; G, G-protein; AC, adenylate cyclase. (Modified from Shepherd, 1988.)

maximum elaboration of circuit connections within a minimum of neural space. A second general property is that the junction is *oriented* from the *pre*synaptic process to the *post*synaptic process. A third property is that, at a chemical synapse, the presynaptic process liberates a *transmitter* substance that acts on the postsynaptic process. Thus, from an operational point of view, a synapse converts a presynaptic electrical signal into a chemical signal and back into a postsynaptic electrical signal. In the language of the electrical engineer, such an element is a nonreciprocal two-port (Koch and Poggio, 1987).

The mechanism for the action of a synapse involves a series of steps, which are summarized in Fig. 1.2. These include depolarization of the presynaptic membrane (1); influx of Ca ions into the presynaptic terminal (2); a series of steps (3–5) leading to fusion of a synaptic vesicle with the plasma membrane; release of a packet (quantum) of transmitter molecules (6); diffusion of the transmitter molecules across the narrow synaptic cleft separating the presynaptic and postsynaptic processes (7); action of the transmitter molecules on receptor molecules in the postsynaptic membrane (9), leading in some cases to direct gating of a membrane conductance (10), which in turn causes a change in the membrane potential (11), and thus in the excitability, of the postsynaptic process. Alternatively, the receptor (9) may activate a second messenger system (10a) which indirectly gates a membrane conductance, as well as having other metabolic effects (11a,12). The transmitter may also act on autoreceptors on the presynaptic process (Fig. 1.2, step 9a) to activate systems that influence the subsequent release of transmitter.

The mechanisms described above are concerned primarily with rapid (2–20 msec) and relatively rapid (20 msec to 2 sec) transmission. Many cellular mechanisms impinge on synaptic transmission over longer time periods. These include the steps involved in axonal and dendritic transport; storage of transmitters and peptides; and corelease of peptides and direct modulation of transmitter responses (see Neuromodulation in Fig. 1.2). These effects are slow (seconds to minutes) or very slow (hours and longer); the slowest processes merge with mechanisms of development, aging, and hormonal effects.

From these properties one can appreciate that the synapse is admirably suited to be a unit for building circuits. The multiple steps of its mechanism confer a considerable flexibility of function, by means of different transmitters and modulators, different types of receptors, and different second-messenger systems linked to the different kinds of machinery in the cell: electrical, mechanical, metabolic, and genetic. Several mechanisms, with different time courses, can exist at the same synapse, conferring on the individual synapse the ability to coordinate rapid activity with the slower changes that maintain the behavioral stability of the organism over time.

In view of this tremendous potential for functional diversity, it is remarkable that synapses throughout the nervous system show such a high degree of morphological uniformity. One of the clearest differences has been the finding that synapses in the brain tend to fall into two groups, those with asymmetrical densification of their pre- and postsynaptic membranes, and those with symmetrical densification. Gray (1959) termed these type 1 and type 2, respectively. Depending on the histological fixatives used, type 1 is usually associated with small, round, clear synaptic vesicles. In a number of cases, type 1 synapses have been implicated in excitatory actions. By contrast, type 2 is usually associated with small, clear, flattened or pleomorphic vesicles, and is implicated in inhibitory synaptic actions. Many examples are shown throughout this book. There are well recognized exceptions to these structure–function relations—for example, inhibitory actions by synapses that do not have type 2 morphology (cf. cerebellar basket cells, Chap. 7). Thus, although the type 1 and type 2 designations are a useful working hypothesis, and will be often invoked in later chapters, there is always the clear understanding that this is only a first step in classifying synaptic structure and function.

In addition to its ability to mediate different specific functions, a second important property of the synapse is its small size. The area of contact has a diameter of 0.5–2.0 μm, and the presynaptic terminal (bouton) has a diameter that characteristically is only slightly larger. These small sizes mean that large numbers of synapses can be packed into the limited space available within the brain. For example, in the cat visual cortex, 1 mm^3 of gray matter contains approximately 50,000 neurons, each of which gives rise on average to some 6,000 synapses, making a total of 300 million (300×10^6) synapses (Beaulieu and Colonnier, 1983). It has been estimated that 84% of these are type 1 and 16% are type 2. If the cortical area of one hemisphere in the human is approximately 100,000 mm^2, there must be on the order of 10 billion cells in the human cortex, and 60 trillion (60×10^{12}) synapses.

Like the national debt, these numbers are so large that they lose meaning. The important point is that the number of synapses amplifies the number of neurons by several orders of magnitude, providing a rich substrate for the construction of microcircuits within the packed confines of the brain. During early development an exuberance of synapses is generated throughout the nervous system, of which as much as one half are pruned away as an animal reaches maturity (Rakic, 1986).

The importance of the synapse in the construction of synaptic circuits has justified this introduction to its properties. The physiological actions of different types of synapses will be discussed in the following chapter.

LEVELS OF ORGANIZATION OF SYNAPTIC CIRCUITS

It might seem that one could simply connect neurons together by means of synapses and make networks that mediate behavior, but this is not the way nature does it. A general principle of biology is that any given behavior of an organism depends on a hierarchy of levels of organization, with spatial and temporal scales spanning many orders of magnitude. This is nowhere more apparent than in the construction of the brain. As applied to synaptic circuits, it means that one needs to identify the main levels of organization in order to provide a framework for understanding the principles underlying their construction and function.

The analysis of local regions over the past two decades has led to the recognition of several important levels of circuit organization (see Fig. 1.3). If we start at the most fundamental level with the synapse, the most local patterns of synaptic connection and interaction, involving small clusters of synapses, are termed *microcircuits* (Shepherd, 1978). The smallest microcircuits have sizes measured in microns; their fastest speed of operation is measured in milliseconds. These are grouped to form *dendritic subunits* (Shepherd, 1972; Koch et al., 1982) within the *dendritic trees* of individual neurons. The whole *neuron*, containing its several dendritic subunits, is the next level of complexity. Interactions between neurons of similar or different properties constitute *local circuits* (Rakic, 1976); these perform the operations characteristic of a particular region. Above this

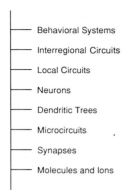

— Behavioral Systems

— Interregional Circuits

— Local Circuits

— Neurons

— Dendritic Trees

— Microcircuits

— Synapses

— Molecules and Ions

FIG. 1.3. Some of the main levels of organization in the nervous system. This book focuses on the levels from *Synapses* to *Local Circuits*, as a basis for understanding the expression of *Molecules and Ions* in an integrative context, and for understanding the circuit basis of *Behavioral Systems*.

level are the interregional *pathways,* columns, laminae, and topographical maps, involving multiple regions in different parts of the brain, that mediate specific types of behavior. It is important to note that these many interwoven levels of organization are a trait of the brain not shared by its closest artificial cousin, the digital computer, in which few intermediate modular structures exist between the individual transistor, on the one hand, and a functional system, such as a random-access memory chip, on the other.

An important aim of the study of synaptic organization is to identify the types of circuits and the functional operations they perform at each of these organizational levels. In the rest of this chapter, we will consider examples at each of these levels. Subsequent chapters show how, in each region, the nervous system rings the changes on these basic themes, expressing variations of circuits exquisitely adapted for the specific operations and computations carried out by that region on its particular input information.

MICROCIRCUITS

Excitation and inhibition by single synapses usually have little behavioral significance by themselves; it is the assembly of synapses into *patterns of connectivity* during development that produces functionally significant operations. The process can be likened to the assembly of transistors onto chips to form microcircuits in computers. By analogy, we refer to these most local synaptic patterns as neuronal microcircuits. Let us consider several basic types.

SYNAPTIC DIVERGENCE

One of the simplest types of patterns provides for divergence of output from a single terminal. In neural network terminology, this would be called "fan-out." As indicated in Fig. 1.4A, a presynaptic terminal (a) has excitatory synapses onto postsynaptic dendrites (b–f). An action potential (ap) invading the presynaptic terminal can thus cause simultaneous excitatory postsynaptic potentials (EPSPs) in many postsynaptic dendrites. The advantage of this arrangement is that activity in a single axon is amplified into activity in many postsynaptic

A. SYNAPTIC
DIVERGENCE

B. SYNAPTIC
CONVERGENCE

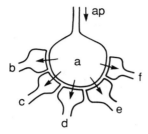

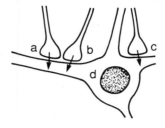

C. PRESYNAPTIC
INHIBITION

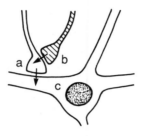

FIG. 1.4. The simplest types of microcircuits. **A.** Synaptic divergence, from a single large axonal terminal (a) onto multiple postsynaptic dendrites (b–f). Synapses are indicated by arrows. ap, Action potential. **B.** Synaptic convergence, of several axons (a–c) onto a single postsynaptic neuron (d). **C.** Presynaptic inhibition, by axon (b) onto axon (a), which is presynaptic to axon (c). See text.

neurons, conferring a high gain upon the system. The activity is simultaneous, thus retaining the timing of the input, and it is of the same sign: excitation → excitation.

Comparison of this arrangement with the well-known model of the neuromuscular junction (NMJ) is of interest. The NMJ also consists of a large presynaptic terminal with many release sites, in this case made onto the muscle. It is known that of 1000 or so release sites, only 100–200 are actually activated by invasion of a single impulse into the presynaptic terminal. Thus, there is a probability of only .1–.2 that a given site will release transmitter when depolarized by an impulse. Active zones of central synapses also appear to have release probabilities. Thus, in the example of Fig. 1.4A, presynaptic depolarization causes some synapses to release transmitter (e.g., b,c) but not others (e.g., d). This makes the important point that morphological studies can identify the pattern of synaptic connections, but their actual use is physiological and probabilistic (see Korn and Faber, 1987). If release probability is dependent on the amount of activity, it provides an effective mechanism for adjusting the relative proportions of "silent synapses" in a given microcircuit.

An important point to note is that, whereas all active zones at a NMJ connect to the same postsynaptic cell, in the example of Fig. 1.4A they connect to different cells; this is the way that divergence is achieved at the level of a single terminal. Such an arrangement is found in many parts of the nervous system. A single mossy fiber terminal in the cerebellum, for example, may make synapses onto dendrites of as many as 100 or more granule cells. Single terminals with more modest divergence factors are made by sensory afferents in thalamic relay nuclei and the substantia gelatinosa of the dorsal horn. In addition to divergence from a single terminal, there is of course also divergence of axons as they branch to give rise to multiple terminals. This is a more common means of achieving multiple synaptic outputs, as exemplified by the several thousand synapses of a typical cortical cell mentioned above.

SYNAPTIC CONVERGENCE

The considerable divergence that characterizes the output of a neuron is matched by convergence of many inputs. In neural network terminology, this is called "fan-in." The essence of this convergence at the microcircuit level is depicted in Fig. 1.4B, where two terminals (a and b) make synapses onto a postsynaptic dendrite (d).

Let us consider first the case in which both terminals are excitatory. Spread of an impulse into terminal (a) sets up an EPSP; slightly later, spread of an impulse into terminal (b) sets up an EPSP which summates with that of (a). This is termed *temporal summation*. Note that although the impulses in (a) and (b) may be asynchronous, their EPSPs nonetheless can summate. For relatively fast EPSPs, the prolongation that makes temporal summation possible is due mainly to the membrane capacitance, which slows the dissipation of charge across the postsynaptic membrane (see Appendix). For slower EPSPs, the time course is controlled by biochemical processes, such as second messengers.

Although it might appear that temporal summation involves simple addition of PSPs, in general this is not the case. This is because PSPs are generated by changes in membrane conductance to specific ions and not by current injection (see Chap. 2); the conductances act to shunt, or short-circuit, each other, so that the combined amplitude of a PSP will be less than the sum of its parts. This means that synaptic summation is essentially a nonlinear process (Rall, 1964, 1977; see Chap. 2 and Appendix).

Synaptic convergence also involves summation of excitatory and inhibitory PSPs. This process lies at the heart of the integrative mechanisms of neurons. Consider, for example, in Fig. 1.4B, that (b) is inhibitory. Activation sets up an IPSP which opposes the EPSP set up by (a), and repolarizes the membrane toward the reversal potential for the inhibitory conductance (see Chap. 2). If the reversal potential is near the resting membrane potential, the IPSP by itself will not be apparent as a change in membrane potential; this is called "silent" or "shunting" inhibition. Obviously, integration of excitatory and inhibitory synaptic responses is nonlinear and complex (Rall, 1964; Koch et al., 1983).

It remains to note that inputs are characteristically distributed over the entire dendritic surface of a neuron [see (c) in Fig. 1.4B]. This means that, in addition to temporal summation, there is *spatial summation* of responses arising in different parts of a dendrite, as well as different parts of the whole dendritic tree. Spatial summation allows for the combining of many inputs into one integrated postsynaptic response. The separation of PSPs reduces some of the nonlinear interactions between synaptic conductances, making the summation more linear. However, it also increases the possibilities for local active mechanisms and the generation of nonlinear sequences of activation from one site in the dendritic tree to the next. Examples are considered in the Appendix.

Presynaptic inhibition is a final type of simple synaptic combination involving convergence. In this arrangement (Fig. 1.4C), a presynaptic terminal (a) is itself postsynaptic to another terminal (b). The presynaptic action may involve a conventional type of IPSP produced by (b) in the presynaptic terminal (a). Alternatively, there may be a maintained depolarization of the presynaptic terminal, reducing the amplitude of an invading impulse and with it the amount of transmitter released from the terminal. The essential operating characteristic of this microcircuit is that the effect of an input (a) on a cell (c) can be reduced or abolished (by b) without there being any direct action of b on the cell (c) itself. Control of the input (a) to the dendrite or cell body can thus be much more specific.

Presynaptic control may be exerted by either axon terminals or presynaptic dendrites. It is important to note that the effect is presynaptic only with regard to the response of the postsynaptic cell. From the point of view of the presynaptic terminal, the effect is postsynaptic. There are many situations in the nervous system, involving multiple synapses between axonal and/or dendritic processes, in which sequences of pre- and postsynaptic effects can occur (see Chaps. 4–6 and 8, on the Spinal Cord, Olfactory Bulb, Retina, and Thalamus).

INHIBITORY OPERATIONS

The combinations considered thus far provide for elementary excitatory and inhibitory operations. Let us next consider arrangements that carry out operations that are building blocks for specific functions.

Feedforward inhibition. Sensory processing commonly involves an inhibitory "shaping" of excitatory events. The synaptic basis has been explored, and a pattern of synaptic connections has been identified in several nuclei of the thalamus that could play a role in this type of *feedforward* or *afferent inhibition* (see Chap. 8). As illustrated in Fig. 1.5A, the arrangement consists of an afferent terminal (a) which makes synapses onto the dendrites of both a relay neuron (b) and an interneuron (c). The dendrites of both neurons respond by generating EPSPs. However, the interneuron also has inhibitory dendrodendritic synapses onto the relay neuron; the EPSP activates these synapses, producing an inhibition of the relay neuron. The extra synapse in this pathway helps to delay the inhibitory input, so that the combined effect in (b) is an excitatory—inhibitory sequence. This type of sequence is common in many sensory pathways. By restricting the

A. FEEDFORWARD INHIBITION   B. RECURRENT INHIBITION

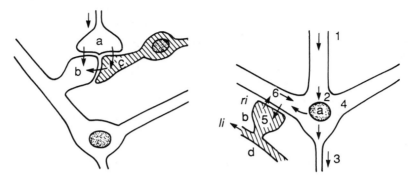

FIG. 1.5. Microcircuits that mediate different types of postsynaptic inhibition. **A.** Feedforward inhibition, in which presynaptic axon (a) directly excites relay neuron dendrite (b) and interneuron dendrite (c), which feeds-forward inhibition onto (b). **B.** Recurrent inhibition, in which a relay neuron (a) is both pre- and postsynaptic to the dendrite (d) of an inhibitory interneuron (b). This microcircuit mediates both recurrent and lateral inhibition, through the series of steps indicated by 1–6. See text.

excitation of relay neurons to the onset of an excitatory input, it serves to enhance the sensitivity to changing stimulation, and thus performs a kind of *temporal differentiation* on changing sensory states (Koch, 1985). By spread of responses through dendritic trees, it may also contribute to the enhancement of spatial contrast through *lateral inhibition* (see below).

The student will note that the microcircuit in Fig. 1.5A is built of all three elementary patterns discussed above and depicted in Fig. 1.4. Thus, it combines *divergence* from terminal (a) with *convergence* of (a) and (c) onto (b) and *presynaptic* control by (a) of (c). Many variations of this circuit occur throughout the CNS, the more common involving feedforward inhibition through more widely dispersed synapses and/or terminals than those illustrated in Fig. 1.5A.

Feedback inhibition. A very common type of operation in the nervous system is one in which the excitation of a neuron leads to inhibition of that neuron and/or of neighboring neurons. This is called *feedback or recurrent inhibition*, to distinguish it from feedforward, afferent inhibition. It can be mediated by several types of circuit, the most local type of which involves reciprocal dendrodendritic synapses.

This mechanism has been worked out most clearly at the synaptic level in the olfactory bulb, and is illustrated in Fig. 1.5B (Rall and Shepherd, 1968). The output neurons of the olfactory bulb are mitral and tufted cells (a). They are activated by EPSPs which spread through a primary dendrite (1) to the cell body (2) to set up an impulse which propagates into the axon (3). The impulse also spreads into secondary dendrites (4), where it activates output synapses which are excitatory to spines of granule cell dendrites (5). The EPSP in the spine then activates a reciprocal inhibitory synapse back into the mitral cell dendrite (6); the IPSP spreads through the neuron to inhibit further impulse output.

Reciprocal synapses thus form an effective microcircuit module carrying out an elementary computation, in this case for recurrent inhibitory feedback of an activated neuron. In addition, the EPSP in the granule cell spine spreads through the dendritic branch to other spines, activating inhibitory output onto neighboring, less active, mitral cells. This provides for lateral inhibition which enhances the contrast between the activity of an excited neuron and its less stimulated neighbors, a fundamental mechanism of sensory processing.

Reciprocal synapses are found in a number of regions of the nervous system; in addition to the olfactory bulb, these include the dorsal horn of the spinal cord, retina, thalamus, and suprachiasmatic nucleus. They appear to be largely absent from the basal ganglia and the different regions of cortex. It is interesting that, although they are absent from the local circuits of the neocortex, their presence in the different nuclei of the thalamus means that they play a role in the inter-regional circuits that control cortical operations. One may conclude that they are not special for a particular region or a particular modality. Rather, they are an excellent example of a microcircuit module that can perform several operations of information processing, and can be assembled during development for that purpose wherever it is needed. At this very local level of organization, full-blown propagating action potentials may not be needed; synaptic output can be controlled by graded electrotonic spread of nearby EPSPs (as in the case of the granule cell) or by partial voltage-dependent "boosting" effects.

The recurrent inhibition illustrated in Fig. 1.5B is dependent on excitation of an intermediary inhibitory element. The most common arrangement for this general type of circuit involves an inhibitory short-axon cell; a well-known example is the Renshaw cell circuit in the spinal cord (Chap. 4). This type of inhibitory interneuron is present in many parts of the brain, and is one of the main building blocks of local circuits (see examples below).

DENDRITIC INTEGRATION AND DENDRITIC SUBUNITS

We take for granted that the extensive branching of neuronal dendrites increases their surface area for receiving synaptic inputs. If this were its only function, there would be some justification for representing the neuron as a single, spatially lumped node, and representing all inputs as converging onto this node. This is a common practice in neuroanatomical textbooks, and it is the common assumption underlying the vast majority of neural network simulations, which consider individual nerve cells to be single-node, linear integration devices. In both cases, the effect of dendritic morphology and synaptic architecture on the function of individual cells is totally neglected. In fact, the patterns of dendritic branching, unique to different types of neurons, impose geometrical constraints which separate the activity in different branches from each other. Furthermore, the geometry of the branches and the sites of specific inputs combine with the electrotonic properties to ensure that parts of a dendritic tree can function semiin-dependently of one another. If one adds the fact that voltage-gated channels can confer excitable properties onto local dendritic regions, it is clear that the dendrites, far from being negligible appendages of a cell body, are the substrate for generating a rich repertoire of computation, all still below the level of the whole neuron. It is evident that single-node network models ignore several levels of dendritic organization responsible for much of the computational complexity of the real nervous system.

We may summarize at this point by noting that four factors—dendritic branching, synaptic placement, and passive and active membrane properties—must be taken into account in assessing the nature of the integrative activity of dendrites. Characterization of the electrotonic spread of potentials is difficult because of the complex branching patterns of many dendrites. An introduction to one-dimensional passive cable theory is provided in the Appendix. The ways that active conductances can contribute to dendritic activity are considered in the following chapter. This expression of dendritic activity is a common theme running through the accounts of most of the brain regions considered in this book. Here we provide a brief introduction to the nature of dendritic integration and the ways that functional compartments are created at several levels of dendritic organization.

DENDRITIC COMPUTATION

In assessing the nature of dendritic integration, it is increasingly fashionable to use the computational metaphor. This obscures many functional roles of dendrites that are not strictly "computational" (e.g., mechanisms involved in development, maturation, activity-dependent changes, etc.), but it has the advantage of providing a useful framework within which the capacity for dendrites to carry out specific types of operations can be assessed.

The importance of the sites and types of synaptic inputs on a dendritic branch can be illustrated by using the paradigm of logic operations. In the diagram of Fig. 1.6A, alternating excitatory and inhibitory synapses are arranged along a dendritic branch. Given the nonlinear interactions between these synapses, an inhibitory synapse (i1, i2, i3) with a synaptic reversal potential close to the resting potential of the cell ("shunting" or "silent" synapse, Chap. 2) can effectively oppose ("veto") the ability of a membrane potential change generated by any more distal excitatory synapse to spread to the soma and generate impulses there. By contrast, an inhibitory synapse has little effect in vetoing the voltage change initiated by more proximal excitatory synapses. This operation is an analog form of a digital AND-NOT gate (e *and not* more proximal i), and has been postulated to be a mechanism underlying various computations, such as direction selectivity in retinal ganglion cells (Koch et al., 1983).

This type of synaptic arrangement can also be found in more localized parts of dendritic trees. Figure 1.6B depicts a case in which a dendrite has numerous distal branches, each with an excitatory and an inhibitory synapse. The same "on-path" role still applies: An inhibitory synapse effectively vetoes a more distal excitatory synapse on the same branch, but has little effect in opposing excitatory responses originating anywhere else in the dendritic tree, which are effectively sited more proximally to the soma. Thus, the combination of dendritic morphology in conjunction with synaptic placement enables the cell to "synthesize" analog versions of logical, boolean operations.

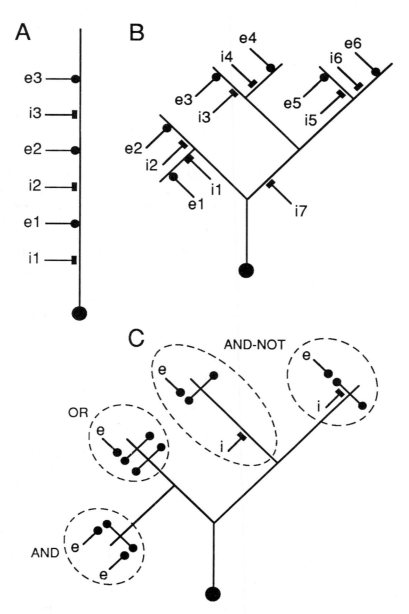

FIG. 1.6. Possible arrangements of synapses that have the potential for subserving logic operations. **A.** A single dendrite receives excitatory (e1–e3) and inhibitory (i1–i3) synapses. An inhibitory input can effectively veto only more distal excitatory responses; this approximates an AND-NOT logic operation, e.g., [e2 AND NOT i1 or i2]. **B.** Branching dendritic tree with arrangements of excitatory and inhibitory synapses. As in A, inhibitory inputs effectively veto only the excitatory response more distal to it, e.g., {[e5 AND NOT i5] AND NOT i7}. **C.** Branching dendritic tree with excitatory synapses on spines, and inhibitory synapses either on spine necks or on dendritic branches. Different types of logic operations arising out of these arrangements are indicated. In all cases (A–C), inhibition is of the shunting type. See text. (A,B from Koch et al., 1983; C based on Shepherd and Brayton, 1987.)

DENDRITIC SPINE UNITS

The smallest compartment within a dendritic tree is the dendritic spine, a small (1–2 μm), thornlike, dendritic protuberance. It is already evident from Fig. 1.5 that spines are an important component in many kinds of microcircuits. An electron micrograph of a spine in the cerebral cortex is shown in Fig. 1.7. Spines are extremely numerous on many kinds of dendrites; in fact, they account for the majority of postsynaptic sites in the brain. They are especially prominent in the cerebellar cortex, basal ganglia, and cerebral cortex. On dendrites of pyramidal cells, spine densities may reach several spines per micrometer of dendrite. Because spines are characteristically located on dendrites at some distance from the cell body, experimental evidence regarding their physiological properties is still limited. However, their obvious importance has stimulated considerable interest. Within the cerebral cortex, about 79% of all excitatory synapses are made onto spines and the rest directly onto dendrites, whereas 31% of all inhibitory synapses are made onto spines. The latter type of spine always carries both an excitatory as well as an inhibitory synaptic profile (Beaulieu and Colonnier, 1983). Given the dominance of excitatory synapses, about 15% of all dendritic spines carry both excitatory and inhibitory synaptic profiles.

Subsequent chapters will give abundant testimony to the importance and interest of spines; the electrotonic properties that underlie their function are considered in the Appendix. Here we will pursue the metaphor of logic operations, and inquire to what extent it might apply to interactions between spines. The diagram in Fig. 1.6C represents a dendritic tree with its distal branches covered by spines. Let us assume that there are patches of active membrane in the distal dendrites, and that these give rise to an impulse if there is sufficient depolarization by an excitatory synaptic response. One possible arrangement is that the impulse would fire if any one of several spines in a cluster should receive an excitatory input; this would be equivalent to an OR gate in the logic paradigm. Alternatively, two simultaneous inputs might be required; this would constitute an AND gate. Finally, one might have AND-NOT gates. Depending on the placement of the inhibition, the gate might be localized to an individual spine, or might involve a dendritic branch containing a cluster of spines. These possibilities can all be traced in the diagram of Fig. 1.6C. Experimental studies suggest that these simple combinations of excitatory and inhibitory interactions do occur in natural activity, and computer simulations have shown that the logic operations arise readily out of these arrangements (Shepherd and Brayton, 1987; Shepherd et al., 1989; see also Appendix).

The point of these studies is not to suggest that the brain is constructed like a digital computer based on binary logic. Rather, it is to explore the properties of interactions in the smallest compartments of the nervous system—branched dendrites and dendritic spines—and show that they may be capable of powerful and precise types of information processing. A further interest is that, through sequential activation of active sites in the dendritic tree, synaptic responses initiated in the most distal parts of the tree nonetheless can exert precise control over the generation of impulses in the cell body and initial axonal segment.

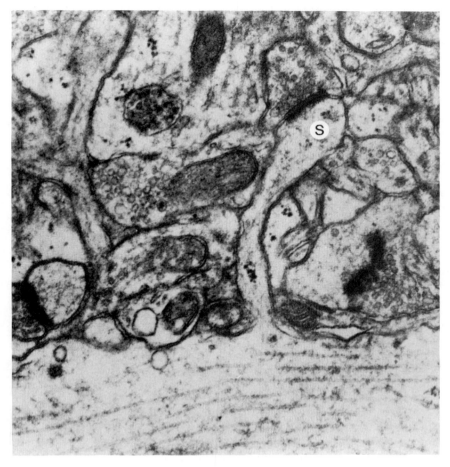

FIG. 1.7. The fine structure of a dendritic spine. This electron micrograph shows, at the bottom, a longitudinally cut dendrite from which arises a spine (s). The spine is approximately 1.5 μm in length, and 0.1 μm at its narrowest width. At its head it receives a synapse, which has the round vesicles and asymmetrical density characteristic of Gray's type 1. In the neck and head are small clumps of ribosomes; in the dendrite are longitudinally cut microtubules. (From Feldman, 1984.)

Associative learning. Spine interactions may also be a mechanism for *associative learning.* In recent years considerable attention has been focused on the possibility that long-term potentiation (LTP) of cortical neurons may underly learning and memory (reviewed in Brown et al., 1988b). This involves a long-term (hours to weeks) increase in synaptic efficiency in response to a presynaptic input volley. There is growing evidence of anatomical, biochemical, and physiological changes in dendritic spines of these cells during LTP. It has been shown, for example, that sufficient depolarization of a spine will increase calcium ion conductance; the calcium ions are then available to bring about biochemical and structural changes in the spine that could function in the storage of information (see Fig. 1.2 above; these mechanisms are discussed in detail in Chap. 11). To the extent that these changes involve thresholds and nonlinear properties, they can be incorporated into the logic paradigm of spine interactions illustrated in Fig. 1.6C.

DENDRITIC SUBUNITS

Functional compartmentalization of dendrite trees can be created in various ways. The interactions between excitatory and inhibitory responses described above define relatively small, activity-dependent subunits. At the other extreme are larger compartments built into the branching structure during development.

The mitral cell of the olfactory bulb provides a clear example of this level of organization. As shown in Fig. 1.8A (left), each mitral cell has a primary dendrite, divided into two subunits: a terminal tuft (t) and a primary dendritic shaft (1°). The function of the terminal tuft is to receive the sensory input through the olfactory nerves and process the responses through dendrodendritic interactions (see insert). The function of the primary dendritic shaft is to pass on this integrated response to the cell body. The third dendritic subunit in this cell consists of the secondary (2°) dendrites, which take part in dendrodendritic interactions with the granule cells and thereby control the output from the cell body (these have been described above; see Fig. 1.5B). Thus, the mitral cell dendritic tree is fractionated into three large subunits, each with a distinct function that is carried out semiindependently of the others (see Chap. 5 for further details).

Another example of this level of organization is provided by the starburst amacrine cell of the retina. This cell (see Fig. 1.8B) has a widely radiating dendritic tree, which at first glance would seem to be diffusely arranged. Like olfactory granule cells, amacrine cells lack axons; the distal dendritic branches are the sites of synaptic output, whereas synaptic inputs are present both distally and proximally (see inserts). Thus, each dendrite appears to function as a relatively independent input–output unit (Koch et al., 1982; Miller and Bloomfield, 1983). These dendritic subunits appear to be part of the circuitry for computing the direction of a moving stimulus in the vertebrate retina (see below, and Chap. 6 for further details).

The starburst cells synthesize and release acetylcholine (ACh), providing excitatory input to direction-selective ganglion cells. Pharmacological evidence suggests that the most common inhibitory neurotransmitter, gamma-aminobutyric acid (GABA), provides the inhibitory input in the cell's null direction. Thus, an excitatory bipolar cell input to the amacrine cell could, in conjunction with GABAergic input from inhibitory bipolar or inhibitory cells, function as an AND-NOT gate, in analogy with the corresponding arrangement illustrated in Fig. 1.6B. However, paradoxically, starburst amacrine cells also appear to synthesize, store, and release GABA (see Chap. 6). Until recently, such a colocalization of two fast-acting neurotransmitters was thought not to exist. Its presence

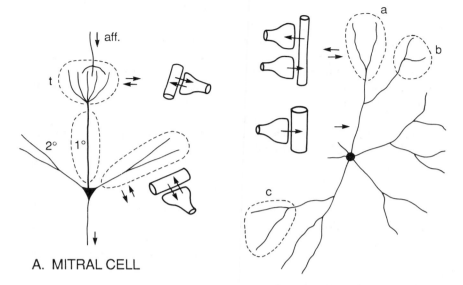

A. MITRAL CELL

B. STARBURST AMACRINE CELL

FIG. 1.8. Organization of subunits within dendritic trees. **A.** Mitral cell of the olfactory bulb, showing division of the dendritic tree into three main subunits. Abbreviations: aff., afferent; t, dendritic tuft; 1°, 2°, primary and secondary dendrites. Synaptic microcircuits are indicated in insets. **B.** Starburst amacrine cell in the retina, showing division of dendritic tree into functional subunits, as exemplified by *a–c*. Microcircuits are indicated in the insets. (A after Shepherd, 1979; B based in part on Koch et al., 1982.)

obviously increases the opportunities for more complex synaptic interactions at the local level.

THE NEURON AS AN INTEGRATIVE UNIT

We have seen that a neuron generally contains several levels of organization within it, starting with the synapse as the basic functional unit. Many different configurations of synapses, coupled with passive and active membrane properties and the geometry of the dendrites, provide a rich substrate for carrying out neuronal computations. The time scale of these computations varies greatly, from the fraction of a millisecond required for inhibition to suppress EPSPs in dendritic spines to many hundreds of milliseconds or seconds in the case of the slowly acting effects of neuropeptides on the electrical properties of neurons.

A detailed description of the way that different types of membrane conductances, each with a characteristic distribution in the cell body and dendrite, combine to control the flow of information through the neuron is given in the next chapter. In Table 1.1, we provide a brief compendium of some elementary synaptic circuits and biophysical mechanisms underlying specific computation in the nervous system. In addition to their interest for neuroscience, these operations are of considerable potential relevance in computer science, where current work on the "physics of computation" attempts to characterize the physical mechanisms that can be exploited to perform elementary information processing operations in artificial neural systems (Mead and Conway, 1980). These mechanisms constrain in turn the types of operations that can be exploited for computing. We believe that a "biophysics of computation" is now needed for understanding the role of membranes, synapses, neurons, and synaptic circuits in information processing in biological systems, in order to bridge the gap between computational theories and neurobiological data (Koch and Poggio, 1987; Shepherd, 1990). This knowledge will also enable us to understand the fundamental limitations in terms of noise, accuracy, and irreversibility on neuronal information processing.

The vast majority of neural network simulations—in particular, connectionist models—consider individual nerve cells to be single-node, linear integration devices. They thus neglect the effect of dendritic, synaptic, and intrinsic membrane properties on the function of individual cells. An important goal of the study of synaptic organization is therefore to identify the specific operations (such as those summarized in Table 1.1) that arise from these properties and incorporate them into more realistic network simulations of specific brain regions.

LOCAL CIRCUITS

No matter how complicated a single neuron may be, it cannot play a role in the processing of information without interacting with other neurons. The circuits that mediate interactions within a region are called *intrinsic,* or local, circuits. They include all of the levels of organization we have considered previously, plus the longer-distance connections made by axons and axon collaterals within a given brain region. In turning our attention to these more extensive circuits, we will continue to distinguish between simple excitatory and inhibitory synaptic actions. Although the types of neurons and their intraregional connections appear to be distinctive for each region, we will see that the operations they carry out can be grouped into several basic types.

EXCITATORY OPERATIONS

Excitatory operations can be grouped into two main types: feedforward excitation in input pathways, and feedback excitation through axon collaterals. We will discuss these operations in relation to the organization of the cerebral cortex.

Feedforward excitation. In many regions of the nervous system, the input fibers (arising from output cells in other regions) are excitatory. Thus, all the long-range projections to, from, and within cortex are excitatory, as are the projections to and from the specific thalamic nuclei. The rules of connectivity for the targets of this excitatory input vary, however, in different regions.

A common pattern is that the input goes directly to the output neurons of that

Table 1.1. Some Neuronal Operations and Their Underlying Biophysical Mechanisms

Biophysical mechanism	Neuronal operation	Example of computation	Time scale
Action potential initiation	Threshold, one-bit analog-to-digital converter		0.5–5 msec
Action potentials in dendritic spines	Binary OR, AND, AND-NOT gate	[a]	0.1–5 msec
Nonlinear interaction between excitatory and inhibitory synapses	Analog AND-NOT veto operation	Retinal directional selectivity[b]	2–20 msec
Spine–triadic synaptic circuit	Temporal differentiation high-pass filter	Contrast gain control in the LGN[c]	1–5 msec
Reciprocal synapses	Negative feedback	Lateral inhibition in olfactory bulb[d]	1–5 msec
Low, threshold calcium current (I_T)	Triggers oscillations	Gating of sensory information in thalamic cells[e]	5–15 Hz
NMDA receptor	AND-NOT gate	Associative LTP[f]	0.1–0.5 sec
Transient potassium current (I_A)	Temporal delay	Escape reflex circuit in Tritonia[g]	10–400 msec
Regulation of potassium currents (I_M, I_{AHP}) via neurotransmitter	Gain control	Spike frequency accommodation in sympathetic ganglion[h] and hippocampal pyramidal cells[i]	0.1–2 sec
Long-distance action of neurotransmitters	Routing and addressing of information	[j]	1–100 sec
Dendritic spines	Postsynaptic modification of functional connectivity	Memory storage[k]	∞

Note: The time scales are only approximate. LGN, lateral geniculate nucleus; LTP, long-term potentiation; NMDA, N-methyl-D-aspartate.

Sources: Includes the chapter in which the mechanism is discussed and the original reference.

[a]Chap. 1; see also Shepherd and Brayton, 1987.
[b]Chap. 1; see also Koch et al., 1982, 1983.
[c]Chap. 8; see also Koch, 1985.
[d]Chap. 5; see also Rall and Shepherd, 1968.
[e]Chaps. 2 and 8; see also Jahnsen and Llinás, 1984a,b.
[f]Chap. 2; Jahr and Stevens, 1986.
[g]See Getting, 1983.
[h]Chap. 3; see also Adams et al., 1986.
[i]Chap. 11; see also Madison and Nicoll, 1982.
[j]See Koch and Poggio, 1987.
[k]Chaps. 1 and 12; see also Rall, 1974a,b.

region. As shown in Fig. 1.9 for the case of an area of cerebral cortex, the target may be the distal dendrites (AFF$_1$), or the proximal dendrites (AFF$_2$) of pyramidal output neurons (P$_1$). Although it is widely believed that the distal sites can provide only for weak, background modulation of more specific proximal responses, this is not generally true. There are many regions, such as the olfactory cortex (Chap. 10), where specific inputs in fact preferentially and exclusively make connection only to the most distal dendrites. We have seen above that these distal sites provide a fertile ground for local signal processing.

Another common pattern is that input fibers connect to interneurons which make excitatory connections within the region in question. These provide circuits for feedforward excitation. In the cerebellar cortex, for example, mossy fibers make excitatory synapses onto granule cells, which then make excitatory connections onto Purkinje cells. The advantage of this arrangement is that it gives the opportunity for additional patterns of convergence and divergence at each synaptic

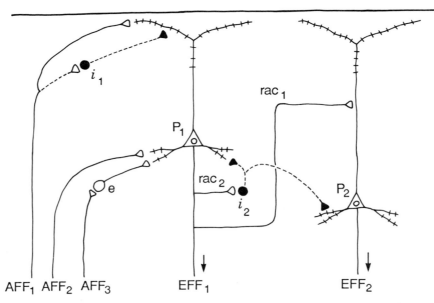

FIG. 1.9. Basic circuit for the synaptic organization of cerebral cortex. The elements comprising this basic circuit include (1) pyramidal (P) output neurons, with apical and basal dendrites; (2) differentiation of P cells into different populations in different layers or areas; (3) dendritic spines, which receive excitatory inputs; (4) input fibers which connect to P cells in specific ways—direct excitation (AFF$_1$ and AFF$_2$), feedforward excitation (AFF$_3$, e), feedforward inhibition (AFF$_1$, i$_1$); (5) intrinsic recurrent axon collaterals (rac) for feedback excitation of P cells (rac$_1$); (6) intrinsic rac for feedback and lateral inhibition of P cells (rac$_2$, i$_2$); (7) lamination of different inputs at different levels of the P cell dendritic tree, for complex gating of input responses; (8) different targets of different populations of P cells (EFF$_1$, EFF$_2$). Taken together, these constitute a unique set of circuit elements, which is adapted and elaborated in different cortical areas to generate the operations unique for each area. (After Shepherd, 1988b.)

relay. In cerebellar cortex, these complex patterns are complemented by the direct access of excitatory climbing fiber inputs to the Purkinje cells (see Chap. 7).

In the cerebral cortex, excitatory interneurons in input pathways are exemplified by the spiny stellate cells of laminar IV, found in sensory areas and association areas of granular cortex. As shown in Fig. 1.9 (AFF_3), this provides for an intracortical feedforward excitatory relay from the afferents onto the output neurons (P_1). This gives the opportunity for an additional step in the processing of afferent information, by convergence–divergence patterns of connections as mentioned above. In addition to serial sequences, there are also parallel input pathways, including direct afferent inputs to the pyramidal neurons, which contribute to the abstraction of receptive field properties. Inhibitory interactions also contribute very importantly to these properties (Ferster and Koch, 1987; see below).

Feedback excitation. Excitation fed back through axon collaterals of an excitatory output neuron onto that neuron and neighboring output neurons is a special property found only in certain regions of the brain. Recurrent excitation, where it exists, rarely is fed back through an excitatory interneuron. An obvious reason is that this would create a loop for positive feedback that would lead to powerful and widespread excessive excitation and seizure activity. Thus feedback excitation is usually mediated only by direct connections of the recurrent collaterals onto other principal neurons.

The clearest examples of this type of excitation are to be found in the vertebrate cerebral cortex. The simplest case is olfactory cortex, where the pyramidal neurons give off recurrent collaterals which feed back excitation onto the basal dendrites of nearby pyramidal neurons and a site midway out on the shafts of the apical dendrites of distant pyramidal neurons (Chap. 10). Another example is the Schaffer collateral system of the hippocampus, in which recurrent collaterals from pyramidal neurons of CA3 make excitatory connections onto the apical shafts of pyramidal neurons in CA3 and CA1 (Chap. 11). Evidence for direct feedback excitation has also been obtained in reptilian dorsal general cortex, which is regarded as a model for the evolutionary precursor of mammalian neocortex (Kriegstein and Connors, 1986). In the neocortex itself, pyramidal neurons have well-developed recurrent axon collateral systems, and there has long been evidence for excitatory actions attributable to them. A very important and massive interregional feedback system originates among the pyramidal cells in the lower layers of cortex and projects back to those specific thalamic nuclei that provide the input to the cortex (Chap. 12).

The significance of the local feedback connections is that activated pyramidal neurons can respond to an initial excitatory input with subsequent waves of reexcitation. Through this means, a subset of activated pyramidal neurons imposes a subsequent pattern of activation onto an overlapping subset of pyramidal neurons in the same region. It is believed that this is a powerful mechanism for achieving combinatorial patterns of activation reflecting both the pattern of the input signal and the experience-dependent patterns stored within the distributed connections of the local circuits of the region in question (see Haberly, 1985; Granger et al., 1988; Wilson and Bower, 1988).

Comparisons between the local circuits in the different types of cortex, as discussed above, give a different perspective on cortical organization from that derived from traditional views based on cortical cytoarchitectonics. From this new perspective, what is striking is the local circuit elements that are common to the different types. We have thus far emphasized the pyramidal neuron and the intrinsic circuits for feedforward and feedback excitation. These are paralleled by circuits for feedforward and feedback inhibition, respectively. Inhibitory circuits and their functions will be discussed in detail below. Here we note that 20% of cortical cells stain for the inhibitory neurotransmitter GABA; these can be subdivided into a large number of subclasses, according to their morphology, size, location, postsynaptic targets, and functions (Fitzpatrick et al., 1987). Thus, cortical inhibitory circuits are likely to subserve a number of different cortical operations (see below).

The full set of local circuits depicted in Fig. 1.9 and listed in the legend is in fact a unique property of the vertebrate cerebral cortex. It constitutes *a basic circuit for cerebral cortical organization,* which is adapted and elaborated in the different types of cortex and the different cortical regions in order to carry out the set of operations characteristic of each (Shepherd, 1974, 1988). It shares many common elements and properties with the "canonical cortical circuit" of Douglas and Martin (Chap. 12). A full exposition of these circuit elements and their roles in different cortical functions is given in Chaps. 10–12. The general properties of basic circuits are discussed further below.

As is already clear, inhibitory circuits play large roles in determining the types of operations generated within a region. This is supported by studies showing that if synaptically mediated inhibition is blocked pharmacologically, cells lose most of their distinguishing features, such as their sensitivity to bars of a certain orientation or moving in a particular direction (see Chap. 12).

We will examine here three examples of inhibitory local circuits. Our aim, first, is to show how inhibitory circuits can give rise to specific functions. A second aim is to consider these circuits within a comparative framework. This will allow us to see that each circuit employs an inhibitory neuron in a slightly different way, so that distinctly different operations can be generated by relatively fine tuning of the connections of a generic inhibitory interneuron.

Rhythmic generation. Rhythmic activity can be generated by intrinsic membrane properties, such as in the pacemaker neurons in central pattern generator circuits controlling breathing, walking and other highly stereotyped behaviors in invertebrates.

In fact, since the early 1980s, research carried out on brain slices has shown that a large number of cells in the central nervous system, such as thalamic neurons (Chap. 8) and neurons in the inferior olive, possess very complex and highly nonlinear ionic conductances that endow these cells with the ability to respond to inputs with oscillations at various frequencies (Llinás, 1988). Thus, intrinsic oscillatory neurons may be far more common in the CNS than previously thought, enhancing the computational power of the system. An introduction to these ionic conductances is provided in Chap. 2.

Rhythmic activity can also be a property of local circuit interactions. A common model, shown in Fig. 1.10A, consists of output neurons (b) connected through axon collaterals to inhibitory neurons (i), which in turn connect back into the output neurons. When an input (a) activates the output neurons, they begin to generate impulses, which leads to activation of the interneurons. This activation leads to feedback inhibition of the output neurons, which now can no longer respond to the input and thus also deprives the interneurons of their source of activation; they are, in a sense, presynaptically inhibited by themselves. Both populations, therefore, are silent until the IPSPs in the output neurons wear off and the cycle is ready to be repeated. The degree of synchronization of a region will obviously depend on the extensiveness of the connections. The circuit could thus be laid down during development by a simple rule for the interneurons to make extensive random connections on the output cell populations.

Rhythmic activity can also be generated by dendrodendritic microcircuits, as discussed above (see Fig. 1.5B). Although the neural elements are different, the principles underlying the interactions are similar. This illustrates an important concept, that similar functions can be mediated by different neural substrates. Conversely, similar substrates can mediate different functions, by specific adaptations of general mechanisms.

High-frequency oscillations (40–60 Hz) of a large proportion of the neuronal population in a given area appear to be a common occurrence in the awake, behaving animal. Evidence in both the olfactory bulb and olfactory cortex (Freeman, 1983) as well as in the cat visual cortex relates these oscillations to perception and behavior. Thus, both at the single-cell and at the network level, neurons show very complex dynamic patterns of neuronal excitability which are essential for the neural computations which they carry out. This is another property of real neurons that needs to be incorporated into network simulation and connectionist models.

Receptive field organization. These concepts are further exemplified by the role of inhibitory circuits in mediating enhancement of spatial contrast, a common property of receptive field organization in many sensory systems. This property is illustrated in Fig. 1.10B, where there is strong stimulation of input (a_1) and all elements to the left in the diagram (not shown), and weaker stimulation of input (a_2) and elements lying to the right. The responses of b_1 and b_2 would start out being proportional; however, the stronger inhibition by b_1 and i_1 suppresses b_2 more than the suppression of b_1 by b_2 and i_2, thereby enhancing the difference in

A. RHYTHM GENERATION

C. DIRECTION SELECTIVITY

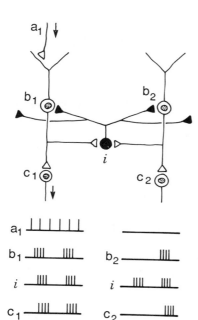

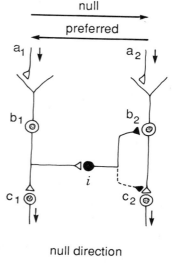

B. SPATIAL CONTRAST

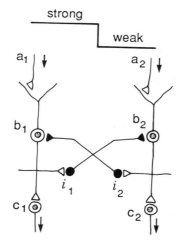

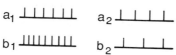

FIG. 1.10. Intrinsic inhibitory circuits are organized to mediate different types of functional operations characteristic of a given region. **A.** Rhythm generation (a_1, input; b_1, b_2, relay neurons; c_1, c_2 targets of relay neurons; i, inhibitory interneuron). Impulse firing patterns are shown below. **B.** Spatial contrast. Stimulation consists of strong and weak areas of stimulation, with a sharp edge between them. **C.** Direction selectivity. Arrows indicate movement of a stimulus in the null and the preferred direction. See text.

26

firing rates of b_1 and b_2. This effects falls off the further away the elements are from the border, thereby giving the effect of enhancing the contrast between strong and weak simulation at the border.

This is the basis for the classical description of Mach bands in the visual system, as first demonstrated in the *Limulus* eye (see Ratliff, 1965). Paradoxically, the circuit for mediating this effect in *Limulus* appears to involve dendrodendritic connections without intervention of an inhibitory interneuron (Fahrenbach, 1985). In mammalian sensory systems, however, it characteristically involves inhibitory interneurons, interacting with the output neurons through the type of local circuits indicated in Fig. 1.10B, or through dendrodendritic microcircuits.

Directional selectivity. A third type of local circuit in which an inhibitory interneuron plays an essential role is in direction selectivity. The best elaborated model is the retina, where ganglion cells in most vertebrate species show selective activation by stimuli moving in one direction. The proposed model is shown schematically in Fig. 1.10C. The essential element is an inhibitory interneuron whose connections are made in one direction, *opposite* to the preferred direction. This means that stimuli moving in that direction (called the "null" direction) activate the inhibitory connections in that direction, so that cells further along cannot respond. In the opposite, preferred, direction, by contrast, the cells are free to respond.

Several types of circuit connections might mediate this selectivity. In the retina the most likely involves connections onto the dendrite and somata of ganglion cells (dashed line in Fig. 1.10C). The starburst amacrine cell has been proposed to play this role. The organization of this cell has been discussed above (see Fig. 1.8B). The recent discovery that these cells synthesize, store, and release the two most common fast-acting excitatory (ACh) and inhibitory (GABA) neurotransmitters suggests that one and the same cell could potentially act as both an excitatory and an inhibitory circuit element. Directionally selective cells are also found in visual cortex, where this same property is mediated by a combination of excitatory inputs and inhibitory interneurons (Ferster and Koch, 1987; see above), whose connections are mainly onto cell bodies and proximal dendrites (solid line in Fig. 1.10C). This site is less selective, but might be easier to target during development. Blocking the action of the inhibitory cells by an appropriate chemical substance leads to the almost total loss of direction selectivity (Sillito, 1980). Interestingly, interneurons, whether excitatory or inhibitory, appear to be absent or inoperant in the very young, immature cortex, and begin to exert their specific effect only after the development of the extrinsic projection neurons (Jacobson, 1978).

THE BASIC CIRCUIT OF A BRAIN REGION

From this overview it can be seen that each brain region contains a plethora of types of circuits at successive levels of organization. How can one represent all this information in a way that is not merely a catalog, but rather gives insight into the functions of each region?

A useful way to do this is by means of a *basic circuit*. This can be defined in a general way as a diagram representing the main patterns of synaptic connections and interactions most characteristic of a given region. Such a representation is useful in several ways: for identifying the principles of circuit organization of a region; for better understanding of relations between synaptic actions and dendritic properties; for making comparisons between the organization of different regions; finally, for identifying those aspects that must be included if a network simulation is to have validity as an accurate representation of that region.

An example of a basic circuit has already been presented in Fig. 1.9, which represents the key elements and connections that are found in the cerebral cortex. This particular example emphasizes those aspects that are common to different types of cortex, as discussed in the text. For closer examination of a given type, one could introduce the particular adaptations and elaborations that underlie its unique properties. The basic circuit is thus a flexible tool, not rigidly defined; the purpose is to represent the minimum of elements and connections that will capture the essence of the functional operations of a region.

The utility of this concept in comparing different regions is further illustrated in Fig. 1.11. On the left is shown a basic circuit of the retina; on the right, the olfactory bulb. Despite the fact that these regions process entirely different sensory modalities, the basic circuits are similar in outline and in several details. In each case, there are parallel vertical pathways for straight-through transmission of information. In addition, there are horizontal connections, arranged in two main levels, for processing of information by lateral interactions. Within this framework are further similarities, such as reciprocal synapses and interneurons that lack axons.

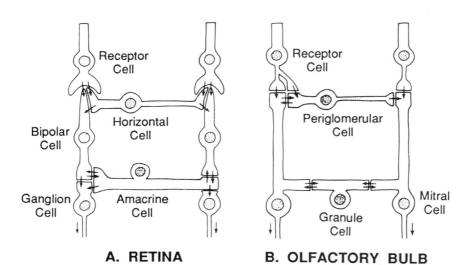

A. RETINA   **B. OLFACTORY BULB**

FIG. 1.11. Comparison between basic circuits for the retina (**A**) and olfactory bulb (**B**). (From Shepherd, 1988a.)

The purpose of such a comparison is not, of course, to claim that the two regions are identical; rather, it is to be able to identify more clearly the principles that are common across regions, so that one can better analyze and understand the adaptations that make each region unique. Among the common principles are the notions that in each region, there is an initial stage of *input processing,* a second stage of *intrinsic operations* within the synaptic circuits of the region, and a final stage of *output control.* These three levels can be seen most clearly in the basic circuits of tightly organized and highly laminar regions like the olfactory bulb and retina (see Fig. 1.11), but also are evident in more spread-out regions like the cerebral cortex (Fig. 1.9). This approach thus provides a useful starting point for categorizing the overall organization of a region.

A possible objection to the idea of a basic circuit is that it does not adequately represent the rich diversity of neural elements and synaptic connections that can be found in most brain regions. But the problem with this diversity is that it obscures the crucial issue, of determining which properties are essential for which operations. This issue is critical, not only for experimentalists attempting to analyze these operations, but also for theorists who seek to incorporate these essential properties into network simulations.

FROM SYNAPTIC CIRCUITS TO NETWORK PROPERTIES

Now that we have examined the individual circuit components making up most brain regions, and have studied some of the operations they can carry out, we are confronted with the challenge of understanding the resulting neural network, consisting usually of hundreds of thousands of nerve cells interconnected by millions of synapses. These networks show emergent behaviors that may not be apparent or implied in their constituent elements. Furthermore, an understanding of the way the components are organized through their synaptic connections will be essential to understanding the functional properties and operating ranges of these behaviors.

An example of such an emergent property is shown in Fig. 1.12. Here are illustrated three connection schemes possibly underlying orientation selectivity in the mammalian visual cortex. Whereas cells in the lateral geniculate nucleus have circular symmetrical receptive fields, their targets in primary visual cortex have elongated receptive fields, responding best to an oriented bar (Fig. 1.12a). A large number of different wiring schemes have been postulated to explain this phenomenon, starting with the influential study of Hubel and Wiesel (1962), who postulated that a number of geniculate cells, with their receptive field centers arranged in a row, converge onto a cortical cell (Fig. 1.12b); for an overview, see Rose and Dobson (1985). Another class of models involves the use of the feedforward type of inhibition to prevent the cortical cell from firing if the visual stimulus falls on a neighboring region, outside the bar-shaped receptive field (Fig. 1.12c; "iso-orientation inhibition"; Sillito, 1980; Heggelund, 1981). So far, however, no significant population of nonoriented cells that could provide such inhibition has been reported in cat visual cortex. The same operation, however, can be accomplished by using a massively interconnected network of inhibitory cells, as indicated in Fig. 1.12d. If a vertical bar stimulus falls on the

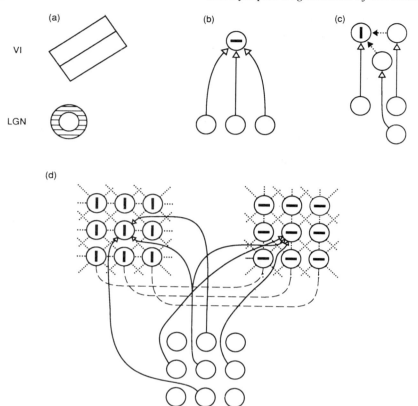

FIG. 1.12. Illustration of the way by which a selective response (e.g., orientation selectivity) arises as a network property. See text. (From Ferster and Koch, 1987.)

receptive fields of the central row of three geniculate cells, cortical cells in the vertical orientation column (left side in Fig. 1.12d) will be optimally activated and the cell will fire. These cortical neurons then also inhibit cells (dashed lines) in the horizontal orientation column (right side in Fig. 1.12d). This sort of inhibition is referred to as cross-orientation inhibition. If the bar is now tilted by 45°, such as to excite the diagonally located geniculate cells, local inhibition among similarly oriented cortical neurons (dotted lines) will prevent, or at least reduce, the excitability of the cortical cells in the vertical orientation column. Similarly, if the stimulus bar is vertical but broad, covering more than the central row of geniculate cells, the cortical response will be muted due to this instance of lateral inhibition.

In the model of Fig. 1.12d, orientation selectivity is a collective property of the network due to the massive asymmetrical inhibition among cortical cells, and not a property of two or three cells. This, together with a superimposed Hubel-and-Wiesel-like excitatory scheme would ensure selectivity to oriented objects. An important consequence is that no single neuron or biophysical mechanism is solely responsible for this orientation tuning, since they act synergistically. Thus, if any one component fails, the overall system will perform, albeit at a somewhat reduced level of performance (referred to as "graceful degradation" in engineering terms). This will probably turn out to be a major feature of most biological computations. A further point is that even though all the different circuits in Fig. 1.12 lead to orientation selectivity, we will never discover which ones are correct by mere theoretical considerations. Although many models have been proposed, all able in some measure to demonstrate orientation selectivity, it is only the knowledge of the local synaptic circuits, the location and strength of inhibitory synapses, the detailed pharmacology, etc., which will enable us to identify the correct circuitry underlying even this rather simple computation.

2

MEMBRANE PROPERTIES AND NEUROTRANSMITTER ACTIONS

DAVID A. McCORMICK

Information processing depends not only on the anatomical substrates of synaptic circuits but also on the electrophysiological properties of neurons and neuronal elements, and how these properties are altered and tuned by the plethora of neuroactive substances impinging upon them. Even if two neurons in different regions of the nervous system should possess identical morphological features, they may respond to the same synaptic input in very different ways, owing to each cell's intrinsic properties. To understand synaptic organization and function in different regions of the nervous system therefore requires an understanding of the electrophysiological and pharmacological properties of each of the constituent neuronal elements.

The electrophysiological behavior of a neuron is determined by the presence and distribution of different ionic currents in that cell, and the ability of various neurotransmitters either to increase or to decrease the amplitude of these currents. This chapter will provide a general overview of neuronal currents known to exist in brain cells, how they may be modulated by neurotransmitters, and how the interplay between the two can result in complicated patterns of activity in synaptic circuits. For a more detailed introduction to the biophysical mechanisms of ionic currents in neurons, the reader is referred to Hille (1984) and Kuffler et al. (1984).

MEMBRANES AND IONIC CURRENTS

Neurons, like cells elsewhere in the body, are bounded by a lipid bilayer membrane which contains a large number of protein macromolecules. The lipid bilayer allows the composition of the medium on either side to be very different. Of particular importance for electrical signaling is the fact that certain key ions have different concentrations on the inside and outside of the neuron (Fig. 2.1). On the outside, Na^+, Ca^{2+}, and Cl^- exist in much higher concentrations; by contrast, K^+ ions and membrane-impermeant anions (denoted as A^-) are concentrated on the inside.

32

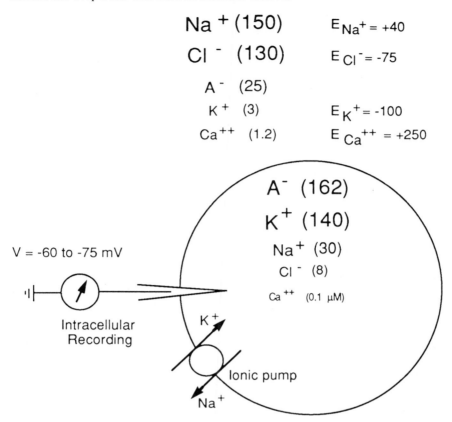

FIG. 2.1. Distribution of ions across neuronal membranes (concentrations in mM) and their equilibrium potentials (expressed as mV). At rest the cell membrane is permeable to K^+, Na^+, and Cl^- and exhibits a voltage difference (inside vs. outside) of approximately -60 to -75 mV, as seen by an intracellular recording electrode.

Protein macromolecules in the membrane subserve a variety of functions. Those that underly electrical signaling are large molecules that form ionic channels. Membrane channels possess a number of important features, including the presence of a water-filled pore through which ions flow; selectivity for one or more types of ions (e.g., K^+, Na^+, Cl^-, Ca^{2+}); sensitivity to (i.e., opening or closing in response to) the electrical potential across the membrane or to a neurotransmitter substance, or both; and the ability to be modified by a variety of intracellular biochemical signals.

Because ions are unequally distributed across the membrane, they tend to diffuse down their concentration gradient through ionic channels. This tendency arises from the fact that the intrinsic movements of ions in a solution tend to disperse them from regions of higher to lower concentration. However, because ions are electrically charged molecules, their movements are dictated not only by concentration gradients but also by the voltage difference across the membrane.

For example, if the membrane of a neuron were made permeable to K^+ ions by opening K^+ channels, the higher concentration of these ions on the inside versus the outside would make it more probable that K^+ ions would leave, rather than enter, the cell. As K^+ ions exit the cell, they carry positive charge with them, thereby leaving behind a net negative charge (made up in part by impermeant anions; Fig. 2.1). However, this negative charge (expressed as a *voltage difference*) on the inside versus the outside of the cell will attract the K^+ ions and slow down the rate at which they leave. At some point, the tendency for K^+ to flow out of the cell will be offset exactly by the attraction of the negative charge left inside the cell. The voltage difference at which this occurs is known as the *equilibrium potential* (denoted as E) and is different for each ionic species (see Figs. 2.1 and 2.2). It will be seen later in this and other chapters that the equilibrium potential is important for determining the effect of activation (synaptic or intrinsic) of an ionic current.

The flow of ions across the membrane obeys physical laws in a consistent and reproducible manner. Therefore, the equilibrium potential for any ion can be calculated from its intracellular and extracellular concentrations by using the Nernst equation (see Hille, 1984; Kuffler et al., 1984; Shepherd, 1988a). In the CNS, for example, K^+ is approximately 140 mM on the inside and 3 mM on the outside of neurons, corresponding, as shown in Fig. 2.1, to an equilibrium potential (E_K) of approximately -100 mV (inside vs. outside). In this manner, the tendency for K^+ to flow across the membrane down its concentration gradient (i.e., out of the neuron) is equally offset by the attractive nature of the negative membrane potential inside the cell only when the voltage difference is -100 mV.

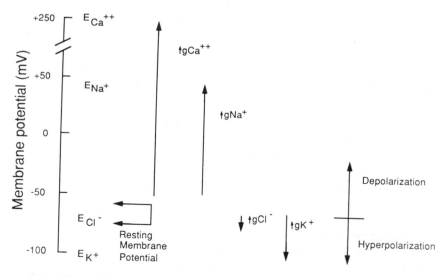

FIG. 2.2. Effect of increasing membrane conductance (denoted as g) to Ca^{2+}, Na^+, Cl^-, or K^+. Increases in gCa^{2+} or gNa^+ bring the membrane potential toward more positive values (*depolarization*), whereas increases in gCl^- or gK^+ bring it toward more negative values (*hyperpolarization*).

The same mechanism underlies the equilibrium potential of other ions (see Fig. 2.2).

The ease with which an ion diffuses across the membrane is expressed as the ion's *permeability*. Increasing the permeability of the membrane to a particular ionic species (e.g., by increasing the probability that membrane channels which conduct that ion will be open) increases the electrical conductance and brings the membrane potential of the cell closer to the equilibrium potential of that ion. This is true whether the membrane potential subsequently becomes more negative (i.e., hyperpolarized) or more positive (i.e., depolarized). Of course, if the membrane potential is already at the equilibrium potential, its value will not change in response to a further increase in conductance. In this circumstance, the most significant change will be that the ability of other currents to move the membrane potential away from its present potential will be diminished. For example, if the membrane were only slightly permeable to Cl^- ions, then increases in membrane conductance to other ionic species could easily move the membrane potential away from E_{Cl^-}. However, if the permeability to Cl^- was greatly increased, the membrane potential would be effectively "clamped" close to E_{Cl^-}. In this circumstance, movements of other ions into or out of the cell would now largely be offset by compensating movements of Cl^- ions, thereby keeping the membrane potential close to E_{Cl^-}. As we shall see below, this type of "shunting" of the membrane potential near E_{Cl^-} is important in the actions of some types of inhibitory neurotransmitters.

RESTING MEMBRANE POTENTIAL

When there is no synaptic input, in other words when the neuron is "at rest," the cellular membrane is dominated by its permeability to K^+. This permeability to potassium ions draws the membrane potential of the cell toward approximately -100 mV (see Figs. 2.1 and 2.2). If the membrane were only permeable to K^+, then the membrane potential would be equal to E_K. However, even at rest, neuronal membranes are also permeable to other ions, Na^+ and Cl^- in particular, so that the membrane potential is also pulled toward E_{Na^+} ($+40$ mV) and E_{Cl^-} (-75 mV). The point at which the movements of these varied ions come into equilibrium such that there is no *net* current (denoted as I) flow across the membrane corresponds to the resting membrane potential and is typically between -60 to -80 mV (see Fig. 2.2). In this manner, the weighted mixture of all of the ionic currents flowing across the membrane determines the resting membrane potential, as well as the membrane potential during nearly all types of activity. This principle allows the calculation of the membrane potential at any given point in time by use of the *Goldman–Hodgkin–Katz* equation based upon the concentration gradient and membrane permeability of each ion (Goldman, 1943; Hodgkin and Katz, 1949). A consequence of this relationship of ionic currents and membrane potential is that, in general, the membrane potential of the cell will be closest to the equilibrium potential of the ion to which the membrane is most permeable.

The exact value of the resting membrane potential varies between different types of neurons and is very important in determining the manner in which a particular neuron behaves both spontaneously and in response to extrinsic inputs. For example, cortical pyramidal cells (Chap. 12) have a resting membrane potential in the absence of synaptic input of approximately -75 mV, thalamic relay neurons (Chap. 8) are at approximately -65 mV at rest, and retinal photoreceptor cells (Chap. 6) have a resting membrane potential of approximately -40 mV. Some types of neurons do not have a true "resting" membrane potential because they are spontaneously active even during the lack of all synaptic input (see below).

ACTION POTENTIAL

Rapid signaling in nerve cells is accomplished by brief changes in the membrane potential. Traditionally, the most characteristic type of signal has been considered to be the action potential, or nerve impulse (also referred to as a "spike"). Action potentials represent a rapid depolarization of the membrane potential from threshold (approximately -55 mV) to positive to 0 mV. One of their main functions is to communicate information from the soma to the terminals of a cell's axon. Local action potentials in patches of dendritic membrane can serve as boosters for the spread of synaptic potentials to the soma (as discussed in Chap. 1; see also Appendix).

The basic changes in membrane ionic permeability that underlie the action potential were first well characterized by Alan Hodgkin and Andrew Huxley (1952a–d) using the squid giant axon preparation. The squid giant axon is very large, approximately 1 mm in diameter, as its name implies. This allowed these investigators to thread a wire into the axon, giving them the ability to control accurately the membrane potential by a procedure known as *voltage clamp.* In this procedure, the amount of current injected into the cell is adjusted so that the voltage across the membrane is kept constant (i.e., the voltage is "clamped"). This technique allows one not only to observe directly the transmembrane currents responsible for the electrical behavior of the cell, but also to measure the current's kinetics and sensitivity to membrane potential. The isolation of the squid giant axon in vitro meant that Hodgkin and Huxley could also control the ionic composition of the medium surrounding the axon. These experiments revealed that the rapid upswing of the action potential is mediated by a regenerative increase in a transient Na^+ current, which we denote as $I_{Na,t}$ (Fig. 2.3). Because $I_{Na,t}$ is rapidly activated by depolarization and is itself a depolarizing influence, it forms a positive feedback loop, as shown diagrammatically in Fig. 2.3A. Depolarization of the membrane causes a rapid increase in the number of Na^+ channels that are open, thereby allowing more Na^+ ions to enter the cell, depolarizing that site even more, and so on.

Membrane currents that change their amplitude in response to changes in the membrane potential in a "nonlinear" manner, such as $I_{Na,t}$, are referred to as *voltage sensitive,* whereas the ionic channels that underlie these currents are said to be *voltage gated.* The depolarization caused by entry of Na^+ ions into a localized region of membrane spreads to neighboring membrane by *electrotonic* current flow (see Appendix). The depolarization of this new region of membrane

THE ACTION POTENTIAL

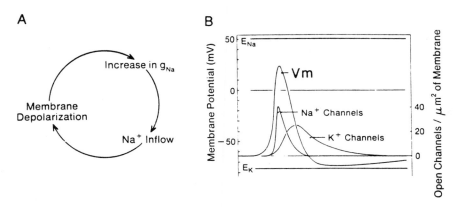

FIG. 2.3. **A.** Regenerative relation among membrane depolarization, increase in membrane conductance to Na$^+$ (g_{Na}), and Na$^+$ current that underlies the action potential. **B.** Reconstruction of changes in ionic conductance underlying the action potential in squid giant axon; scale for the membrane potential (V_m) is shown on the left. The equilibrium potentials for Na (E_{Na}) and K (E_K) are also indicated on the left. Changes in Na and K ionic conductances are scaled on the right in terms of calculated open channels per square micrometer of membrane. (Adapted from Hodgkin and Huxley, 1952a–d; Hille, 1984.)

may then activate the same regenerative mechanism and thus bring about the propagation of the action potential along the axon.

Repolarization of the action potential is very important not only because of the obvious need to be able to generate more than one impulse during the life of the cell, but also for determining the way the cell responds to repetitive inputs. Two processes are important for repolarization of the action potential: rapid inactivation of $I_{Na,t}$ and activation of K$^+$ currents. The rate of *inactivation* (termed "inactivation kinetics") of $I_{Na,t}$ is only slightly slower than the kinetics of its activation. Even during the rising phase of the action potential, the available Na$^+$ current is gradually reduced as a result of inactivation. Simultaneously, but with a slower time course, a delayed K$^+$ current, known as I_K, is activated by the membrane depolarization associated with the action potential, allowing K$^+$ to leave the cell. These two currents, flowing through their respective ionic channels, are indicated in Fig. 2.3B. At some point, the hyperpolarizing influence of K$^+$ leaving overcomes the depolarizing influence of Na$^+$ entering, thereby terminating the action potential and repolarizing the membrane.

The triggering of an action potential occurs when the membrane potential of the neuron is depolarized sufficiently to reach action potential *threshold*. In many cells this potential is approximately -50 to -60 mV. *Action potential threshold* is the membrane potential at which the regenerative activation of depolarizing currents (e.g., $I_{Na,t}$) is strong enough to overcome the inactivation of these currents as well as the activation of other currents that hyperpolarize the neuron back toward rest. At threshold, the generation of an action potential is an "all-or-nothing" event. If threshold is surpassed, an action potential is generated and the information is transferred down the axon to cause the release of neurotransmitter at synapses. If firing threshold is not reached by a depolarizing event, an action potential is not produced and the event is not relayed to other cells. However, as we shall see, the depolarization can still serve to modify the probability that other postsynaptic potentials in the neuron may do so.

Since 1952 it has become apparent that $I_{Na,t}$ is the dominant depolarizing current in the generation of action potentials in axons and cell bodies. However, in somatic and dendritic regions, voltage-gated Ca^{2+} currents are also involved, as documented below and in ensuing chapters (e.g., see Fig. 2.5B,C, and E). In mammalian somata and dendrites, in contrast to squid giant axon, repolarization of action potentials is accomplished not only by I_K, but also by a complicated array of different K^+ currents (see sections on I_K, I_C, and I_A below).

IONIC PUMPS

The quantity of ions that enter and exit the cell during electrical activity is actually very small in comparison with the number of ions present. However, even these small exchanges of ions across the membrane coupled with a constant "leak" at rest can eventually destroy the correct ionic distribution and thereby render a neuron nonfunctional. To compensate for this "rundown," neuronal membranes possess specialized protein macromolecules known as *ionic pumps*. Ionic pumps maintain the correct distribution of all of the ions involved in electrical activity by actively transporting these ions "upstream" against their concentration gradient. The energy required to perform this task is obtained in some cases through the hydrolysis of ATP (adenosine triphosphate). The ionic pump that has been best characterized is the "electrogenic sodium–potassium" pump (see Kuffler et al., 1984; Shepherd, 1988). This ionic pump carries approximately three Na^+ ions out for every two K^+ ions it brings in, thereby generating an electric current (Fig. 2.1). The exact amplitude of this current depends upon the rate at which the pump is active, which is in turn related to the intracellular concentration of Na^+ and the extracellular concentration of K^+. Ionic pumps are essential and important constituents of neurons and neuronal membranes. Their time scale of action is seconds and minutes, and they are therefore thought of as being more involved in long-term, rather than short-term, neuronal processing.

TYPES OF IONIC CURRENTS

Neurons in the mammalian nervous system do not simply lie at rest and occasionally generate an action potential. Rather, neuronal membranes are in a constant state of flux, due to the presence of a remarkable variety of ionic currents (Table 2.1). These currents are distinguished not only by the ions that they conduct (e.g., K^+, Na^+, Ca^{2+}, Cl^-), but also by their time course, sensitivity to membrane potential, and sensitivity to neurotransmitters and other chemical agents (for review, see Hille, 1984; Adams and Galvan, 1986; Jahnsen, 1986; Rudy, 1988; Tsien et al., 1988). As the various ionic currents were discovered, they were divided into two general categories: those that are sensitive to changes

Table 2.1. Neuronal Ionic Currents

Current	Description	Function
Na+		
$I_{Na,t}$	Transient; rapidly inactivating	Action potentials
$I_{Na,p}$	Persistent; non-inactivating	Enhances depolarizations; contributes to steady-state firing
Ca²+		
I_T, low threshold	"Transient"; rapidly inactivating; threshold negative to −65 mV	Underlies rhythmic burst firing
I_L, high threshold	"Long-lasting"; slowly inactivating; threshold around −20 mV	Underlies Ca²+ spikes that are prominent in dendrites; involved in synaptic transmission
I_N	"Neither"; rapidly inactivating; threshold around −20 mV	Underlies Ca²+ spikes that are prominent in dendrites; involved in synaptic transmission
K+		
I_K	Activated by strong depolarization	Repolarization of action potential
I_C	Activated by increases in $[Ca^{2+}]_i$	Action potential repolarization and interspike interval
I_{AHP}	Slow afterhyperpolarization; sensitive to increases in $[Ca^{2+}]_i$	Slow adaptation of action potential discharge
I_A	Transient, inactivating	Delayed onset of firing; interspike interval; action potential repolarization
I_M	"Muscarine" sensitive; activated by depolarization	Contributes to spike frequency adaptation
I_Q, I_h	"Queer"; activated by hyperpolarization/mixed cation current	Prevents strong hyperpolarization
$I_{K,leak}$	Contributes to neuronal resting "leak" conductance	Helps determine resting membrane potential

in membrane potential and those that are altered by neurotransmitters. However, with the recent discovery of a number of voltage-sensitive ionic channels that are also gated by neurotransmitters, and vice versa, it has become apparent that there is substantial overlap between these two groups. The currents that possess both voltage and neurotransmitter sensitivity have received much recent attention because of their ability to modulate the electrical behavior of neurons in unusual and interesting ways (see section on Chemical Synapses, below).

Most currents sensitive to membrane potential are turned on (i.e., *activated*) by depolarization. The rate at which they activate, and the membrane potential at which they start to become active (i.e., *threshold*), are important characteristics. Many voltage-dependent currents do not stay on once they are activated, even during a constant shift in membrane potential. The process by which they turn off despite a stable level of membrane potential in their activation range is known as *inactivation*. Inactivation is a state of the current and ionic channels that is distinct from simple channel closure. Once a current becomes inactive, this inactivation must be removed before it can again be activated. *Removal of inactivation* is generally achieved by repolarization of the membrane potential. Like the process of activation, inactivation and removal of inactivation are time and membrane potential dependent. Together, all of these characteristics define the temporal and voltage domain over which the current influences the electrical activity of the neuron.

The names given to each ionic current often reflect a property that distinguishes that current from the others. If the current is activated by relatively small deviations in the membrane potential (denoted as V_m) from rest, then it may be referred to as *low threshold* (e.g., low-threshold Ca^{2+} current), whereas if the current is activated only at levels that are substantially positive (depolarized) from rest, the current may be referred to as *high threshold* (e.g., high-threshold Ca^{2+} current). In addition, if activation of the current through a constant and steady change in membrane potential (i.e., under voltage-clamp conditions in which the membrane potential is held constant) leads to only a transient response, then it is referred to as *transient* or *rapidly inactivating* (examples are the A-current and the T-current). Likewise, a current that persists during constant activation (i.e., is *noninactivating*) is known as *sustained, persistent* or *long-lasting* (e.g., persistent Na^+ current and the L-current).

The ionic currents that determine the neuronal firing behavior of neurons in different regions of the nervous system have been intensively investigated. To date, at least a dozen distinct currents, many of which are common to neurons at all levels of the neuraxis, have been identified (Table 2.1). We will briefly summarize these currents and the unique contribution that each makes to the firing behavior of neurons (see Fig. 2.4).

SODIUM (Na) CURRENTS

Two sodium currents, $I_{Na,t}$ (transient) and $I_{Na,p}$ (persistent), are widely distributed in neurons from different regions of the nervous system. These two currents are distinguished from each other by their rate of inactivation, their threshold for activation, and their amplitude.

Transient sodium ($I_{Na,t}$). As we have noted, the transient Na$^+$ ($I_{Na,t}$) current rapidly activates and then inactivates within a few milliseconds during steady depolarization. All central neurons studied to date possess a large $I_{Na,t}$, whereas $I_{Na,p}$ is considerably smaller in amplitude. The rapid activation and inactivation properties of $I_{Na,t}$ make this current ideal for its role in the generation of trains of action potentials (Fig. 2.4A).

Persistent sodium ($I_{Na,p}$). In contrast to $I_{Na,t}$, the persistent Na$^+$ ($I_{Na,p}$) current shows little, if any, inactivation. This current is also rapidly activated by membrane depolarization, but its noninactivating nature and lower threshold allow it to serve a very different role in neuronal function (Hotson et al., 1979; Llinas, 1981; Stafstrom et al., 1985). A large percentage of neuronal computations occur in a narrow range of membrane potential between approximately -75 and -50 mV. This range is between resting membrane potential and a level of depolarization at which the neuron is firing repeatedly at a high rate. The nature of $I_{Na,p}$ is such that it is activated by depolarizations, such as synaptic potentials, which bring the membrane potential from rest to near action potential firing threshold. The added depolarizing influence of the influx of Na$^+$ ions resulting from the activation of $I_{Na,p}$ serves to enhance the response of the neuron to excitatory inputs, as illustrated in Fig. 2.4B. The amplitude and cellular distribution of $I_{Na,p}$ can therefore have an important role in determining the responsiveness of neurons.

The persistent nature of $I_{Na,p}$ allows this current also to participate in the determination of the baseline firing rate of neurons. $I_{Na,p}$ appears especially important to the ability of some neurons to maintain intrinsic "pacemaker" activity (e.g., the generation of action potentials in a repeated temporal pattern in the absence of synaptic input). In these cells, the steady influx of Na$^+$ ions into the neuron depolarizes the cell to above firing threshold, thereby triggering baseline activity. The membrane potential of these cells is in a state of constant change, cycling through the generation of an action potential to the repolarization of the cell (see K currents, below) to again the generation of an action potential. Examples of such neurons in the CNS include those of the locus coeruleus, dorsal raphe, and medial habenula (see Fig. 2.5F, below) (Vandermaelen and Aghajanian, 1983; Williams et al., 1984; McCormick and Prince, 1987b).

CALCIUM (Ca) CURRENTS

Calcium ions play a dual role in neurons, as modulators of neuronal firing pattern and as intermediaries of a number of nonelectrical cellular activities including neurotransmitter release, enzyme activation, metabolism, and perhaps even gene expression. The diverse involvement of Ca^{2+} ions, as well as the ubiquitous nature of Ca^{2+} currents, has led to intensive investigation of these currents.

Like Na$^+$ currents, Ca^{2+} currents depolarize neurons and are distinguished by either their persistent or their transient nature. However, Ca^{2+} currents also differ markedly in the membrane potential at which they become active. Recent nomenclature has separated neuronal Ca^{2+} currents into three groups: I_T (transient), I_L (long-lasting), and I_N (neither) (reviewed by Tsien et al., 1988).

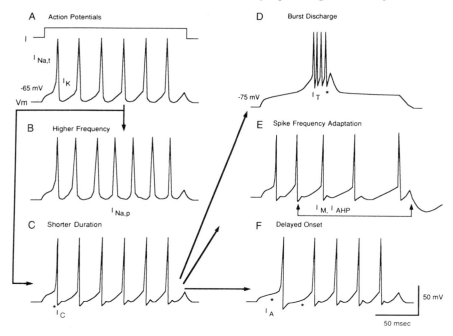

FIG. 2.4. Effect of various ionic currents on the pattern of neuronal firing generated by a model neuron. **A.** With only $I_{Na,t}$ and I_K, the neuron responds to intracellular injection of depolarizing current (A, top trace; labeled I) with a train of action potentials (V_m). **B.** Addition of the persistent Na$^+$ current, $I_{Na,p}$, results in an increase in response to the depolarizing input such that cell fires at a higher frequency. **C.** In contrast, addition of Ca^{2+} currents (I_L and I_N) and the fast Ca^{2+}-activated K current, I_C, adds a short duration hyperpolarization after each action potential. This additional K current decreases the duration of each action potential even though Ca^{2+} currents are now active and may lengthen the interval between action potentials. **D.** Addition of I_T to the neuron of C allows the cell to generate a discrete burst of fast Na$^+$-dependent action potentials due to the generation of a low-threshold Ca^{2+} "spike" (cell now held at -75 mV). **E.** Addition of I_M and I_{AHP} to the neuron of C gives rise to spike frequency adaptation. **F.** Addition of I_A causes a delay in onset of the first two action potentials. Asterisks in each case indicate the unique contribution of the current to the cell's electrophysiological behavior. Cerebral cortical pyramidal neurons generally behave like that illustrated in E, unless I_M and I_{AHP} are blocked (e.g., by acetylcholine), in which case they behave more like that of C. Thalamic relay neurons behave like those illustrated in C and D, depending upon the membrane potential (burst firing at -75 mV, single spike activity at -65 mV).

Transient and low threshold (I_T). I_T, which is also known as the "low threshold" Ca^{2+} current, is activated by small depolarizations of the membrane potential positive to approximately -65 mV (Carbonne and Lux, 1984; Nowycky et al., 1985). The transient nature of I_T, which is qualitatively similar to but much slower than $I_{Na,t}$, allows activation of this current to generate a slow "spike" or hump (Fig. 2.4D). This Ca^{2+}-mediated depolarizing spike can in turn trigger a number of short-duration Na$^+$-mediated action potentials in a high-frequency barrage known as a *"burst discharge."* After its activation, I_T inactivates, such that if V_m is tonically held positive to approximately -60 mV, I_T cannot be activated by further depolarization, as illustrated in the burst of Fig. 2.4D. I_T can be activated only if this inactivation is removed by again hyperpolarizing the membrane potential below approximately -65 mV for a sufficient period of time (approximately 50–100 msec). These properties of I_T endow the neuron with the interesting and important feature of being able to generate burst discharges at relatively hyperpolarized membrane potentials negative to approximately -65 mV (Figs. 2.4D and 2.5E), whereas slow repetitive action potentials are generated at relatively depolarized membrane potentials positive to approximately -55 mV (see Figs. 2.4C and 2.5D). The significance of these two firing modes will be discussed further below.

Recent investigations have shown that I_T is a prominent current in determining firing behavior of neurons at many different levels of the nervous system, including thalamus, medial septum, superior colliculus, inferior olive, as well as other brainstem and forebrain regions (e.g., Llinás and Yarom, 1981a,b; Jahnsen and Llinás, 1984a,b). The temporal and voltage characteristics of I_T make it a prime candidate for one of the dominant currents underlying rhythmic burst firing in these neurons. This rhythmic activity serves multiple roles in the nervous system, one of which is reflected in synchronization of the electroencephalogram during drowsiness and slow-wave sleep (Steriade and Deschênes, 1984).

Long lasting and high threshold (I_L). I_L, which is also known as the "high threshold" Ca^{2+} current, is activated by large depolarizations of the membrane potential positive to (more depolarized than) approximately -20 mV, such as those associated with action potentials. High threshold calcium currents are found in the dendrites of Purkinje cells of the cerebellum (Llinás, 1981), pyramidal cells of the cerebral cortex (Connors et al., 1982) and hippocampus (Wong et al., 1979), and inferior olivary cells of the brainstem (Llinás and Yarom, 1981a,b). These Ca^{2+} currents have profound effects on neuronal firing behavior. For example, in Purkinje cells dendritic high-threshold Ca^{2+} "spikes" can greatly modulate the pattern of discharge of the faster, more stereotyped Na^+-mediated action potentials generated in or near the cell body (see Fig. 2.5C below and Chap. 7).

Calcium influx may influence the duration of each action potential and the pattern in which they are generated by adding an additional depolarizing influence and by activating potassium currents that are sensitive to the intracellular concentration of Ca^{2+} (Fig. 2.4C; see potassium currents below). I_L, along with I_N, may provide the calcium influx that mediates this important function.

Transient and high threshold (I_N). I_N is characterized by a high threshold (approximately -20 mV) like I_L, but is transient in nature, like I_T. Thus, large sustained depolarizations only transiently activate I_N. I_N is localized in many parts of the neuron (e.g., soma, dendrites, axon terminals) and may play an important role in the presynaptic release of neurotransmitters (Tsien et al., 1988).

POTASSIUM (K) CURRENTS

Neuronal potassium currents (which are hyperpolarizing in nature) form a large and diversified group. They are intimately involved in determining the pattern of activity generated by neurons. They are responsible not only for the repolarization of the action potential, but also for the determination of the *probability* of generation of an action potential at any given point in time. As with other neuronal currents, potassium currents are distinguished by their voltage and time dependency, as well as by pharmacological techniques (for review, see Rudy, 1988).

Delayed rectifier (I_K). As we have seen, the early studies in the squid giant axon not only defined the role of the transient Na^+ current in the generation of the action potential, but also identified an equally important outward potassium current known as the *delayed rectifier* or I_K. The activation kinetics of I_K are slower than those of the transient sodium current, and therefore, it appears somewhat "delayed" (Fig. 2.3B). This potassium current is voltage sensitive, being activated at membrane potentials positive to approximately -40 mV, and only slowly inactivates. I_K is found in neurons throughout the nervous system and is believed to contribute to the repolarization of action potentials and the hyperpolarization that follows them (Figs. 2.3B and 2.4A).

Calcium-activated potassium currents. An additional class of potassium currents that are important for determining the firing behavior of neurons includes those that are Ca^{2+} sensitive (denoted $I_{K,Ca}$). This family of potassium currents is activated by increases in the intracellular concentration of unbound Ca^{2+} ($[Ca^{2+}]_i$). Two $I_{K,Ca}$s have been widely identified in neurons: I_C and I_{AHP} (see Adams and Galvan, 1986; Rudy, 1988). I_C is not only sensitive to increases in $[Ca^{2+}]_i$ in the micromolar range but is also strongly voltage dependent, becoming larger with depolarization. I_C contributes to control of the frequency of action potential generation during a steady depolarization by causing a marked hyperpolarization after the occurrence of each spike (Fig. 2.4C). I_C may even be important in some neurons in repolarization of the action potential. The voltage dependence of I_C results in its rapid inactivation once the membrane potential is repolarized. This inactivation constrains the influence of I_C in the temporal domain to tens of milliseconds or less.

I_{AHP}, in contrast to I_C, is much slower in time course and not very voltage dependent. Its influence on the membrane potential of the cell is best seen after the generation of a number of action potentials as a prolonged *afterhyperpolarization*, for which it is named. This potassium current contributes significantly to the tendency of the firing frequency of some types of neurons (e.g., cortical and hippocampal pyramidal neurons) to decrease during maintained depolarizations, a process known as *spike frequency adaptation* (Fig. 2.4E; see below).

The generation of action potentials, by increasing $[Ca^{2+}]_i$ through L- or N-type Ca^{2+} channels, triggers I_C and I_{AHP}. The hyperpolarization of the membrane potential resulting from K^+ leaving the cell during these currents regulates the rate at which the neuron fires. Due to its brief time course, I_C contributes substantially to short interspike intervals. By contrast, because of its slow activation and prolonged time course, I_{AHP} contributes more to the overall pattern of spike activity. The relative voltage-independence of I_{AHP} means that the influence of this current on the membrane potential is more closely related to changes in $[Ca^{2+}]_i$ than is I_C. Importantly, the amplitude of I_{AHP} appears to be under the control of a number of putative neurotransmitters (see Decrease of I_{AHP}, below).

Transient potassium currents. The first of a family of potassium currents that are activated by membrane depolarization and that undergo relatively rapid inactivation was discovered in molluscan neurons (Connor and Stevens, 1971) and later termed I_A. The A-current is a transient K^+ current: After its activation by depolarization of the membrane potential positive to approximately -60 mV, it rapidly inactivates. Like other transient and voltage-activated currents (e.g. $I_{Na,t}$ and I_T), this inactivation is removed by repolarization of the membrane potential. I_A is involved in the response of neurons to a sudden depolarization from hyperpolarized membrane potentials and serves to delay the onset of the generation of the first action potential (Fig. 2.4F). I_A can also slow a neuron's firing frequency during a maintained depolarization and help to repolarize the action potential. For example, in a spontaneously active neuron, the hyperpolarization that occurs after the generation of an action potential will remove some of the inactivation of I_A. As the membrane potential depolarizes back toward firing threshold, I_A will be activated and slow down the rate of depolarization. Once firing threshold is reached and an action potential is generated, the rapid depolarization may activate more of I_A, which then helps to repolarize the cell. In this manner, I_A can be an important current in the determination of firing behavior of neurons.

Muscarine-sensitive potassium current (I_M). Recently, a new type of potassium current was discovered in sympathetic ganglion neurons of bullfrogs by David Brown and Paul Adams (1980). This potassium current is activated by depolarization of the membrane potential above approximately -65 mV, does not inactivate with time, and is blocked by stimulation of muscarinic cholinergic receptors (hence its name, I_M). I_M is found in neurons throughout the nervous system, including pyramidal cells of the cerebral cortex and hippocampus. Depolarizations that are large enough to result in the generation of action potentials also cause the activation of I_M. However, because of its relatively slow kinetics and modest amplitude, I_M probably does not substantially affect the waveform of a single action potential, but rather contributes to the slow adaptation of spike frequency seen during a maintained depolarization (Fig. 2.4E).

Currents activated by hyperpolarization. Hyperpolarization of neurons in many regions of the nervous system results in the activation of a current that brings the membrane potential toward more positive values (e.g., back toward rest). This current, or family of currents, has been variously named I_Q ("queer"), I_h ("hyperpolarization-activated"), and I_f ("funny") (Halliwell and Adams, 1982; Di-

Francesco, 1985; Crepel and Penit-Soria, 1986). These currents carried by both Na^+ and K^+ ions are relatively slow in time course, although this varies widely between different cell types. The functions of these currents at this time are unclear. They may serve to stabilize the membrane potential toward rest by preventing prolonged hyperpolarizations, or they may participate in the generation of rhythmic neuronal activity.

SUMMARY OF INTRINSIC MEMBRANE PROPERTIES

Neurons possess a virtual cornucopia of different ionic currents. The magnitude, cellular distribution, and sensitivity to pharmacological manipulation of each of these ionic currents are different for every major neuronal region in the central and peripheral nervous systems. These differences result in widely varying electrophysiological properties and patterns of neuronal activity generated by cells in different parts of the brain. Each class of neuron is exquisitely "tuned" to do its particular task in the nervous system through its own special mixture of the basic ionic currents available and by the precise modulation of these currents by neuroactive substances. An analogy to this situation would be "nature versus nurture" in determining human behavior. The cells are endowed with a particular mixture of ionic currents through genetic programing (nature) which can then be modified on either a short- or long-term basis through development or the actions of a number of substances impinging upon the cell (nurture).

Examples of the different electrophysiological "behaviors" of neurons due to different combinations of ionic currents are illustrated in Fig. 2.5. Cortical pyramidal neurons respond to a depolarizing current pulse with a train (Fig. 2.5A) or a burst (Fig. 2.5B) of action potentials (Connors et al., 1982; McCormick et al., 1985). The spike frequency adaptation of cortical pyramidal neurons (Fig. 2.5A) is due to the presence I_{AHP} and I_M. In contrast to neocortical pyramidal neurons, the major output cell of the cerebellar cortex, the Purkinje cell, responds to a depolarizing current pulse with a high-frequency discharge of short-duration action potentials (Fig. 2.5C). This high-frequency discharge is modulated by dendritic calcium spikes (Fig. 2.5C asterisks) as well as by prolonged sodium ($I_{Na,p}$) and calcium currents (Fig. 2.5C, arrow heads).

Thalamic relay neurons are unusual in that they possess two distinct modes of action potential generation: single spike activity when depolarized above -65 mV (Fig. 2.5D) and burst firing when depolarized at or negative to -75 mV (Fig. 2.5E). Thalamic neurons respond with a burst of action potentials at -75 mV, due to the presence of a large I_T, which is completely inactivated at membrane potentials above -65 mV.

Some neurons display spontaneous activity in a regular and stereotyped manner, even in the lack of all synaptic input, such as the medial habenular neuron illustrated in Fig. 2.5F. These cells appear to possess prolonged and complicated spike afterhyperpolarizations (arrows) which help determine the rate at which the action potentials are generated.

Although the electrophysiological behaviors of neurons can be markedly changed by the neurotransmitter "environment," they also remain distinct in that

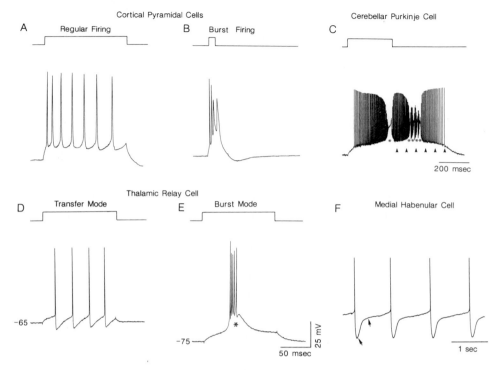

FIG. 2.5. Electrophysiological behavior of neurons in different regions of the mammalian brain. **A.** Example of a "regular" firing cortical pyramidal neuron. Intracellular injection of a depolarizing current pulse (top trace) results in the generation of a train of action potentials which occur at progressively slower frequencies (spike frequency adaptation). **B.** In contrast, intracellular injection of depolarizing current pulses in a "burst"-generating cortical pyramidal neuron results in the clustering of action potentials together on top of a slow potential. **C.** Electrical activity of a cerebellar Purkinje cell in response to intracellular injection of a depolarizing current pulse. The cell generates initially a high-frequency discharge of fast Na^+-dependent action potentials (generated in the soma). This discharge is modulated by the occurrence of dendritic Ca^{2+} spikes (asterisks). The discharge outlasts the duration of the intracellular depolarizing pulse (top trace) because of the presence of a plateau potential mediated by $I_{Na,p}$ and calcium currents (arrow heads). **D.** Depolarization of a thalamic relay neuron results in the generation of a train of four action potentials if the membrane potential is positive to approximately −65 mV, but a burst of action potentials if the cell is at or negative to −75 mV (**E**). The low-threshold Ca^{2+} spike underlying this burst discharge is indicated by an asterisk. **F.** Example of a neuron in the medial habenula which generates an intrinsic "pacemaker" discharge. Intracellular recording reveals the presence of large hyperpolarizations after each action potential which are complicated in time course and help determine the rate at which the neuron fires (arrows).

47

it generally is not possible to cause one class of neuron (e.g., cortical pyramidal neuron) to behave *identically*, electrophysiologically, to another (e.g., cerebellar Purkinje cell). However, substantial and interesting transformations take place in response to neuron-to-neuron communication.

TYPES OF NEURONAL COMMUNICATION

Communication from one neuron to another in the nervous system occurs through at least three different mechanisms: (1) gap junctions; (2) ephaptic interactions; and (3) the release of neuroactive substances.

GAP JUNCTIONS

Gap junctions are actual physical connections between neighboring neurons made by large macromolecules that extend through the membranes of both cells and contain water-filled pores (Fig. 2.6). Gap junctions allow for the direct exchange of ions and other small molecules between cells. Ionic current through these channels directly couples the electrical activity of one cell to that of the other. Although in some cases gap junctions can be viewed as simple linearly conducting connections, in many other cases they are known to *rectify* (i.e., pass

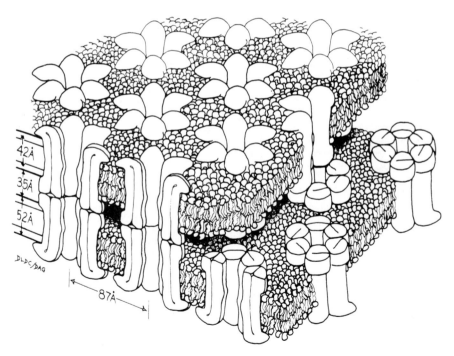

FIG. 2.6. Diagram of direct electrical connection between cells (gap junction). Channels provide for cell-to-cell exchange of low-molecular-weight substances and electric ionic current. (From Makowski et al., 1977.)

current in one direction better than in the other). Gap junctions are known to be a prominent feature of neuron-to-neuron communication in many submammalian species, but in only a small number of regions in the mammalian nervous system (e.g., retina, inferior olive, vestibular nucleus, and the mesencephalic nucleus of the fifth cranial nerve). Gap junctions in these regions serve to synchronize the activity of individual elements with those of their neighbors. The ability of neurotransmitters to alter the conducting properties of gap junctions in some regions (e.g., retina) gives additional complexity to this system of communication (reviewed by Rogawski, 1987).

EPHAPTIC INTERACTIONS

Ephaptic interactions refer to interactions between neurons based largely upon their close physical proximity (Fig. 2.7). The flow of ions into and out of one neuron sets up local electrical currents which can partially pass through neighboring neurons. The degree to which a neuron can be influenced by the activity of its neighbor is determined in part by the proximity of the cells and their processes (i.e., dendrites, cell bodies, and axons). In regions that possess closely spaced neuronal elements, such as the close packing of cell bodies in hippocampus and cerebellum or the bundling of dendrites in the cerebral cortex, there is the possibility of significant ephaptic interaction. Ephaptic interactions, like gap junctions, serve to synchronize local neuronal activity and may influence the general firing pattern of functionally related neurons (e.g., Taylor and Dudek, 1984a,b).

CHEMICAL SYNAPSES

The release of neuroactive substances at the specialized connections called *synapses* is by far the most common method by which neurons influence other neurons. Some neuroactive substances can also diffuse over rather long distances to activate extrasynaptic sites, although it is not yet clear how common this type of transmission is.

As discussed in Chap. 1, neurotransmitters are released by neurons through exocytosis of packets (*vesicles*) of the substance from synaptic specializations into the space (*synaptic cleft*) between the cells. Examples of two of the most prevalent types of synapses are shown in Fig. 2.8. The release of transmitter is triggered by the entry of Ca^{2+} into the presynaptic terminal. This Ca^{2+} entry results from the depolarization associated with the arrival of the action potential. Once the neurotransmitter is released, it rapidly traverses the short distance between the neurons and binds to specific proteins (*receptor molecules*) on the postsynaptic cell. The activation of the receptors by the neurotransmitter may then cause a myriad of postsynaptic responses, many of which are expressed as an altering of the probability that a particular type of ionic channel will be open.

The actual receptor binding site may be part of, or separate from, the macromolecule making up the ionic channel. Examples of ionic channels to which the neurotransmitter directly binds include the glutamate and GABA-activated channels (Fig. 2.8), and the nicotinic cholinergic receptor. The latter is activated

Ephaptic Interactions

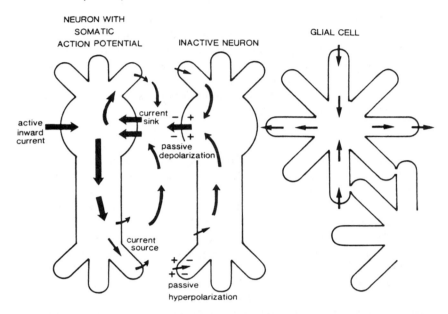

FIG. 2.7. Schematic diagram of current flow proposed to underlie excitatory electrical field effects between pyramidal neurons in the hippocampus (an example of ephaptic interactions). Arrows denote current flow of positive charges. The driving force of the ephaptic electrical field effect is the flow of positive current into somata produced by the synchronous firing of a population of hippocampal pyramidal neurons (left). Positive current then flows passively out of dendrites of active cells and returns through extracellular space. The relative decrease in positive charge in the extracellular space at the cell body layer causes the voltage on the inside of inactive cells (center) to appear relatively more positive (i.e., depolarized) than before. Likewise, the addition of positive current to the extracellular space at the levels of the dendrites by the neuronal activity causes the intracellular potentials of inactive dendrites to appear more negative (hyperpolarized) than before. Depolarization of the neuronal somata increases the probability that neighboring cells will generate action potentials in synchrony. Passive glial cell also develops transmembrane current flow within electrical field (right). (From Taylor and Dudek, 1984b.)

by acetylcholine (ACh) at the neuromuscular junction, in sympathetic ganglion neurons (see Chap. 3), and in many other regions of the nervous system. The binding of ACh to the nicotinic postsynaptic receptor induces a conformational change in the ionic channel, thereby opening the "gate" and allowing ions (in this case, Na^+, Ca^{2+}, and K^+) to flow through the pore (reviewed by Hille, 1984).

An example of a receptor site that appears to be separate from the channel molecule is the muscarinic receptor in the heart, which, when activated by acetylcholine, results in an increase in membrane potassium conductance. This

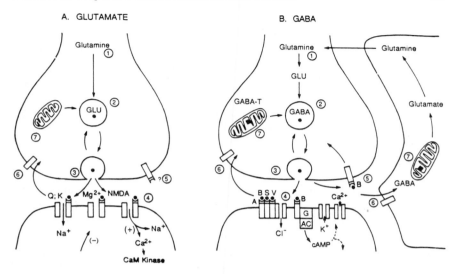

FIG. 2.8. Molecular mechanisms of amino acid synapses. **A.** Glutaminergic synapses: (1) synthesis of glutamate (GLU) from glutamine; (2) transport and storage; (3) release of GLU by exocytosis; (4) binding of GLU to quisqualate (Q), kainate (K), and NMDA receptors. The Q and K receptors gate Na$^+$ and K$^+$ flux; the NMDA receptor also allows Ca^{2+} entry when the membrane potential is depolarized (+). When the membrane potential is hyperpolarized (−), Mg^{2+} blocks this channel. The release of GLU may be regulated by presynaptic receptors (?5). Once GLU is released, it is removed from the synaptic cleft by reuptake (6) and processed intracellularly (7). **B.** GABAergic synapse: (1) synthesis of GABA from glutamine; (2) transport and storage of GABA; (3) release of GABA by exocytosis; (4) binding to a GABA$_A$ receptor which can be blocked by bicuculline (B), picrotoxin, or strychnine (S) and can also be modified by benzodiazepines, such as valium (V); GABA$_B$ receptors, by contrast, are linked via a G-protein to K$^+$ and Ca^{2+} channels which are blocked by GABA; (5) release of GABA is under the control of presynaptic GABA$_B$ receptors; GABA is removed from the synaptic cleft by uptake into terminals or glia (6); (7) processing of GABA back to glutamine. (A from Shepherd, 1988, based upon Cooper et al., 1987; Jahr and Stevens, 1987; Cull-Candy and Usowicz, 1987. B from Shepherd, 1988a; modified from Cooper et al., 1987; Aghajanian and Rasmussen, 1988; Nicoll, 1982.)

response to ACh is associated with the receptor-mediated activation of an intracellular second messenger known as a G-protein. G-proteins are a class of molecule that requires the binding of guanyl nucleotides in order to be active. The active component (catalytic subunit) of the G-protein is then thought to act as an intermediary between the receptor molecule and the ionic channel (reviewed by Neer and Clapman, 1988).

Once a neurotransmitter is released, the length of time that it is present in the synaptic cleft is controlled, alternatively, by hydrolysis of the transmitter, reuptake into the presynaptic terminal, uptake into neighboring cells, or simple diffusion out of the cleft.

NEUROTRANSMISSION VERSUS NEUROMODULATION

Neuroactive substances in the nervous system have often been classified as either putative "neurotransmitters" or "neuromodulators," according to the duration and the functional implications of their actions. Substances released by neurons that have typical neurotransmitter roles cause phasic postsynaptic responses which are both quick in onset (e.g. $<$ a few milliseconds) and relatively short in duration (e.g., $<$ tens of milliseconds). The summation of phasic excitatory and inhibitory postsynaptic potentials, and the way they interact with the intrinsic electrophysiological and morphological properties of the neuron, forms to a large extent the manner in which neuronal computations occur.

By contrast, modulatory actions of neuroactive substances are characterized by their prolonged duration and the ability to *modulate* the response of the neuron to other, perhaps more phasic, inputs. Although the distinction between these two types of neurotransmitter action is not always easy, it is nonetheless useful.

It is probably safe to say that most neurons in the brain are under the influence of as many as a dozen or more neuroactive substances (see Table 2.2). The wide range of cellular responses (ionic as well as biochemical) to these substances adds great depth and richness to the possible behavior of individual neurons and consequently to a neuronal circuit as a whole. For example, it is the job of

Table 2.2. Common Neurotransmitter Responses in the CNS

Response	Neurotransmitter	Receptor
$\uparrow I_{Na}$, $\uparrow I_{K}$	Glutamate	Quisqualate/Kainate
$\uparrow I_{Na}$, $\uparrow I_{K}$, $\uparrow I_{Ca}$	Glutamate	N-Methyl-D-aspartate
	Acetylcholine	Nicotinic
$\uparrow I_{Cl}$	Gamma-aminobutyric acid	$GABA_A$
$\uparrow I_{K,r}$	Acetylcholine	M_2
	Norepinephrine	$Alpha_2$
	Serotonin	5-HT_{1A}
	Gamma-aminobutyric acid	$GABA_B$
	Dopamine	D_2
	Adenosine	A_1
	Somatostatin	
	Enkephalins	mu, delta
$\downarrow I_{AHP}$	Acetylcholine	Muscarinic
	Norepinephrine	$Beta_1$
	Serotonin	(?)
	Histamine	H_2
$\downarrow I_{M}$	Acetylcholine	Muscarinic
$\downarrow I_{K,l}$	Acetylcholine	Muscarinic
	Norepinephrine	$Alpha_1$
	Serotonin	(?)
$\downarrow I_{A}$	Norepinephrine	$Alpha_1$
$\downarrow I_{Ca}$	Multiple transmitters	

neurotransmitters not only to allow neurons to communicate accurately and quickly the exact details of a complicated visual scene (e.g., the reading of this page), but also to control the proper level of arousal of the nervous system (e.g., awake and attentive) for efficient and accurate processing of the information, as well as to cause the generation of relatively permanent cellular changes (memory) through which the contents of the written page can be recalled. Considering the wide range of involvement of neurotransmitters in simple (e.g., reflexes) as well as complicated (e.g., psychiatric disorders) behavioral attributes, it is not surprising that one should find that there is an equally wide range of neurotransmitter actions on single neurons.

IONIC ACTIONS OF NEUROTRANSMITTERS

A large number of substances are thought to be released by neurons in order to modify the electrophysiological properties of other neurons (Table 2.2). Many of these substances can cause more than one postsynaptic response. Most, if not all, of these various responses are mediated by pharmacologically distinct receptor molecules. In this manner, a neuroactive substance released onto a pyramidal neuron in the cerebral cortex may have a very different effect from the same neurotransmitter released onto a relay neuron in the thalamus (see below). Indeed, the same neurotransmitter may have very different, or even opposite, postsynaptic effects on neighboring neurons in the same neuronal region, depending upon the particular function of the neuron in the local circuit.

Many of the ionic currents in neurons are under the control of neuroactive substances. Recently, it has become apparent that different neurotransmitters, each acting through its own distinct class of receptor molecules, can modify the same ionic current. For this reason, I will review here the more common postsynaptic responses to neurotransmitters in terms of the physiological action rather than the type of neurotransmitter.

FAST POSTSYNAPTIC POTENTIALS

The classical postsynaptic potential (PSP) occurs through a temporally (e.g., milliseconds) and spatially (i.e., local) limited increase in membrane ionic conductance. The relatively brief time course of these postsynaptic potentials allows neurons to perform a large number of computations within short time periods, limiting the interactions between events which are widely separated in time. Synaptic potentials, especially those brief in duration, are usually classified by whether they increase (excitatory) or decrease (inhibitory) the probability of action potential discharge. However, it is always better to know the actual biophysical and biochemical actions of a neuroactive substance than to refer to it as being just "excitatory" or "inhibitory," especially when considering the *modulatory* actions of many putative neuroactive substances (see below).

Fast excitatory postsynaptic potentials. Two main types of brief duration excitatory PSPs have been identified in the nervous system: those due to the activation of nicotinic receptors by ACh, and those caused by excitatory amino acids.

Nicotinic Cholinergic Responses. Fast nicotinic excitatory PSPs have been demonstrated in the spinal cord, peripheral nervous system, and skeletal muscle. Although numerous regions of the CNS, particularly subcortical areas, also possess functional nicotinic cholinergic receptors, the *synaptic* activation of these receptors has not yet been shown. The activation of the nicotinic receptor–ionic channel complex by ACh results in a conformational change in the shape of critical portions of this macromolecule which function as a "gate," thereby allowing ions to flow through. The nicotinic ionic channel is a "nonselective" cation channel, meaning that positively charged ions (e.g., Na^+, Ca^{2+}, and K^+) pass through the channel with about equal proficiency. Because of the mixed nature of the ions flowing through the nicotinic channel, the equilibrium (reversal) potential of the nicotinic response, approximately -5 mV, lies between the equilibrium potentials of the various cations (see Fig. 2.2). The actions of ACh through nicotinic receptors in the nervous system are of particular interest since nicotine, in the form of tobacco products, is one of the most widely used drugs of addiction.

Excitatory Amino Acid Responses. A substantial proportion of the fast excitatory PSPs in the brain, particularly those in the cerebral cortex and hippocampus, are believed to be due to the release of an excitatory amino acid such as glutamate or aspartate. Postsynaptic receptors for glutamate have been categorized according to their affinity for three different exogenous agonists: quisqualate, kainate, and *N*-methyl-D-aspartate (NMDA) (reviewed by McLennan, 1983; Watkins and Olverman, 1987). Activation of these receptors is believed to underlie fast glutaminergic EPSPs. The postsynaptic potentials mediated by quisqualate and kainate receptors, like those associated with nicotinic channels, are caused by an increase in a mixed cation conductance (mainly Na^+ and K^+) such that the reversal potential is approximately 0 mV (see McDermott and Dale, 1987). These synaptic potentials have a very short delay from the arrival of the action potential at the presynaptic terminal to the appearance of the postsynaptic potential, and a rapid rate of rise. The falling phase is much slower, being determined in large part by the membrane properties of the neuron (see Fig. 2.9B).

In contrast to the fast PSPs mediated by quisqualate/kainate receptors, the action of glutamate through NMDA receptors is much more complicated (reviewed by Ascher and Nowak, 1987). Stimulation of NMDA receptors results in the activation of a voltage-dependent current that is carried not only by Na^+ and K^+ but also importantly by Ca^{2+}. The voltage-dependent nature of this NMDA receptor-mediated current is due to the differential block of the ionic channel by magnesium ions (Mg^{2+}) at different membrane potentials (Mayer et al., 1984). At resting membrane potential (e.g., -75 mV), the driving force on Mg^{2+}, which is concentrated on the outside of the cell, to enter the neuron is quite high. Because of this, magnesium ions compete with Ca^{2+} and Na^+ ions for access to the pore of the channel. Because Mg^{2+} ions cannot flow through the pore, the channel is effectively blocked whenever one of the ions enters, thereby reducing the time during which the channel is open and conducting (see Fig. 2.9C). When the cell is depolarized, the tendency for Mg^{2+} to fill the pore is

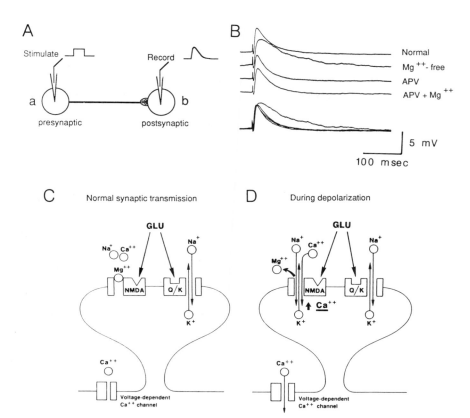

FIG. 2.9. Synaptic potentials mediated by the release of glutamate. **A.** Schematic diagram of experimental protocol in which the actions and pharmacology of monosynaptic connections between cultured cortical pyramidal cells is investigated. Intracellular recordings are used to stimulate a generator cell (a) which is monosynaptically connected to a follower cell (b). **B.** Activation of an action potential in the generator cell (a) causes a monosynaptic EPSP in the follower cell (b) through the stimulation of quisqualate and kainate receptors (top trace; normal). Removal of Mg^{2+} from the medium bathing the cultures enhances the late components of this EPSP (second trace; Mg^{2+}-free). Addition of the NMDA receptor antagonist 2-amino-5-phosphovalerate (APV) abolishes this late component, indicating that it was due to the activation of NMDA receptors (third trace; APV). Returning Mg^{2+} to the bathing medium now has no additional effect on the EPSP (fourth trace; APV + Mg^{2+}). At the bottom of B the traces are superimposed for comparison. The data illustrate that the release of glutamate can activate quisqualate/kainate and NMDA receptors, and that the NMDA, but not the quisqualate/kainate, ionic channel can be blocked by Mg^{2+} ions. **C.** Schematic summary diagram illustrating that glutamate release from the presynaptic terminal at a low frequency ("normal synaptic transmission") acts on both the NMDA and quisqualate/kainate type of receptors. Na^+ and K^+ flow through the quisqualate/kainate channel, but not through the NMDA receptor channel owing to Mg^{2+} block. **D.** Depolarization of the membrane potential, or activation of the glutaminergic inputs at a high frequency, relieves the Mg^{2+} block of the NMDA channel, thereby allowing Na^+, K^+, and, importantly, Ca^{2+} to flow through the channel. Depolarization due to the synaptic potential now also activates other voltage-dependent channels, such as those that conduct Ca^{2+}. (B from Huettner and Baughman, 1988; C,D from Nicoll et al., 1988.)

substantially reduced, thereby lessening the block and allowing a larger $Na^+/Ca^{2+}/K^+$ current to flow. Because of this voltage dependence, activation of a glutaminergic synapse onto a neuron at resting membrane potential may result in a fast EPSP mediated through the activation of kainate and quisqualate receptors with little contribution of NMDA receptor-mediated current, even though glutamate may be binding to these receptors (Fig. 2.9C). However, repetitive activation of the same synapse may cause a large depolarization of the cell through temporal summation of the unitary PSPs. The more these PSPs depolarize the cell, the more the degree of magnesium block will be removed, and thus the larger the activation of the NMDA current (Fig. 2.9D). Because NMDA channels conduct Ca^{2+} as well as Na^+ and K^+, calcium will flow into the postsynaptic cell and, by activating further biochemical mechanisms, can result in a *potentiation* of the strength of the unitary excitatory PSP. This enhancement of the PSP can last for prolonged periods (hours, days, longer?) and therefore is known as *long-term potentiation* (LTP) (see Collingridge and Bliss, 1987, and Nicoll et al., 1988, for review). LTP is currently one of the leading models of the mechanisms by which synapses change their efficacy in order to participate in the encoding of memories in the nervous system (see Chap. 11, Hippocampus).

Fast inhibitory postsynaptic potentials. Postsynaptic potentials that are quick in onset and inhibit the postsynaptic activity of the neuron are known to be mediated by two different neurotransmitters in the CNS: gamma-aminobutyric acid (GABA) and glycine.

GABA-Mediated IPSPs. Gamma-aminobutyric acid (GABA) is the major inhibitory neurotransmitter of the nervous system. GABA-releasing cells are present throughout all levels of the neuraxis. In the cerebral cortex and thalamus, they account for approximately 20–30% of all neurons. Neurons utilizing GABA as a neurotransmitter form a diverse group, with different morphologies specific for their roles in neuronal processing. They are instrumental in defining and confining the response properties not only of single neurons, but also of large neuronal circuits. They figure prominently as interneurons in the types of inhibitory circuits illustrated previously in Chap. 1. It would be fair to say that without GABAergic neurons the nervous system would not function in any logical manner.

There are two major types of GABA receptors, which are referred to as $GABA_A$ and $GABA_B$ (Bowery et al., 1987). We consider first the $GABA_A$ receptor. Many fast inhibitory PSPs in the brain are believed to result from the release of GABA acting upon the $GABA_A$ subclass of receptor (see early IPSP, Fig. 2.10). Binding of GABA to this class of receptor opens ion channels that are selective for Cl^- ions, and therefore the reversal potential of $GABA_A$-mediated responses is at the equilibrium potential for chloride (i.e., approximately -75 mV). Like the fast EPSPs in the nervous system, fast $GABA_A$-mediated IPSPs possess a rapid rising phase and a slower decay. These IPSPs are only tens of milliseconds in duration and are involved in rapid computations by neuronal networks (see Chap. 1).

Glycine-Mediated IPSPs. Glycinergic interneurons are believed to exist in the spinal cord and perhaps the brainstem. Glycine inhibits neuronal activity by increasing a Cl^- conductance similar to that activated by GABA (reviewed in Krnjević, 1974; Hamill et al., 1983). Indeed, it has been proposed that glycine and GABA receptors may couple to the same Cl^- ionic channels (Hamill et al., 1983). Although glycine probably does not serve as a classical neurotransmitter in the forebrain, recent evidence that very low doses of glycine can greatly potentiate the action of glutamate at NMDA receptors has renewed interest in this amino acid (Johnson and Ascher, 1987). This potentiating action occurs at low enough doses that even the concentrations of glycine occurring in the extracellular fluid are large enough to have a significant effect.

SLOW SYNAPTIC POTENTIALS

Like fast PSPs, slow PSPs are found at all levels of the nervous system. They have a large variety of sizes, shapes, and effects on the functional properties of neurons and neuronal circuits. Because of their delayed onset and prolonged duration, these PSPs are probably more involved in the regulation of the underlying *excitability* of single neurons and neuronal circuits as opposed to the relatively high-frequency transfer of information (see Hartzell, 1981; Adams and Galvan, 1986).

Increase in potassium conductance. Applications onto neurons of a large variety of putative neurotransmitters, including acetylcholine, norepinephrine, serotonin, GABA, dopamine, and various peptides, have been found to cause an increase in membrane potassium conductance (g_K; Fig. 2.10 and Table 2.2) (reviewed by North, 1987; Nicoll, 1988). This occurs through a specific subtype of neuronal receptor for each neuroactive substance. Although all of these substances have the ability to increase potassium conductance in some region of the nervous system, the nonuniform distribution of receptors that mediate this response means that some neurons exhibit it and others do not. For example, application of acetylcholine to GABAergic interneurons in the feline thalamus results in an *increase* in g_K, whereas in neighboring thalamocortical relay cells it causes a *decrease* in g_K (McCormick and Prince, 1987a; McCormick and Pape, 1988). Furthermore, in many regions of the nervous system there is convergence of different neuroactive substances, each one generating an increase in g_K in the same postsynaptic neuron. For example, hippocampal pyramidal cells respond to serotonin, GABA (through $GABA_B$ receptors), and adenosine with an increase in the same potassium conductance (see Nicoll, 1988, and Chap. 11). In this manner, a variety of neuroactive substances can activate or inactivate the same ionic currents in a given neuron and perhaps even converge onto the same ionic channel molecules.

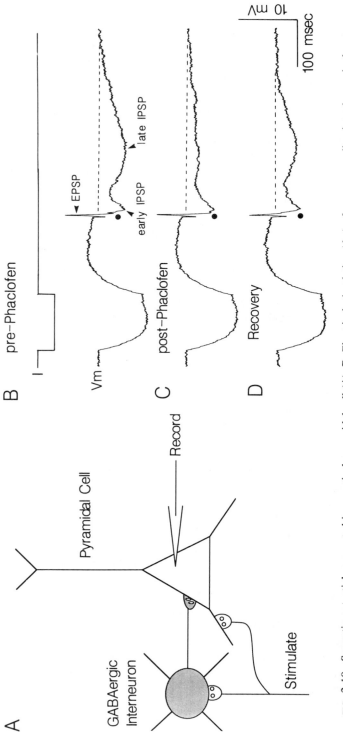

FIG. 2.10. Synaptic potentials generated in a cortical pyramidal cell (**A**). **B**. Electrical stimulation (dot) of axons ascending into the cerebral cortex results in the generation of a fast EPSP followed by an early (fast) and late (slow) IPSP. Activation of the axons excites a local GABAergic interneuron (**A**) which subsequently releases GABA onto the recorded pyramidal cell. GABA then activates both $GABA_A$ and $GABA_B$ receptors. Activation of $GABA_A$ receptors causes an increase in Cl− conductance and underlies the early IPSP, whereas activation of $GABA_B$ receptors causes an increase in K^+ conductance and is responsible for the generation of the late IPSP. **C**. Local application of the $GABA_B$-specific antagonist phaclofen substantially reduces the late IPSP, confirming that this PSP is due to the activation of $GABA_B$ receptors. **D**. The effect of phaclofen is reversible. (B from McCormick, 1989.)

Functionally an increase in membrane potassium conductance is considered inhibitory because it usually decreases the probability of action potential discharge. It is important to note that the inhibitory nature of an increase in g_K is very different from that of an increase in g_{Cl}, and this can have important functional consequences. For example, GABA can cause an increase in both g_{Cl} (through $GABA_A$ receptors) and g_K (through $GABA_B$ receptors); the result is a fast $GABA_A$-mediated increase in g_{Cl} followed by a slow $GABA_B$-mediated increase in g_K in the postsynaptic neuron (early and late IPSP, Fig. 2.10) (Newberry and Nicoll, 1985; Dutar and Nicoll, 1988). In addition, there are many differences between the fast and slow GABA-mediated IPSPs other than just their time course. The conductance increase associated with the late IPSP is much smaller than that associated with the fast IPSP, even though the amplitude of the voltage deviation associated with each may be similar. Indeed, if the membrane potential is negative to E_{Cl}, the fast IPSP will be *depolarizing* (although it is still inhibitory), whereas the late IPSP remains hyperpolarizing. In addition, the GABA-activated late IPSP is mediated through a second-messenger system (G-proteins), whereas the fast IPSP is probably the result of GABA binding to a receptor located directly on the ion channel.

These physiological differences make fast IPSPs a primarily *shunting* inhibition (i.e., the membrane potential of the cell is held close to E_{Cl}, and the input resistance of the cell is "shunted"), whereas the late IPSP is a *hyperpolarizing* inhibition. Fast IPSPs are useful for local (e.g., particular subparts of the cell) "yes–no" decisions, whereas the late IPSP is useful for the *modulation* of the overall excitability of the neuron. The restricted time and space domains of the fast IPSPs allow them to participate in relatively high-frequency neuronal processing, whereas the slow IPSP is important for setting a particular level of excitability in the neuron for more prolonged periods of time.

The postsynaptic morphological locations of the IPSPs are also very important in determining their consequences for processing within synaptic circuits. Many types of GABAergic neurons form synaptic contacts at specific locations on the postsynaptic neuron. For example, *chandelier cells* of the cerebral cortex give rise to chains of synaptic terminals on the axon hillocks of cortical pyramidal cells (see Chap. 12), whereas *basket cells* give rise to a "basket" or "pericellular nest" of terminals around the cell bodies of pyramidal neurons. In this way, both of these inputs have a powerful "veto" ability on the output of the entire neuron. It may even be possible for the chandelier cell to prevent the propagation of an action potential down the axon after its generation in the cell body and/or dendrite.

The opposite extreme of the above two examples of a rather global inhibition by GABA of the output of the neuron is found in the very localized processing that occurs in synaptic microcircuits (see Chap. 1, Fig. 1.5). At this level of organization, individual GABAergic terminals may have effects that are relatively independent of one another, as well as independent of the output activity of the neuron itself. In these situations, the GABAergic process may affect only a particular portion of the postsynaptic dendritic tree, or perhaps only particular

synaptic terminals. Numerous examples of GABAergic contributions to processing in synaptic glomeruli, dendritic trees, and other types of microcircuits will be
discussed in subsequent chapters.

Decrease in potassium currents. Neuroactive substances can not only increase,
but also decrease, neuronal potassium currents. To date, four different potassium
currents have been identified that can be decreased in amplitude by the actions of
various neurotransmitters: I_{AHP}, I_M, I_A, and a resting "leak" potassium current
which we denote as $I_{K,l}$.

Decrease of I_{AHP}. I_{AHP} has been shown to be decreased by a number of putative
neurotransmitters (norepinephrine, acetylcholine, serotonin, histamine, etc.),
much in the same way that various neurotransmitters converge onto the activation of potassium currents (Haas and Konnerth, 1983; Madison and Nicoll,
1986a; Andrade and Nicoll, 1987). In the case of norepinephrine, the decrease in
I_{AHP} is achieved through an increase in the intracellular activity of a second
messenger, cyclic adenosine monophosphate (cAMP) (Madison and Nicoll,
1986b).

As stated previously, I_{AHP} contributes substantially to spike frequency adaptation. Therefore, block of this current greatly reduces the tendency for cells to
slow down their firing rate during maintained depolarization (Fig. 2.11). This is
an important effect, for it allows a neurotransmitter to increase the response of a
cell to barrages of excitatory PSPs with little or no change either in the resting
membrane potential, or in the response to inhibitory PSPs. Indeed, if the putative
neurotransmitter simultaneously increases membrane conductance to K^+ or Cl^-
as well as blocking I_{AHP}, the result may actually be an increase in "signal-to-
noise" ratio. The baseline spontaneous firing of the cell will be reduced by the
increase in potassium and/or chloride currents, whereas the response of the cell
to barrages of large EPSPs may actually be enhanced by the decrease in I_{AHP} (see
below).

Decrease of I_M. As stated above, I_M is a potassium current that is slowly (tens of
milliseconds) activated by depolarization of the membrane potential above approximately -65 mV (Brown and Adams, 1980; Brown, 1988a,b). This current
has been shown to be potently reduced by stimulation of muscarinic receptors by
acetylcholine. Like I_{AHP}, I_M contributes to spike frequency adaptation; blocking
it subsequently increases the response of a neuron to barrages of excitatory PSPs.
Because I_M is active only at depolarized potentials, its blockade may have little
effect on the cell's resting membrane potential or response to IPSPs. In the
peripheral nervous system, I_M is blocked not only by acetylcholine, but also by
some neuropeptides, such as leutinizing hormone-releasing hormone (LHRH)-
like peptide, which apparently work through extrasynaptic receptors (see Chap.
3). Furthermore, recent data suggest that some putative neurotransmitters may
actually be capable of *increasing* I_M (Sims et al., 1988), thereby giving the
system additional complexity.

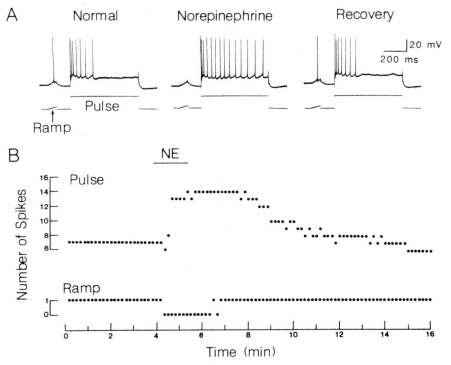

FIG. 2.11. Effect of norepinephrine on the excitability of cortical pyramidal neurons. The response of this hippocampal pyramidal neuron to two different types of inputs was examined: a small depolarizing ramp (to mimic weak EPSPs) and a prolonged depolarization (to mimic a train of strong EPSPs). **A.** In the normal condition, the small ramp input causes the generation of a single action potential, whereas the prolonged depolarization results in a train of action potentials that show strong spike frequency adaptation (A, left). Addition of norepinephrine to the bathing medium results in a small hyperpolarization of the membrane potential (not shown). During the hyperpolarization, the small depolarizing input no longer generates an action potential, whereas the response to the prolonged input is actually potentiated, owing to a block of spike frequency adaptation (A, NE). The reduction in spike frequency adaptation is a secondary effect due to the block of I_{AHP} (not shown). This effect of norepinephrine is fully reversible (A, recovery). **B.** Graphic representation of the data in A. The generation of an action potential by the small ramp input is blocked, whereas the response to the prolonged input is greatly enhanced. In this manner, norepinephrine can increase the "signal-to-noise" ratio of the cell. (From Madison and Nicoll, 1986a.)

Decrease of I_A. Reduction of I_A by neurotransmitters is a recent finding (Aghajanian, 1985). Because I_A contributes to an increase in the interval between action potentials during certain types of neuronal activity, the block of I_A tends to enhance the response of the neuron by increasing the frequency of action potential discharge.

Decrease in Calcium Currents. Numerous putative neurotransmitters, including acetylcholine, norepinephrine, serotonin, and GABA, can reduce the flow of Ca^{2+} across the membrane (see Tsien et al., 1988). The functional consequences of neurotransmitter suppression of calcium currents have not been well studied in the CNS. One possible effect is related to the actions of neurotransmitters at presynaptic terminals. The amount of transmitter released after the invasion of a terminal by an action potential is under the control of neuroactive agents binding to receptors located on these terminals. In most (perhaps all) systems, the binding of a neurotransmitter to receptors on its own presynaptic terminal *reduces* the quantity released by subsequent action potentials. This *autoinhibition* then forms a negative feedback loop which is useful for regulating the concentration of transmitter in the area of the synaptic cleft. The ionic mechanisms of this negative feedback are not known. However, because neurotransmitter release is highly dependent upon Ca^{2+} entry, transmitter-mediated decreases in Ca^{2+} currents may be involved.

Possible gating actions of neurotransmitters. As discussed previously, many different types of neurons in the nervous system possess two intrinsic and physiologically distinct firing modes: *single spike* and *burst* activity (e.g., Llinás and Yarom, 1981a,b; Jahnsen and Llinás, 1984a,b). The cell's membrane potential determines in part which of these two firing patterns the neuron will exhibit. Burst firing occurs in response to excitatory inputs whenever the membrane potential is negative to approximately −65 mV, whereas single spike activity occurs at membrane potentials positive to approximately −55 mV (Fig. 2.5D and E). Therefore, a neuroactive substance that activates a potassium conductance can actually increase the probability of a neuron firing by hyperpolarizing the cell into the burst firing mode of action potential generation (e.g., from −60 to −70 mV). In this situation, the increase in membrane conductance is acting more as a "switching" or modulatory mechanism than as a strict "yes–no" inhibition (McCormick and Prince, 1986). Likewise, decreasing resting conductance to K^+ is an effective mechanism by which a neuron can be tonically depolarized out of the burst firing mode and brought closer to threshold for generation of the more unmodulated single spike discharge (Fig. 2.12). Such changes in membrane potential have been found to occur during shifts in arousal (Hirsch et al., 1983) and may underlie the well-known shift in the characteristics of the electroencephalogram (EEG) from synchronized slow waves to desynchronized, higher frequencies during increases in arousal (e.g., Moruzzi and Magoun, 1949).

INTRINSIC AND SYNAPTIC CURRENTS: PUTTING IT ALL TOGETHER

With our new armament of knowledge of the intrinsic properties of neurons and how they may be affected by neurotransmitters, we can proceed (with due caution) to propose a scenario of how synaptic computations may be implemented and modulated in a representative neuron. We take as our example one of the most abundant and important neuronal cell types in the human brain: the cerebral cortical pyramidal cell.

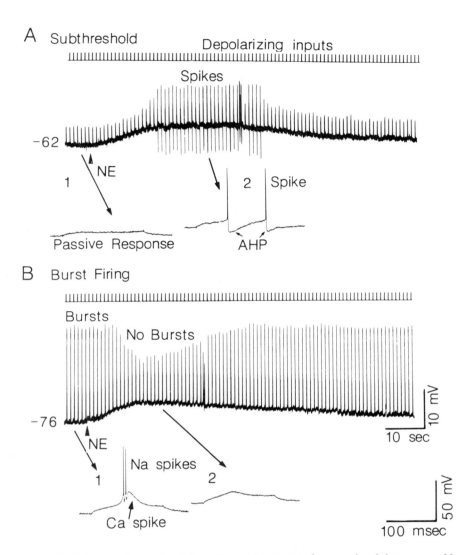

FIG. 2.12. Effect of norepinephrine (NE) on the pattern of neuronal activity generated by thalamic relay neurons. The response properties of the neuron were checked every 2 sec through the injection of a depolarizing current pulse (A,1 and B,1; pre-NE). The effect of NE was investigated with the neuron in two different states: just below single spike firing threshold (A) and in the burst firing mode (B). **A.** Application of NE resulted in the generation of a slow depolarization by reducing the membrane conductance to K + ions. During the slow depolarization, the response to a previously subthreshold depolarizing input (A,1) became suprathreshold and generated two action potentials (A,2). A_1 and A_2 are expanded traces of the response of the neuron to the depolarizing input (note different time bases in B). **B.** In contrast, application of NE while the cell was in the burst firing mode (B,1) resulted in a slow depolarization that potently inhibited the generation of the low-threshold Ca^{2+} spike-mediated burst (B2). These data illustrate the *modulatory* influence of NE in that it decreases the probability of burst discharge, while increasing the probability of single spike activity.

63

Cortical pyramidal cells, like neurons in most other parts of the brain, receive excitatory, inhibitory, and modulatory inputs from a variety of sources. Putative *glutaminergic* synapses, which have typical fast excitatory actions, are found on the spines of apical and basilar dendrites (Fig. 2.13). Notable sources of these excitatory inputs are other pyramidal cells (located in neighboring or distant cortical regions), spiny stellate neurons of layer IV, and inputs from the thalamus (see Chap. 12). In contrast to excitatory inputs, *GABAergic* inhibitory synapses are found on the soma, proximal and distal dendrites, and initial segment of the axon; they arise largely from intrinsic cortical interneurons, which are morphologically and functionally heterogeneous. Putative *neuromodulatory* substances arrive from a variety of subcortical (cholinergic, noradrenergic, serotonergic) and intracortical (cholinergic and peptidergic) neurons. Their synaptic contacts on pyramidal neurons are found largely on dendrites. Some types of GABAergic neurons also contain, and may release, one or more peptides. The

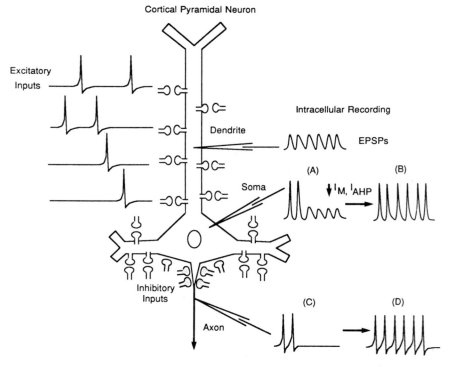

FIG. 2.13. Effect of activation of excitatory inputs to a cortical pyramidal cell. A train of action potentials arriving at different synaptic endings on the apical dendrite of the pyramidal cell results in the generation of a train of EPSPs. The first two EPSPs generate action potentials in the somatic region, whereas the last four fail, owing to activation of I_M and I_{AHP} (A). This is further reflected in the axonal output of the neuron (C). Block of these two currents reduces spike frequency adaptation and allows all six EPSPs to generate action potentials (B and D). See text for details.

ionic actions of these peptides and how they interact with the actions of GABA are not yet known.

Let us imagine that our cortical pyramidal cell is in the visual cortex and that, although the animal is awake and attentive, the cell is not yet receiving any specific visual input. The resting potential of our hypothetical cell will probably be somewhere around −65 mV, depending upon the state of input from the slowly acting neurotransmitters, especially those (e.g., acetylcholine) that can alter the level of resting potassium conductance. This resting potential is about 10 mV below (more hyperpolarized than) the threshold of around −55 mV for the generation of action potentials by a cortical pyramidal neuron.

Now let us stimulate the visual receptive field of our cell with an adequately adjusted light stimulus to the retina (e.g., a moving bar of light). This input will first cause excitation of thalamic relay neurons. Because the animal is awake and attentive, the thalamic neurons respond to the input in a one-spike-out/one-spike-in fashion (e.g., Fig. 2.5D) and in turn give rise to a train of action potentials that reach some of the presynaptic terminals onto our cell. Each action potential causes an increase in $[Ca^{2+}]_i$ in the presynaptic terminal, which in turn causes the release of an excitatory transmitter from a variable number of synaptic vesicles. The transmitter diffuses across the synaptic cleft and binds to specific receptor molecules on the postsynaptic spine, increasing the probability that certain ionic channels (assume that they conduct Na^+ and K^+ ions) will be in the open and conducting state. In this manner, each presynaptic spike will cause an EPSP in the postsynaptic dendrite (Fig. 2.13). The exact amplitude and time course of each EPSP depends upon a large number of factors, including the amount of transmitter released, the density of postsynaptic receptor molecules, the sensitivity of the postsynaptic element to the transmitter, the size and shape of the postsynaptic element, and finally the amplitude and distribution of active currents that the postsynaptic element possesses. Indeed, the "efficacy" of each synaptic connection is not a static number, since it is probably modified during the acquisition of new information, as well as by new strategies to analyze that information, perhaps through a process similar to LTP (see *Fast Excitatory Postsynaptic Potentials*, above).

In order for the barrage of EPSPs generated by the train of inputs from the thalamus to cause our cell to fire, it must cause the output decision point of the cell (the cell body and *axon hillock* in this case) to rise above firing threshold (e.g., −55 mV). To do this, the EPSPs must spread from their points of generation in the dendrites, through the cell body, to the axon hillock. What happens to these EPSPs as they make this trip is determined by the intrinsic properties of our cell and the actions of other neuroactive substances impinging upon it. The dendritic EPSPs will probably be large enough to activate $I_{Na,p}$, or a Ca^{2+}, current and thereby receive an extra "boost" from these depolarizing currents. This enhancement is needed to help overcome the fact that cell membranes are not perfect insulators and some of the current will leak out, thereby reducing the size of the EPSP as it spreads toward the cell body (see Appendix, on electrotonic conduction). If the train of EPSPs comes at a high enough frequency, they will exhibit temporal summation, whereas EPSPs that arise from more than one point in the cell will exhibit, in addition, spatial summation. If the summated EPSP is large enough, it may be capable of causing the generation of a dendritic Ca^{2+}-mediated action potential which will, of course, greatly enhance and transform the response of the cell to the synaptic input (see Fig. 2.5C). However, for simplicity, assume that the threshold for the generation of a dendritic Ca^{2+} spike is not reached.

Now consider the situation in which many of the EPSPs in the train are large enough to cause the generation of an action potential in the cell body and axon hillock. In this circumstance, the initial EPSPs in the train will be more likely to cause the generation of spikes than the later ones, owing to the progressive activation of I_{AHP} and I_M, both of which contribute to spike frequency adaptation (Fig. 2.13A and C). Thus, although the cell may fire in response to the initial few EPSPs, the later ones will not reach firing threshold, and the cell's firing will cease. This is where modulatory transmitters come into play. If we were to arouse our animal such that there were an increase in the release of, for example, norepinephrine and acetylcholine, then I_{AHP} and I_M (and perhaps $I_{K,1}$) would be reduced. Reduction of these potassium currents would enhance the response of the neuron by reducing spike frequency adaptation as well as by moving the cell's membrane potential closer to firing threshold (Fig. 2.13B and D).

As the visual stimulus moves out of the cell's excitatory receptive field and into those of neighboring cortical neurons, our pyramidal cell may now be actively inhibited through the connections of intrinsic GABAergic neurons. These barrages of IPSPs will meet with many of the constraints as did the previous EPSPs, although they may occur in a portion of the membrane potential where other conductances are not activated (i.e., between -65 and -75 mV). The fast GABAergic IPSPs will be important in terminating the residual excitation from the previous barrage of EPSPs by causing an increase in Cl^- conductance. The influential position of the IPSPs on or near the soma and initial portion of the axon (axon hillock) make them particularly effective.

Many of the properties outlined for our hypothetical cortical pyramidal neuron can be generalized to neurons in all regions of the nervous system. However, each type of neuron is an individual, and generalizations must be used with caution so as not to neglect the important features of each neuronal type that allow it to perform its own unique brand of cellular processing and thereby make its specific contributions to the synaptic circuits of which it is a part.

3

PERIPHERAL GANGLIA

PAUL R. ADAMS AND CHRISTOF KOCH

Peripheral ganglia are clusters of nerve cells wrapped in connective tissue capsules to form small nodules. They lie outside the central nervous system, distributed in various places in the body. They are part of the peripheral nervous system, which is composed of two divisions, named somatic and autonomic. The *somatic* system is composed of the axons of spinal motoneurons that directly innervate the skeletal muscles. There are no ganglia in this system, hence there is no possibility of information processing, and each impulse leaving the CNS reaches the neuromuscular junction relatively unchanged. In invertebrates, there can be complex synaptic interactions and activity-dependent plasticity at neuromuscular junctions, but this level of processing is largely absent in vertebrates.

The *autonomic* nervous system is organized into three divisions: *sympathetic, parasympathetic,* and *enteric.* Each of the three divisions is characterized by the presence of peripheral neurons, and therefore is capable not only of relaying information but also of processing and modifying that information. In the first two divisions the peripheral neurons are located in clusters called ganglia. In the third division the neurons are found in sheets, called *plexi,* in the gut wall. Neuronal circuitry in the enteric system is complex and not well understood (cf. Gershon, 1987), and will not be considered here.

The basic plans of sympathetic and parasympathetic ganglia are similar (Fig. 3.1). Motoneurons located in the spinal cord send their preganglionic, lightly myelinated axons to the autonomic ganglia, where they synapse on principal neurons called ganglion cells. The ganglion cells send their postganglionic axons out to various target tissues, notably smooth muscle and glands.

Traditionally, interest in the autonomic nervous system has focused almost entirely on its role in regulating the physiological state of the organism through its effects on glandular secretion, cardiovascular function, intestinal motility, and many other properties. In this role, the ganglia have been considered to be merely relay stations with no significant functions. In recent years, however, it has been realized that the ganglia offer many advantages as simple systems for the analysis of neuronal membrane properties and synaptic transmission. The results of this analysis have contributed significantly to the much enlarged view

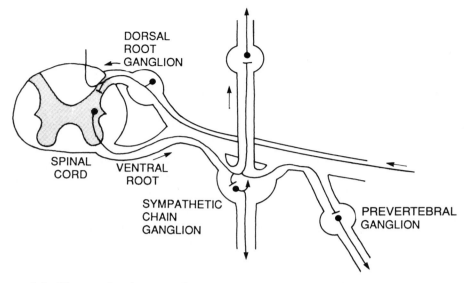

FIG. 3.1. Diagram of major types of ganglia in the sympathetic nervous system and their relation to the spinal cord and dorsal root ganglion. Note that motoneurons in the lateral horn of the spinal cord send their axons through the ventral root to provide the preganglionic fibers to the sympathetic ganglia.

of the complexity of the information processing that takes place in individual neurons and their local circuits, as outlined in Chap. 2 above.

In this chapter, we will focus on the array of membrane conductances in these neurons and the mechanisms by which they can be gated and modulated by neurotransmitters and neuromodulators over a wide range of time periods. We will also describe how the ganglia provide the opportunity to make direct observations of neurons while they undergo structural modification as synapses are formed and local circuit connections are laid down and modified during development and activity. The ganglia thus constitute simple systems, permitting direct information on many aspects of neuronal function and synaptic organization that are important for understanding more complex central systems.

There is a good deal of variation between different autonomic ganglia of a given species, as well as between corresponding autonomic ganglia of different species. In this chapter, we will focus on two different sympathetic ganglia that have been studied particularly intensively: the most caudal sympathetic ganglia of amphibians, the lumbar sympathetic ganglia, and the most rostral sympathetic ganglia of mammals, the superior cervical ganglion. The main account of neuronal properties and synaptic organization will concern amphibian ganglia; it will be followed by consideration of specific aspects of dendritic function in the mammalian ganglion.

NEURONAL ELEMENTS

INPUTS

The sympathetic ganglia lie in two chains on either side of the spinal column. The ganglia receive their main inputs from motoneurons of the intermediolateral column of the spinal cord, in the thoracic region. By contrast, the ganglia of the parasympathetic system lie close to or within the organs they innervate, and receive their inputs from motoneurons in the brainstem and the cervical and lumbar regions of the spinal cord.

The input axons from the spinal motoneurons are termed *preganglionic fibers*, or simply *preganglionics*. In the frog, most of the electrophysiological studies to be described below have been carried out on the largest, most caudally located ganglia (numbered IX and X) in the sympathetic chain (Weight, 1983). The preganglionics to these ganglia are of two types. Type B axons enter the sympathetic chain rostral to the VI ganglion; they then course along the chain, traversing the intervening ganglia to reach ganglia IX and X. These axons are of medium size and are lightly myelinated, with diameters of 2–4 μm. They conduct impulses moderately rapidly, at rates of 2–10 m/sec. Type C axons enter the sympathetic chain at more caudal locations. These are fine, unmyelinated fibers, with diameters of less than 1 μm and conduction velocities of less than 1 m/sec (Nishi et al., 1965; Dodd and Horn, 1983a).

PRINCIPAL NEURONS

The output neurons of the ganglia (Weitsen and Weight, 1977) are called, appropriately, *ganglion cells*. An attractive aspect of the organization of the lumbar ganglia of the frog is that there are two main types of ganglion cell, each selectively innervated by one of the two types of preganglionic fiber.

The larger of the two types is a *B cell,* so-called because it receives B preganglionics. These have a relatively large cell body (40–60 μm in diameter in the bullfrog). A striking feature of these cells is that they are virtually devoid of dendrites, and therefore appear as a unipolar cells, as depicted in Figs. 3.2 and 3.3. In this respect, they are exceptions to the rule that most neurons give rise to elaborate dendritic trees; however, this rule applies to ganglion cells in other species, as we shall see below. The type B preganglionics wrap around the axon hillock and their branches splay out on the surface of the soma, giving rise to varicosities and terminal boutons in the manner illustrated in Fig. 3.3. There are typically 40 boutons per cell, though this appears to vary somewhat with cell size, so that the synaptic density remains relatively constant.

The smaller ganglion cells are mostly *type C;* as one might expect, this terminology indicates that they are innervated by type C preganglionics. These cells have soma diameters of 20–40 μm, and thus fall into the medium-to-large size range relative to neurons elsewhere in the nervous system. Their innervation pattern is similar to that depicted in Fig. 3.3. The B and C cells are anatomically intermingled within the ganglion, but nevertheless show almost complete func-

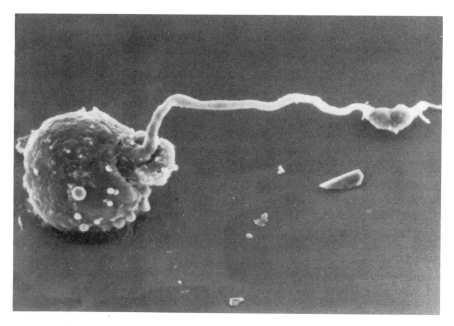

FIG. 3.2. Electron micrograph of an isolated bullfrog sympathetic ganglion "B" type cell with its associated axon (×1600). The spherical shape of the soma, the complete absence of dendritic processes, and a single nonbranching axon make this cell an excellent preparation for electrophysiological study. The diameter of the cell body is approximately 30 μm. (Courtesy of Barry Burbach. From Yamada et al., 1989.)

tional independence, reflecting the fact that they receive separate innervation and in turn innervate different peripheral target cells. The functional independence is not complete, however, as we shall see.

The ganglion cell (whether B or C) gives rise to axons that join to form the nerves that innervate peripheral targets such as the vasculature or intestines. These are termed *postganglionic fibers,* or simply postganglionics. The axons of B cells appear to acquire myelin on the way to their postsynaptic targets; the C axons remain unmyelinated. It is unusual in the nervous system that such large cell bodies give rise to such thin (less than 4-μm diameter) unmyelinated axons. Traditionally it has been thought that the axons usually do not give rise to recurrent collaterals. However, a recent study by Forehand and Konopka (1989) using HRP injections has shown that 19% of B cells and 65% of C cells give off axon collaterals within the ganglion. The significance of these collaterals for information processing is not yet known.

INTRINSIC NEURONS

Autonomic ganglia also usually contain scattered groups of small, neuronlike cells that show intense fluorescence after formaldehyde treatment. These "small, intensely fluorescing" cells (SIF cells; Eränkö and Harkonen, 1965) appear to have only short, dendritelike processes and no axons. They are thus classified as

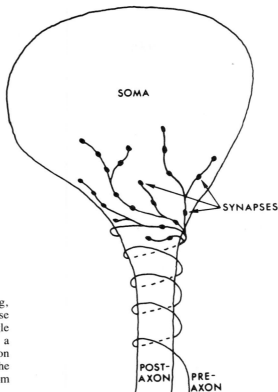

FIG. 3.3. Schematic drawing, based on horseradish peroxidase (HRP) staining, of how a single preganglionic fiber innervates a bullfrog sympathetic ganglion cell by wrapping itself around the axon hillock. (Modified from Weitsen and Weight, 1977.)

intrinsic neurons. There is little evidence that these cells play a significant role in modifying impulse traffic through the ganglia that are the subject of this chapter, and we will therefore not discuss them further.

BASIC CIRCUIT

As noted in Chap. 1, it is convenient to summarize the main patterns of synaptic connection in a local region by a diagram called a basic circuit. This serves several purposes: identifying the principles of circuit organization; characterizing the main computational and functional operations carried out by that region; summarizing synaptic actions and neuronal membrane properties; and aiding comparisons between regions. It should be emphasized that a basic circuit does not, nor can it possibly, include all connections and circuits within a region; rather, it serves as a starting point for assessing those that are most essential.

A basic circuit for the frog lumbar sympathetic ganglion is shown in Fig. 3.4. It is the simplest among those considered in this book, and certainly one of the simplest for any region of the nervous system. This is due in large part to several of the features of the neuronal elements noted in the previous section. The most important among these are the separate innervations of B and C cell populations; the lack of recurrent axon collaterals feeding-back information within the gang-

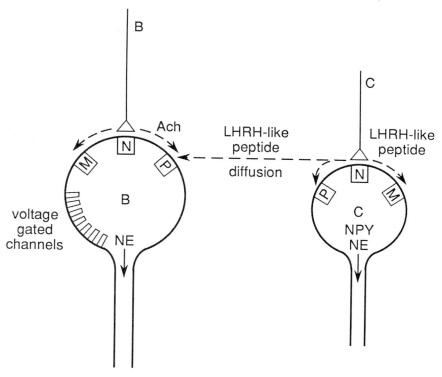

Fig. 3.4. Basic circuit diagram for the bullfrog sympathetic ganglion cell. Synapses of preganglionic axons onto B cells release only acetylcholine (ACh) which binds to postsynaptic nicotinic (N) as well as muscarinic (M) receptors. Synapses onto C cells release both ACh and a LHRH-like neuropeptide. This neuropeptide can diffuse and bind to peptidergic receptors (P) on C as well as B cells (spillover). B cells release norepinephrine (NE) onto their postsynaptic targets, whereas C cells release both NE and a neuropeptide (NPY).

lia; and the absence of interneurons as a substrate for local processing circuits. In addition, there is no evidence for axons feeding-back information from the output targets of the principal neurons, as is common within the central nervous system.

The simple neural elements of Fig. 3.4 of course represent populations of elements. Most B cells usually receive input from only a single preganglionic fiber. A second or third input axon, if activated, has only a small additional effect. B cell preganglionic fibers are not thought to diverge to innervate several ganglion cells. Furthermore, the "unitary" (single fiber) fast EPSP generated by an action potential in a preganglionic axon is normally large enough by itself to reach threshold and initiate an impulse in the ganglion cell. Thus, B cells appear to function as an almost one-to-one automatic relay station, reliably converting presynaptic spikes into postsynaptic spikes. However, C cells typically receive two or three preganglionic axon inputs, and the unitary EPSP evoked by only one of these input axons may not reach threshold.

NEUROTRANSMITTERS

Classical pharmacology of the nervous system was based to a great extent on studies of the autonomic system, because of its accessibility to experimental analysis. These studies established that acetylcholine (ACh) is the transmitter released by all autonomic preganglionic axons; ACh is also the transmitter released by postganglionic axons in the parasympathetic system, whereas norepinephrine (NE; also referred to as noradrenaline) is released by sympathetic postganglionic axons.

The classical studies also established that ACh has two kinds of effects: a rapid effect, mimicked by the substance nicotine and referred to therefore as the nicotinic cholinergic action; and a slow effect, mimicked by the substance muscarine, referred to as the muscarinic cholinergic action. This has turned out to be a useful way to characterize cholinergic synapses elsewhere in the nervous system (cf. Chap. 2). The molecular receptors for these two actions have been identified and cloned (Kubo et al., 1986).

The localizations of ACh and NE in the lumbar sympathetic ganglion are indicated in the basic circuit diagram of Fig. 3.4, together with a representation of the nicotinic (N) and muscarinic (M) receptors. The physiological properties of these receptors will be considered in the following section. It may be noted that the genetic mechanisms controlling the expression of transmitter phenotype by ganglion cells during the early embryonic development of the animal have come under study in recent years. It has been shown that if sympathetic ganglion neurons are maintained in cell cultures by themselves they express the adrenergic phenotype; that is, they take up, store, synthesize, and release NE. By contrast, if they are cocultured with certain types of target cells, they can be induced to express cholinergic properties (see Potter et al., 1981). There is, in fact, a period during these experiments when they have dual transmitter functions. Present studies are aimed at identifying the mechanisms by which substances released from target cells can control gene expression of the appropriate neurotransmitter machinery.

In addition to the classical transmitters, sympathetic ganglia have provided opportunities to analyze the presence and actions of neuropeptides. In the amphibian, Stephen Kuffler and his co-workers (Jan and Jan, 1982; Kuffler and Sejnowski, 1983) provided evidence that a luteinizing hormone-releasing hormone-like (LHRH-like) peptide mediates a very slow depolarization in the type C ganglion cell. The mechanism of this action will be discussed in the next section. Here it may be noted that the evidence includes the facts that an LHRH-like peptide is released from stimulated preganglionic type C fibers in a calcium-dependent manner, and that the very slow depolarization can be mimicked by application of LHRH and LHRH agonists and blocked by LHRH antagonists. Furthermore, only synaptic terminals on C cells stain immunocytochemically for LHRH. Recent studies implicate T-LHRH or a very similar peptide as the substance that produces the very slow depolarization (Jan and Jan, 1982; S. W. Jones, 1985).

The presence of an LHRH-like peptide in the type C preganglionics is indicated in the basic circuit diagram of Fig. 3.4. Also indicated is the fact that type C ganglion cells have been shown to contain the candidate neurotransmitter neuropeptide Y (NPY) (Horn and Stofer, 1988). By contrast, type B preganglionics and type B ganglion cells appear not to contain neuropeptides. This suggests that the type C pathway is specifically subject to modulation by neuropeptides. However, it appears that there is some degree of interaction with the type B pathway, as indicated in the diagram and discussed in the next section.

SYNAPTIC AND MEMBRANE PROPERTIES

Type B bullfrog sympathetic ganglion cells have proven themselves in recent years to be unusually favorable objects for electrophysiological investigations. A primary reason is that they present few space-clamp problems because dendrites are absent, synapses are formed on or near the cell body, and dissociated cells even lack visible axon stumps. Moreover, the cells are robust enough to allow application of the entire gamut of microelectrode methods by themselves or in combination: single and two-electrode voltage clamp, whole-cell and single-channel patch recording, and intracellular injection.

In ganglion cells, as in all neurons, the ionic channels in the postsynaptic membrane can be divided into two broad categories: ligand-gated and voltage-gated conductances. Traditionally, these have been associated with synaptic input and with the generation and propagation of action potentials, respectively. However, as we have seen in Chap. 2, there is a great deal of overlap in these properties, and nowhere in this more evident than in gangion cells. In order to deal with the complex interrelationships between these conductances, we will first describe the fast synaptic response followed by the range of voltage-gated conductances; we will then consider the two main types of slow synaptic responses that modulate these conductances.

THE FAST EXCITATORY POSTSYNAPTIC POTENTIAL (EPSP)

The major transmitter released from preganglionic axon terminals onto ganglion cells is ACh. Acetylcholine diffuses from these terminals in the immediate wake of the presynaptic action potential and binds to the postsynaptic nicotinic ACh receptors in the postsynaptic membrane of the ganglion cell (Marshall, 1981), where it causes the opening of the associated ionic channels. The resulting influx of sodium ions through these open channels into the cell leads to a brief depolarization of the ganglion cell, called a fast EPSP (Fig. 3.5; Blackman et al., 1963; Kuba and Minota, 1986). If the fast EPSP is large enough, it will reach impulse threshold, triggering a spike in the cell which is propagated down the postganglionic axon. On reaching the postganglionic terminals, this spike in turn initiates release of the neurotransmitter NE.

If the ganglion cells are voltage-clamped, the current generating the fast EPSP, called the fast EPSC, can be recorded without the complication of a superim-

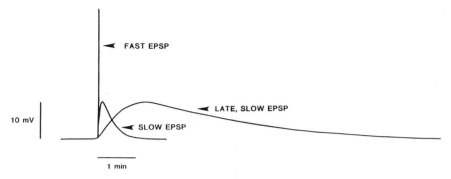

FIG. 3.5. Schematic drawing of the three types of synaptic potentials that can be observed in type B bullfrog sympathetic ganglion cells. The fast EPSP (with a rise time of 3–5 msec) is mediated by ACh binding to nicotinic receptors, and the slow EPSP is mediated via ACh binding to muscarinic receptors. The late, slow EPSP results from a neuropeptide being released by neighboring type C cells and diffusing to receptors on the B cell surface.

posed action potential (Kuba and Nishi, 1979). The fast EPSC fluctuates randomly in amplitude with successive nerve stimuli, because it is composed of a variable number of smaller units, called miniature EPSCs (MEPSCs) (McLachlan, 1978; Kuba and Minota, 1986). These events also occur individually at a low rate even in the absence of nerve stimulation. They are therefore also sometimes called spontaneous EPSCs. A typical preganglionic nerve action potential causes the almost simultaneous occurrence of 20–50 MEPSCs. Because each MEPSC reflects a peak conductance increase of 300 picosiemens (pS), which decays exponentially with a time constant of a few milliseconds, the single nerve-evoked EPSC represents a peak conductance increase of about 6–15 nS at about 2.5 msec, decaying exponentially with a time constant of several milliseconds. (For orientation to the relation between membrane conductance and membrane current, see the Appendix.) In fact, the time course of the postsynaptic conductance change can be fitted reasonably accurately by a so-called alpha function: $g(t) = const \cdot t e^{-t/t_{peak}}$, with $t_{peak} = 2.5$ msec (Yamada et al., 1989; Chap. 2).

Any factor that changes the average number of released MEPSCs, or the size of individual MEPSCs, will therefore change the size of the final EPSC. But since even the smallest EPSC is large enough normally to evoke a spike, only fairly drastic decreases in EPSC size are likely to cause failure of the normally faithful one-to-one transmission through the ganglion. It is far from clear whether the occurrence of a single MEPSC represents the release of a single vesicle by transmitter by the presynaptic axon, or all-or-nothing release from a single bouton (e.g., see Korn et al., 1987). If a single bouton can release only one vesicle at a time, these alternatives would not be separable.

Perhaps the most important physiological factor that influences EPSC size is the frequency of presynaptic impulses. Impulses that occur more closely together than some minimum frequency interact, producing complex patterns of increase and decrease of the EPSP. It is thought that all these effects very likely reflect presynaptic mechanisms, probably dependent on intracellular calcium accumulation.

The action potentials of B and C cells differ in waveform, the C-cell spike afterhyperpolarization (AHP) being briefer and smoother. Little is known about the voltage-dependent currents of C cells, and thus the origin of this difference is uncertain. However, the voltage-dependent currents of B cells have intensively studied, and a detailed picture of the mechanism underlying the B cell action potential is available (see Adams and Galvan, 1986; Adams et al., 1986). B cell somata possess many of the currents described in Chap. 2: a rapidly activating sodium current, I_{Na}; a rapidly activating but slowly inactivating calcium current, I_{Ca} (this current most closely resembles $I_{Ca,1}$; Chap. 2); and five outward potassium currents: I_K, I_C, I_{AHP}, I_A, and I_M. The delayed-rectifier current, I_K, and the fast calcium- and voltage-dependent I_C current appear to play complementary roles. In the normal ganglion cell, I_C is the current that causes spike repolarization, with only a small contribution from I_K. If I_C is blocked by certain experimental manipulations, I_K appears to take over the function of repolarization. I_{AHP} and I_M together regulate cell firing during prolonged excitation. Because no natural hyperpolarizing input to B cells is known, I_A is usually inactivated and its functional significance is unknown.

The accessibility of the ganglion cell has enabled these different currents to be characterized rather completely, thus providing a basis for constructing computational models of each one. Furthermore, the absence of dendrites in these cells means that the complex electrotonic properties of dendrites, as described in the Appendix, are not present, and the cell can be modeled by a single electrical compartment, consisting of the membrane capacitance, the passive membrane resistance, and the models for each of the seven voltage-dependent membrane conductances. This model (Yamada et al., 1989) then reconstructs the voltage trajectory for arbitrary synaptic and current input using a Hodgkin–Huxley-like formalism, where the state of each conductance is described by rate constants whose values depend on the membrane potential as well as on intracellular calcium concentration. In order to describe these cells adequately, the diffusion and buffering of free calcium ions throughout the cell is simulated by using a series of 20 or so concentric shells. This model closely mimics the behavior of type B ganglion cells under a variety of different experimental protocols and can be used to study the function of the various channels in isolation (Adams et al., 1986; Yamada et al., 1989).

The membrane events occurring during and subsequent to the application of a short current pulse are documented in Figs. 3.6 and 3.7. It is important to point out that the conductance change that generates the fast EPSC itself shunts to some extent the spike it triggers. Thus the synaptically evoked spike differs somewhat from the spike generated by a "pure" current impulse. The injected current passively charges up the membrane potential (V_m in Fig. 3.6a), until I_{Na} is activated (b in Fig. 3.6) and leads to the rapid depolarization of the membrane level to 0 mV and beyond. This voltage change causes the calcium channel to

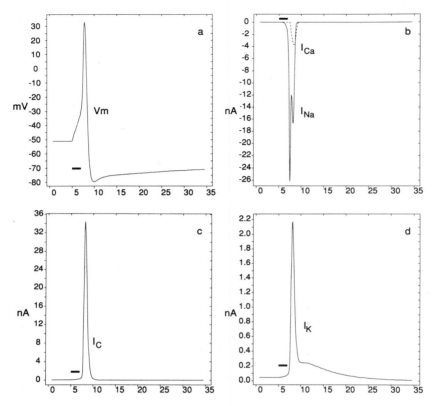

FIG. 3.6. Simulated response of the model type B bullfrog cell at rest to a brief, 2-msec, depolarizing current pulse of 1.75-nA amplitude (indicated on all plots by black bar). This panorama depicts the fast components of the cell's response. The voltage trajectory is shown in (a). Notice the two pronounced phases in the afterhyperpolarization (AHP) following the action potential. Panel (b) illustrates the two fast currents—inward sodium (I_{Na}) and calcium (I_{Ca}). Panel (c) shows the fast, calcium- and voltage-dependent potassium currents (I_C). and panel (d) the delayed rectifier—potassium current (I_K). Because I_C is much larger than I_K, it is primarily responsible for recharging the cell's membrane potential back to rest. (From Yamada et al., 1989.)

open (I_{Ca} in Fig. 3.6b), leading to an influx of calcium ions. The role of the calcium current in depolarizing the cell body is slight, because only about 9% of the incoming charge is being carried by Ca^{2+} ions. However, the intracellular presence of these ions activates I_C (Fig. 3.6c), which very rapidly repolarizes the cell membrane to below the resting level. Because the peak amplitude of I_K (Fig. 3.6d) is about ten times smaller than the peak amplitude of I_C, it does not appear to play a dominant role in healthy cells. The influx of calcium also activates the noninactivating, voltage-independent I_{AHP} (Fig. 3.7a) that is responsible for the second, long-lasting, phase of the afterhyperpolarization. I_{AHP} remains activated far beyond the duration of the action potential, whereas the purely voltage-dependent potassium currents, I_K and I_M, are deactivated within 5–10 msec

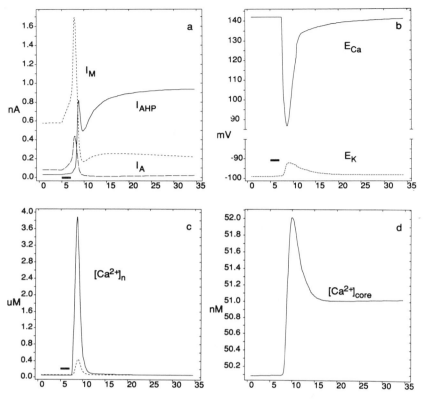

FIG. 3.7. This panoramic view illustrates the slower components of the model cell responding to a brief current stimulus (as in Fig. 3.6). The three, "small" potassium currents are plotted in panel (a). I_{AHP} and, to a lesser extent, I_M, are responsible for the slow phase of the AHP following action potentials (see Fig. 3.6a). The changes in the Nernst reversal potential for the calcium and the potassium batteries following the action potential are plotted in (b). The dynamics of free intracellular calcium ions in the shell just below the membrane [solid line in (c)], in a shell approximately 2 μm removed from the membrane [dotted line in (c)] as well as in the innermost core compartment of the spherical soma (d), are illustrated here. The concentration of Ca^{2+} just below the membrane of the cell controls the strength of the two, calcium-dependent potassium currents, I_C and I_{AHP}. (From Yamada et al., 1989.)

following the spike (Fig. 3.7a). The dynamics of free intracellular Ca^{2+} ions in a thin shell just below the membrane (Fig. 3.7c) reflect the rapid inrush of calcium ions through the open I_{Ca} channel and the equally rapid binding to intracellular calcium buffers such as calmodulin, diffusion of calcium into the interior of the cell, and the extrusion of calcium into the extracellular environment via pumps in the membrane.

The role of calcium in patterning the action potential is clearly revealed in the experiment illustrated in Fig. 3.8, which compares the model with experimental data. The standard action potential is superimposed on the experimental trace

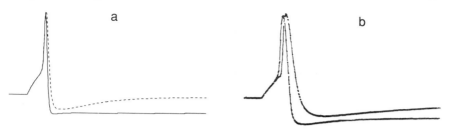

FIG. 3.8. Effect of blocking calcium entry on simulated and observed action potentials. Panel (a) shows the superimposed computed action potential with the normal value (116 nS; solid line) and 5% of the normal value (5.8 nS; dotted line) of the maximal calcium conductance. Panel (b) indicates that experimental recordings before and after application of cadmium to block calcium channels show a very similar result. Both traces are 50 msec in duration. (From Yamada et al., 1989.)

obtained by blocking 95% of the calcium conductance. Both the observed and the simulated action potentials are broader, and show a less pronounced AHP.

Studies of calcium dynamics in bullfrog neurons have shown that calcium movement within the cell following initial entry through membrane channels is governed by simple passive diffusion, once the binding of Ca^{2+} ions to intracellular buffers has been accounted for. Such a scheme is incorporated into our model shown in Figs. 3.6 and 3.7. Thus, intracellular calcium acts as a second messenger, coupling calcium channel activity to potassium channel opening. Of course, calcium also plays a number of other trigger roles in neurons, such as initiating transmitter release and long-term potentiation. Ultimately, then, influx of Ca^{2+} during the action potential provides the crucial link between electrical activity and the initiation of metabolic events.

Spike frequency adaptation. Research on bullfrog cells has also very clearly revealed one particular mechanism for spike frequency adaptation, a property shown by a large proportion of neurons in the central nervous system. This is most clearly seen in experiments in which the source of the excitation is a microelectrode supplying a maintained depolarizing current (Fig. 3.9). The squid giant axon, as well as the standard Hodgkin–Huxley equation, responds to such a stimulus—if above threshold—by a steady stream of action potentials, until accumulation of extracellular potassium reduces the spike frequency. However, the normal and healthy bullfrog sympathetic ganglion cell fires only one or a few action potentials and then falls silent as the two small but persistent potassium currents, I_M and I_{AHP}, are recruited (Fig. 3.9; Jones and Adams, 1987; Brown, 1988a,b). These hyperpolarizing currents can be thought of as simply counterbalancing the effect of the depolarizing current injected into the cell. Thus, the effective current seen by the cell is the depolarizing current supplied by the microelectrode minus the currents through the I_M and I_{AHP} channels. The decrease or cessation of action potential discharge is referred to as adaptation. If either I_M or I_{AHP} is blocked pharmacologically, some loss of adaptation is seen both experimentally (Fig. 3.9) and in our simulations (Fig. 3.10). Only when

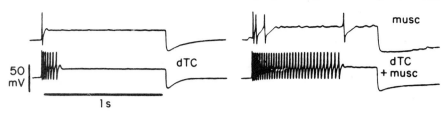

FIG. 3.9. *Upper left*: Experimental record showing the response of a healthy bullfrog cell to a suprathreshold current step. The cell adapts to this stimulus and stops responding. Some loss in adaptation can be seen upon applying muscarine, which blocks a fraction of M channels (*upper right*), or *d*-tubocurare, which blocks AHP channels (*lower left*). The synergistic action of I_M and I_{AHP} is seen in *lower right,* where simultaneous application of muscarine and *d*-tubocurare leads to an almost complete block of spike frequency application.

both currents are blocked (Fig. 3.9, lower right, and Fig. 3.10D) does adaptation of neuronal firing cease. Interestingly, adaptation appears to be at least partially controlled by presynaptic input, because both I_M and I_{AHP} become partially blocked during the slow EPSP (see next section), caused by the binding of acetylcholine to muscarinic receptors in the ganglionic membrane (Figs. 3.4 and 3.5 above). Thus, the control of these small but persistent currents via "slow" synaptic input represents an important neuronal operation.

Some of the currents in these cells (e.g., I_K and particularly I_A) do not seem to be significantly activated by physiological patterns of membrane depolarization caused by synaptic potentials or action potentials. One way to assess the possible significance of a current is to block it pharmacologically. However, there are two problems with this approach. First, the available drugs may not be completely selective. Second, even a highly selective drug may have little effect because small changes in electrical behavior caused by removing a particular current may cause other currents to become more prominent, counteracting the drug effect. One approach to this problem is to perform computer simulations embodying quantitative empirical descriptions of the known currents, to assess to what extent the real cell behavior can be reproduced. This approach is exceedingly difficult in central neurons with many poorly characterized currents and complex dendritic geometry, but it has worked well in bullfrog neurons.

THE SLOW EPSP

In addition to binding to nicotinic receptors to generate a fast EPSC, the ACh released by the preganglionic axon making synapses onto the ganglion cells also binds to postsynaptic muscarinic receptors and elicits a slow EPSP (see Fig. 3.4). The slow EPSP is not large, 10 mV at most, and thus does not exceed threshold and trigger action potentials. However, the total charge (i.e., the current integrated over time) moved across the membrane during the fast and slow EPSPs is quite comparable. This slow EPSP, after an initial electrically silent latency of about 100–200 msec, rises to a peak in 1–2 sec, and may last for a minute (see Fig. 3.5). During the slow EPSP the excitability of the cell is radically modified.

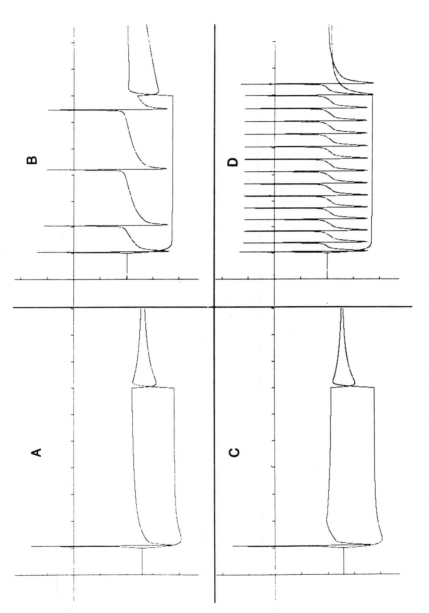

FIG. 3.10. The same phenomena are observed in the computer simulation, where a 1.25-nA, suprathreshold current leads only to a single action potential (**A**), since recruitment of I_M and I_{AHP} counteracts the depolarizing effect of the current stimulus. After block of I_M (**B**) or I_{AHP} (**C**), adaptation is still obvious. Only simultaneous block of both I_M and I_{AHP} abolishes adaptation (**D**). Superimposed on all four figures is the computed response to a 1.25-nA hyperpolarizing current step.

As mentioned above, much of the adaptation normally present in these cells disappears during the slow EPSP. This removal of adaptation has a much more pronounced effect on the cell's excitability than the small potential changes witnessed during the EPSP. Note that in this case, ACh acts both as a classical neurotransmitter, generating the fast EPSP via nicotinic receptors, and as a neuromodulator, generating the slow EPSP via muscarinic receptors present on the ganglion cells.

The mechanism of this cholinergic modulation has been studied in some detail. It appears that activation of muscarinic receptors leads to activation of an unidentified G-protein, which in turn, either directly or via an unknown intracellular second messenger, leads to inhibition of the currents, I_M and I_{AHP}, responsible for adaptation (Pfaffinger, 1988; Lopez and Adams, 1989).

The interaction between the fast and slow EPSPs described above is an example of *homosynaptic modulation,* since one transmitter released from one set of presynaptic boutons is responsible for both classical transmission and neuromodulation at those same synapses. Homosynaptic modulation can be effective only if the transmitter is released over a period of time, either phasically or tonically, so that the transmitter progressively modifies its own effect. A more interesting situation, *heterosynaptic modulation,* is also exhibited in bullfrog ganglia (Kuffler and Sejnowski, 1983; Jan and Jan, 1982). As already mentioned, preganglionic terminals in C cells also release a peptide, akin to and perhaps identical with chicken II LHRH, when invaded by action potentials. This peptide passively diffuses and binds to peptide receptors located on both B and C cells. The diffusion of the neuromodulator from synapses in close conjunction with the postsynaptic membrane of C cells to receptors on neighboring B cells is termed "spillover." Activation of these peptide receptors (again via a G-protein) also leads to inhibition of the slow potassium currents I_M and (in B cells) I_{AHP}, thereby attenuating spike frequency adaptation. This effect is even slower than the muscarinic slow EPSP, generating the so-called "late slow EPSP," which can last up to 10 min.

There are several important points to note about this arrangement:

1. Preganglionic terminals on C cells can make and release two distinct transmitter substances: One, the classical transmitter ACh, is rapidly broken down by an enzyme, acetylcholinesterase, and thus cannot travel far in the ganglion. The other, the peptide, is only very slowly broken down, and continues to diffuse in the ganglion for minutes. Because it is thus distributed over a large volume, it must be able to act at very low concentrations.

2. The sphere of influence of these modulatory substances is not confined to the synaptic cleft, but may well extend beyond the 10-μm range. Therefore, the physiological postsynaptic target of neuromodulators may not correspond to the anatomically postsynaptic structure. The speed of action of these neuromodulatory substances is limited, in the absence of any active transport system, by diffusion; that is, distance is proportional to the square root of the elapsed time.

3. At other synapses, and possibly also in bullfrog sympathetic ganglia, colocalized classical and peptide transmitters may be differentially released, for example, by different patterns of presynaptic activity (cf. Hökfelt et al., 1984).

4. Classical transmitters produce rapid increases in conductance that directly initiate, or inhibit, action potential firing. Neuromodulators produce slower changes (both decreases and increases) in conductances that participate in the shaping of individual action potentials or groups of action potentials.

It may well be that this pattern of neuromodulators controlling cellular properties over long times and relatively large distances may repeat itself throughout the central nervous system and subserve a very important role in routing information among neurons (Koch and Poggio, 1987).

SYNAPTIC INHIBITION

So far we have discussed the various excitatory synaptic events, both fast and slow, exhibited by bullfrog ganglia. C cells, but not B cells, also possess a slow inhibitory synaptic event, the slow IPSP. Acetylcholine released by terminals on C cells acts on C cell muscarinic receptors to initiate a potassium conductance increase, lasting several hundred milliseconds, that hyperpolarizes the membrane and inhibits impulse generation (Dodd and Horn, 1983a,b). In other situations— for example, the heart—such muscarinically activated increases in potassium conductance are mediated by a G-protein, which probably acts directly on the potassium channel in the GTP-bound form (Brown and Birnbaumer, 1988). Hydrolysis of the bound GTP by the GTPase activity of the G-protein then leads to termination of the muscarinic action, the receptor probably having lost its bound ACh somewhat earlier.

Thus, in B and C cells, different transmitters can have either convergent effects (decrease of potassium conductance) or divergent effects (decrease in one potassium conductance and increase in another). Also, the same transmitter (ACh) can have opposite effects in different postsynaptic cells. Opportunism and flexibility are as prevalent in the nervous system as elsewhere in biology.

MAMMALIAN SUPERIOR CERVICAL GANGLIA

We turn now from the most caudal of the sympathetic ganglia to the most rostral, and from amphibians to mammals. The most significant difference between these ganglia is that superior cervical ganglion (SCG) cells possess dendrites, which extend several hundred microns beyond the cell body. Interestingly, the number of these cells, as well as the extent of complexity of their dendritic arbors, seem to be related to the size of the animal (Purves et al., 1986b). Thus, over a 65-fold range of average adult weight of several mammals (mouse, hamster, rat, guinea pig, and rabbit), the number of SCG cells increases by about a factor of 4 whereas the number of preganglionic cells increases by 2. The average diameter of the SCG cell bodies also increases, and, much more strikingly, ganglion cells in larger animals bear progressively more complex dendritic arbors (Fig. 3.11).

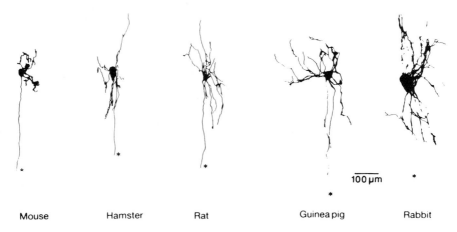

Mouse   Hamster   Rat   Guinea pig   Rabbit

FIG. 3.11. Comparison of homologous sympathetic ganglion cells in several related adult mammals that differ in body size. Both the overall length of dendrites and the complexity of dendritic branching parallel body size. The asterisks indicate the axon of each cell, which extends much farther than shown here. (From Purves and Lichtman, 1985.)

Synaptic boutons are found mostly on dendrites. For example, in the rat SCG there are typically about 250 boutons on the dendrites, and only 50 on the cell body (Purves et al., 1986b). These boutons are supplied by several preganglionic axons—as many as 25 in the rabbit, and 6–8 in the mouse. Indeed, the number of preganglionic axons innervating a given cell seems to reflect strongly the size of the available postsynaptic surface, dominated by the dendrites. And, as noted above, the size and complexity of the ganglion cells in turn correlate strongly with the size of the animal. Mammalian autonomic ganglion cells that lack dendrites are, like frog cells, typically innervated by only one axon. The boutons arising from different presynaptic fibers are not found randomly over the cell surface. They tend to focus on a subset of the dendritic tree (Purves et al., 1986b).

Most of the preganglionic axons that innervate a SCG cell can, individually, generate an EPSP large enough to reach threshold. This arrangement of convergent, potent inputs may mean that the same output cell can be time-shared by several different inputs. Alternatively, regulating the number of asynchronously active inputs may be a way of varying the degree of activation of individual ganglion cells.

Early in development, SCG cells are innervated by many more axons than in the mature condition. It is believed that synapse elimination proceeds during development as a result of competition among axons, and it may well be that such competition is less severe when a larger postsynaptic target is available, explaining the greater convergence in more complex cells and larger species. However, despite elimination of preganglionic inputs, the total number of boutons actually increases during development, the remaining successful axons forming synapses (Purves and Lichtman, 1985).

Each different preganglionic axon may have a different conduction velocity, so that the compound EPSPs generated by preganglionic volleys can have quite complicated, serrated time courses. However, the underlying single fiber ("unitary") EPSPs is very similar to that already described for the frog. Because of this scatter, only antidromic conduction velocity can be used to classify SCG cells. It appears that separate B (myelinated) and C (unmyelinated) populations exist, the latter type predominating. However, there has been no real attempt to correlate slow synaptic potentials (sEPSP, sIPSP, etc.) with cell type. Indeed, slow synaptic potentials, of various types, have been more difficult to record intracellularly in mammalian ganglia than in frogs, and have been less well studied. However, it is likely that the basic mechanisms reviewed above for frogs also apply here.

The presence of dendrites on mammalian ganglion cells has hampered the application of voltage-clamp techniques to the analysis of excitability mechanisms in these cells. However, it appears that rat SCG cells possess all of the conductances found in bullfrog neurons, and that action potential and repetitive firing mechanisms are basically the same. The main difference is that the A current is active at less negative potentials, and plays a significant role in spike organization.

Miniature EPSP amplitudes in SCG cells show a good deal of scatter in size, but a similar range is also seen in frog cells, and is thus probably not related to variable electrotonic locations of the boutons over the cell surface. The compactness of the dendrites probably means that all synapes are roughly equivalent electrotonically, the ratio of dendritic conductance to somatic conductance being considerably less than 1 (Rall, 1977).

In addition to the convergence of several preganglionic axons onto single ganglion cells, each preganglionic axon can innervate numerous target cells— varying from 60 in the mouse to over 400 in the rabbit. The target cells for an individual axon are found scattered throughout the ganglion, there being little tendency for neighboring cells to be innervated by the same axon. This must be related to the observation that there is little topographical organization, cells projecting to the same end organs being found throughout the ganglion.

Recently, repeated observation of SCG cells in situ, with the use of fluorescent dyes, in young adult, living mice, has shown that the dendritic morphology of postsynaptic elements undergoes rearrangements over a time scale of weeks to months (Purves et al., 1986a). Dendrites of the same identified cells showed no significant difference if inspected within 3–10 days. Neurons visualized at intervals of 2–3 weeks, however, showed retraction of some branches and extension of others. Over months (Fig. 3.12), these changes can be quite significant as the dendrites become longer and branch more profusely. Because about 90% of all synapses from preganglionic axons are made onto the dendritic arbor (Forehand, 1985), this change in morphology must go hand in hand with a corresponding change in synaptic connectivity (which is experimentally much more difficult to show). The distribution and number of the boutons on the cell surface also continuously vary. Neither of these processes seems to be simply growth or

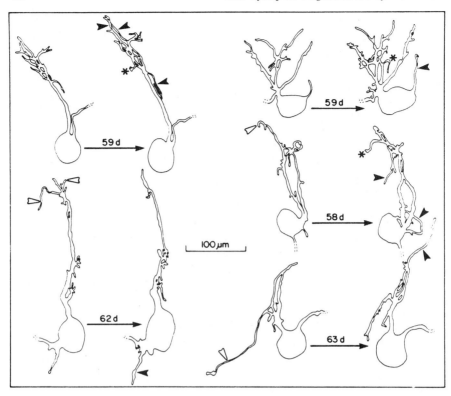

FIG. 3.12. Reconstructions of selected portions of the dendritic arbor of several representative neurons in the superior cervical ganglion of living mice visualized over an interval
of about two months. Open arrowheads point to branches that appear to have retracted;
closed arrowheads point to dendritic branches that appear to have extended; and asterisks
indicate branches that have formed de novo. Dashed lines indicate processes that continue
but were not chosen for study. (From Purves et al., 1986a.)

elaboration, but are more akin to structural turnover. It will be very interesting to
see if the synaptic input contributed by individual axons waxes or wanes in a
similar manner, perhaps in correlation with their separate activities.

Thus far there is little direct evidence for such dynamic remodeling of neurons
in the adult mammalian brain. However, the elegant studies of Merzenich and his
colleagues (1984) in the somatosensory cortex of primate indicate that cortical
neurons sensitive to touch, heat, or pain in well-specified areas for the arm and
hand can change their receptive fields within hours. It therefore appears likely
that the neural and synaptic architecture underlying information processing in the
brain continuously changes in response to external stimuli. This, of course, is in
stark contrast to all computers and other artificial computational devices built so
far whose electronic structure remains invariant over time.

THE FUNCTIONS OF AUTONOMIC GANGLIA

As already noted, the main postganglionic transmitter in the parasympathetic nervous system is ACh. ACh acts on muscarinic receptors in postsynaptic smooth muscle cells or gland cells to produce characteristic parasympathetic effects such as relaxation or secretion, leading to slowing of the heart or increased motility of the intestine. In the sympathetic nervous system, the main transmitter released from postganglionic nerve terminals is NE. This acts on a variety of different types of adrenergic receptors to produce effects such as decreased gut mobility, increased heart rate, dilatation of the pupil and increased sweat production.

Although autonomic ganglion cells are among the most intensively studied of all neurons, amazingly little is know about their function, in that, as far we know, not much would change if effector organs were directly innervated by preganglionic axons, without the intercalation of ganglia. In frog lumbar ganglia, B cells are thought to innervate mostly cutaneous slime glands, whereas C cells innervate mostly hind limb vasculature. SCG neurons innervate a variety of cephalic targets such as hair, blood vessels, and ocular structures such as the iris. Obviously, much more needs to be known about natural patterns of impulse flow in the sympathetic nervous system, in order to understand the functional significance of the ganglia and of the synaptic and membrane properties that control these activity patterns.

4

SPINAL CORD: VENTRAL HORN

ROBERT E. BURKE

A great deal of detailed information about synaptic organization within the central nervous system has come from studies of the spinal cord. There are several reasons for this. For example, the input/output relations of the spinal cord—that is, reflexes—are reproducible and easily elicited in normal animals as well as in reduced preparations. Sensory inputs to and motor outputs from the spinal cord are physically accessible and can be defined in functionally meaningful terms. And, most important for the present discussion, the relative proximity of functionally defined sensory and motor elements makes it possible to attach behavioral significance to the organization of intrinsic synaptic systems—an elusive goal in most other parts of the CNS.

The spinal cord is a remarkably complex system of neurons that subserves sensory, motor, and autonomic functions. The diversity of its systems and the wealth of information about many of them preclude any attempt at comprehensive discussion here. Rather, as in earlier editions of this book, the emphasis in this chapter will be on the organization of neuronal elements and synaptic interconnections in the ventral horn that are relevant to the control of movement. Much attention will be paid to motoneurons, which belong to perhaps the most intensively studied cell type in the mammalian CNS. For better or worse (e.g., see Llinás, 1988), the motoneuron is the basis for many fundamental notions about neuronal function in general.

This chapter will first introduce the general types of neurons in the spinal cord and then deal with some aspects of their interconnections in identifiable circuits. Once the background is established on this "macro" level, emphasis will shift to a consideration of synaptic action at a subcellular level and the influence of dendritic electrotonus on the propagation of synaptic potentials within individual neurons. The chapter ends with a discussion of the process of motor unit recruitment, which involves an integration of factors from the micro to the macro level.

NEURONAL ELEMENTS

There are three major categories of neuronal elements in the spinal cord: (1) neurons that carry sensory information into the cord (primary afferents and descending fibers); (2) neurons with axons that leave the cord to innervate skeletal muscle fibers (motoneurons) or autonomic ganglia (preganglionic neurons); and (3) neurons with axons that terminate exclusively within the CNS, either relatively locally in spinal segments (interneurons) or in more distant, more rostral sites (tract cells). Although the cell bodies of primary afferent neurons lie in the dorsal root ganglia outside the spinal cord, their central projections are crucial to understanding synaptic organization within the spinal cord.

The diagram in Fig. 4.1 illustrates some aspects of the structure of the spinal cord viewed in cross section. Unlike the rest of the brain, the outermost part of the spinal cord—the white matter—consists of the axons of spinal and supraspinal neurons running parallel to the long axis of the cord. The inner gray matter core contains the cell bodies of spinal neurons and most of the synaptic neuropil in which they interact. The two major gray matter divisions, the dorsal and ventral horns, can be subdivided into layers, numbered I through X, on the basis of cell numbers, shapes, and densities ("cytoarchitectonic" divisions; Rexed, 1952; left half of Fig. 4.1). Particular classes of spinal interneurons ("Ia/Ib INs," "Ia INs," "Renshaw cells") and motoneurons (MNs) occupy reasonably well-defined parts of the gray matter (Fig. 4.1). Their cell bodies and dendrites

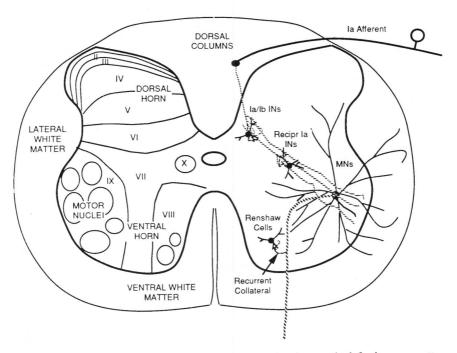

FIG. 4.1. Diagram of the spinal cord in cross section, showing, on the left, the gray matter divisions and lamination determined by neuron densities and sizes ("cytoarchitectonics"; Rexed, 1952), and, on the right, the trajectory of a group Ia afferent, the approximate extent of motoneuron dendrites, and the location of certain identified groups of interneurons discussed in the text.

receive synaptic input from sensory afferents and from each other largely within the gray matter. These interactions will be the main topic of this chapter.

PRIMARY AFFERENTS

A wide variety of primary afferent types have been identified on anatomical and functional grounds. Some schemes depend on the tissue innervated (muscle, skin, joints, and other deep tissues, and the viscera) and the characteristics of the receptor organ giving rise to them. The size (diameter) of the peripheral afferent axon and whether or not it is myelineated are also important considerations. The peripheral receptor organ governs an afferent's response to various types of natural stimulation, and its diameter and myelination control the speed with which its information can be conducted to the spinal cord. As might be expected, there are systematic interrelations among all of these characteristics (for reviews, see Darian-Smith, 1984; Fyffe, 1984). There is also recent evidence that different classes of afferent neurons exhibit distinctive intrinsic membrane properties (Koerber et al., 1988). Brown (1981) has provided detailed descriptions of the morphology and physiology of the central projections of primary afferents.

Other afferent classification schemes are more general—for example, the dichotomy between "exteroceptive" and "proprioceptive" afferents. The former are viewed as primarily responsive to events in the environment as sensed by the skin (touch, temperature, pain, etc.). Proprioceptors, on the other hand, are activated mainly by the animal's own movements, as sensed by sensory structures in muscles, joints, and deep tissues of the trunk and limb. This dichotomy is useful and well entrenched, but it is useful to keep in mind that movements can, and do, activate skin afferents as well as those from muscle and deep tissues. Still other systems group sensory afferents according to the type of response they produce when activated. An example of this is the "flexor reflex afferents," an assortment of joint, muscle, and skin afferents with relatively high electrical threshold that, when stimulated electrically, tend to activate flexor muscles and inhibit extensors (i.e., the "flexor reflex"; Eccles and Lundberg, 1959; see Baldissera et al., 1981, for review).

One species of primary afferent—the group Ia, or primary muscle spindle, afferents that arise within muscle spindle stretch receptors—is of special interest here because its synaptic organization has been extensively studied. Group Ia afferents arise from specialized end organs called muscle spindles and signal changes in muscle length whereas their first cousins, the group Ib muscle afferents (so called because they conduct at slightly slower conduction velocities), arise from Golgi tendon organs and signal muscle tension. Many fundamental ideas about synaptic organization within the CNS are based on information derived from studies of the direct (monosynaptic) excitatory synaptic contacts made by group Ia afferents on motoneurons and interneurons within the spinal cord.

In mammals, group Ia afferents are among the largest diameter and fastest conducting primary afferents that enter the spinal cord. The group Ia afferents from a given muscle enter the cord in the same region that contains their target motoneurons—ordinarily those that innervate the same muscle from which the afferents originate and those of its mechanical and functional synergists (Eccles et al., 1957). After entry, each Ia afferent divides into a large ascending and a smaller descending branch, which travel rostrally or caudally in the ipsilateral dorsal column. These in turn give off collateral branches, at roughly 1- to 2-mm intervals, which descend into the gray matter (Ishizuka et al., 1979; Brown, 1981). Each Ia collateral gives rise to preterminal arborizations, and associated synaptic boutons, in three definable locations: (1) the intermediate nucleus (the medial part of Rexed laminae V and VI, where they contact a variety of spinal interneurons (Czarkowska et al., 1981); (2) in lamina VII, just dorsomedial to lamina IX, where they make contacts with a more specific group of inhibitory interneurons (Jankowska and Lindström, 1972, see below); and (3) in lamina IX where they contact the somata and dendrites of alpha motoneurons (Burke et al., 1979; Brown and Fyffe, 1981; see Fig. 4.1). Normally, every muscle spindle in a given muscle gives rise to one group Ia afferent (e.g., there are about 70 in the cat medial gastrocnemius muscle nerve; Boyd and Davey, 1968), so that one can envision the interstices between collaterals from any given afferent as filled by the collaterals of others. Thus, the terminal fields containing Ia boutons from a given muscle can be viewed as longitudinal columns, or perhaps better, clouds of synaptic boutons in the three gray matter loci noted above.

Motoneurons in a given motor nucleus receive functional Ia contacts from nearly all of the group Ia afferents originating in the innervated muscle (the "homonymous" connection; Mendell and Henneman, 1971; Fleshman et al., 1981b). The projection frequency is somewhat smaller but still considerable to motoneurons that innervate muscles that act synergistically at the same joint ("heteronymous" connections). Group Ia connections can also occur between muscles acting at different joints (e.g., from the knee extensor, quadriceps, to the ankle extensor, soleus; Eccles et al., 1957). Presumably, such Ia connections facilitate the coordinated action of different muscles during movement (Eccles and Lundberg, 1958). Muscles that are linked by Ia interconnections are sometimes referred to as a "myotatic unit" (see Lloyd, 1960).

Anatomical studies in which the tracer substance horseradish peroxidase (HRP) was injected into both identified group Ia afferents and alpha motoneurons presumed or identified to be postsynaptic to them have demonstrated putative contacts on the cell somata and throughout the dendritic tree, including quite distal regions (Burke et al., 1979; Brown and Fyffe, 1981; Redman and Walmsley, 1983a). The wide spatial distribution of these contacts confirmed earlier inferences that were based purely on electrophysiological information (Rall et al., 1967; Jack et al., 1971). The available data suggest that an individual Ia afferent can make between 3 and 35 contacts with homonymous motoneurons, averaging about 10 per afferent/motoneuron pair, and a somewhat smaller number on heteronymous cells (Burke et al., 1979; Brown and Fyffe, 1981; Burke and Glenn, unpublished). The total number of group Ia boutons on a lumbosacral cat motoneuron can be estimated, using these averages and the total number of Ia afferents in the appropriate muscle nerves, as between 500 and 1000, or less than 1% of the more than 100,000 total synaptic boutons that probably end on these cells.

MOTONEURONS

Motoneurons (or motor neurons, if you prefer) are the only CNS neurons that make synaptic contacts on nonneural tissue (i.e., skeletal muscle fibers). They also are one of the relatively few classes of CNS neurons with a precisely defined functional role—to activate muscle. Motoneuron cell bodies and their extensive dendritic extensions lie within the ventrolateral gray matter of the spinal cord (Fig. 4.1) and in certain cranial nerve nuclei in the brainstem. The cells that innervate a given muscle in the limbs and trunk lie clustered together in circumscribed "motor nuclei," which are elongated, cigar-shaped collections of cell bodies spreading along the rostrocaudal axis of the ventral horn (see Burke et al., 1977; Burke, 1981). The five motoneurons shown in Fig. 4.2 were labeled by intracellular injection with HRP and came from the triceps surae motor nucleus of a cat. The intricate feltwork of intertwined dendrites of even these few cells gives some idea of the complexity of the neuropil, which is filled with the cell bodies and dendrites of many more motoneurons, in addition to afferent axons that are not shown.

The relative positions of the motor nuclei that innervate functionally related muscles exhibit fundamental similarities throughout the vertebrate series, from amphibia to man (Romanes, 1951; Sharrard, 1955; Landmesser, 1978; Fetcho, 1987). Motoneuron dendrites, however, usually extend for considerable distances (Fig. 4.1; up to 2 mm; see Cullheim et al., 1987) outside the confines of the nuclear cell column, to invade adjacent regions of both gray and white matter, where they receive synaptic contacts (Rose and Richmond, 1981).

There are two distinct types of motoneurons in mammals, alpha and gamma, which were so named initially on the basis of the fact that alpha-motoneuron axons have faster conduction velocities (generally >60 m/sec, in the "alpha peak" of the compound action potential recorded from muscle nerves) than the gamma motoneurons (conduction velocities <40 m/sec; Matthews, 1972, 1981). The alpha motoneurons innervate the large "extrafusal" striated muscle fibers that make up the major bulk of muscles and produce output force. In contrast, the two species of gamma motoneurons ("static" and "dynamic") exclusively innervate one or more of the three kinds of small, highly specialized "intrafusal" muscle fibers that exist only within the muscle spindle stretch receptors. Their action is to modulate the sensitivity of the two types of muscle spindle afferents, group Ia and group II, to static muscle length or to changes in length. The details of the operation of this remarkable sensory system can be found elsewhere (Matthews, 1981; Hasan and Stuart, 1984).

The somata of alpha motoneurons are larger and their dendritic trees are more numerous and highly branched than those of gamma cells (Ulfhake and Cullheim, 1981; Ulfhake and Kellerth, 1981; Burke et al., 1982; Cullheim et al., 1987). Although the two neuron types are quite thoroughly intermixed within motor nuclei (Burke et al., 1977), they do not receive identical synaptic inputs.

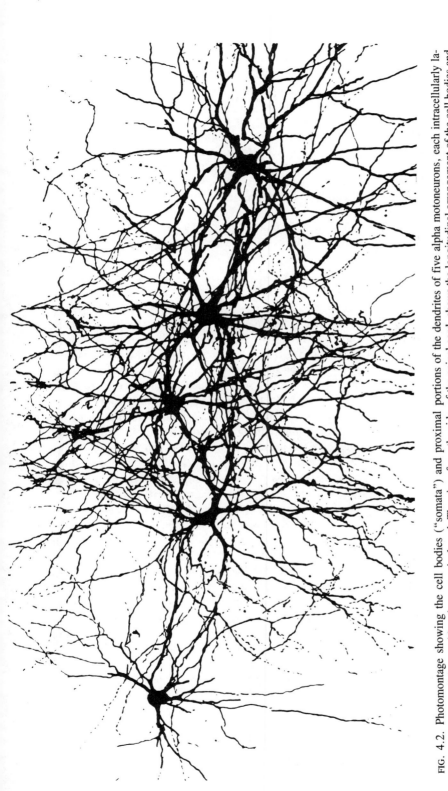

FIG. 4.2. Photomontage showing the cell bodies ("somata") and proximal portions of the dendrites of five alpha motoneurons, each intracellularly labeled with horseradish peroxidase in a cat spinal cord. The view, from the side of the spinal cord, shows the longitudinal arrangement of the cell bodies and the remarkably complex interdigitation of their dendrites. The area shown is approximately 2.1 × 1.15 mm. (Adapted from Burke, 1981.)

In particular, gamma motoneurons receive no monosynaptic excitation from group Ia muscle spindle afferents (Eccles et al., 1960a), whereas virtually all alpha motoneurons do (Eccles et al., 1957; for review see Burke and Rudomin, 1977). The lack of direct excitation of gamma motoneurons by spindle afferents presumably prevents a "positive feedback" loop that might produce instability in the spindle servo-control system.

This simple dichotomy became more complicated when Emonet-Denand and co-workers (1975) showed that some motoneurons in cats innervate *both* extrafusal and intrafusal muscle fibers. Such "beta motoneurons" are quite common in certain limb muscles in cats (Jami et al., 1982), and they can also be found in primates (Murthy et al., 1982). In contrast to their gamma motoneuron half-brothers, it is probable that most, if not all, beta motoneurons receive monosynaptic excitation from group Ia afferents (Burke and Tsairis, 1977). This would appear to form a positive feedback system with unknown functional consequences.

INTERNEURONS AND TRACT CELLS

Strictly speaking, all neurons with axons confined entirely within the CNS are interneurons. However, for many reasons it is convenient in the spinal cord to define "interneurons" as nerve cells with axons that terminate exclusively relatively close to the parent cell body—that is, within the same spinal segment or in nearby segments (see INs and Renshaw cells in Fig. 4.1). Interneurons with axons that end at greater distances but still within the spinal cord itself are sometimes called "propriospinal" cells. Propriospinal neurons have obvious utility in coordinating the activities in many spinal segments, such as ensuring correct movements of fore- and hindlimbs in four-footed animals and, perhaps less obviously, the actions of trunk muscles in maintaining balance and posture while providing a stable platform for limb movements.

Spinal neurons that send their axons primarily to supraspinal destinations can be conveniently distinguished as "tract cells." Examples include the various divisions of the spinocerebellar tracts, ending as mossy fibers in the cerebellar cortex, and neurons of the spinothalamic tract that end in the thalamus. The most obvious role for tract neurons is to deliver information to supraspinal regions of the brain that are specifically associated with sensory perception, but here again the situation is not necessarily so clear-cut. For example, the distinction between tract cells and interneurons is sometimes blurred, because some tract cells have axon collaterals that end locally within the spinal cord (Brown et al., 1977). Such neurons may therefore also act as "interneurons" in certain reflex actions. Furthermore, systems such as the spinocerebellar tract carry information from both deep (i.e., "proprioceptive") and cutaneous receptive fields to supraspinal structures ordinarily associated with motor functions (e.g., cerebellum) rather than with sensory perception. As will be discussed below, some tract cell systems have synaptic organizations which suggest that they signal information about the "state of affairs" in spinal reflex pathways rather than simply relaying incoming signals, underscoring their connection with motor control functions.

In the spinal cord, there are relatively few morphological clues to guide a functional taxonomy of interneurons and tract cells, other than a few cell groups with characteristic morphology, such as Clarke's column in the upper lumbar spinal segments, which contains neurons that project to the cerebellum in the dorsal spinocerebellar tract (DSCT). The laminae present in the dorsal horn (Rexed's laminae I through IV; Fig. 4.1) have been shown to contain neuronal organizations that process incoming afferent information, especially from cutaneous sources, in orderly sequences that also display topographical relations to the receptive field of particular afferent species (Brown, 1981; Darian-Smith, 1984). The large neurons making up lamina IX and VIII in the ventral horn (see Fig. 4.1) are mainly the output cells—motoneurons that innervate limb and trunk muscles, respectively. The functional identities of the many interneurons in the intermediate portion of the spinal gray (Rexed's laminae V–VIII) are much less clearly delineated by spatial location or morphology alone.

REFLEXES AND BEYOND: THE SYNAPTIC ORGANIZATION OF SPINAL CIRCUITS

The central nervous system is a vast assemblage of neurons with highly complex but quite specific patterns of interconnection. As a first approximation, an "understanding" of the function of any given region (e.g., an anatomically identifiable "nucleus") is usually inferred from information about its inputs and outputs. For example, the nuclear groups that receive direct input from particular sense organs [e.g., the lateral geniculate nucleus (see Chap. 8), which receives input directly from retinal ganglion cells (see Chap. 6)], clearly must be involved in processing sensory signals. Conversely, regions that project directly to motoneurons, autonomic preganglionic neurons, or hypothalamic/hypophyseal sites of hormone secretion, can be assumed to have motor, autonomic, or humoral functions, respectively. By extension, the inputs to and the outputs from any particular neuron serve to establish not only its position in the CNS circuitry but also its probable function, to the extent that those inputs and outputs are functionally definable.

As mentioned at the beginning of this chapter, one of the great advantages of the spinal cord for functional analysis is the fact that primary afferents and motoneuron outputs, the two linchpins of functional identification in the motor system, are both locally present, separately accessible, and readily identifiable. A set of interneurons that receives direct (monosynaptic) input from a particular afferent system and projects, directly or indirectly, to motoneurons can be defined in terms of spinal cord circuitry and in probable function. Such aggregates form the "reflex" pathways.

REFLEX PATHWAYS

Systematic research on the organization of the spinal cord began with studies of "reflexes"—the predictable output patterns in muscles or autonomic effector organs that are produced by a particular input stimulus. One example of a simple reflex is the withdrawal of a limb that is produced by noxious stimulation (the "flexion reflex"). Another is the contraction of a muscle (sometimes of its synergists), produced by suddenly stretching it (the "stretch reflex"). Noxious stimulation can also produce autonomic reflexes such as constriction of arterioles, raising blood pressure. There are also more elaborate and coordinated reflexes that involve many muscles, such as the rhythmic scratching movements by a dog's hindleg that are produced when its ear is tickled (the "scratch reflex"). For over a century, physiologists have used spinal reflexes as test systems in attempts to define the neuronal circuitry and synaptic interactions that produce them. One cannot overstate the importance of reflexes in the development of ideas about CNS function that we now take for granted (e.g., see Brazier, 1960).

Until relatively recently, the various reflexes have been regarded as distinct entities, resulting from the operation of separate bits of neural machinery. In addition, "voluntary" or "willed" movement is often viewed as separate and distinct from "reflex" or "automatic" actions. However, over the past three decades, research on the synaptic organization of the spinal cord has produced a synthesis among these elements. The spinal cord can now be seen as a full-fledged part of the functioning "brain" that integrates incoming sensory information with "motor command" signals descending from supraspinal regions, combining the neural organizations of reflex and voluntary movement control. The following section stresses two fundamental points: (1) The spinal pathways of nominally different "reflexes" in fact interact extensively, often by sharing common interneurons; and (2) control of movement (i.e., of motoneurons) by the supraspinal brain is mediated mainly through interneurons in "reflex" pathways. This appears to be true even in primates which, unlike most other species, have a significant direct pathway from the motor cortex to motoneurons (see Phillips, 1969). The classical Sherringtonian notion of the motoneuron as the "final common path" has been amended to include spinal interneurons that function as multimodal nodes upon which afferent and descending control converge.

DIRECT ACTION OF AFFERENTS ON MOTONEURONS: THE STRETCH REFLEX

The monosynaptic stretch reflex is the most basic reflex organization in the spinal cord. Using relatively simple but ingenious methods, David Lloyd (see Lloyd, 1960) was able to show that the stretch reflex is generated by the fastest-conducting (i.e., group Ia) afferents from muscle spindles and that the time delay ("central latency") between the arrival of a volley of action potentials in group Ia afferents at the spinal cord and the activation of motoneurons is so short (less than 1.0 msec) that the connection between Ia afferents and motoneurons must be direct, as shown in the diagram in Fig. 4.3. The subsequent development of intracellular recording from individual motoneurons (Brock et al., 1952) demonstrated the production of excitatory postsynaptic potentials (EPSPs) with central latencies of about 0.8 msec (Fig. 4.3, record B) after the Ia volley enters the cord (record A). The delay after the arrival of the Ia volley in the ventral horn, signaled by a small deflection known as the "terminal potential" (Fig. 4.3, record B, arrow), is even shorter (about 0.2 msec) and this is regarded as the "synaptic delay" between arrival of action potentials at the synapses and the onset of postsynaptic action.

It is now known that the slower conducting group II spindle afferents also

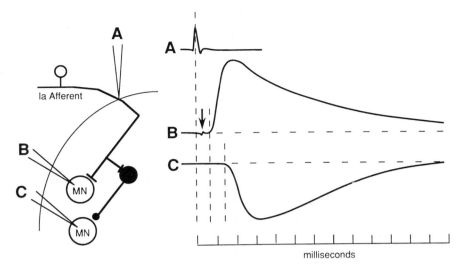

milliseconds

FIG. 4.3. *Left:* Diagram showing the direct, monosynaptic projection and an indirect, disynaptic projection from a group Ia afferent to two motoneurons. The interneuron in the disynaptic pathway is shown as a filled circle, to denote its inhibitory action on the motoneuron. *Right:* Diagram showing the potentials that would be found at recording site A (on the dorsal surface of the spinal cord), and at B and C (intracellularly in the motoneurons). The arrival of the group Ia volley in the ventral horn, at a slight delay after the main volley enters the cord (A), is indicated by the arrow in trace B. The synaptic delay between this arrival and the onset of the EPSP is about 0.2 msec. Note that the disynaptic Ia IPSP (trace C) has a longer central latency than the EPSP.

produce monosynaptic excitation in many motoneurons (see Sypert and Munson, 1984). These are the only primary afferent systems that make direct connections onto motoneurons. All other motor reflex arcs exhibit longer central latencies, enabling one to infer that one or more levels of interneurons are interposed between the input afferents and motoneurons. An example that will be discussed in more detail below is the inhibition produced by group Ia afferents in motoneurons of antagonist muscles. The central latency of these group Ia inhibitory postsynaptic potentials (IPSPs) is about 1.4 msec, about 0.6 msec longer than the monosynaptic EPSP (Fig. 4.3, record C). Although brief, this additional delay signals the presence of one level of interneurons interposed between the afferents and motoneuron, as shown in the diagram in Fig. 4.3.

The monosynaptic stretch reflex is obviously designed for speed. Its fundamental purpose is to facilitate muscle activity to counteract imposed stretch, tending to restore the muscle to its original length (a "length-servo" function). The apparent simplicity of the monosynaptic pathway is deceptive, however, since the sensitivity of stretch receptor afferents can be controlled by the CNS through the gamma motoneurons. In addition, both Ia and group II spindle afferents exert significant effects indirectly through interneuron pathways (see Lundberg et al., 1987). A variety of functions have been proposed for the muscle spindle servo, and none are mutually exclusive (see Houk and Rymer, 1981, and Rymer, 1984, for reviews). This subject continues to be a matter of intense research interest.

MULTISYNAPTIC REFLEX ARCS

The monosynaptic reflex arc has certain disadvantages, one of which is that it cannot be completely disabled. As will be discussed below, the synaptic effects of group Ia afferents on motoneurons can be modulated by the mechanism of "presynaptic inhibition," but this does not completely suppress Ia synaptic action on motoneurons. However, when interneurons are interposed between an afferent system and motoneurons (a "polysynaptic" pathway), two significant advantages accrue. First, the sign of the effect at motoneurons can be changed. So far as is known, all primary afferents excite the neurons to which they project, but an interposed interneuron can produce an inhibitory synaptic effect. Second, transmission in a polysynaptic reflex pathway can vary from zero to considerable amplification, by virtue of excitatory and inhibitory effects converging onto the interposed interneurons. Multisynaptic circuits can thus function as logical elements (in effect, digital gates) as well as analog signal amplifiers.

A DISYNAPTIC REFLEX PATHWAY: RECURRENT INHIBITION

The first spinal interneurons to be functionally identified as individual cells were the "Renshaw cells," so called by John Eccles and co-workers (1954) in honor of Birdsey Renshaw, an American neurophysiologist who first described inhibition of motoneurons following antidromic activation of motor axons ("recurrent inhibition"). Renshaw cells are monosynaptically excited by synapses from collaterals that arise along the course of motoneuron axons before they exit via ventral spinal roots. In turn, they project to the same and related motoneurons (see Fig. 4.4; for reviews, see Burke and Rudomin, 1977; Baldissera et al., 1981). The central latency of onset (approximately 1.5 msec) for recurrent inhibitory postsynaptic potentials (IPSP) in motoneurons indicated that this connection was disynaptic—that is, that a single level of interneurons was interposed between the recurrent motor axon collaterals and the recipient motoneurons. The key to functional identification of Renshaw interneurons was that the input and output sources were clearly identified; indeed, they were the same cells (i.e., motoneurons).

A further happy chance led to the early identification of individual Renshaw cells using microelectrode recording (Eccles et al., 1954). The excitatory postsynaptic potentials (EPSPs) produced in Renshaw cells by motor axon collaterals are both powerful and of relatively long duration (Walmsley and Tracey, 1981), generating very high-frequency (up to 1000 Hz) repetitive action potentials (Fig. 4.4A). These distinctive bursts after ventral root stimulation could be identified in microelectrode recordings with considerable confidence (Eccles et al., 1954), without having to perform the much more difficult technical feat of proving that an individual cell indeed projects monosynaptically to motoneurons (Van Keulen, 1981). Individual Renshaw cells have now been studied mor-

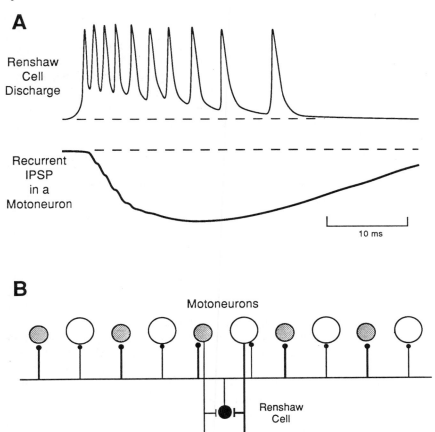

FIG. 4.4. **A.** Drawing of records typical of those obtained inside a Renshaw interneuron during synaptic activation from motor axon collaterals (upper trace). The powerful EPSP produced by antidromic stimulation of motor axons generates high-frequency, repetitive firing of Renshaw cells. The corresponding IPSP produced by Renshaw cells in many motoneurons is shown in the lower trace. The prolonged time course of the IPSP is due in part to the repetitive firing of the Renshaw cells responsible for it. **B.** Diagram of a side view of the spinal cord, to illustrate the restricted spatial distribution of input to a Renshaw cell and the much wider longitudinal distribution of its output. The larger motoneurons of fast-twitch motor units (denoted by open circles) tend to produce more powerful excitation of Renshaw cells than the smaller motoneurons of slow-twitch units (stippled circles), whereas the synaptic efficacy of the Renshaw cells on motoneurons is just the reverse (denoted by the light and boldface lines).

phologically and their projections confirmed directly (Lägerbach and Kellerth, 1985a,b). Although Renshaw cells are located deep in Rexed's lamina VII, ventro-medially adjacent to the motor nuclei to which they project (Fig. 4.1), their axons do not take the shortest route to the adjacent motor nuclei. Rather, the axons enter the ventral and lateral white matter and then drop finer collaterals back into the gray matter. This "funicular" pattern of axonal trajectory appears to be a common trait among the types of spinal interneurons that have been studied morphologically.

The basic circuit diagram of the Renshaw system (see Fig. 4.4 and below) is relatively simple, but one must appreciate that many neurons are symbolized by the single element labeled "Renshaw cell." Dozens or perhaps hundreds of Renshaw cells may be interposed between the axon collaterals from a given motor nucleus and the motoneurons that receive recurrent inhibition. It is also clear that a given Renshaw cell receives input from many motoneurons ("input convergence") and almost certainly projects to many individual motoneurons ("output divergence"; see Baldissera et al., 1981). Before rushing to interpret the function of recurrent inhibition in terms of the obvious negative feedback configuration implicit in the Renshaw cell circuit, one should consider a few interesting complexities in their synaptic organization.

A given Renshaw interneuron is activated only by collaterals of nearby motoneurons, because motor axon collaterals spread for only about 1 mm in the rostrocaudal direction (Cullheim and Kellerth, 1978a). Since most motor nuclei in the cat spinal cord are 7–10 mm in length, this means that only a fraction of a given nucleus can contribute input to any local group of Renshaw cells. However, Renshaw cell axons can project over 12 mm rostrocaudally, extending their influence to large fractions of the motor nuclei (Jankowska and Smith, 1973). The diagram in Fig. 4.4B is intended to illustrate this situation. In addition, recurrent collaterals from the motoneurons that innervate fast contracting muscle units (most notably, type FF units; see below) are more luxuriant than those of cells innervating slow-twitch muscle units (type S; Cullheim and Kellerth, 1978b) and appear to have greater functional input to Renshaw cells than those from type S motoneurons (see Baldissera et al., 1981). On the output side, in contrast, recurrent IPSPs are largest among type S motoneurons and smallest in type FF (Friedman et al., 1981). These points are shown diagrammatically in Fig. 4.4B by the width of the lines interconnecting the Renshaw cell and large, fast motoneurons (open circles) and smaller, slow cells (stippled circles). Some distal motor nuclei do not have recurrent axon collaterals at all and thus cannot activate Renshaw cells (Cullheim et al., 1977; Cullheim and Kellerth, 1978a). Thus the motoneurons that excite Renshaw cells most strongly receive the weakest recurrent inhibition, and vice versa. There are other rather clear patterns in the distribution of recurrent inhibition. Particular motor nuclei can be joined as input–output partners in recurrent inhibition even though separated by considerable distances, whereas near neighbors may not be. As a general rule, recurrent inhibition is not found between motor nuclei of muscles that are strict functional antagonists at a particular joint (Baldissera et al., 1981).

Last but certainly not least, motor axon collaterals are not the only source of synaptic input to Renshaw cells. There is strong evidence that Renshaw interneurons receive both excitatory and inhibitory input from a variety of other sources, including both muscle and skin afferents, and from a variety of supraspinal regions as well (Baldissera et al., 1981). Thus, it is clear that Renshaw cells do not constitute a "private pathway" exclusive to motor axon collaterals.

On the output side, Renshaw cells also inhibit one another, accounting for the phenomenon of "recurrent excitation" (Ryall et al., 1971). Renshaw cells also inhibit interneurons in the reciprocal disynaptic pathway between group Ia afferents and motoneurons (see below). All of this evidence clearly suggests that Renshaw interneurons probably subserve functions that go well beyond simple negative feedback. The existence of supraspinal projections onto Renshaw cells suggests that recurrent inhibition can be modulated by descending motor "commands." This extended discussion is intended to convey the message that what appears to be at first glance a rather simple neuronal organization has, on closer inspection, remarkable complexity, with functional implications that remain to be clarified.

DISYNAPTIC GROUP Ia RECIPROCAL INHIBITION

The above message can be amplified by considering another "simple" reflex pathway—that mediating the inhibition between muscle spindle group Ia afferents and motoneurons of antagonist motor nuclei (Laporte and Lloyd, 1952). Eccles and co-workers (1956) showed that the time from the first arrival of the Ia volley at the spinal cord to the onset of the resulting IPSP—that is, the "central latency"—in antagonist motoneurons was sufficiently long (1.2–1.8 msec, or about 0.6–0.8 msec longer than that of the monosynaptic EPSP) to indicate the presence of one level of interposed interneurons (see also Fig. 4.3). This reciprocal inhibition (so named because each member of an agonist–antagonist pair inhibits the other) is thus a "disynaptic" pathway, like the Renshaw cell pathway, with two synaptic stages before the motoneuron. However, the individual interneurons in the reciprocal Ia inhibitory pathway were not as easily identified as were the Renshaw cells. Many spinal interneurons receive monosynaptic group Ia excitation (Eccles et al., 1960b), and it was not initially clear how one could prove that any given cell also directly inhibited antagonist motoneurons.

The key to this problem was found by Hultborn and co-workers (1971a,b), who demonstrated that reciprocal group Ia inhibition of motoneurons is suppressed by none other than recurrent inhibition from Renshaw cells. They used the technique of "spatial facilitation" of reflex effects, as recorded intracellularly from motoneurons. This powerful method has been used by Lundberg and his colleagues (see Lundberg, 1975; Jankowska and Lundberg, 1981) to make inferences about the organization of many spinal reflex pathways. It relies on the observation that stimulation of a second input system (the "conditioning" input), when combined with stimulation of the test reflex pathway, can enhance the amplitude of the EPSPs or IPSPs produced by the test reflex pathway in motoneurons, or even reveal PSPs that are not detectable when either input is activated in isolation. With appropriate controls, it is possible to infer from such observations that both the test and conditioning input pathways excite common interneurons in the test pathway, as illustrated in Fig. 4.5.

Of all the inhibitory reflex pathways tested by Hultborn and co-workers, disynaptic reciprocal Ia inhibition alone was subject to recurrent inhibition. Thus, if an individual interneuron was both excited by group Ia afferents and

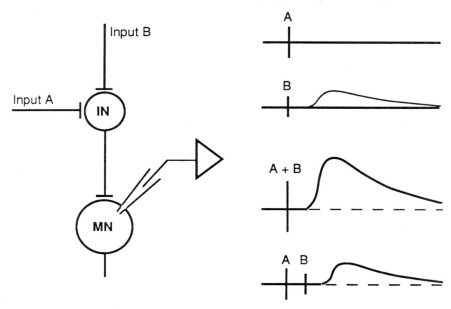

FIG. 4.5. Diagram to show the method of spatial facilitation that has been used to investigate synaptic organization in spinal reflex pathways. In this hypothetical example, input A generates no synaptic potential when stimulated alone, and input B generates an intermittent, small EPSP. However, when both are stimulated such that their effects arrive synchronously at the interneuron, a large EPSP is recorded in the motoneuron (MN; "A + B"). Delaying one or the other input can still produce an augmented EPSP, but of smaller amplitude ("A B"). (Adapted from Lundberg, 1975.)

inhibited after ventral root stimulation, it could be inferred to belong to the reciprocal Ia inhibitory reflex pathway. Subsequent elegant work by Jankowska and Roberts (1972a,b) showed directly, using the technique of spike-triggered averaging, that such interneurons indeed directly inhibit the appropriate motoneurons. The Ia inhibitory interneurons have now been studied morphologically by intracellular injection of tracer substances (Jankowska and Lindström, 1972). Like Renshaw cells, Ia inhibitory interneurons have axons that travel in the white matter, dropping collaterals back into the gray matter to make synaptic contacts with both alpha and gamma motoneurons.

The disynaptic reciprocal Ia inhibitory pathway provides an even more striking example of convergent control than the Renshaw cells. Using spatial facilitation and direct interneuron recordings, Hultborn, Jankowska, and their colleagues (for review, see Baldissera et al., 1981) have shown that the reciprocal Ia inhibitory pathway is modulated by a wide variety of input systems in addition to Ia afferents and Renshaw cells. These other input sources include other primary afferent systems, such as ipsilateral group Ib muscle afferents, and cutaneous and joint afferents from both sides of the body (the "flexor reflex afferents," or FRA; see above), as well as a variety of identifiable systems that descend to the spinal cord from the supraspinal brain (e.g., the corticospinal, rubrospinal, vestibulospinal, and reticulospinal tracts). Since the Ia inhibitory pathway is disynaptic, all of these systems can be inferred to converge, directly or indirectly, onto pathway interneurons, as shown diagrammatically in Fig. 4.6. Note also that, like Renshaw cells, groups of Ia inhibitory interneurons also inhibit one another.

Although seemingly bewildering in complexity, the synaptic organization of reciprocal Ia inhibitory interneurons displays patterns that suggest important functional correlates. For example, descending systems like the vestibulospinal tract directly excite certain extensor motoneurons as well as Ia inhibitory interneurons that receive Ia input from the same muscle and project to its flexor antagonists. These same Ia inhibitory interneurons also inhibit the Ia inhibitory interneurons that receive flexor Ia input and inhibit the extensor motoneurons (Fig. 4.6; see Baldissera et al., 1981). Working through this organization, it becomes apparent that activation of an extensor muscle by, for example, descending vestibulospinal "commands" would at the same time reduce any ongoing extensor inhibition due to "flexor" Ia interneurons but increase inhibition of the antagonist flexor nucleus. If the flexor muscle were passively stretched by extensor contraction, the usual central effects of the resulting flexor Ia afferent input (activation of flexor motoneurons and disynaptic inhibition of extensors) would thus tend to be suppressed. This antagonist suppression presumably can be kept within bounds, at least to some extent, by the organization of recurrent inhibition (Fig. 4.6). Activation of the agonist (extensor) motor nucleus would, through recurrent inhibition, suppress "its" Ia interneurons and remove the antagonist suppression. Flexibility is, however, ensured by the existence of convergent control of both the Ia inhibitory and Renshaw interneurons (see above). A thorough discussion of many of the possible options is given in a review by Illert and colleagues (1981).

At this point, one is justified in viewing the Ia reciprocal inhibitory pathway as still, after all, just a "reflex" system (i.e., dependent on afferent information for activation), albeit one with elegant and quite sophisticated interconnections. All of the converging input systems depicted in Fig. 4.6 could exist simply to modulate muscle stretch reflex effects to suit the needs at hand. This is probably quite true, but there is more to the story—another piece of evidence to further thicken the stew. Group Ia reciprocal inhibitory interneurons are phasically modulated by the spinal central pattern generation system (CPG) that produces the basic muscle activity patterns underlying locomotion (reviewed by Grillner, 1981, and Stein, 1984). In fact, the centrally generated rhythm produced by the CPG can drive Ia inhibitory interneurons *without* any phasic contribution from Ia afferents themselves (Feldman and Orlovksy, 1975)—as it were, coopting them to generate "locomotor" inhibition rather than "stretch reflex" inhibition. In this situation, they might be called "CPG inhibitory interneurons." What if, say, the descending rubrospinal system could also activate them without help from primary afferents? How would we then categorize the same set of interneurons? There is evidence that "Ia reciprocal inhibitory interneurons" can also participate in synaptic effects produced by cutaneous afferents (see Baldissera et al., 1981). It is worth remembering that the neat pigeonholes produced by our need to

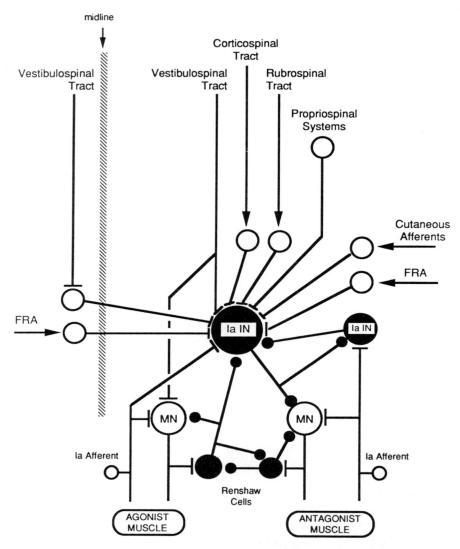

FIG. 4.6. Schematic diagram showing the organization of descending and afferent synaptic input to an interneuron (Ia IN) in the disynaptic reciprocal inhibitory pathway between group Ia afferents and antagonist motoneurons (MN). The organization of Renshaw interneurons in the recurrent inhibitory pathway is also partially diagramed. By convention, excitatory interneurons are denoted by open circles and excitatory synapses by bars, whereas inhibitory cells and synapses are indicated by filled circles. (Adapted from Baldissera et al., 1981.)

organize and communicate information do not always fit the flexibility evident in reality.

The extensive convergence of multiple input systems onto Renshaw cells and reciprocal Ia inhibitory interneurons illustrates a critical point about spinal reflex pathways in mammals: Few if any of them are "private" to a particular category of input or output (see Lundberg, 1975, 1979; Baldissera et al., 1981). It seems quite possible that the spinal interneurons that are interposed in the long-recognized spinal reflex pathways and dominated by a particular kind of primary afferent under some conditions, may under other conditions subserve quite different functions and obey other masters (Jankowska and Lundberg, 1981).

OTHER REFLEX SYSTEMS

There are a great many other spinal reflex organizations that have been identified, some dominated by muscle afferent input and others by cutaneous or deep afferent systems. For example, spindle group Ia and group II afferents can produce di- and multisynaptic excitation of some motoneurons and the Group Ib Golgi tendon organ afferents that sense muscle tension also generate a wide variety of excitatory and inhibitory reflex effects (see Baldissera et al., 1981). There are also disynaptic excitatory pathways from skin afferent to some motoneurons (Illert et al., 1976; Fleshman et al., 1988a,b), although most cutaneous reflexes operate through at least two levels of interneurons (referred to simply as "multisynaptic").

It is very difficult to make inferences about the synaptic organization in multisynaptic reflex pathways using the spatial facilitation approach because the elements that receive direct afferent input are not the same as those that project to the output motoneurons. Nevertheless, many of these more complex pathways appear to exhibit the same kind of convergent control that is characteristic of the disynaptic systems discussed above. The great flexibility thus conferred on the spinal cord suggests that this phylogenetically old part of the CNS retains mechanisms capable of sophisticated integration of sensory information and descending motor commands to produce appropriate motor outputs. It seems not too far-fetched to imagine that the segmental "reflex" systems are organized to enable descending motor commands to take advantage of the fact that spinal interneurons have immediate access to the sensory information which signals, for example, current limb position, muscle lengths and tensions, and the possible existence of external objects. By operating through interneurons of segmental reflex pathways, descending motor commands may be "filtered" according to the existing "state-of-affairs" in the limb before they reach the motoneurons. This organization is essentially a "feedforward" mechanism that operates in a predictive manner, in contrast to "feedback" mechanisms, which depend on sensory signals about the results of a movement. The latter are inherently slower because of conduction and synaptic delays. If all sensory information about the results of movement had to be transmitted to the supraspinal brain in order to adjust motor "commands," the resultant delays would be quite unsuited to control movements of even modest speed (see Rack, 1981). This viewpoint seems quite compatible with the fact that the majority of descending motor systems

operate through segmental interneurons of the reflex pathways rather than projecting directly to spinal motoneurons. This is true even in the primates, where
direct projections from the motor cortex and red nucleus to motoneurons occur
only in particular motor nuclei controlling relatively distal musculature (reviewed
in Phillips, 1969; Porter, 1973, 1987; Burke and Rudomin, 1977).

THE SYNAPTIC ORGANIZATION OF ASCENDING TRACTS

The synaptic organization of some ascending tracts fits quite well with the ideas
put forward above. As mentioned previously, the basic function of some ascending tract neurons is clearly to relay sensory information from primary sensory
afferents to regions of the brain where the information produces conscious sensation and/or guides the formation of appropriate motor commands. However,
other ascending systems receive mixtures of sensory modalities that appear to
reflect a more "integrative" function at the spinal cord level. An instructive
example of the contrast between "relay" and "integrative" functions is apparent
when comparing the input organization to the two major spinocerebellar systems
that project from the lumbosacral enlargements of carnivores and primates to the
cerebellar cortex, as well as to certain brainstem nuclei.

The dorsal spinocerebellar tract (DSCT) originates from cells in Clarke's
column, one of the few anatomically distinct groups of neurons in the upper
lumber segments of the cord. Individual DSCT neurons receive powerful input
from particular afferent species, either from muscle afferents (e.g., group Ia or
group Ib afferents from one muscle or a functionally related group of muscles) or
from cutaneous and some high-threshold muscle afferents (for review, see
Bloedel and Courville, 1981). In both cell types, the firing of a given DSCT cell
is tightly coupled to, and relays with reasonable accuracy, the input from the
afferents that project to it (see Kröller and Grüsser, 1982). The "relay" analogy
is not perfect because there is evidence that DSCT neurons also receive input
from segmental and local interneurons, and possibly some descending input as
well. Nevertheless, DSCT neurons appear generally to provide the cerebellum
with a relatively accurate, "unprocessed" view of particular afferent inputs.

In contrast, individual neurons of the ventral spinocerebellar tract (VSCT)
receive a complex mixture of inputs, some directly from primary afferents and
some indirectly via segmental interneurons (Lundberg and Weight, 1971). Unlike
DSCT neurons, there is often convergence of effects from muscle and cutaneous
afferent systems onto individual VSCT cells. It is not uncommon, for example,
to encounter VSCT cells that receive monosynaptic input from group Ia and/or Ib
muscle afferents, as well as mixed polysynaptic excitation and inhibition from
low- and high-threshold skin and joint afferents. Stimulation of primary afferent
systems can produce inhibition as well as excitation of VSCT neurons. In general, no individual input system dominates the input pattern, and the synaptic
inputs to individual VSCT cells are sufficiently diverse so that each appears to be
unique in the constellation of information that it conveys (see Lundberg, 1971).
Some VSCT cells (the "spinal border cells") in fact appear rather like
motoneurons in the organization of their synaptic inputs and, indeed, are located

within the ventral horn of upper lumber spinal segments, near to and even among motoneurons (Burke et al., 1971a; Matsushita and Hosoya, 1979).

What use can the cerebellum make of the output of VSCT neurons that receive such diverse inputs? Lundberg and Weight (1971; see also Lundberg, 1971) have suggested an intriguing hypothesis, based in part on observations such as the fact that some VSCT cells receive monosynaptic excitation from group Ia afferents from a particular muscle nerve and, at the same time, disynaptic inhibition apparently produced by the same Ia afferents (Fig. 4.7). The probability of discharge from such cells is increased by direct Ia excitation and simultaneously reduced by disynaptic Ia inhibition from the same source. VSCT neurons with this organization would carry garbled messages about group Ia input alone, but the net result of competing excitation and inhibition could well signal whether or not the disynaptic Ia inhibitory pathways was active at the time. When looking at Fig. 4.7, recall the complexity of input to the reciprocal Ia inhibitory inter-neurons shown in Fig. 4.6.

It would seem of significance to the supraspinal brain, including the cere-bellum, to have some reflection of the "state of affairs" at the segmental level, when so many of the reflex pathways discussed above are subject to complex control from many sources. It seems quite plausible that at least some elements of the VSCT inform the cerebellar cortex about the activity in spinal reflex path-

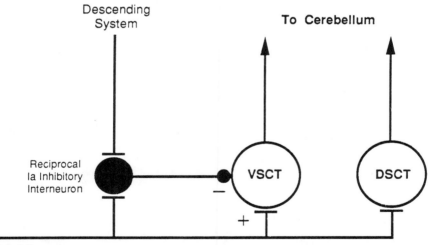

FIG. 4.7. Diagram to illustrate a pattern of synaptic organization found in some neurons of the dorsal (DSCT) and ventral spinocerebellar (VSCT) tracts. Many DSCT neurons receive relatively "pure" and powerful group Ia input, whereas some VSCT cells receive both monosynaptic excitation from group Ia afferents in a given muscle nerve and dis-ynaptic inhibition produced by the same Ia afferents. At least some of the inhibitory interneurons are in the reciprocal disynaptic pathway shown in Fig. 4.6; these also receive descending control inputs. This organization can, in principle, signal information about transmission in spinal reflex pathways (see discussion in the text). (Adapted from Lundberg, 1971.)

ways that are subject to descending control (Lundberg, 1971). The fact that the DSCT relay provides the cerebellum with a much more accurate, unprocessed version of, for example, Ia input would give the opportunity to make additional comparisons of input to, and output from, particular spinal interneuron pathways. Although much further work is required to validate this hypothesis, it is, in principle, testable and suggests a further rationale for the remarkable complexity of synaptic organization at the spinal cord level (see Baldissera et al., 1981, for further discussion).

SYNAPTIC TYPES IN THE SPINAL CORD

The above material addresses some features of the "macroscopic" organization of the spinal ventral horn. The sections to follow focus more closely on the level of individual synapses in order to examine some aspects at a subcellular level. The types of synaptic contacts found in ultrastructural studies of the ventral horn resemble those found in other parts of the CNS (Fig. 4.8). There are three basic types: (1) type S boutons, which are 0.5–2.0 μm in diameter and contain spherical synaptic vesicles; (2) type F boutons, which are about the same size but contain vesicles that appear oval or flattened with many types of fixation, and

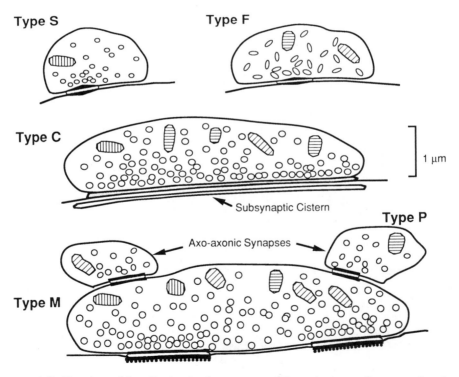

FIG. 4.8. Drawings of the ultrastructural appearance of the major types of synapses found in the ventral horn of the spinal cord. Type S boutons also receive axoaxonic (type P) synapses. (Adapted from Conradi et al., 1979a.)

which have thin layers of postsynaptic density; and (3) type C boutons, which are larger (2–7 μm in diameter) and are characterized by an extensive, thin membranous "cistern" immediately beneath the region of synaptic apposition (see Conradi et al., 1979a). A subtype of the S bouton, the type M synapse, is found in the ventral horn; it is quite large (3–5 μm in diameter) and has particularly prominent and multiple postsynaptic densities.

Both the S and M bouton types often have synapses that are *presynaptic* to them (the "axoaxonic" or "P" synaptic type). An example is shown in Fig. 4.8. These P boutons are relatively small (about 1 μm in diameter) and have spherical vesicles. They are associated with synaptic specializations that indicate functional synaptic connections onto the larger S and M boutons on which they end. Such "axoaxonic contacts" are believed to modulate the release of transmitter from the recipient (i.e., postsynaptic) boutons, producing "presynaptic inhibition" (of which more later).

About 30–50% of the somatic membrane surface of ventral horn motoneurons lies beneath synapses while the synaptic covering is higher on the proximal dendrites (up to 80%; see Conradi et al., 1979a,b). Over half of this area is beneath type F boutons, which make up about 75% of the synapses on the motoneuron soma. Type S and type C boutons comprise about 14% and 8% of the synapses on the motoneuron soma, respectively (Conradi et al., 1979a,b). The large type M boutons are infrequent and are found mainly on the dendrites.

There is now direct evidence that the monosynaptic contacts established by group Ia afferents on alpha motoneurons include type S boutons (and possibly the "M" boutons) of varying size, many of which also exhibit axoaxonic P boutons (Conradi et al., 1983; Fyffe and Light, 1984). As discussed in Chap. 1, synapses with spherical vesicles (Gray's type 1) are associated with excitatory synapses in many parts of the CNS. Conversely, boutons with flattened vesicles (type F, resembling Gray's type 2) are often associated with inhibitory synaptic effects, and there is indirect evidence that this is the case in the spinal ventral horn. The relatively high proportion of type F synapses on and near the motoneuron soma fits with the notion that at least some types of inhibitory input are concentrated near the site of action potential generation in the axon initial segment arising from the soma, where the inhibition can effectively prevent cell discharge irrespective of the source of excitatory input. The origin of the type C boutons in the ventral horn remains unknown, but they are of interest because, during the process of synaptic remodeling during early development, motoneurons apparently can themselves prune them by phagocytosis (Ronnevi, 1979).

Although the direct dendrodendritic synaptic arrangements that are found in some specialized CNS regions (see Chap. 5) are not present in the spinal ventral horn, there is some evidence for types of cell-to-cell interactions beyond those associated with "plain vanilla" synapses. For example, Nelson (1966; see also Gogan et al., 1977) found evidence that there is weak electrical (nonsynaptic, or "ephaptic") interaction between motoneurons in the cat spinal cord, a phenomenon that is much more potent in the frog, where there are gap junction contacts between motoneuron dendrites (see Sonnhof et al., 1977). There are sites of apparent close apposition, without membrane specializations, between cat motoneuron dendrites, especially in regions where dendrites are "bundled" closely together (Matthews et al., 1971). Electrical coupling between motoneurons has been demonstrated to produce sharply timed synchronous discharge of motoneurons in electric fish (see Bennett, 1977) and this might be of use to frogs. However, there is little evidence, nor much functional reason, for this type of tight coupling among mammalian spinal motoneurons.

Oddly enough, however, motoneurons apparently interact directly via synaptic connections from motor axons collaterals, which have been shown to end monosynaptically on motoneurons as well as on Renshaw interneurons (Cullheim and Kellerth, 1978a,b). It has been difficult to demonstrate a clear synaptic effect that can be attributed to such direct motoneuron-to-motoneuron contacts (see Gogan et al., 1977), and it seems possible that such direct intermotoneuron connections might have some other function.

MECHANISMS OF SYNAPTIC ACTION IN MOTONEURONS

EXCITATORY SYNAPTIC MECHANISMS: Ia EPSPS

It is useful to define three levels of anatomical and functional hierarchy in the group Ia input to motoneurons: (1) *single-bouton* (SB) EPSPs produced by an individual Ia synaptic bouton; (2) *single-fiber* (SF) EPSPs produced by all of the boutons belonging to an individual Ia afferent; and (3) *composite* EPSPs produced by synchronous action in multiple Ia afferents. This hierarchy is illustrated diagrammatically in Fig. 4.9. The three levels of synaptic input are fundamental for understanding the functional relations between synaptic inputs and postsynaptic responses in all of the CNS neurons considered in this book.

The earliest intracellular recordings, by Eccles and co-workers, of synaptic potentials in the CNS were of composite Ia EPSPs in alpha motoneurons following electrical stimulation of a muscle nerve (Brock et al., 1952; Coombs et al., 1955). Later, Kuno (1964) and others (Burke, 1967; Jack et al., 1971; Mendell and Henneman, 1971) described single-fiber EPSPs produced in motoneurons by individual Ia afferents. The varying shape of these EPSPs fit the theoretical predictions made by Rall (1967) for synapses that are widely dispersed throughout the dendritic tree (see Figs. 4.10 and 4.11; see also Appendix). This wide dispersion of group Ia boutons has now been confirmed directly by anatomical studies of HRP-labeled Ia afferents ending on labeled motoneurons (Burke et al., 1979; Brown and Fyffe, 1981; Redman and Walmsley, 1983a).

Despite some early enthusiasm for the idea that Ia EPSPs result from direct transmission of the electrical currents from the presynaptic action potentials into the postsynaptic neurons ("electrical" synapses; see chap. 1 in Eccles, 1964), it has been accepted for many years that, in mammals, the Ia synapse operates by an exclusively "chemical" mechanism (see Chaps. 1 and 2). The synapses between muscle afferents and motoneurons in amphibians, however, have both electrical and chemical components—an interesting combination for some research purposes (Shapovalov and Shiraev, 1980). The chemical component of amphibian Ia EPSPs can be blocked by increasing the external concentration of Mg^{2+} ions, leaving the initial electrical component unaffected. In mammals,

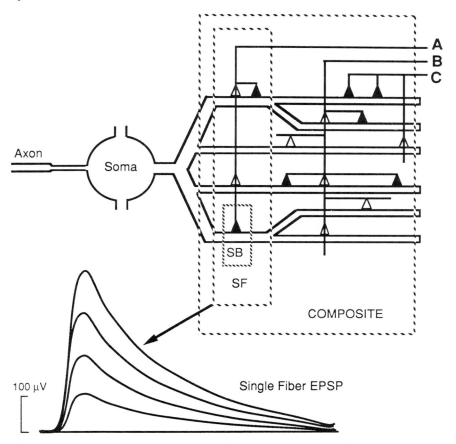

FIG. 4.9. Diagram showing the anatomical hierarchy of synapses from group Ia afferents to a motoneuron, which is typical of the structure of many input systems within the CNS. Multiple group Ia axons (A, B, and C) make contact at various locations in the dendritic tree. Some synaptic boutons occur along the course of the afferent axon ("boutons en passant"; open triangles) whereas others occur at the terminations of axonal branches ("terminal boutons"; filled triangles). The anatomical substrate of single-bouton (SB), single-afferent (SF), and multiafferent (composite) PSPs discussed in the text are indicated by dashed lines. The lower traces show the PSP components that might be expected in an intracellular record of EPSPs produced by afferent A if transmitter release at each of the four boutons sometimes fails (i.e., the probability of release is less than unity). Transmission failure at all four boutons would produce no EPSP, whereas the detectable EPSPs could be composed of one, two, three, or all four unitary, or "quantal" components. (Adapted from Burke, 1987.)

however, the entire Ia EPSP behaves as expected for a purely chemically mediated event.

"As expected" means that composite Ia EPSPs exhibit an equilibrium potential ($E_{eq,syn}$; see Chap. 2) near 0 mV and reverse to hyperpolarizing polarity with more positive membrane potentials (Smith et al., 1967; Engberg and Marshall, 1979). This synaptic E_{eq} suggests that Na$^+$ ($E_{eq,Na}$ of about +40 mV), K+

($E_{eq,K}$ of about -90 mV), and possibly Ca^{2+} ($E_{eq,Ca}$ about $+145$ mV) pass through the activated channels, since the observed $E_{eq,syn}$ depends on the net balance of the equilibrium potentials of relevant ions and their respective conductivities (see Puil, 1983 and Chap. 2). The reversal potential of Ia EPSPs was very difficult to demonstrate experimentally (see Burke and Rudomin, 1977), due in part to the fact that a majority of Ia boutons contact relatively distal dendritic regions where they are electrically isolated from current injected into the soma (Rall, 1967). An additional problem is the fact that motoneurons can exhibit sudden changes in membrane resistivity when strongly depolarized (Engberg and Marshall, 1979), limiting the ability of injected currents to alter the membrane potential further. Similar problems have been encountered in other parts of the CNS (e.g., the spiny neuron of the neostriatum, Chap. 9; see also Appendix).

QUANTIZATION OF SYNAPTIC ACTION

The anatomical organization of an input system like Ia afferents suggests that EPSPs produced by one or more afferents must be composed of tens or even hundreds of single-bouton EPSPs. Single-fiber Ia EPSPs fluctuate in amplitude, and sometimes apparently fail to occur, during repetitive activation (Kuno, 1964; Jack et al., 1981a,b; Redman and Walmsley, 1983a). This is illustrated in Fig. 4.9 and suggests that individual synaptic boutons vary in their participation in particular single-fiber PSPs. Such variations could be produced by failure of bouton activation due to failure of action potentials to invade the complex branching systems of Ia collaterals (see Burke and Rudomin, 1977; Lüscher et al., 1983) or, alternatively, could result from failure of transmitter liberation from fully activated synaptic boutons, as occurs at the peripheral neuromuscular synapse. The mechanisms are not in fact mutually exclusive. Although the weight of evidence available at this time favors the latter explanation (see Jack et al., 1981a; Lev-Tov et al., 1983, 1988), the branch-point failure hypothesis has not been ruled out.

In exceptional cases, all-or-none unitary, or "quantal," components have been observed in single-fiber Ia EPSPs (Kuno, 1964; Burke, 1967). However, in most cases, indirect statistical analysis of single-fiber EPSP amplitude fluctuations is required to infer the number and sizes of single-fiber EPSP components because the background synaptic noise is as large as or larger than the EPSP signals being analyzed (see Jack et al., 1981a,b). For example, the mathematical technique of "deconvolution" has been used to show that Ia EPSPs are in fact quantized in the same way as neuromuscular synaptic potentials, and to estimate the number and individual amplitudes of unitary EPSPs (Redman, 1979; Jack et al., 1981a). In the few cases studied both physiologically and morphologically, the maximum numbers of quantal EPSPs match reasonably well the numbers of HRP-labeled boutons observed anatomically for single Ia afferents (Redman and Walmsley, 1983b), making it likely that "quantal" Ia EPSPs represent events associated with single synaptic boutons. There is little or no detectable fluctuation in *quantal* EPSP sizes, and it is therefore assumed that they may be generated by release of individual packets of neurotransmitter.

The probability of transmitter release at any given Ia synaptic bouton can vary between zero and one. Some boutons do not appear to liberate transmitter at all for long periods of time, whereas others do so with some or all incoming action potentials. Henneman and co-workers (1984) have evidence that these probabilities are not fixed but rather vary with time. The peak amplitudes of unitary Ia EPSPs found in most analyses are in the range of 80–180 μV. Remarkably, these peak amplitudes do not seem to vary with the shape of the potentials, from which the spatial location of the active synapses can be inferred (Jack et al., 1981a,b; see below). This might indicate the presence of some amplification of synaptic potentials generated in distal dendrites, but the indirect nature of the statistical analyses used to obtain these data prevents their firm interpretation at present.

TRANSMITTER RELEASE AT Ia SYNAPSES

For some time, the excitatory amino acid (EAA) L-glutamate has been the leading candidate as the neurotransmitter substance released at group Ia synapses in the mammalian spinal cord (e.g., Krnjević, 1981). Recent in vitro studies using pharmacological agonists and antagonists to EAAs, which are essentially impossible to perform in the intact animal in vivo, provide support for this suggestion. Group Ia EPSPs can be blocked by the EAA antagonist kynurenate but not by 2-amino-5-phosphovalerate, a potent inhibitor of N-methyl-D-asparate (NMDA) receptors (Jahr and Yoshioka, 1986). It thus seems likely that the postsynaptic receptor involved is a non-NMDA glutamate receptor (see Chap. 2).

In addition to the trial-to-trial variations discussed above, the amplitude of composite group Ia EPSPs can also be modulated by the history of presynaptic action potentials. For example, during repetitive activation of Ia afferents at moderate to high frequencies (25–80 Hz), EPSP amplitudes are depressed (Curtis and Eccles, 1960), owing to an apparent depletion of the amount of transmitter immediately available for release (see Barrett and Magleby, 1976). However, Ia EPSPs generated after a high-frequency tetanus are usually considerably larger than pretetanic control responses for periods of many minutes. This "posttetanic potentiation" (PTP) ordinarily rises to maximum over several tens of seconds and then decays to control amplitudes over a period of tens of minutes. There is no true "long-term potentiation" in this synaptic system, as found in other areas of the brain (see Chaps. 2 and 11). The time course of PTP is complicated by the fact that EPSPs that occur within a few seconds after the end of a tetanus are smaller than control values ("posttetanic depression"), and are succeeded by potentiation to greater than control amplitudes (see Lev-Tov et al., 1983).

The coexistence of simultaneous posttetanic depression and potentiation is revealed by the administration of the drug l-baclofen, which activates $GABA_B$ receptors (see Chap. 2). Among its several effects, l-baclofen reduces Ca^{2+} entry into synaptic terminals (Dunlap and Fischbach, 1981; Dolphin and Scott, 1986), reducing total transmitter output and consequently eliminating posttetanic depression due to transmitter depletion (Lev-Tov et al., 1988). The net effect is to reveal uncontaminated PTP, in which EPSPs immediately after the end of the conditioning tetanus can be up to six times larger than the pretetanic EPSPs. The drug 4-aminopyridine, which prolongs the afferent action potential and allows greater influx of Ca^{2+}, also produces marked enhancement of Ia EPSP amplitudes (Jankowska et al., 1977) and increases the average probability of single-fiber quantal Ia EPSPs without changing their size (Jack et al., 1981b). In all of the above respects, the behavior of group Ia EPSPs within the CNS closely resembles that of the neuromuscular junction (see Martin, 1977), which is undoubtedly the best understood (certainly the most intensively studied) chemical synapse. These results strongly suggest that transmitter release at group Ia synapses within the intact CNS and at the peripheral neuromuscular junction depend on the same basic mechanism—increased intracellular $[Ca^{2+}]_i$ secondary to the entry of Ca^{2+} during synaptic depolarization. Despite the great anatomical differences between the consolidated neuromuscular junction and the distributed, individual synaptic boutons belonging to a single Ia afferent (see Fig. 4.9), the synaptic endings of both systems appear to operate with the same basic mechanism (see Chap. 1, especially Fig. 1.2).

OTHER EXCITATORY SYSTEMS

Obviously, the group Ia afferent system is only one of many synaptic systems that control activity within the spinal cord ventral horn. Many segmental interneurons make excitatory synapses on one another and on motoneurons. These are not well characterized because of technical limitations. However, some functionally defined systems that descend from the brain have been studied with experimental methods analogous to those applied to the group Ia system. Perhaps the best known are the corticospinal and rubrospinal tracts, which contain fibers that make direct, monosynaptic excitatory contact with certain species of alpha motoneurons in primates (see reviews by Phillips, 1969, and Porter, 1973, 1987). Recent anatomical studies have shown that both types of afferents have terminal arborizations within the cervical spinal cord of monkeys that are quite different from those of Ia afferents (Lawrence et al., 1985), in that their collaterals are more widely spaced than Ia collaterals in the cat, and their arborizations within the gray matter tend to spread in a rostrocaudal direction for several millimeters, making fewer bouton contacts on individual motoneurons than do Ia afferents.

Corticospinal EPSPs recorded in primate motoneurons also differ from Ia EPSPs, in that they tend to be smaller in amplitude and exhibit remarkable intratetanic facilitation during double-pulse or short-train repetitive activation (Muir and Porter, 1973). This is not characteristic of all descending systems, however, since the vestibulospinal tract produces monosynaptic EPSPs in motoneurons that behave much like those of Ia afferents during repetitive activation (see Burke and Rudomin, 1977). The morphology of vestibulospinal fibers is also more like that of group Ia afferents (Shinoda et al., 1986). Although electrical stimulation of descending tracts is much less easily controlled than that of Ia afferents in peripheral nerves, the evidence at hand suggests that the transmitter and/or receptor kinetics may well be different for these various species of spinal afferent systems. The identities of the transmitter substances for these descending systems are unknown.

INHIBITORY SYNAPTIC MECHANISMS

Inhibitory synaptic mechanisms are extremely important in controlling CNS activities. All primary afferents from the periphery make excitatory synapses with first-order spinal neurons and most descending systems also excite spinal neurons. Thus, inhibition in the spinal ventral horn is largely produced by segmental interneurons, making it correspondingly difficult to analyze the synaptic mechanisms of spinal inhibition. The best known example of spinal inhibition is that produced by the group Ia reciprocal inhibitory interneurons discussed above.

At normal motoneuron resting potentials (about -70 mV), group Ia disynaptic IPSPs are hyperpolarizing (see Fig. 4.3), indicating that the equilibrium potential for the process is more negative than the resting potential. Additional polarization produced by injecting current into the cell causes apparent reduction in IPSP amplitude and eventual reversal to depolarizing synaptic potentials when the transmembrane potential is more negative than E_{eq} (usually about -75mV to -80mV). Injection of Cl^- ions into the motoneuron also produces rapid and marked changes in Ia IPSPs, reversing them to depolarizing events at normal resting potential (Coombs et al., 1955) and it is generally accepted that Cl^- is the major ionic species involved in this inhibition (see Bührle and Sonnhof, 1985; see also Chap. 2). Because group Ia IPSPs are so readily influenced by injection of electrical current or Cl^- ions into the motoneuron soma, it seems likely that the synapses of Ia inhibitory interneurons are located on and near the cell soma (Burke et al., 1971b). Motoneuron responses to direct application of the amino acid glycine behave in exactly the same manner as Ia IPSPs, having the same reversal potential and sensitivity to changes in intracellular Cl^- (reviewed by Young and Macdonald, 1983). Such responses are also blocked by the convulsant drug strychnine, which also blocks the production of Ia IPSPs. Thus it seems likely that glycine is the transmitter at this synapse.

The recurrent IPSPs produced in motoneurons after ventral root stimulation are longer and less sharply peaked than group Ia IPSPs (Fig. 4.4) because Renshaw cells usually fire repetitively at high frequency (up to 1000 Hz; see Burke and Rudomin, 1977). In fact, recurrent IPSPs can be seen to be composed of more or less synchronous wavelets (see Burke and Rudomin, 1977), each generated by the relatively synchronized spikes in the Renshaw cells that converge onto that cell. The reversal potential for recurrent IPSPs is similar to that of disynaptic Ia IPSPs, but recurrent IPSPs are only partially blocked by the glycine antagonist strychnine (Cullheim and Kellerth, 1981). The strychnine-resistant remnant is instead blocked by the drug picrotoxin, which specifically blocks the action of the inhibitory transmitter gamma-aminobutyric acid (GABA). This suggests that there may be two populations of Renshaw interneurons, one that secretes glycine and the other GABA. It seems much less likely, although certainly not impossible, that individual Renshaw interneurons could secrete both inhibitory transmitters.

Interestingly, recurrent IPSPs in motoneurons are less readily reversed by small injections of Cl^- into the motoneuron soma than are Ia IPSPs, although both are about equally affected by current injected at the soma (Burke et al., 1971b). This suggests that Renshaw cell synapses end mainly on proximal motoneuron dendrites, where they are relatively isolated from small alterations of intrasomatic Cl⁻ concentration but not from voltage perturbations. If Renshaw cell synapses are located on proximal motoneuron dendrites, it is possible that they might strategically reduce the effects of synapses located more distally on just those dendrites by mechanisms illustrated in Chap. 1 (Fig. 1.6) and discussed in the Appendix.

PRESYNAPTIC INHIBITION

In 1957, Frank and Fuortes reported that stimulation of certain muscle nerves reduced group Ia EPSPs in motoneurons without producing detectable IPSPs. Such an effect might be produced by conventional inhibitory synapses acting directly on the motoneuron but located in the dendrites at a distance sufficient to attenuate their somatic voltage so as to make them undetectable ("remote" postsynaptic inhibition). Alternatively, they suggested that transmission might be modulated "presynaptically." Since then, a great deal of work has shown that *both* mechanisms are in fact present in the spinal cord.

Conditioning stimuli that produce presynaptic inhibition are invariably associated with signs of depolarization ("primary afferent depolarization," or PAD) in the intraspinal arborizations of primary afferents. The possible mechanism by which PAD can decrease presynaptic release of transmitter is unclear (e.g., Illes, 1986) but it could depend on reduction of voltage-sensitive Ca^{2+} entry into terminals when they are subject to background depolarization (see Burke and Rudomin, 1977). The anatomical substrate necessary for selective depolarization of particular afferent terminals is found in the "axoaxonic" synapses that indeed end on Ia and other primary afferent synapses that are subject to presynaptic inhibition (Fig. 4.8).

The neurotransmitter released by axoaxonic synapses in the spinal cord appears to be GABA (reviewed by Nistri, 1983). PAD and presynaptic inhibition in several systems can be blocked by local application of picrotoxin or bicuculline, both of which block the action of GABA on the $GABA_A$ receptors (see Chap. 2). Direct application of GABA onto primary afferents in the spinal ventral horn depolarizes them, and axoaxonic synapses (Fig. 4.8) contain the enzyme systems required to manufacture GABA (McLaughlin et al., 1975). The depolarizing action of GABA on primary afferent terminals may seem somewhat odd because, when applied to motoneurons and many other types of neurons, GABA usually produces hyperpolarization, not depolarization, by increasing Cl⁻ conductance. The Cl⁻ equilibrium potential in primary afferent terminals is apparently more positive than the resting membrane potential, as is true also in frog motoneurons (Bührle and Sonnhof, 1985).

The recent discovery of $GABA_B$ receptors (see Chap. 2), which are also present on primary afferent terminals (Bowery et al., 1987) and reduce transmitter output when activated (Shapovalov and Shiraev, 1982; Lev-Tov et al., 1988), introduces an interesting complication into this story. Selective activation of

GABA$_B$ receptors by the drug l-baclofen generates little or no detectable change in primary afferent polarization or excitability (Fox et al., 1978; Curtis et al., 1981) but nevertheless markedly depresses transmitter release from spinal afferent synapses, presumably by depressing the voltage-dependent influx of Ca^{2+} on which liberation of chemical transmitters depends (Dolphin and Scott, 1986). Although this evidence suggests that presynaptic inhibition and PAD may not in fact be causally linked, it is also possible that PAD and GABA$_B$ receptor activation act cooperatively to modulate Ca^{2+} entry into afferent terminals, with additive effects on net transmitter output.

DENDRITIC FUNCTION: ELECTROTONIC PROPERTIES

Virtually all neurons within the CNS, and certainly those in the spinal ventral horn, have extensive dendritic trees that in fact contain most of the membrane of the cells. In the case of alpha motoneurons, the dendrites contain over 97% of the nonaxonal cell membrane (Cullheim et al., 1987), and accordingly receive most of the synaptic input. The electrical properties of dendrites are thus critical to the way in which motoneurons (and most other CNS neurons) process synaptic information. Most of our current understanding of the electrotonic properties of dendrites is based on Wilfrid Rall's pioneering application of the theory of electrical cables to membrane cylinders that represent certain classes of dendritic trees (see Rall, 1959a,b, 1977; Jack et al., 1975), as outlined in the Appendix. In this section, I will concentrate on the application of electrotonic cable models to alpha motoneurons, which have served as a major test system on this topic.

DENDRITIC CABLE PROPERTIES

The term "electrotonus" signifies the reaction of excitable cells to the passage of electrical currents that are *subthreshold* for the generation of active responses (e.g., action potentials). Many discussions of dendritic electrotonus assume, explicitly or implicitly, that membrane properties are invariant with time and voltage (i.e., they behave linearly), under the conditions tested. The active membranes characteristic of real neurons are inherently nonlinear and much more difficult to deal with in model simulations. Nevertheless, conceptual and computer models that embody only passive membrane properties have provided very useful first approximations for understanding the role of dendritic electrotonus. They are the essential foundation upon which to build more "realistic" formulations that include the wealth of active conductances present in real neurons (see Chaps. 1 and 2 and the Appendix).

Idealized cable models often adopt the useful simplifying assumption that the electrical properties of cell membranes and cytoplasm (specific membrane resistivity, R_m; capacitance, C_m; and cytoplasmic resistivity, R_i, see Appendix for definitions) are the same throughout the cell. With any given set of these properties, the passive electrotonic properties depend primarily on the morphology of the neuron. Because motoneuron dendrites are, in aggregate, so much larger than the soma, they dominate the electrical properties of the cell. However, the soma is strategically located and it can exert a disproportionate influence on the electrical behavior of the whole neuron if its effective membrane resistivity is significantly different from that elsewhere (see Durand, 1984; Fleshman et al., 1988).

SOME COMPLICATIONS

As noted in the Appendix, branched dendritic trees can be represented electrically by a single, constant-diameter membrane cylinder under certain conditions: (1) The sum of the 3/2 power of the daughter branch diameters at each branch point equals the 3/2 power of the parent branch (sometimes called the "three-halves power rule"); (2) the electrotonic lengths of all dendritic paths are the same; and (3) the boundary conditions at all dendritic terminations are the same (Rall, 1977). Although the three-halves power rule is reasonably well fulfilled in motoneuron dendrites, the individual paths within a dendritic tree terminate at widely varying anatomical distances (e.g., Cullheim et al., 1987). If one assumes the same values for R_m and R_i through the dendritic tree, the electrotonic paths are also disparate. This is illustrated by the branched tree in Fig. 4.10 (Fleshman et al., 1988). The actual lumped electrotonic cable that represents the dendritic tree of an individual motoneuron is roughly cylindrical only in its proximal half, while the distal half tapers. This is apparent in the model neuron shown in Fig. 4.10, where ten of the cell's 11 dendrites were lumped into one equivalent cable (see Fleshman et al., 1988, for details). An analytical theory of tapered membrane cylinders is considerably more complex than that for ideal (i.e., uniform diameter) cylinders, and it is often easier to simulate complex neuronal morphologies by using compartmental models implemented in digital computers (Segev et al., 1985, 1989).

Given accurate data about the morphology of a motoneuron, plus electrophysiological measurements from the same cell, it is possible, at least in principle, to constrain the possible values of R_m, given reasonable estimates for C_m (1.0 μF/cm² in most biological membranes; Cole, 1968) and R_i (about 70 Ω-cm in cat motoneurons; Rall, 1977; Barrett and Crill, 1974). Specific neuronal models developed in this way can be used to gain insight into the distribution of synaptically generated voltages within the neuron and the effect of dendritic location on the shape and amplitude of synaptic potentials at the soma, where synaptic voltages are integrated and output action potentials are generated.

One of the most serious of the many problems in applying electrotonic theory to actual neurons is the assumption that neuronal membranes are passive. Real neuronal membranes contain an array of time- and voltage-dependent conductances in addition to a postulated passive "leak" conductance ($I_{K,leak}$; see Chap. 2 and Appendix). Motoneurons, for example, generate action potentials in the initial axon segment, which propagate antidromically to involve the soma membrane. Although the dendrites of normal mammalian motoneurons do not generate action potentials, their proximal portions appear capable of supporting further limited antidromic spike propagation (see Burke and Rudomin, 1977; Dodge, 1979; Traub and Llinás, 1979). The active conductances that support action potentials can, at least in principle, contribute to net membrane conductivity at "rest," and they may also respond to the voltage perturbations that are used experimentally to measure neuronal input resistance, R_N, and the time constants

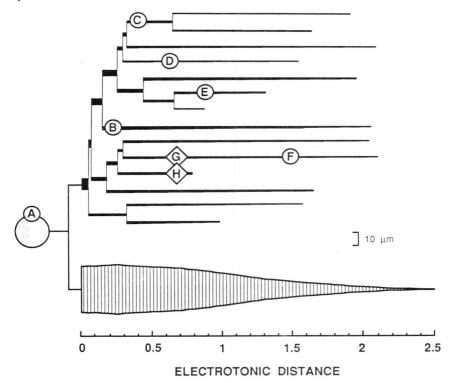

FIG. 4.10. Diagram of a neuron model including the soma (stippled circle), one fully branched dendrite, and a tapered cable structure representing the other ten dendrites of the modeled neuron. The model is based on morphological and electrophysiological data from a cat type FR motoneuron (cell 43/5 in Fleshman et al., 1988). The dendrograms are plotted against electrotonic distance from the soma, calculated from the cell morphology and estimates of the specific electrical properties of the cell membrane and cytoplasm. The diameters of dendritic branches are plotted according to the 10-μm scale shown. The positions of synapses applied to the model neuron in positions typical of group Ia synaptic boutons are shown as open circles or diamonds (cf. Figs. 4.11 and 4.12).

that follow voltage perturbations (Fleshman et al., 1988). There is good reason to suspect that some of these "active" conductances, certainly including those responsible for the action potential, are more concentrated near the initial axon segment and the axon hillock, where action potentials are generated, and are less densely distributed in the dendrites that do not support action potentials (Dodge, 1979). To the extent that active and/or passive conductance channels are distributed nonuniformly in the neuronal membrane, the effective net conductivity of the membrane, G_m, and its inverse, R_m, will also be spatially nonuniform. There is also reason to believe that intracellular penetration with micropipettes can introduce an artifactual "leak" at the site of penetration, lowering the effective R_m of the soma membrane (Jack, 1979; see Appendix). For most types of neurons, we lack sufficient quantitative information about the various active conductance channels and their spatial locations to permit truly realistic simulation. Until such information is available, passive neuronal models can be used to provide useful insights but must be viewed with appropriate caution.

Subject to this caveat, recent studies suggest that alpha motoneurons exhibit spatially nonuniform R_m (Barrett and Crill, 1974; Durand, 1984; Fleshman et al., 1988), with R_m much smaller in the soma than in the dendrites. "Leaky" somata could be produced by micropipettes, but there is no a priori reason to exclude the possibility that spatially nonuniform membrane may be present in uninjured neurons. Estimates (read "best guesses") of R_m in alpha motoneuron dendrites are relatively high (10,000–20,000 Ω-cm^2) in the dendrites ($R_{m,d}$) and 10- to more than 100-fold lower on and near the soma ($R_{m,s}$; see Fleshman et al., 1988). The neuron model shown in Fig. 4.10 is based on morphological and physiological measurements from a cat motoneuron (see Fleshman et al., 1988, cell 43/5). With an estimated $R_{m,d}$ of 11,000 Ω-cm^2 and R_i of 70 Ω-cm, the dendrite paths of this cell were relatively short in electrotonic terms (1–2 electrotonic space constants, λ; see Fig. 4.10 and Appendix). The $R_{m,s}$ of this cell was estimated to be 225 Ω-cm^2, but the longest time constant of voltage decay in the neuron was between 7 and 8 msec, that is, intermediate between the time constants of the dendritic (11 msec with $C_m = 1.0$ μF/cm^2) and somatic (0.2 msec) membranes. Obviously a neuron with nonuniform R_m cannot have a true "membrane" time constant (since there is no unique value of R_m). One of the challenges of current spinal cord neurophysiology is to continue to refine such estimates and to apply new techniques such as whole-cell patch clamping to reduce some of the uncertainties. Further orientation to this problem is provided in the Appendix.

THE EFFECT OF DENDRITIC LOCATION ON SYNAPTIC POTENTIALS

Given current information about the morphology of motoneuron dendrites and the location of group Ia synapses on them (Burke et al., 1979; Brown and Fyffe, 1981; Redman and Walmsley, 1983b), it is of interest to consider the synaptic potentials that "typical" group Ia synapses would generate in a model motoneuron with a fully branched dendritic tree. The labeled circles in Fig. 4.10 denote synaptic locations that are quite compatible with direct anatomical evidence about the distribution of boutons from a single group Ia afferent on alpha motoneurons in the cat (Burke and Glenn, unpublished). Conductances equivalent to those found in voltage-clamp experiments for somatic group Ia synapses (Finkel and Redman, 1983) were generated in the computer model at the individual synaptic locations in order to simulate Ia EPSPs, as illustrated in Fig. 4.11. This conductance (5 nS peaking at 200 μsec) produced a rapidly rising and falling EPSP of about 100 μV when applied across the soma membrane (A), and slower and smaller EPSPs when applied at progressively greater electrotonic distances (B → F), when recorded at soma. This is as expected from equivalent cylinder models (e.g., Rall et al., 1967), with a constant conductance at all synapses (cf. Jack et al., 1981a; see Fig. 13.11 in the Appendix).

The exponential falloff in somatic EPSP peak amplitudes with increasing electrotonic distance is shown by the open diamonds and solid line in the upper

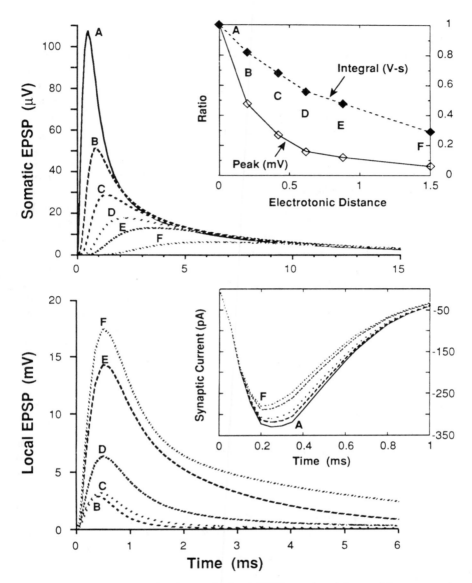

FIG. 4.11. The large graphs show calculated EPSPs produced by model synapses A through F (open circles) in the model shown in Fig. 4.10 when each synapse is individually activated. The upper graph shows EPSPs recorded in the cell soma, whereas the lower graph shows the *local* EPSPs recorded in the dendritic compartments containing the active synapses. The upper inset graph shows the falloff in peak EPSP amplitude and the integral of voltage over 20 msec for the *somatic* EPSPs, both as functions of the electrotonic distance of the respective synapses. The lower inset shows the time course of synaptic current injected into the local compartments receiving the active synapses. Full discussion in the text.

inset graph in Fig. 4.11. However, if one plots the integral of voltage over a relatively long time (20 msec in this case), the falloff is much less steep (filled diamonds, dashed line). This integral reflects the electrical charge delivered by the synapses to the soma (see Edwards et al., 1976). Beyond this, however, is the fact that Ia afferents (and most other input systems) fire repetitively at frequencies (50–200 Hz) that allow effective temporal summation of PSPs that have relatively slow time courses. Thus, distal synapses delivering repetitive EPSPs can still have a significant influence on the somatic potential, and therefore on cell output, relative to their more proximal brethren.

The lower large graph shows that the peak amplitudes of *local* EPSPs, recorded at the site of generation, *increase* markedly with progressively more distant sites of generation (from B to F) and are many fold larger than their somatic reflections (almost 3000-fold larger in case F). These large local EPSPs result from the fact that the local input resistances at the sites of generation in branched dendritic trees can be very high (see Appendix, and Rinzel and Rall, 1974). The local EPSPs in the dendrites are sufficiently large that they can influence the magnitude of synaptic current generated by the constant conductance used in this simulation. The inset graph shows that the peak inward current (hence negative polarity) at synapse F was only about 85% of that at the somatic bouton (A) and fell more rapidly. This can be understood by the fact that synaptic current, I_{syn}, depends not only on the conductance, g_{syn}, that produces it but also on the instantaneous difference between the local transmembrane voltage, $V(t)$, and the constant synaptic equilibrium potential, E_{syn}, for the ionic species involved (see also Chap. 2):

$$I_{syn}(t) = g_{syn}(t) \cdot (V(t) - E_{syn}) \tag{4.1}$$

In the case of Ia EPSPs, E_{syn} is near 0 mV, so that the rising phases of the local EPSPs at synapses E and F represent a substantial fraction of the EPSP driving-potential when they coincide with the conductance transient, despite the latter's brevity.

The complexities of dendritic branching introduce other interesting complications into the interpretation of EPSP shape and amplitude, which are not present in idealized cylinder models. For example, the synaptic sites denoted by diamonds (G and H) in Fig. 4.10 are at approximately the same electronic distance from the soma (about 0.7 λ). However, as shown in the upper graph in Fig. 4.12, the EPSPs at the soma are not identical in amplitude and time course, as they would be if generated at the same distance in ideal membrane cylinders. The local EPSPs (lower graph) are also quite different in peak amplitude, and the current at H is also significantly smaller and shorter than that at G. The explanation for this seemingly paradoxical result lies again in differences in the electrotonic conditions at the two locations. Synapse G was placed in the proximal third of a long dendritic branch, whereas H was near the termination of a short branch. The input resistance at H was much larger than at G, leading to a correspondingly larger local EPSP. The difference in EPSP shape at the soma is likely due to the fact that most of the current generated at H had to spread toward the soma whereas that produced at G spread in both directions, outward as well

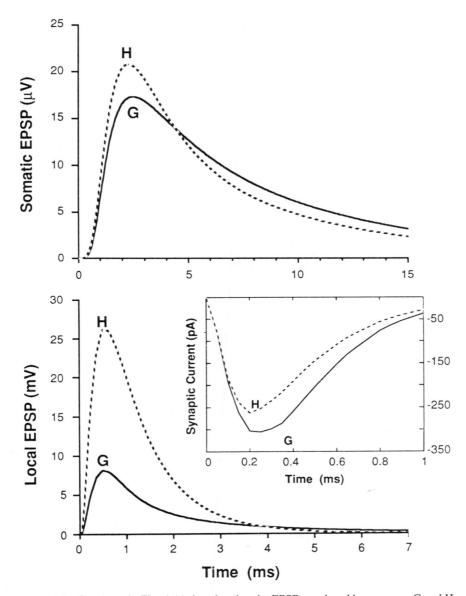

FIG. 4.12. Graphs as in Fig. 4.11, but showing the EPSPs produced by synapses G and H (diamonds in Fig. 4.10) that are at approximately the same electrotonic distance from the soma. The differences in somatic and local EPSPs are accounted for by the location of the active synaptic sites in relation to dendritic path terminations (see text).

as inward. Inferences about the electrotonic locations of active synapses from the shape of a somatic PSP (see Rall et al., 1967) must therefore be interpreted with appropriate caution.

THE SPINAL CORD IN ACTION: RECRUITMENT OF MOTOR UNITS

Aspects of the synaptic organization of ventral spinal cord have been discussed above at the level of individual synapses, recipient neurons, and simple neuronal circuits. This "small system" approach can be used to examine an important aspect of more global spinal cord function—the control of motor unit recruitment. This process is illustrative not only of the mechanisms that operate in motor nuclei but also of the mechanisms that control aggregate output from any functionally related group of CNS neurons.

MOTOR UNITS

Motoneurons are, anatomically and functionally, inseparable from the muscle fibers that they innervate. Among the many key concepts that we owe to Sherrington (Liddell and Sherrington, 1925) is that of the *motor unit*—the combination of a motoneuron (alpha or beta; see above) and the group of muscle fibers that it innervates. Most limb and trunk muscles in mammals contain two fundamental motor unit types: fast twitch (type F) and slow twitch (type S), based primarily on the morphology, biochemistry, and mechanical properties of the innervated muscle fibers (the "muscle unit"). The type F population can be further divided into two major subtypes, one relatively fatigable (type FF) and the other much more resistant to fatigue (type FR), and one minor subtype intermediate between the two (type F(int); see Burke, 1981, for review).

All of the muscle fibers in a muscle unit share the same morphology, chemistry, and presumably mechanical characteristics. There is a large and still growing list of intrinsic motoneuron properties that are systematically related to motor unit type, as defined by the innervated muscle unit (Burke, 1981; Zengel et al., 1985). A further extension of these interrelations has been found in the quantitative and even qualitative organization of synaptic inputs to the various motor unit types (Burke, 1981). Some of these interrelations, as illustrated in the multidimensional plot shown in Fig. 4.13, almost certainly represent functional specializations that enable the different types of motor units to play particular roles during reflex and voluntary movements (reviewed in Burke, 1981, 1986; Henneman and Mendell, 1981).

RECRUITMENT: THE "SIZE PRINCIPLE"

Force output from a muscle is modulated by activating and deactivating motor units ("recruitment" and "derecruitment, respectively), as well as by controlling the firing frequency of active units (a process often referred to as "rate coding"; see Burke, 1981). Because of the wide range of individual motor unit properties (see Fig. 4.13), an understanding of the recruitment process requires that the *identities* as well as the numbers of active motor units be taken into consideration.

It has been known for over half a century that, in many kinds of movements,

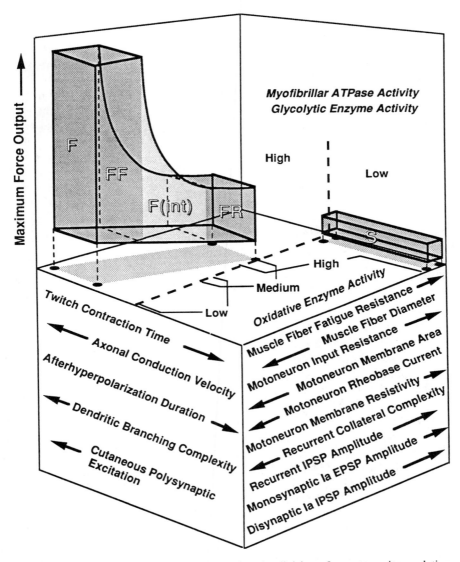

FIG. 4.13. Three-dimensional diagram showing the division of a motor unit population into two basic clusters—fast twitch (type F) and slow twitch (type S)—according to properties of the muscle unit portion. The type F cluster is subdivided into types FF, F(int), and FR on the basis of relative resistance to fatigue. The box outlines denote the loci of data points for individual motor units (see text, and fig. 5 in Burke et al., 1973), plotted against maximum tetanic force (vertical ordinate), twitch contraction time (left horizontal abscissa), and muscle fiber resistance to fatigue (right abscissa). Histochemical enzyme activities characteristic of each muscle unit type are given on the main diagram. The two abscissae also show other properties of muscle units (fiber diameter), motoneurons (motor axon conduction velocity, etc.), and quantitative characteristics of synaptic organization (group Ia EPSP amplitude, etc.), with arrows to indicate increasing magnitudes. Most of the actual distributions of these characteristics are not discontinuous as implied by this summary diagram. (Data for the graph were taken from Burke et al., 1970, 1971a,b, 1973, 1976; Burke, 1967, 1968a,b; Cullheim and Kellerth, 1978a,b; Fleshman et al., 1981a,b; Friedman et al., 1981; Cullheim et al., 1987).

recruitment is a stereotyped and orderly process (Denny-Brown, 1929, 1949). In the mid-1960s, Elwood Henneman and his co-workers (Henneman et al., 1965; Henneman and Olson, 1965) provided indirect but compelling evidence for an orderly relation between recruitment threshold and muscle unit characteristics. For example, recruitment in the stretch reflex begins with small, slow twitch units and ends with very large force, fast units. Derecruitment proceeds in the reverse order. Henneman coined the term "size principle" to describe the recruitment sequence and to encapsulate his idea that the process depends on intrinsic motoneuron properties that are closely related to the anatomical size of the cells, irrespective of synaptic input source.

Subsequent work has confirmed that differences in intrinsic motoneuron properties indeed play a role in recruitment control but have substantiated the early suggestion of Denny-Brown (1929) that the organization of synaptic input is also crucial. There is good evidence that the recruitment sequence within a given motor unit population is not immutable, as would be the case if it depended on motoneuron properties alone, but rather can vary with the nature of the synaptic system driving the action and the movement in question (see Burke, 1981). The CNS has sufficient flexibility so as to produce recruitment patterns that are biomechanically and metabolically tailored to meet an enormous range of functional demand.

INTRINSIC MOTONEURON PROPERTIES

There is experimental evidence for quantitative variations in intrinsic motoneuron properties among the cells that belong to a given "motor pool" (defined here as the motoneurons that innervate muscle units within one anatomically defined muscle). Some of these properties are included schematically in Fig. 4.13 (see also Zengel et al., 1985). For example, the rheobase current (the depolarizing current necessary to bring a motoneuron to its firing threshold), when adjusted for differences in cell input resistance, increases systematically in the sequence types S < type FR < type FF (Fleshman et al., 1981a; Zengel et al., 1985). This implies that type S motoneurons are intrinsically more excitable than type FR, which are in turn more excitable than type FF cells. An intrinsic membrane process called *accommodation* also tends to limit the responsiveness of some type F motoneurons to injected current much more strongly than in type S cells (Burke and Nelson, 1971). Such intercell differences are presumably accounted for by different densities of the active conductance channels, as discussed in Chap. 2.

Motoneurons exhibit limited ranges of firing frequency, largely because of well-developed, relatively long-duration postspike afterhyperpolarizing potentials (e.g., Calvin and Schwindt, 1972), that depend largely on Ca^{2+}-dependent K^+ conductances (Barrett and Barrett, 1976). They do not exhibit the high-frequency "burst-firing" mode described for neurons in many other parts of the CNS (see Chap. 2). This makes excellent physiological sense, since muscle units in limb muscles produce virtually their full range of force output modulation over a relatively limited range of input frequencies (10–50 Hz in most animal muscles; see Burke, 1981). However, alpha motoneurons do show what has been called "bistable" membrane responses in the decerebrate state in cats (Crone et al., 1988) or when exposed to the neurotransmitter serotonin (Hounsgaard and Kiehn, 1985) and to drugs like L-DOPA which cause release of noradrenaline (Hounsgaard et al., 1984, 1988). Under these conditions, sustained firing at moderate frequencies that does not depend on sustained input excitation can be produced in motoneurons by a short-lasting depolarization and is curtailed by sudden membrane hyperpolarization. This activation of an intrinsic membrane conductance (presumably to Ca^{2+}; Hounsgaard and Kiehn, 1985) by neurotransmitters is an apparent example of "neuromodulation" in the mammalian spinal cord (see Chap. 2). It is not known whether there is a systematic difference in this property that is related to motor unit type, but voltage-clamp data suggest that the tendency for self-sustaining firing may be more characteristic of the cells with relatively high input resistance, like type S motoneurons (Schwindt and Crill, 1977; Crill and Schwindt, 1983).

The overall passive membrane resistivity in cat motoneurons, when adjusted for the influence of nonuniform R_m (see above) increases in the sequence FF < FR < S (Fleshman et al., 1988). It is unknown whether there is a corresponding difference in the density of "leak" channels or if evident differences in the density of active conductances also contribute. This factor has complex effects on the generation and intraneuronal spread of synaptic potentials (see above), which can influence the efficacy of synaptic potentials delivered to the spike–trigger zone. In summary, intrinsic motoneuron properties indeed seem important to the control of recruitment order, as envisioned in the original size principle.

SYNAPTIC ORGANIZATION UNDERLYING RECRUITMENT

The average amplitudes of EPSPs produced by group Ia afferents increase in the sequence FF < FR < S (Fig. 4.13; see Burke, 1968a; Burke et al., 1973; Fleshman et al., 1981b), and there is a clear correlation between Ia EPSP amplitude and the functional thresholds of individual motor units in the stretch reflex (Burke, 1968b), as predicted by Henneman and Olson (1965). If all synaptic systems were organized like Ia afferents, one would expect an invariant recruitment process under all conditions. However, departures from this expectation have been known for some time (e.g., Denny-Brown, 1929; Rall and Hunt, 1956).

Among the more dramatic of these departures are the observations that stimulation of distal skin regions can alter the ordering of relative thresholds among motor units responding to stretch reflexes in animals (Kanda et al., 1977) and during voluntary activation in man (Garnett and Stephens, 1981; Datta and Stephens, 1981). Under these conditions, normally high-threshold motor units can be activated before those with otherwise low functional thresholds, some of which may in fact be inhibited. It seems quite possible that such observations can be linked to the finding that low-threshold afferents from distal skin regions in the cat hindlimb produce polysynaptic excitation that is ordered quite differently from Ia input (Fig. 4.13; "cutaneous polysynaptic excitation"). Such synaptic effects are minimal or even undetectable in most type S motoneurons but are present and sometimes quite powerful in all cells of FR and FF units (Burke et al., 1970, 1973). The responsible pathway in the cat includes at least two levels of interneurons between the afferent and the motoneuron and receives convergent supraspinal excitation (Pinter et al., 1982). There is less direct evidence that a qualitatively similar pathway exists in man (Garnett and Stephens, 1980).

The two proposed patterns of synaptic input organization to motoneurons are illustrated schematically in panel A of Fig. 4.14. Input system A has greatest efficacy (denoted by line width) in type S motoneurons but progressively less in

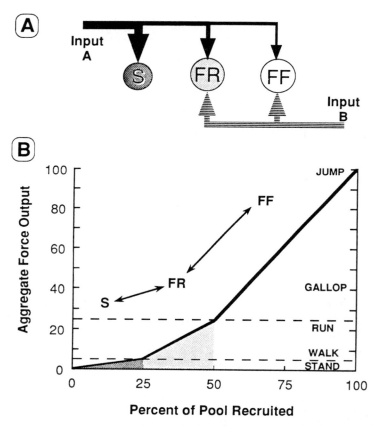

FIG. 4.14. **A.** Two patterns of synaptic organization that have been found in cat motoneurons of different motor unit types. The relative efficacy of input is denoted by the width of the arrows. **B.** A "recruitment model" derived from data on the properties of cat medial gastrocnemius motor units and the force output from this muscle in intact cats during various actions. It is assumed that recruitment in all of these activities begins with type S motor units and that no units of the next unit type are recruited until all of the preceding type are fully active. The nonlinear character of the force output versus recruitment curve is accounted for by the differences in average force output among the motor unit type groups shown in Fig. 4.13. (A adapted from Kanda et al., 1977; B adapted from Walmsley et al., 1978.)

FR and FF cells. The same set of motoneurons also receives another input, B, which is organized to excite only FF and FR units. Activation of the pool by system A would generate the "size principle" recruitment sequence, especially when combined with the intrinsic properties discussed above. If the pool were activated by *both* inputs A and B simultaneously, one would expect a narrowing in net range of functional thresholds (Garnett and Stephens, 1981). This would be useful in promoting synchronous activation of the entire motor pool, such as occurs in very rapid and forceful ("ballistic") movements (Desmedt and Godaux, 1977). Under exceptional conditions, particularly in very rapid alternating movements, an excitatory synaptic system organized like input B could produce selective activation of the rapidly contracting muscle units alone (see Smith et al., 1980). The existence of only two different orderings of synaptic efficacy thus provides the CNS with great flexibility in how a given motor pool might be "used" to provide the enormous range of functional demand placed on the motor systems, from sustained postural maintenance to powerful, ballistic movements like galloping, jumping, and rapid limb shaking.

FUNCTIONAL CONSEQUENCES

The recruitment order envisioned by the size principle implies that the first activated motor units produce small forces and are resistant to fatigue (i.e., type S). These features are advantageous to motor units that are likely to be "used" a good deal of the time as, for example, in the maintenance of posture. Conversely, the unit with the highest functional thresholds are likely to be recruited much less often (low "duty cycle" units), presumably in actions that require large force outputs but, being only occasional, relatively little fatigue resistance. There is a long list of additional biomechanical and metabolic reasons why this ordering of motor unit recruitment is advantageous for most, but not all, motor acts (see Burke, 1981, and Henneman and Mendell, 1981, for details).

Panel B in Fig. 4.14 gives a schematic "recruitment model" based on data from the medial gastrocnemius (MG) motor pool in cats (Walmsley et al., 1978) to illustrate the markedly nonlinear nature of the usual (i.e., "size principle") recruitment process. This muscle and its synergists make up a large part of the fleshy part of the calf and act to extend the ankle. In the cat, these muscles are active to support the weight of the body whenever the animal is standing or walking (the cat essentially stands and walks on its toes). The tendon force measured directly from the MG muscle during quiet standing in intact cats is only 200–400 g, or about 5% of the total aggregate force of which that muscle is capable (8–10 kg when fully tetanized). This modest force can be provided by fully activating all of the type S units in the MG muscle. Although type S units make up about 25% of the MG pool, their aggregate force is small because they individually produce small force outputs (Fig. 4.13).

When the cat begins to walk, larger forces are required from the ankle extensor muscles. Locomotor activities ranging from slow walking to fast-running demand forces that range up to 25% of aggregate MG force (0.5–2.0 kg). This level of output can be provided if the S and FR unit populations (about 50% of the MG motor unit population) work together. The orderly recruitment sequences produced by inputs organized like input A in the upper diagram can easily be envisioned to control recruitment within this range of motor demand.

The only movements that require larger output from the MG muscle occur during rapid, forceful actions such as galloping and jumping. In a sedentary predator like the cat, these actions are performed relatively infrequently and only for short periods of time (for further details, see Walmsley et al., 1978). They involve brief and virtually synchronous activation of the entire MG motor unit pool, which, as noted above, could be produced quite efficiently by conjoint action of synaptic systems organized like inputs A and B in the upper diagram (see Burke, 1981). Whether or not the synaptic organizations found in immobile, anesthetized animals are in fact used during normal movements in the suggested way remains to be demonstrated.

FACTORS THAT CONTROL SYNAPTIC EFFICACY

One measure of synaptic strength, or "efficacy" (denoted by the width of the arrows in Fig. 4.14), is the peak amplitude of synaptic potentials measured by an intracellular electrode in the cell soma. The factors that can account for the large variation in Ia EPSP amplitude among homonymous motoneurons (Burke et al., 1976) are surprisingly numerous and difficult to assess (Table 4.1). Systematic variations in any or all of them can, in principle, produce the experimentally observed difference in group Ia synaptic efficacy (see also Burke, 1987; Burke et al., 1989).

Table 4.1 includes three levels of synaptic organization at which interactions between presynaptic and postsynaptic factors can affect the efficacy of group Ia synapses in motoneurons. At the level of *single synaptic boutons,* one must consider the amount of transmitter (presumably glutamate; see above) released presynaptically, and the sensitivity and density of postsynaptic receptors for it. The interaction between these pre- and postsynaptic factors results in the transmembrane conductance change that generates a single-bouton EPSP (see Fig. 4.9 and Chap. 2). There is some evidence that variations at this level may be systematically related to motoneuron type (Honig et al., 1983).

The next level of analysis is the *synaptic system,* including all of the Ia synapses that impinge on a given motoneuron. Effective synaptic strength can vary from cell to cell as a function of the total number of Ia synapses that release transmitter during any given presynaptic action potential (see above). However, synaptic number alone has little meaning without considering the total size of the postsynaptic cell over which they are distributed. The latter is best evaluated in terms of total cell membrane area. The ratio between synaptic number and membrane area is synaptic density. When all other things are equal, variations in synaptic density should produce equivalent variations in synaptic efficacy.

The last level of analysis concerns *electrotonic interactions* that depend on the spatial distribution of the active boutons and the electrotonic characteristics of the cell (soma and dendrites) on which they are distributed. As discussed in the preceding section, such spatial dispersion profoundly affects the amplitude and shape of synaptic potentials arriving at the cell soma, where they are integrated to

Table 4.1. Factors That Control the Synaptic Potential Amplitude at Chemical Synapses

Presynaptic factors	Interaction	Postsynaptic factors
Single bouton level		
Amount of transmitter liberated	Conductance change per synapse	Receptor sensitivity, density, and kinetics
Synaptic system level		
Number of active synapses	Synaptic density	Postsynaptic membrane area (A_N)
Electrotonic interaction level		
Spatial distribution of active synapses	Local PSP amplitudes	Dendritic morphology
	Nonlinear interactions between different synapses	Specific membrane properties, R_m and C_m, and cytoplasmic resistivity, R_i
	Electrotonic attenuation	PSP driving potential

produce, or fail to produce, action potentials. For example, if most of the Ia synapses to motoneuron A were electrotonically near the soma while those in cell B were mostly more distant, an equal density might still result in quite different EPSP amplitudes (larger and faster in A than in B; see Fig. 4.9). Equivalent spatial distributions of the same number of synapses to cells with very different electrotonic architectures could give the same result.

The problem for the physiologist is to sort out, to the extent possible with current techniques, which of this array of factors in fact accounts for the observed variation in Ia EPSP efficacy found among motoneurons within a single motor nucleus. A definitive answer is not yet available for this or any other synaptic system. However, a combination of electrophysiological, morphological, and computer modeling studies suggests that variations in the density of active Ia synapses (i.e., the synaptic system level) can account for much of the experimental data (see Burke et al., 1989). In principle, the second and third levels of interactions noted in Table 4.1 can be attacked with available technologies in the intact spinal cord of mammals, but the single-bouton level remains inaccessible to analysis in the intact CNS. Perhaps future refinement of experimental techniques and the use of more accessible preparations such as afforded by tissue culture and spinal cord slice preparations will permit more concrete answers.

SUMMARY

This chapter has attempted to present an overview of the neuron types present in the ventral spinal cord, some features of their cellular and synaptic neurobiology, and some of the more significant circuit interactions between them, in order to provide the reader with a starting point for the study of synaptic organization in this part of the CNS. The spinal cord has been studied systematically from the earliest days of scientific investigation of brain function, but it has much still to teach us. It remains one of the few parts of the CNS in which inputs and outputs are both accessible and can be interpreted in functional terms. Given sufficient energy and time, these linchpins can permit definition of neuronal properties and synaptic circuits that can be directly linked to CNS function.

5

OLFACTORY BULB

GORDON M. SHEPHERD and CHARLES A. GREER

The olfactory bulb is an outgrowth of the forebrain. It receives all of the input from the olfactory sensory neurons (see Fig. 5.1), and sends its output directly to the olfactory cortex (see Chap. 10). This basic relationship is seen in the most primitive fish, and has endured throughout the evolution of the vertebrates.

As a region for experimental analysis, the olfactory bulb is attractive for several reasons. In its position in front of the rest of the brain, it is easily accessible. The sensory nerves to the bulb are completely separated from its output fibers to the brain. This enables each to be manipulated individually, whether by electrical stimulation or tracer injection, as in the spinal cord (Chap. 4). Within, the bulb is a distinctly laminated structure, containing sharply differentiated cell types, particularly in terrestrial animals. All of these features facilitate the application of different experimental techniques and the precise interpretation of results.

Historically, these advantages have been put to good use; work on the olfactory bulb by classical histologists contributed to the evidence that led to acceptance of the neuron doctrine, and studies by modern workers have contributed to emerging principles of synaptic organization, as outlined in Chap. 1. A disadvantage has been the limited knowledge available concerning the sensory mechanisms in the olfactory pathway—in particular, the nature of the molecular stimulus and its modes of processing. The work on synaptic organization is therefore of additional significance for the insight it can give into the nature of these sensory mechanisms. An important step has been the realization that many aspects of the synaptic organization of the bulb reflect general principles that apply to other sensory systems, especially the retina (Chap. 6) and thalamus (Chap. 8). The goal of this chapter is to describe the essentials of the synaptic organization of the olfactory bulb within the framework of these general principles.

NEURONAL ELEMENTS

As in other brain regions, the neuronal elements fall into three categories: input, output, and intrinsic. In describing these elements, we will also relate them to the histological layers that they form (see Fig. 5.2). The classical descriptions were based on Golgi-impregnated neurons (cf. Cajal, 1911); they have been confirmed

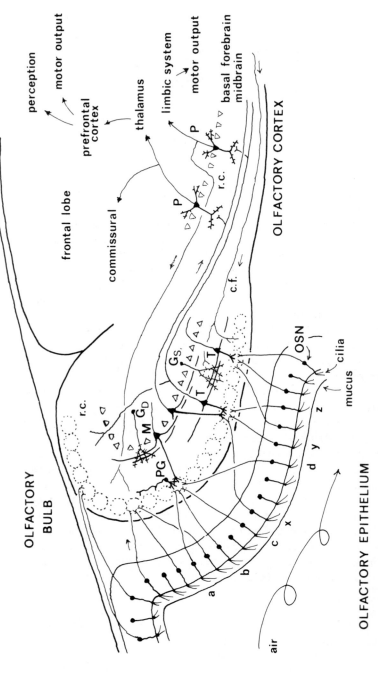

OLFACTORY EPITHELIUM

FIG. 5.1. An overview of the olfactory pathway. The olfactory bulb receives input from the receptor neurons in the olfactory epithelium, and projects to the olfactory cortex. The diagram indicates some essential aspects of the projection patterns between the regions, as well as the main neural elements within the olfactory bulb. Note that the olfactory epithelium is arranged in overlapping populations of receptor neurons (a–d, x–z) which project to individual glomeruli. Some of the central olfactory connections to limbic brain structures are also indicated. Abbreviations: olfactory epithelium—OSN, olfactory sensory neuron. Olfactory bulb—PG, periglomerular cell; M, mitral cell; T, tufted cell; G_S, superficial granule cell; G_D, deep granule cell; r.c., recurrent axon collateral; c.f., centrifugal fiber. Olfactory cortex—P, pyramidal cell; r.c., recurrent axon collateral.

and amplified by more recent studies employing Golgi methods and HRP injections (see below).

INPUTS

Afferents. The sensory input consists of the parallel array of axons from the olfactory sensory neurons in the nasal cavity (Fig. 5.2A). Although a detailed consideration of the sensory neuron is not possible here (see Lancet, 1986, for review), several facts are relevant. Traditionally, it has been believed that the sensory neurons are morphologically homogeneous. However, there is evidence for at least one small population of a morphologically distinct subtype (cf. Jourdan, 1975; Moran et al., 1982), and plentiful evidence for molecular subtypes (see later). The sensory axons are all unmyelinated; their diameters form a unimodal spectrum with a mean of approximately 0.2 μm and a range of 0.1–0.4 μm. Although this morphological uniformity stands in contrast to the variety of input fibers that characterize many regions of the brain, it conceals nonetheless a heterogeneity of physiological specificities of the sensory neurons for different odor molecules. The tight packing of many axons within a single Schwann cell mesaxon provides the opportunity for ephaptic interactions between neighboring axons in which K^+ extruded during impulse activity may play a role (Getchell and Shepherd, 1975; Bliss and Rosenberg, 1979; Gesteland, 1986; Eng and Kocsis, 1987); there is a close similarity in this respect with parallel fibers in the cerebellum.

The olfactory axons are grouped in bundles, which enter the bulb surface where the bundles splay out and interweave. The axons terminate in regions of neuropil called *glomeruli*. In fish and amphibians, the glomeruli are small (20–40 μm in diameter) and not distinctly demarcated; in mammals, they are spherical with sharp borders, and range in size from 30–50 μm in diameter in small mammals (e.g., mice) to 100–200 μm in rabbits or cats (cf. Allison, 1953). The olfactory glomeruli are among the clearest examples in the brain of the principle of grouping of neural elements and synapses. They are analogous to "barrels" and "columns" in cerebral cortex, representing a higher level of organization than the synaptic glomeruli of the cerebellum and thalamus.

The olfactory axons do not branch on their way to the glomeruli, but once inside they ramify to varying extents prior to terminating. During early development in rats, some axons terminate transiently in deeper layers (Monti-Graziadei et al., 1980; Farbman, 1986), but in the adult all sensory terminals are within the glomeruli.

A notable feature of the sensory neurons is that they are continuously replaced from stem cells in the epithelium throughout adult life (Graziadei and Monti-Graziadei, 1979). Thus, the specificity of synaptic connections in the glomeruli is achieved despite a constant remodeling due to turnover of the sensory terminals. This degree of plasticity is unique among the regions of the brain considered in this book.

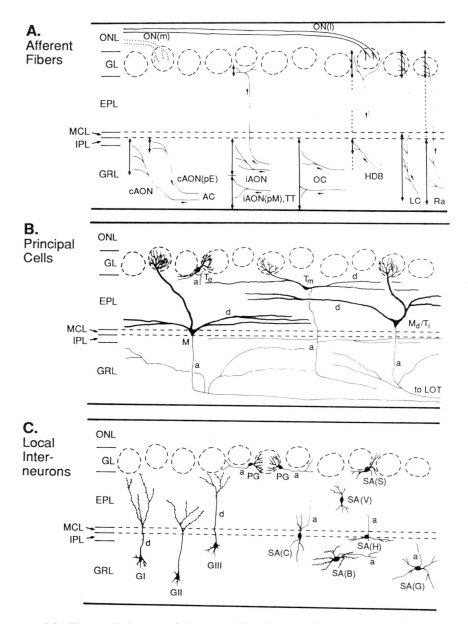

FIG. 5.2. The neural elements of the mammalian olfactory bulb, grouped according to the subdivision into inputs (afferent fibers), principal neurons, and intrinsic neurons (local interneurons). Diagrams based on various studies using the Golgi method and HRP (see text). Abbreviations for layers: ONL, olfactory nerve layer; GL, glomerular layer; EPL, external plexiform layer; MCL, mitral cell layer; IPL, internal plexiform layer; GRL, granule cell layer. **A.** ON(m) and ON(l) indicate medial (m) and lateral (l) subtypes of olfactory nerve fibers. Centrifugal afferents are from the contralateral anterior olfactory nucleus (cAON), ipsilateral anterior olfactory nucleus (iAON), tenia tecta (TT), olfactory cortex (OC), horizontal limb of the diagonal band (HDB), locus coeruleus (LC), and raphe nucleus (Ra). pE, pars externa of the AON; pM, pars medialis of the AON. **B.** The

Central inputs. There are several types of inputs from the brain, each of which has a distinctive laminar pattern of termination (reviewed in Macrides and Davis, 1983; Mori, 1987; Scott and Harrison, 1987). One type consists of axon collaterals from pyramidal neurons in the *olfactory cortex* (OC) (Chap. 10); these end mostly in the granule cell layer (see OC Fig. 5.2A). Extensive connections are made by fibers from different parts of the *anterior olfactory nucleus* (AON); their different laminar projections possibly relate to different populations of granule cells (see below). The nucleus of the *horizontal limb of the diagonal band* (HDB) is one of the basal forebrain cholinergic centers; it sends fibers to both the granule layer and the periglomerular parts of the glomerular layer. From the brainstem, the *locus coeruleus* (LC) and the *raphe nucleus* (Ra) send fibers diffusely to the granule layer and specifically to the interiors of the glomeruli (see Fig. 5.2A).

These central inputs are also referred to as *centrifugal* inputs to indicate their outward orientation from the brain. It is obvious that the olfactory bulb is under extensive and highly differentiated control by the brain. This is true of many other sensory regions; the retina, by contrast, receives few centrifugal fibers.

PRINCIPAL NEURON

The output from the olfactory bulb is carried by the axons of mitral and tufted cells. The morphology of these cells has been the subject of several studies in recent years (Mori et al., 1981a, 1983; Macrides and Schneider, 1982; Kishi et al., 1984; Orona et al., 1984).

Mitral cells. In fish and amphibia, the principal neurons are relatively undifferentiated. In reptiles, birds, and mammals, however, distinctive mitral cell bodies lie in a thin sheet 200–400 μm deep to the glomerular layer (see Fig. 5.2B). The cell bodies are 15–30 μm in diameter, medium-sized for a principal neuron.

In mammals, each mitral cell tends to give rise to a single *primary* dendrite, which traverses the external plexiform layer (EPL) and terminates within a glomerulus in a tuft of branches. The tuft has a diameter of 50–150 μm, extending throughout most of its glomerulus. The diameter of the dendrite ranges from 2 to 12 μm (depending on the size of the cell body from which it arises); the length is 200–800 μm, depending on how much it angles across the EPL.

Each mitral cell also gives rise to several *secondary* (basal) dendrites; laterally directed, they branch sparingly and terminate in the EPL. They are 1–8 μm in diameter, and in mammals extend 500–800 μm. In turtles, HRP-injected mitral cells commonly display two thin (1–2 μm) primary dendrites up to 700 μm in length, and several thin secondary dendrites which extend over 1 mm, and may

---

dendrites and axon collaterals of a mitral cell (M), an internal tufted cell (T_i, or a displaced mitral cell, M_d), a middle tufted cell (T_m), and an external tufted cell (T_e) a, axon; d, dendrite. **C.** Three types of granule cells (GI, GII, GIII); PG, periglomerular cell; SA(B), Blanes' cell; SA(C), Clandins' cell; SA(G), Golgi cell. SA(H), Hensen's cell; SA(S), Schwann cell; SA(V), van Gehuchten cell; LOT, lateral olfactory tract. (Modified from Mori, 1987.)

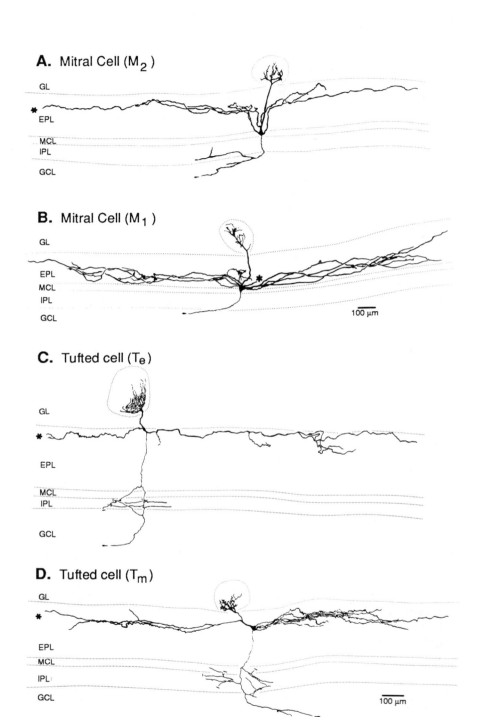

A. Mitral Cell (M₂)

GL

EPL

MCL
IPL

GCL

B. Mitral Cell (M₁)

GL

EPL
MCL
IPL

GCL

100 µm

C. Tufted cell (Tₑ)

GL

EPL

MCL
IPL

GCL

D. Tufted cell (Tₘ)

GL

EPL
MCL

IPL

GCL

100 µm

FIG. 5.3. Different types of principal neurons in the rat olfactory bulb. These reconstructions are based on HRP staining following extracellular microinjections of HRP into the olfactory bulb (injection sites indicated by asterisks). **A.** Mitral cell of type M₂, with secondary dendrites ramifying in a midsuperficial zone of the EPL. **B.** Mitral cell of type

reach up to halfway around the circumference of the EPL (Mori et al., 1981a). In mammals, recent studies (Macrides and Schneider, 1982; Mori et al., 1983; Orona et al., 1984) have identified subtypes of mitral cells on the basis of the branching pattern of their secondary dendrites. As shown in the HRP-stained cells in Fig. 5.3, type I mitral cells send their secondary dendrites into the deepest region of the EPL (see B), whereas type II mitral cells send their secondary dendrites into the middle region of the EPL (see A). These two types form synaptic microcircuits with corresponding subtypes of interneurons (see below). The field of the secondary dendrites may be "disklike" (Mori, 1987) or oriented in the anterior–posterior axis (Shepherd, 1972a).

The primary and secondary dendrites of mitral cells have generally smooth surfaces. They thus are *aspiny* neurons, like motoneurons, but unlike *spiny* principal neurons such as cortical pyramidal cells. The highly differentiated terminal tuft, which segregates primary sensory afferents and their associated microcircuits from the rest of the bulb, is virtually unique among principal neurons, and exemplifies the attractiveness of the olfactory bulb as an experimental model.

The mitral cell axons proceed to the depths of the bulb and then pass caudally to gather at the posterolateral surface to form the lateral olfactory tract (LOT). Within the bulb the axons give off *recurrent collaterals*. According to the classical studies (Cajal, 1911), some collaterals recur to terminate in the EPL, and some remain in the deep granule layer. However, recent studies of cells visualized by intracellular (Kishi et al., 1984) or extracellular (Orona et al., 1984) injections of HRP show that the collaterals remain within the granule layer (GL) and internal plexiform layer (IPL) (Kishi et al., 1984; Orona et al., 1984). Age or species differences may account for some of this discrepancy. The collaterals distribute diffusely within these deeper layers.

The output axons in the LOT give off numerous collateral branches, which terminate in the olfactory cortex, as described in Chap. 10. The distances traveled are relatively short, up to 10–15 mm. This is similar to the projection distances of some other principal neurons, such as cerebellar Purkinje cells or dentate granule cells, but contrasts with the extremely long axons of motoneurons and cortical pyramidal cells.

Tufted cells. Output cells similar to mitral cells but located more superficially in the EPL are called tufted cells (Figs. 5.2B an 5.3C,D). The subgroups and their nomenclature have become quite complex, but for present purposes we can identify three main groups according to their laminar position. The main population, *middle tufted cells* (T_m), lies near the middle of the EPL. They have a cell

---

M_1, with secondary dendrites ramifying in the deep zone of the EPL. **C.** External tufted cell, with cell body and secondary dendrites in the most superficial zone of the EPL, near the border with GL. **D.** Middle tufted cell, with cell body and dendrites in a midsuperficial zone of the EPL. Note luxuriant axon collaterals of tufted cells in the IPL. (Modified from Orona et al., 1984.)

body diameter of 15–20 μm, several thin basal dendrites (300–600 μm), and a primary dendrite (200–300 μm) ending in a relatively confined tuft of branches in a glomerulus (see Fig. 5.3C). The axon gives off collaterals that are mostly confined within the IPL, and then joins the LOT. Its projection sites in olfactory cortex differ from those of mitral cells (see Chap. 10).

There are also several varieties of *external tufted cells* (T_e), whose dendrites have distinctive branching patterns (see contrasting examples in Figs. 5.2B and 5.3C). All of these give off collaterals in the IPL or adjacent GRL, where they constitute a topographically ordered intrabulbar association system (Schoenfeld et al., 1985). Some send an axon into the LOT, whereas others do not and thus should be classified as intrinsic neurons (see below). Finally, some *internal tufted cells* (T_i) overlap in distribution and morphology with outwardly displaced type II mitral cells.

Traditionally tufted cells were considered to be smaller versions of mitral cells (Allison, 1953). However, Cajal (1911) noted that their axon collateral patterns are different, and it was suggested that this could provide for distinctive types of modulation of granule cells (Shepherd, 1972a). More recent research has established that the two types differ genetically (Greer and Shepherd, 1982); in the mouse mutant *pcd,* there is selective degeneration of mitral cells without affecting the tufted cells. The careful studies of Macrides et al. (1965), Orona et al. (1984), and Mori (1987) have documented the detailed morphology of the different subtypes of mitral and tufted cells. Transmitter differences will be noted below. It thus appears that, as in many other regions of the brain, the principal neurons are differentiated into multiple subgroups, on the basis of genetics, position, dendritic morphology, intrabulbar axonal connections, extrabulbar projection sites, and neurotransmitters and modulators.

INTRINSIC NEURONS

There are two main types of intrinsic neuron in the olfactory bulb: periglomerular cells (PG) and granule cells (GC).

Periglomerular cells. Surrounding the glomeruli are the cell bodies of periglomerular (PG) cells (see Fig. 5.2C). The cell body is only 6–8 μm in diameter, among the smallest in the brain. As shown by Cajal (1911) and confirmed by modern studies (see Pinching and Powell, 1971a,b; Schneider and Macrides, 1978), each PG cell has a short bushy dendrite that arborizes to an extent of 50–100 μm within a glomerulus; bitufted PG cells, to two glomeruli, are infrequently seen. The dendritic branches intermingle with the terminals of olfactory axons and the branches of mitral and tufted cells. The PG axon distributes laterally within extraglomerular regions, extending as far as five glomeruli away (Pinching and Powell, 1971a,b). Because the axon remains within the olfactory bulb, the PG cell falls within the category of *short-axon cell.* Morphologically the PG cells appear to be a homogeneous population, but biochemical subtypes containing different transmitters have been identified (see below).

Granule cells. Deep to the layer of mitral cell bodies is a thick layer containing the cell bodies of granule cells (GRL in Fig. 5.2). These cell bodies are also very small (6–8 μm in diameter); they appeared as grains (hence the term "granule") to the early microscopists, who applied this term to many types of small cells in the brain. The cell bodies are grouped in clusters in the granule layer.

Each granule cell gives rise to a superficial process which extends radially toward the surface and ramifies and terminates in the EPL, the branching field extending laterally some 50–200 μm. There is also a deep process which branches sparingly in the granule layer. It was early noted that granule cells located at different depths would have different functional roles to play in intra-bulbar circuits (Shepherd, 1972a). This notion has been greatly amplified by recent studies. Both intra- and extracellular HRP injections have shown that there are three main cell types in rodents (see Figs. 5.2C and 5.4). *Superficial* granule cells (Orona et al., 1983; type G_{III} of Mori et al., 1983) have peripheral dendrites that ramify mainly in the superficial EPL, among the dendrites of tufted cells. *Deep* granule cells (G_{II}) send their dendrites mainly to the deep EPL, among the dendrites of mitral cells. *Intermediate* granule cells (G_I) have dendrites that ramify at all levels of the EPL. It thus appears that mitral and tufted cells have both segregated and overlapping microcircuits through granule cells, as discussed below (see also Macrides et al., 1985). Other subtypes of granule cells have been reported on the basis of light and dark staining of the cell bodies by toluidine blue (Struble and Walters, 1982), and localization of neuropeptides (see below).

The granule cell dendrites are notable for bearing numerous spines (also called gemmules). In general, the spines are larger but less numerous than spines of dendrites of pyramidal cells in the cerebral cortex. In a developmental study, Greer (1984) found that the density of spines increased from 1 per 10 μm of dendritic branch length at birth to a peak of 2–3 at 12 days, settling to an adult value of approximately 2. This is much lower than the spine densities of striatal cells (see Chap. 9), and of pyramidal neurons in the hippocampus (Chap. 11) and neocortex (Chap. 12).

The most notable feature of the granule cell is that it lacks an axon. This was evident in the first studies by Golgi (1885) himself, and has been repeatedly confirmed by use of Golgi methods, HRP, and electron microscopy. The resemblance to amacrine cells in the retina was early recognized. Electron microscopic studies have clearly shown that the granule cell processes are dendrites on the basis of their fine structural features and their close resemblance to cortical dendrites (Price and Powell, 1970a,b). The lack of an axon meant that these cells always stood out as exceptions to the classical "law of dynamic polarization" (Cajal, 1911); this problem was solved by the discovery of the output functions of the dendritic spines of these cells (see Chap. 1 and below).

It remains to note that a third type of intrinsic neuron, *short-axon cell*, is represented sparingly in the glomerular layer and more frequently in the granule layer. The latter consist of several subtypes (Pinching and Powell, 1971a,b; Schneider and Macrides, 1978), with dendritic trees of variable extent, and axons that ramify in the EPL or granule layer (see Fig. 5.2C).

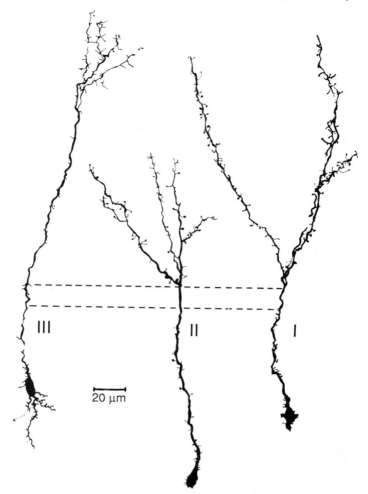

FIG. 5.4. Different types of granule cells in the mouse olfactory bulb. Dashed lines indicate layer of mitral cell bodies. The peripheral dendrites ramify in the EPL, above the mitral body layer, according to type: type I, throughout the EPl; type II, deep zone of EPL; type III, superficial zone of EPL. Note that the dendrites are covered with spines. (Modified from Greer, 1987.)

CELL POPULATIONS

Olfactory sensory neurons in one side of the nose number approximately 50 × 10^6 in the rabbit, giving rise to as many axons entering each olfactory bulb. This is a relatively large array of sensory input channels, exceeded only by the retina. This array converges onto approximately 2,000 glomeruli, to which are connected approximately 50,000 mitral cells and 100,000 tufted cells (see Allison,

1953). Thus, the convergence ratios are very high: onto glomeruli, 25,000 : 1; onto mitral cells, 1,000 : 1; and onto tufted cells, 500 : 1.

The ratios of intrinsic neurons to principal neurons are also high. Some order-of-magnitude estimates for these ratios are: PG to mitral, 20 : 1; granule to mitral, 50–100 : 1; short-axon to mitral, 1 : 1 (Shepherd, 1972a). Better data are needed for different species. However, even these rough estimates suggest an extensive array of intrinsic circuits for information processing in the bulb.

SYNAPTIC CONNECTIONS

The distinct laminae and cell types of the olfactory bulb have greatly simplified the electron microscopical analysis of synaptic connections. In many cases, identification of processes has been further confirmed by serial reconstructions. Because of these advantages, the olfactory bulb was among the first brain regions in which identification of the main patterns of synaptic connection was on a secure basis. These patterns are found in the four main layers of synaptic neuropil in the bulb, which we will describe in sequence.

GLOMERULAR LAYER

Intraglomerular connections. Within the glomeruli the olfactory sensory terminals make axodendritic synapses onto the dendritic tufts of both the relay neurons (mitral and tufted cells) and the intrinsic neurons (PG cells) (Pinching and Powell, 1971a,b). As shown in Fig. 5.5A, the axon terminals are relatively large, especially compared with the thin axons from which they arise. The terminals are filled with small round vesicles. The contacts are type 1 chemical synapses. In at least one strain of mouse (BALB/c), the sensory connection onto the PG cell dendrites is absent. This finding has been documented by White (1972), using serial reconstructions. A possibility is that this strain lacks a PG cell subpopulation that receives the olfactory input. An interesting question is whether this genetic difference in synaptic connectivity is associated with behavioral peculiarities expressed by this strain.

The dendrites within the glomerulus not only receive the sensory input but are themselves presynaptic in position. The most common pattern is dendrodendritic contacts from mitral/tufted cells to PG cells; these are type 1 synapses. The presynaptic dendrites contain only a few synaptic vesicles at the synaptic sites, in contrast to the sensory axon terminals (see above). Another common pattern is dendrodendritic contacts in the opposite direction, from PG cells to mitral/tufted cells; these are type 2 synapses. As indicated in Fig. 5.3, the two types of synapse may be arranged in reciprocal, side-by-side pairs (approximately 25% of the total), or in more widely spaced serial sequences. The type 2 dendrite may in turn receive a type 2 synapse from another, presumably also PG, dendrite. No dendrite has been observed to be presynaptic to an axon terminal [in contrast to dorsal horn (Gobel et al., 1980) and retina (Dowling and Boycott, 1966)]. Patterns of connections form complex microcircuits involving groups of axons and dendrites, sometimes set apart by an incomplete glial wrapping (see below).

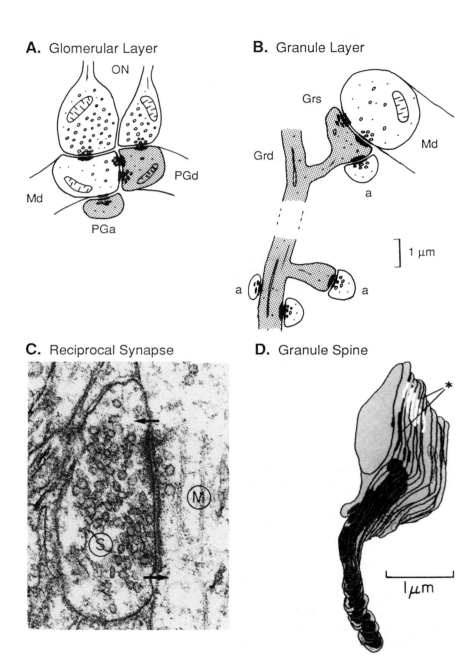

A. Glomerular Layer

ON

Md

PGd

PGa

B. Granule Layer

Grs

Grd

Md

a

a   a

] 1 μm

C. Reciprocal Synapse

M

S

D. Granule Spine

*

1 μm

FIG. 5.5. Synaptic connections in the mammalian olfactory bulb. **A.** Axodendritic and dendrodendritic connections in the olfactory glomerulus. ON, olfactory nerve; Md, mitral dendrite; PGd, periglomerular cell dendrite; PGa, periglomerular cell axon. Note the serial and reciprocal synaptic sequences. **B.** *Above:* Dendrodendritic connections in the external plexiform layer between a mitral secondary dendrite (Md) and a granule cell dendritic spine (Grs); note also the centrifugal axodendritic connection onto the spine. Grd, granule dendrite. *Below:* Axodendritic connections in the granule layer. **C.** Electron micrograph showing the fine structure of a reciprocal synapse between a mitral cell

Finally, there are terminals of certain types of centrifugal fibers (from the dorsal raphe).

Interglomerular connections. In addition to the PG cell bodies there is an extra-glomerular neuropil between the glomeruli. Several types of synaptic connections occur here (Pinching and Powell, 1971a,b). First, the axons of PG cells make type 2 synapses onto the somata and dendrites of other PG cells, and onto the primary dendrites of mitral and tufted cells as they emerge from the glomeruli. Second, there are type 1 synapses from tufted cell axon collaterals onto tufted cell dendrites. Third, there are synaptic terminals of various types of centrifugal fibers.

EXTERNAL PLEXIFORM LAYER (EPL)

In the EPL, the dominant type of synaptic connection is a pair of reciprocal contacts (Hirata, 1964; Andres, 1965; Rall et al., 1966). Serial reconstructions (Rall et al., 1966) established that these contacts occur, as indicated in Fig. 5.5B, between the secondary dendrite (Md) of a mitral/tufted cell and the spine (gemmule) of a granule cell dendrite (Grs). These were the first dendrodendritic synapses identified in the nervous system. An electron micrograph of a typical reciprocal synapse is shown in Fig. 5.5C. A recent reconstruction of a granule cell spine, bearing reciprocal synapses as indicated by (*), is shown in Fig. 5.5D. In the reciprocal pair, the mitral-to-granule synapse is type 1, whereas the granule-to-mitral synapse is type 2 (Price and Powell, 1970a). Over 80% of all synapses in the EPL are involved in such reciprocal pairs. Electron micrographs show the EPL to be a neuropil composed almost entirely of mitral/tufted and granule cell dendrites and their synaptic interconnections (see Reese and Shepherd, 1972). If we consider that there are up to a hundred granule cells for each mitral cell, and that each granule cell has 50–100 spines in the EPL (Greer, 1987; Mori, 1987), it is obvious that these dendrodendritic microcircuits provide for extremely powerful and specific interactions with the mitral cells. Because the basal dendritic fields of most mitral and tufted cells occupy separate zones in the EPL, their microcircuits through granule dendrites are correspondingly separated.

In single EM sections the granule-to-mitral/tufted synapse sometimes appears indistinct or missing, and Ramon-Moliner (1977) suggested that the inhibitory action of granule cells might therefore be mediated by a nonsynaptic mechanism. However, Lieberman and his colleagues carefully reinvestigated the question

---

dendrite (M) and a granule cell spine (S) in the mouse. Note that M → S synapse has round presynaptic vesicles and thick postsynaptic densities, indicative of type 1, whereas S → M synapse has flattened presynaptic vesicles and more symmetrical membrane densities, indicative of type 2. **D.** Computer reconstruction of serial EM sections through a granule spine, showing long, thin neck and larger head which bears a reciprocal synapse (asterisk). (A after Andres, 1965; Reese and Brightman, 1970; Pinching and Powell, 1971a,b; White, 1972; B after Rall et al., 1966; Price and Powell, 1970a; C,D, from Greer et al., 1989.)

(Jackowski et al., 1978) using several EM techniques, and confirmed that the granule-to-mitral synapses are approximately equal partners in the reciprocal pairs, as originally described. This of course does not rule out the possibility of additional nonsynaptic interactions between the cells.

In addition to synapses made by dendrites, the EPL also contains axon terminals from several sources: intrinsic short-axon cells, and centrifugal fibers (as noted above, recurrent mitral collaterals, thought by Cajal to terminate in the EPL, are now believed to be restricted to the IPL). These axon terminals make type 1 synapses predominantly on the presynaptic granule spines (Price and Powell, 1970a); no contacts have been seen on mitral dendrites.

Development and plasticity. How is the dendrodendritic microcircuit assembled during development? The mechanism in fact appears to be relatively simple. According to Hinds (1970), the mitral (or tufted)-to-granule synapse appears first, at about E17 in the mouse, followed a day later by the granule-to-mitral (or tufted) synapse. Whether each neuron has its own genetic timetable of synapse expression, or the earlier mitral synapse induces the granule synapse, requires further study. The later expression of granule synapses is consistent with the general rule that intrinsic neurons develop later than projection neurons (Jacobson, 1978). In the retina, more complex microcircuits also appear to be assembled according to a similar genetic algorithm consisting of sequential expression of individual synaptic types (Nishimura and Rakic, 1987).

Recent studies have revealed a great deal of plasticity in these microcircuits. In the mutant mouse strain *pcd* (i.e., Purkinje cell degeneration), the specific degeneration of mitral, but not tufted, cells occurs at about three months of age (Greer and Shepherd, 1982). The number of reciprocal synapses between granule spines and tufted cell dendrites increases in compensation, suggesting that many denervated granule-to-mitral spines survive and establish new efferent and afferent synapses with tufted cell dendrites (Greer and Halasz, 1987). By contrast, olfactory deprivation causes a reduction in incidence of dendrodendritic synapses, more severe in the granule-to-mitral contacts (Benson et al., 1984). It has been suggested that "the reciprocal pair of synapses exists in a dynamic equilibrium in which each sustains the other through trophic or feedback mechanisms" (Shepherd and Greer, 1988).

GRANULE LAYER

In the granule layer, axon terminals are found on the shafts and spines of the granule cell dendrites (see Fig. 5.5B). The studies of Price and Powell (1970b) showed that these axon terminals derive from both intrinsic and extrinsic (central) inputs. The *intrinsic* sources include the axon collaterals of mitral and tufted cells, and the axons of the deep short-axon cells. There is evidence that the synapses of these terminals are types 1 and 2, respectively. The *extrinsic* sources make connections at different levels of the granule dendritic tree (see Fig. 5.2A). The AC distributes mainly to the deep processes. The AON distributes over the middle part of the dendrites, including the spines in the EPL. The HDB axons distribute mainly to the spines in the EPL; some terminals are also found at the borders of the glomeruli. The synapses made by these inputs from the brain appear to be type 1.

It should be noted that all the synaptic connections in which the granule cell takes part are oriented toward the granule cell, with the sole exception of the dendrodendritic synapses from the granule spines onto the mitral dendrites in the EPL. The latter are, therefore, the only output avenue from the granule cells.

GLIA

The olfactory bulb provides favorable opportunities for analyzing the relations between glial membranes and neuronal elements and synapses. The incoming olfactory axons are gathered into bundles of 100–200 axons contained within one glial (Schwann) cell. The olfactory Schwann cell has several interesting molecular properties. It contains glial fibrillary acidic protein (GFAP), like central astrocytes but unlike Schwann cells of peripheral nerves (Barber and Lindsay, 1982). The olfactory nerve layer in the olfactory bulb is also one of the few sites in the brain where laminin continues to be expressed into adulthood (Liesi, 1985); this may reflect the continued remodeling associated with turnover of the olfactory axons and terminals. Within the glomeruli, a synaptic complex, such as that shown in Fig. 5.5A, is often seen to be surrounded by one or more loose folds of glial membrane; this is similar to, though not nearly as distinct as, the synaptic glomeruli of the cerebellum and thalamus (Pinching and Powell, 1971a,b). Glial folds are also sometimes seen around the reciprocal synapses in the EPL.

Within the EPL, in most vertebrate species, several loose folds of glial membrane surround the primary dendrites of mitral and tufted cells near the glomerular boundary. In mice and primates, it has been found that the folds of membrane are packed into typical myelin, which may surround not only the primary dendrite but even extend to the cell body in the case of tufted cells (Pinching, 1971; Burd, 1980). This shows that a dendrite may be myelinated and that myelin is not exclusively associated with axons. The function of the myelin around dendrites is not known.

BASIC CIRCUIT

The synaptic organization of the olfactory bulb is summarized in the basic circuit diagram of Fig. 5.6. The overall plan depends on the fact that the mitral cell spans the layers, receiving the sensory input in its glomerular tuft and giving rise to the bulbar output from its cell body. The two main functions of *input processing* and *output control,* that characterize all local regions of the brain (Chap. 1), are therefore separated into two distinct levels. We will summarize the organization at these two levels.

INPUT PROCESSING

Intraglomerular microcircuits. The olfactory glomeruli are the most characteristic feature of the olfactory bulb in the vertebrate series. Within the glomeruli, three neuronal elements come together: olfactory axon, mitral/tufted

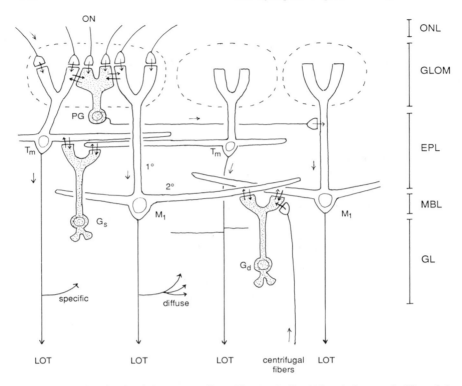

FIG. 5.6. Basic circuit of the mammalian olfactory bulb. Abbreviations as in Figs. 5.1 and 5.2.

cell dendrite, and PG cell dendrite. In more general terms, the three elements are input fiber, principal neuron, and intrinsic neuron. These are the basic elements in the synaptic organization of the local regions of the brain, and as noted (Chap. 1) we refer to them as a *synaptic triad*. In general, the synapses between the input and principal elements provide the necessary basis for input–output transmission, whereas synapses between the principal and intrinsic elements provide for elaboration and control of the input–output transfer.

The synaptic triad in the glomerulus involves synapses from the input onto both the principal and (in most species) the intrinsic elements. This is a common pattern in the brain; the same type of arrangement is found, for example, in the retina, cerebellum, and thalamus (Shepherd, 1979). In the latter regions (see relevant chapters), the synapses onto the principal and intrinsic elements arise from a single large input terminal, whereas, in the olfactory glomeruli, the synapses are made by separate terminals. The arrangement in the olfactory bulb appears to permit considerable combinatorial complexity. One needs to know if the separate terminals arise from separate olfactory axons; if so, do some olfactory receptors project only to the principal neuron, others only to the intrinsic neuron?

After the initial input to the principal and intrinsic elements, further processing takes place within the glomerulus through the dendrodendritic microcircuits. There is increasing evidence that a glomerulus may function to some extent as a functional unit. This has been long suspected on anatomical grounds (see Clark, 1957). Tract-tracing studies indicate that there may be several levels of organization within a single olfactory glomerulus (Land and Shepherd, 1974). There is physiological evidence for glomerular specificity for different olfactory stimuli (Leveteau and MacLeod, 1966). Activity mapping with the 2-deoxyglucose (2DG) technique indicates strongly that there are groups of glomeruli with similar specificities for a given odor (Sharp et al., 1977), and high-resolution 2-DG methods show a relatively homogeneous distribution of activity throughout a glomerulus during odor stimulation (Benson et al., 1985). If glomeruli have this functional specificity, then the group of mitral, tufted, and PG cells with dendrites connected to a particular glomerulus will all share this specificity. This implies a horizontal constraint on the organization of functionally related neurons in the bulb, which may be analogous to the functional columns of the cerebral cortex (Shepherd, 1972). The significance of the glomerulus in processing information about sensory molecules will be discussed further below.

Interglomerular microcircuits. Activity in one glomerulus can affect other glomeruli through interglomerular connections. The main route is PG cell axons, which through their terminals can affect the transmission of information out of neighboring glomeruli (see Fig. 5.6). There is evidence that these interglomerular actions may be excitatory (Shepherd, 1963; Freeman, 1974) or inhibitory (Getchell and Shepherd, 1975a,b). If glomeruli function as units, then one action of the interglomerular microcircuits could be to enhance the contrast between glomeruli of different specificities, in analogy with intercolumnar interactions in neocortex (see Chap. 12).

OUTPUT CONTROL

At the second level of organization, the main type of microcircuit is the reciprocal dendrodendritic synapse between mitral/tufted cells and granule cells. At this level, the mitral/tufted primary dendrite functions as the afferent element of the triad, conveying the input directly to the soma and secondary dendrites, which are the principal neuron component. The triad is completed by the intrinsic element, the granule cell spine.

As we shall soon see, there is strong evidence that in the reciprocal microcircuit, the mitral-to-granule synapse is excitatory and the granule-to-mitral synapse is inhibitory. Because the granule-to-mitral synapse is the sole output of the granule cell, the inhibition it delivers is very powerful, and is the main means for mediating control of output from the olfactory bulb. Although the reciprocal microcircuit seems to be a simple and inflexible arrangement, in fact it can generate several types of functions. The most obvious are self- and lateral-inhibition of the mitral/tufted cells, as will be discussed further below.

The spatial constraints on the circuits controlling output obviously contrast with those involved in input processing. As we have seen, input processing is organized according to distinct glomeruli, each presumably limited to processing some specific stimulus domain. By contrast, output control is mediated through mitral/tufted secondary dendrites, whose fields are extensive and overlapping. A given output neuron can therefore be modulated according to a graded summation of effects from other output neurons. This modulation may be important in determining not only patterns of excitation and inhibition in space, but also sequences of excitation and inhibition in time.

Parallel output pathways. As the diagram of Fig. 5.6 indicates, the mitral and tufted cells form two distinct pathways through the olfactory bulb. It is not known whether, at the level of input processing within the glomeruli, the dendritic tufts of the two types receive input from different receptor cell axons, or if they interact with common (as shown) or different PG cell dendrites. However, it is now known that at the level of output control in the EPL, each type is dominated by different subpopulations of granule cells: Superficial granule cells (G_S) control superficial and middle tufted cells (T_M), and deep granule cells (G_D) control mitral cells (M_1); granule cells forming a third subpopulation appear to interact with both tufted and mitral cells. These cell types can be identified in the drawings of Figs. 5.2–5.4.

When differing projection sites of mitral and tufted cells in olfactory cortical areas were first recognized, it was suggested by Skeen and Hall (1977) that there might be an analogy in this regard with the different classes of retinal ganglion cells. The differing morphologies of the dendritic trees of these cells further support that analogy (Macrides and Schneider, 1982). In discussing that analogy, Orona et al. (1984) noted that, in the retina, the particular sublamina of dendritic ramification of a ganglion cell has been found to be the main morphological feature correlated with the physiological type of its response. The fact that both mitral and tufted cells are further divided into subclasses on the basis of dendritic morphology indicates that multiple parallel pathways exist, which may be important in the mediation of different types of information about molecular stimuli. Thus, within the main olfactory bulb are embedded multiple parallel pathways. Also, in many mammals, a modified glomerular complex forms a separate pathway within the main olfactory pathway, believed to mediate information concerning odor cues related to suckling in young animals (Pedersen et al., 1987). In most vertebrates there is, in addition, a vomeronasal–accessory olfactory pathway which acts in parallel with the main olfactory pathway. In the adult, this pathway has been implicated in mating behavior. Utilization of parallel pathways is therefore an important principle in processing olfactory information.

CENTRIFUGAL MODULATION

In addition to processing sensory information, the bulbar microcircuits are also involved in gating and modulating that information by the brain. A key site for this control is the dendritic spine of the granule cell. As can be seen in Fig. 5.6, a synaptic triad is formed by the centrifugal fiber terminal, the granule spine, and the mitral/tufted dendrite. Through this connection, the centrifugal fiber can exert direct and exquisite control over the function of the reciprocal microcircuit. The nature of that control will be discussed further below.

From these considerations it appears that mitral and granule cell synapses are concerned both with olfactory processing and with integration of information passing forward from the brain through the granule cell. Some of the information from the brain may be in the form of feedback through long loops from the olfactory projection areas. Some of it, however, may be in the form of nonolfactory signals from hypothalamic and limbic structures. The granule-to-mitral synapse is, therefore, of interest as a specific site at which there is an overlap of functions. One may characterize it in this regard as a *multifunctional*, or *multiplex*, synapse.

A second level of centrifugal control occurs in the glomerular layer. Most of the centrifugal fibers that reach this layer have connections restricted to the interglomerular neuropil. However, the serotonergic fibers from the dorsal raphe enjoy a special privilege of being allowed to enter and ramify within the glomeruli. This presumably enables these fibers to modulate in a very specific way the initial processing of the olfactory input to the bulb.

SYNAPTIC ACTIONS

The physiological actions of synapses within local circuits can be revealed in two ways, by analyzing their responses to electric shock stimulation or to natural sensory stimuli. The separate input and output pathways of the olfactory bulb are attractive for precise electrical stimulation, allowing sequences of synaptic actions to be studied. We will first describe these results, as a basis for understanding the properties of local circuits as they relate to processing of natural stimuli.

Modern analysis of synaptic actions in most brain regions relies on in vitro preparations. The isolated olfactory bulb of the turtle has been useful for this purpose; indeed, the entire brain can be removed and maintained in a dish of oxygenated Ringer solution for many hours (Nowycky et al., 1978). In the whole-bulb preparation, one does not have access to cells in different layers under direct observation, as in slice preparations (see Chap. 2), but, on the other hand, there is the great advantage that one can analyze synaptic actions without cutting through nerve processes and disturbing local circuits.

GLOMERULAR SYNAPTIC ACTIONS

A single shock stimulus of the olfactory nerves gives rise to a series of orthodromic potentials in the mitral cell. As shown in Fig. 5.7A(a), this consists of an initial excitatory–inhibitory sequence $(E_1–I_1)$, followed by a second $E_2–I_2$ sequence. These potentials are graded in amplitude with stimulus strength, and represent synaptic potentials. E_1 and E_2 reflect the excitatory synaptic response in the glomerulus. I_1 and I_2 are associated with an increased conductance, and have reversal potentials near resting potentials (Mori et al., 1981b). They appear to be due to feedback and lateral inhibition mediated by the dendrodendritic synaptic interactions between mitral and granule cells. Both of the mechanisms will be discussed further below.

When sufficiently strong, E_1 gives rise to an action potential (Fig. 5.7A,b). The action potential at the cell body has been shown to be generated by both Na^+ and Ca^{2+} currents (Mori et al., 1981a; Jahr and Nicoll, 1982). As described in

A. Fast Orthodromic Response

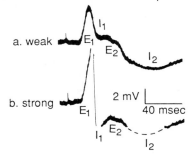

a. weak

I_1
E_1
E_2 I_2

b. strong

E_1

2 mV
40 msec

I_1 E_2
I_2

B. Fast Antidromic Response

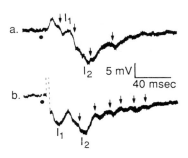

a.

I_1

I_2 5 mV
40 msec

b.

I_1
I_2

C. Slow Inhibitory Response

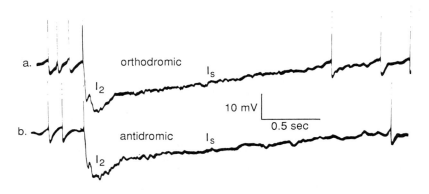

a.

orthodromic
I_s

I_2

10 mV
0.5 sec

b.

antidromic I_s

I_2

D. Slow Excitatory Response (Orthodromic)

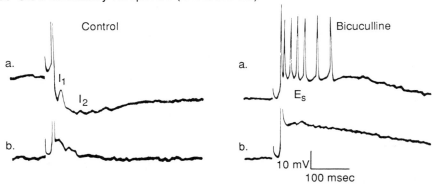

Control

a.

I_1
I_2

b.

Bicuculline

a.

E_s

b.

10 mV
100 msec

FIG. 5.7. The main types of synaptic actions on mitral cells. These are intracellular recordings from the in vitro turtle olfactory bulb, showing responses to a single volley in the olfactory nerves (orthodromic activation) or lateral olfactory tract (LOT) (antidromic activation). **A.** The fast orthodromic response, just below (a) and above (b) threshold for impulse generation. The synaptic potentials consist of excitatory (E_1 and E_2) and inhibitory (I_1 and I_2) components. **B.** The fast antidromic response, similarly just below (a) and

152

Chap. 2, voltage-gated Ca^{2+} influx is an important means of activating K^+ conductances involved in regulating impulse frequency. In the mitral cell, it has a further significance in view of the role of Ca^{2+} in controlling neurotransmitter release from the presynaptic soma and dendrites (cf. Chap. 1; see further, below).

In order to identify the components of this complex response that are due to glomerular synapses, we can make a comparison with the response to antidromic activation. A single shock stimulus to the lateral olfactory tract sets up an impulse volley which invades the mitral cells antidromically. Responses just below and just above threshold for the axon of the cell from which one is recording are shown in Fig. 5.7B. This antidromic response lacks the excitatory synaptic potentials E_1 and E_2, but shows a sequence of inhibitory synaptic potentials I_1 and I_2 that are similar to those in the orthodromic response. This is to be expected if these inhibitory potentials are generated in the secondary dendrites of the mitral cells by the actions of granule cells (see below). In favorable recordings, small hyperpolarizing potentials are seen (downward arrows in Fig. 5.7B), which may represent unitary inhibitory potentials.

Following these early synaptic events, there is a long-lasting hyperpolarization, referred to as I_s. As shown in Fig. 5.7C, it is elicited by both antidromic and orthodromic stimulation. It may last up to 10 sec or more, and the effects of the inhibition may be seen for up to 1 min. The later part of this potential is not associated with an increased membrane conductance, and a reversed potential cannot be obtained (Mori et al., 1981c). This potential appears also to be generated in mitral cell secondary dendrites by granule cells, though by a separate transmitter mechanism (see Transmitters, below).

Further analysis of the glomerular components of the orthodromic synaptic responses has had to rely on soma recordings, because intracellular recordings from the glomerular tuft have not yet been possible. When inhibitory potentials are blocked by either bicuculline or low Cl^-, a prolonged EPSP (E_s) is revealed (Fig. 5.7D). This response is not seen when the cell is activated antidromically, and has therefore been ascribed to a large EPSP generated in the glomerular tuft of the mitral cell. When unopposed by inhibition, this EPSP renders the mitral cell hyperexcitable so that it responds with a burst of impulses (Fig. 5.7D,a). This is similar to findings in many other regions, and demonstrates how important inhibition is in controlling and shaping the responses of a neuron.

Studies of this kind have suggested several types of synaptic actions within the glomeruli. These include (1) the large EPSP (E_s) set up by the olfactory nerve

---

above (b) threshold for impulse generation in the LOT axon. Arrows indicate small, unitary hyperpolarizing potentials. **C.** The slow inhibitory response (I_s), elicited by orthodromic (a) and antidromic (b) stimulation. **D.** A slow excitatory response (E_s) to orthodromic input is revealed by adding bicuculline, a $GABA_A$ blocker, to the medium. The control shows the response at the normal resting potential (a) and during hyperpolarizing current injection, which reversed the I_1 and I_2 components and reduced but did not reverse I_s. The bicuculline traces show the hyperpolarizing current injection, which increased the amplitude of E_s. (A,B from Mori et al., 1981b; C from Mori et al., 1981c; D from Nowycky et al., 1981b.)

synapses onto mitral/tufted cells; (2) prolongation of the excitatory response, possibly by autoreceptors; (3) inhibition of the glomerular tuft, possibly by serial and reciprocal microcircuits through synapses from inhibitory PG cell dendrites; and (4) additional prolongation of the excitatory response, possibly by intrinsic axon terminals on mitral/tufted dendrites. It can be seen that all of these actions are correlated with the synaptic connections identified in anatomical studies (see above). They remain to be verified by combined anatomical and physiological studies in the same animal.

An important general point is the wealth of synaptic integration, involving both excitation and inhibition, taking place in the most distal part of the mitral/tufted dendritic tree. This is the level of input processing mentioned above. It reemphasizes that inhibition is important in distal dendrites as well as at the cell body, and that a significant amount of synaptic integration and sensory processing occurs locally in distal dendritic branches before transmission to the site of impulse output in the cell body.

DENDRODENDRITIC INHIBITION BY GRANULE CELLS

We turn next to consider synaptic actions at the level of the cell body and secondary dendrites of mitral/tufted cells—that is, at the level of output control. The predominant property is inhibition, mediated by the reciprocal synapses between mitral/tufted and granule cells.

The evidence for the reciprocal microcircuit and its physiological actions came in several steps. First, intracellular and extracellular unit recordings showed that antidromic invasion of mitral cells is followed by long-lasting inhibition, and it was suggested that this is mediated by the granule cell dendrites operating in an output mode (Phillips et al., 1963). The next step (Rall et al., 1966; Rall and Shepherd, 1968) was a computational model of the intracellular potentials and the extracellular field potentials generated by the mitral and granule cells (Rall et al., 1966; Rall and Shepherd, 1968). This indicated that the mitral secondary dendrites are likely to be the pathway for synaptic excitation of the granule cell spines, which then mediate the long-lasting inhibition of the mitral cells. The final step was an electron micrographic study, using serial reconstructions, which showed that the mitral secondary dendrites and granule spines are interconnected by reciprocal synapses appropriately located for mediating the interactions predicted by the model (Rall et al., 1966). The excitatory mitral-to-granule synapse was shown to be type 1 and the inhibitory granule-to-mitral synapse type 2 by Price and Powell (1970a) (see Fig. 5.5 above).

The function of the reciprocal synapses as a microcircuit module was previously described in Chap. 1. The sequence of synaptic actions, based on the computational model of Rall and Shepherd (1968), is illustrated in greater detail in Fig. 5.8. In A (diagram 1), depolarization (D) of the mitral cell dendrite, by spread of an antidromic or orthodromic impulse from the cell body, activates the excitatory ($\mathcal{E}$) synapse onto the granule spine (shaded). The EPSP in the spine activates the inhibitory ($\mathcal{I}$) synapse back onto the mitral dendrite (diagram 2). This causes a hyperpolarizing (H) IPSP in the mitral dendrite (diagram 3), which suppresses the excitability of the mitral cell until the inhibitory action has worn off. The long-lasting nature of this inhibitory action, in the absence of impulse activity, was early recognized (Phillips et al., 1963; Rall and Shepherd, 1968). Most of reciprocal synapses are located on the secondary mitral/tufted cell dendrites, where they are activated by the spread of either antidromic or orthodromic impulses from the cell body, as shown in Fig. 5.8.

Physiological testing of the model. The isolated turtle brain preparation, with its preservation of intrinsic circuits, has permitted direct physiological testing of the model. Jahr and Nicoll (1982) carried out a particularly nice experiment to verify the presynaptic role of mitral cell dendrites. As shown in Fig. 5.9A, in a normal mitral cell, injected depolarizing current elicited a fast spike (left, above), which was followed by a slow IPSP (left, below). When tetrodotoxin (TTX) was added to the bath to block Na$^+$ conductance (Fig. 5.9B), the cell responded with a smaller, slower spike (above), presumably due to Ca^{2+} ions. This spike was still

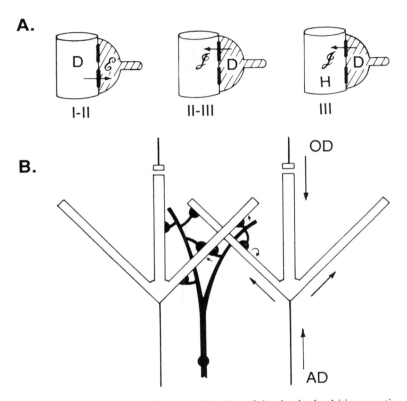

FIG. 5.8. **A.** Postulated mechanisms of action of the dendrodendritic synaptic pathway between mitral (open) and granule (shaded) cells, during successive time periods I, II, and III following an antidromic volley. D, depolarization; H, hyperpolarization; $\mathcal{E}$, excitation; $\mathcal{I}$, inhibition. **B.** Diagram of the pathways for self- and lateral-inhibition through dendrodendritic connections. OD, orthodromic (normal) activation; AD, antidromic activation. (From Rall and Shepherd, 1968.)

A. Control   **B.** TTX,TEA   **C.** TTX, TEA, BMI

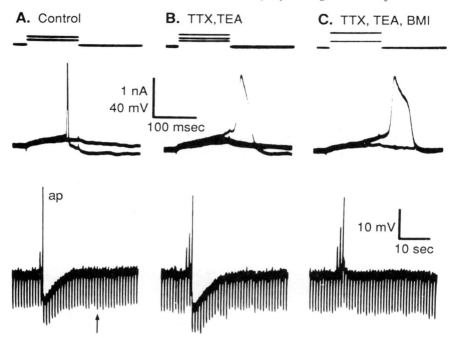

FIG. 5.9. An electrophysiological test of the reciprocal dendrodendritic microcircuit between mitral and granule cells. These tracings are intracellular recordings from a mitral
cell in the in vitro turtle olfactory bulb. The top trace is a monitor of the depolarizing
pulses injected into the cell. The middle trace is the intracellular recording of the response
of the cell on a fast sweep. The bottom trace is a slow trace, showing the intracellular
impulse response (ap) and the hyperpolarizing pulses used to monitor input resistance
(arrow). **A.** Control responses, showing rapid spike (middle) followed inhibition (bottom). **B.** In TTX (to block Na$^+$ conductances) and TEA (to block K$^+$ conductances,
which would shunt Ca^{2+} currents), the spike is broader, but is still followed by inhibition.
C. Addition of bicuculline (BMI), a GABA$_A$ blocker, removes the inhibition, presumably
by blocking granule-to-mitral inhibition. See text. (Modified from Jahr and Nicoll, 1982.)

followed by the slow IPSP (below). [Tetraethylammonium (TEA) was also added
to block K$^+$ conductance, which otherwise shunts out the depolarizing Ca^{2+}
conductance.] When bicuculline (GABA$_A$ blocker) was added to the bath, the
mitral cell still responded with an impulse, but the IPSP was eliminated (Fig.
5.9C). Because the only active mitral cell was the one injected, this experiment
shows that the mitral cell acts as a presynaptic terminal to the granule cell; that
the circuit is recurrent onto the injected cell; and that the inhibitory transmitter is
GABA.

As we have seen, regenerative calcium conductances are important properties
of dendritic membranes. In many neurons they play key roles in promoting
intradendritic transmission and synaptic integration (see Chap. 2, and cerebellar
Purkinje cells, Chap. 6). The experiments of Figs. 5.8 and 5.9 show that in
addition, they may control Ca^{2+} entry and synaptic transmitter release in pre-

synaptic cell bodies and dendrites. In its presynaptic role, the mitral cell soma-dendrite can be considered analogous to the nerve terminal at the neuromuscular junction (NMJ). The NMJ has multiple active zones which are activated by an impulse to release ACh into the postsynaptic muscle endplate membrane; the mitral somadendrite has multiple synapses which are activated by an impulse to release its neurotransmitter onto postsynaptic granule cell spines. The mitral cell is therefore an attractive model for analysis of presynaptic release mechanisms at a central synapse.

Rhythmic activity. The sequence of dendrodendritic interactions described above is of additional interest in that it provides the basis for the generation of rhythmic activity in the neuronal populations of the bulb (Rall and Shepherd, 1968). The proposed mechanism will be briefly described.

As illustrated in Fig. 5.10, the sequence begins with a long-lasting EPSP in the mitral dendritic tufts (MT) in the glomeruli, due to the olfactory nerve input or the intrinsic activity at the glomerular level. The first mitral impulse generated by the EPSP (MC) synchronously activates all the granule cells with which that mitral cell has synaptic connections (GR). These deliver feedback inhibition of the activated mitral cell, and feedforward inhibition of neighboring inactive mitral cells, in the way already described relative to Fig. 5.8. As the mitral IPSP subsides, a point is reached at which the EPSP is again at threshold; an impulse is again initiated, and the cycle repeats itself. Through the extensive interconnec-tions between the mitral and granule cells, a steady input in the glomeruli is converted into a rhythmic impulse output in the mitral cell population, locked to a rhythmic activation of the granule cell population.

The activity in these populations generates electric current, which spreads through the cells according to the electrotonic properties to be described below.

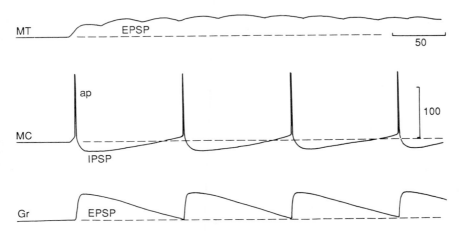

FIG. 5.10. Postulated mechanism whereby the dendrodendritic pathway may provide for rhythmic activity in the olfactory bulb. Postulated intracellular potentials are shown for the mitral dendritic tuft in the glomerulus (MT), mitral cell body (MC), and granule cell (Gr). See text. (From Shepherd, 1979.)

The current paths of the individual neurons summate in the extracellular spaces in and around the olfactory bulb and give rise, thereby, to summed extracellular potentials, which are recorded by an electroencephalograph (EEG). Such rhythmic EEG potentials are a prominent characteristic of the olfactory bulb in the resting state as well as during olfactory-induced activity (Adrian, 1950), and explicit models have been developed (Freeman, 1975).

NEUROTRANSMITTERS

One might suppose that the olfactory bulb, with its relatively stereotyped architecture, would have a simple set of neurotransmitters to mediate its synaptic interactions. In fact, the opposite is true: The bulb is enormously rich in neuroactive substances, rivaling any other brain region in this respect. Why this should be so is not understood, but it is likely related to the fact that the bulb mediates information about crucial behaviors such as feeding, social organization, and reproduction, which are dependent on a number of behavioral state variables.

A summary of the putative neurotransmitter and neuromodulator substances of the main type of neurons is shown in Fig. 5.11. We will discuss briefly the evidence as it relates to the main types of synaptic connections within the basic circuit (reviewed in Halasz and Shepherd, 1983; Macrides and Davis, 1983).

OLFACTORY SENSORY NEURONS

The olfactory sensory neurons are of special interest because they continue to turn over from stem cells in the neuroepithelium during adult life (Graziadei and Monti-Graziadei, 1979). They are thus one of the few examples in the mammal of continued genesis of neurons that make synaptic connections in the central nervous system. The molecular basis of this property is not yet known. The sensory neurons contain a special peptide (olfactory marker protein; OMP) and have a high concentration of the dipeptide carnosine (reviewed in Margolis, 1988). Electrophysiological experiments show that the olfactory axons are excitatory (see above), but neither OMP nor carnosine has been shown to be neuroactive at this synapse. Numerous studies have generated monoclonal antibodies to all or part of the sensory neuron population, indicating the presence of molecules that may be involved in the sensory transduction of olfactory molecules as well as the specificity of synaptic connections (reviewed in Morgan, 1988). Some of these molecules appear to be unique to subsystems of olfactory neurons (cf. Akeson, 1988).

PERIGLOMERULAR CELLS

The first definitive study of glomerular transmitters showed, by histofluorescence and EM autoradiography, that some periglomerular (PG) cells are positive for dopamine-synthesizing enzymes (Halasz et al., 1977). The dopamine is transneuronally regulated; degeneration and regeneration of olfactory nerves following olfactory nerve transection are paralleled by a decrease and increase of DA and dihydroxyphenylacetic acid (DOPAC) levels in the olfactory bulb (Baker et al., 1983).

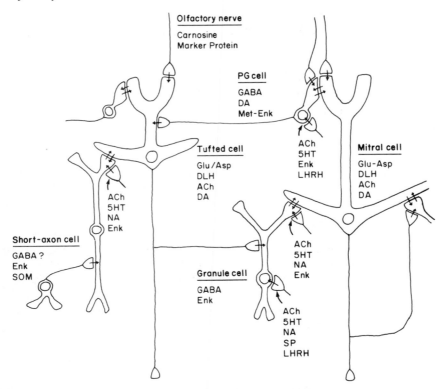

FIG. 5.11. Schematic diagram of the olfactory bulb summarizing the synaptic circuits and the main neurotransmitter and neuromodulator substances that have been proposed for the several types of neurons. The identified neuronal elements are, at least in the case of the PG cells, fractionated into subtypes on the basis of their different neuroactive substances. The identified terminals associated with several substances are meant to indicate centrifugal fibers from the forebrain and midbrain, each type presumably containing a different substance. Figure is based on numerous sources (see text). ACh, acetylcholine; DA, dopamine; DLH, DL-homocysteate; Enk, Met-enkephalin; Glu-Asp, glutamate–aspartate; 5HT, 5-hydroxytryptamine; LHRH, luteinizing hormone releasing hormone; NA, noradrenaline; SOM, somatostatin; SP, substance P. (From Halasz and Shepherd, 1983.)

It has also been found that some PG cells and their dendrites contain glutamic acid decarboxylase (GAD), the GABA-synthesizing enzyme, as well as taking up GABA (Ribak et al., 1977). Comparative studies have shown that DA and GABA may be colocalized in the same PG cell in some species and restricted to separate PG cell subpopulations in others (Mugnaini et al., 1984; Baker, 1988; see Fig. 5.11). This demonstrates two principles: (1) A single morphological cell type may be fractionated into more than one neurotransmitter subtype, and (2) the mix of neurotransmitters in a single morphological cell type is phylogenetically flexible. Dopaminergic and GABAergic subpopulations of intrinsic neurons are also seen in other parts of the nervous system (cf. retina). As pointed out by Oertel et al. (1982), this may reflect an important principle of functional organization of synaptic circuits. PG cells have been postulated to have either excitatory or inhibitory synaptic actions at their dendrites and axons (see above).

MITRAL CELLS

Studies of the mitral-to-granule excitatory synapse have emphasized the importance of knowledge of synaptic organization in interpreting neuropharmacological results. For example, microionophoresis of amino acids, particularly aspartate and DL-homocysteate, tends to inhibit mitral cell activity. At first, this appeared to be confusing, in view of the usual excitatory effects of these amino acids on other neurons. However, closer consideration suggested that these substances, when released from the micropipette, act to excite the granule cell spines, which then inhibit the mitral cells (McLennan, 1971; Nicoll, 1971). This serves as a model for the local interactions that must be taken into account in interpreting microionophoresis studies.

The candidacy of glutamate as the transmitter of mitral cell somadendritic output has been strengthened by the finding of NMDA receptors in the EPL of the olfactory bulb (Cotman et al., 1987). It is also supported by the evidence for glutamate and aspartate as the transmitter of the mitral axons in the olfactory cortex (see Chap. 10). Because the mitral cell has synaptic output from both its dendrites and axon, it furnishes a useful model for testing whether the same transmitter is released from all synapses of a neuron, a possibility known as Dale's law (Dale, 1935).

TUFTED CELLS

It is generally believed that tufted cells share the same neurotransmitters at their dendritic and axonal output synapses with mitral cells. However, some tufted cells appear to be dopaminergic, which makes them more similar to PG cells (Halasz et al., 1977). This may be correlated with other differences between mitral and tufted cells, as noted above.

GRANULE CELLS

The best evidence regarding the identity of a neurotransmitter in the olfactory bulb relates to the axonless granule cell. Numerous studies point to GABA as the neurotransmitter released by the dendrodendritic synapses of the granule spine onto the mitral/tufted cell dendrites. Granule cells take up GABA (Halasz et al., 1978). The GABA-synthesizing enzyme GAD has been localized to the granule cells and their dendritic spines by EM immunohistochemistry (Ribak et al., 1977).

In the in vitro preparation of the turtle olfactory bulb, the early hyperpolarizing components of the IPSP in a mitral cell have several GABAergic properties: They are blocked by bicuculline and low Cl^- in the bathing medium, are reversed by Cl^- filled electrodes, are associated with an increased conductance, and have clear reversal potentials (Mori et al., 1981b; Nowycky et al., 1981a,b; Jahr and Nicoll, 1982). These reflect properties of $GABA_A$ receptors. The later, very slow, inhibitory potential does not have a reversal potential, suggesting that it may be mediated by a different synaptic receptor (possibly $GABA_B$ receptors) or at a more distant locus on the mitral cell (Mori et al., 1981c).

PEPTIDES

The olfactory bulb offers a smorgasbord of delectable peptides. Besides carnosine and OMP, it is especially rich in taurine, thyroid hormone-releasing hormone (TRH), insulin, and cholecystokinin (CCK). However, absolute levels are not the only measure of the significance of a neuroactive substance. Location at a critical site in a synaptic circuit is even more significant. Examples of this are substance P, present in tufted cells, and enkephalins, present in both PG cells and granule cells. Nicoll et al. (1980) studied the effect of a stable enkephalin analogue D-Ala-Met[5]-enkephalin (DALA) in the in vitro turtle olfactory bulb. DALA in the bathing medium reduced the IPSPs induced in mitral cells, especially the recurrent inhibition elicited by intracellular activation of a mitral cell. They suggested that the primary action of enkephalins is to suppress inhibitory interneurons, thereby producing indirectly an increase in excitability of the principal neurons. This would be an example of disinhibition within a synaptic circuit (see Chap. 1), and illustrates again how an understanding of synaptic organization is essential for interpreting pharmacological actions.

Peptides are colocalized with neurotransmitters at most synapses in the olfactory bulb, but the significance is not yet understood. With regard to enkephalins, for example, there is evidence that some enkephalin is contained in granule cells, suggesting that it may be coreleased with GABA and have a direct action on mitral cells (Bogan et al., 1982; Davis et al., 1982). One possibility is that it produces the slow inhibitory potential (see above). There are, however, many types of interactions between transmitters and peptides, as discussed in Chaps. 2 and 3.

CENTRIFUGAL FIBERS

As noted previously, there are three main types of centrifugal fiber, each associated with a specific classical neurotransmitter.

Noradrenaline-containing fibers arrive from the locus coeruleus, and distribute mostly within the granule layer and IPL, as well as the glomerular layer (see Fig. 5.2 above). The actions of NA on mitral cells are complex. One action suggested by the ionophoretic studies of Jahr and Nicoll (1982) is that NA acts on granule cells to reduce their release of GABA.

Serotonin-containing fibers arrive from the dorsal raphe. They distribute preferentially to different laminae in different species (Takeuchi et al., 1982). Of special interest are the fibers that terminate within the glomeruli, thus permitting a brainstem system direct access to the initial level of input processing of olfactory signals. These could be significant in mediating behavioral-state variables set by brainstem systems involved in hunger, satiety, arousal, and sleep. Judging from ionophoresis experiments, the action of serotonin on mitral cells is inhibitory (Bloom et al., 1971).

Cholinergic fibers arrive from the basal forebrain, and distribute relatively evenly through the laminae. According to Rotter et al. (1979), the external plexiform layer "has the highest concentration of muscarinic receptors in the brain . . . and high levels occur in the glomerular layer, mitral cell layer, and the granule cell layer" as well. Because these fibers terminate mainly on granule cell spines, they are well placed to modulate the dendrodendritic inhibition of mitral cells at the level of output control. Ionophoresis of ACh in vivo has mostly depressant effects on mitral cell firing (Bloom et al., 1971), but ACh action has not yet been studied in vitro nor interpreted in relation to the specific details of synaptic organization.

DENDRITIC PROPERTIES

We have considered the anatomy of the neurons, the physiology of their synaptic actions, and the neurotransmitters mediating those actions, and are now in a position to consider how these properties are involved in the processing of information within the synaptic circuits of the olfactory bulb. Processing can be divided into two parts: synaptic integration that occurs within neurons, and synaptic interactions that occur between neurons. Synaptic integration involves *intra*neuronal transmission within dendrites, which we now consider; we then take up *inter*neuronal synaptic interactions in the final section. Each of the three main cell types of the bulb provides a model for illustrating important principles of dendritic signal processing.

MITRAL CELL

In studying the mitral cell, it soon becomes apparent that its dendritic tree is not one homogeneous unit. Rather, it is divided into several distinct anatomical entities, and each entity has its own distinct function. The *glomerular dendritic tuft* is concerned primarily with reception and processing of the olfactory input; it is analogous in this respect to the entire dendritic tree of a thalamic relay neuron. The *primary dendritic shaft* has as its main function the transfer of information from the glomerular tuft to the cell body; it is analogous in this respect to a retinal bipolar cell. The *secondary dendritic branches,* finally, are exclusively concerned with controlling bulbar output through interactions with the granule cells. These divisions are so distinct that one can regard the mitral cell as not one but three cells, transfer between them taking place through *intra*neuronal continuity rather than *inter*neuronal synapses. This means that we must assess dendritic properties in relation to the different functions of each of these entities.

Glomerular tuft. We have seen that the glomerular tuft forms a small dendritic tree within an olfactory glomerulus. The trunks of the tuft have relatively small diameters of 1–3 μm. Let us assume that each dendritic trunk divides in such a way as to conform to the 3/2 power constraint on the diameter, as discussed in the Appendix. Each trunk will thus give rise to a small equivalent cylinder; taken together, they will form an equivalent cylinder for the entire tuft. Assuming a range of values for electrical parameters that is typical of neurons, we can obtain an estimate from the graph of Fig. 13.2 (see Appendix) of a characteristic length of 150–600 μm for the case of 1-μm diameter trunks and 300–1000 μm for 3-μm diameter trunks.

These estimates are considerably higher than the actual extents of the tufts, which range from 150 to 200 μm. The electrotonic length ($L = x/\lambda$) of an equivalent cylinder for the tuft might, therefore, be estimated at less than 1, and possibly less than 0.5 μm. Thus, the smaller branches of the tuft are counterbalanced by their shorter lengths, an expression of the scaling principle (see Appendix). Because of the short electrotonic length of the tuft, current flow through the tuft must be relatively effective by passive means alone, and synaptic responses to sensory inputs can spread effectively to the primary dendrite.

Primary dendrite. This is a single unbranched process, and therefore, it is easy to make a model for it. From the graph of Fig. 13.2, it can be estimated that, for the mammal, likely estimates for the characteristic length of a typical primary dendrite of 6-μm diameter fall in the range of 300–1500 μm. Since a primary dendrite has a length of some 400 μm, it appears that the electrotonic length of this example is less than 1, and perhaps even less than 0.5. Electrotonic spread should, therefore, be relatively effective through such a process (Rall and Shepherd, 1968). In the turtle, primary dendrites are thinner (2–3 μm) and longer (500–700 μm), but this is offset by a higher specific membrane resistance (in the range of 50,000 Ω-cm²), promoting better signal spread (Mori et al., 1981a). We have noted that myelin has been observed wrapped around the distal primary dendrites some species. Possibly the myelin is associated with active properties of the dendrites; alternatively, it may serve to enhance passive spread. More work is needed.

Secondary dendrites. The key question with regard to this compartment of the mitral dendritic tree has been the extent to which an impulse at the cell body would invade the secondary dendrites and activate the mitral-to-granule synapses. In the course of investigating the extracellular potential fields in the olfactory bulb, a biophysical model was developed for the secondary dendrites that is relevant to just this problem. This investigation provided the basis for postulating dendrodendritic synaptic interactions in the bulb (Rall et al., 1966; Rall and Shepherd, 1968).

The steps for modeling the secondary dendrites follow those already outlined. By these steps, an equivalent cylinder for the tree of dendrites is obtained; it is illustrated in Fig. 5.12. Individual secondary dendrites are 2–6 μm in diameter and 400–600 μm in length. Their electrotonic lengths, as well as the values for the equivalent cylinder for the entire tree, have been estimated to lie in the range of 0.5–1.0.

In the investigation of the properties of secondary dendrites, a model for the action potential was also developed, so that it was possible to simulate an experiment in which an impulse propagates into the cell body and spreads into the dendrites (Rall and Shepherd, 1968). Computational experiments were carried out in which different assumptions were made about the electrotonic lengths of the dendrites and their active and passive properties. The use of biophysical

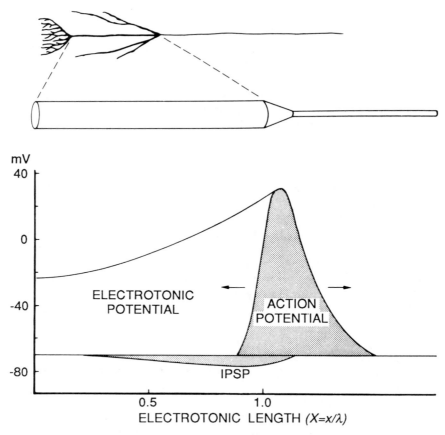

FIG. 5.12. Electrotonic model of the mitral cell, illustrating the spread of potentials in secondary dendrites. *Above:* Schematic diagram of a mitral cell. *Middle:* Equivalent cylinder for the 1° and 2° dendrites together with soma and axon. *Bottom:* Theoretical spatial distribution of intracellular potential at the peak of an action potential, and during recurrent inhibition. (From Shepherd, 1979.)

models to perform experiments that simulate situations, often inaccessible in the biological preparation itself, is a powerful approach that was introduced by Rall (1964); the widespread use of such simulations is evident in most of the chapters in this book.

The main result obtained from the model was that impulse spread into the secondary dendrites is very effective by passive means alone. This is illustrated schematically in the diagram of Fig. 5.12. Passive spread is, in fact, so effective that it was difficult in the computations to distinguish a passively spreading impulse from an actively propagating one, in some cases; this was also true for the primary dendrites. Although this does not answer the question of whether the secondary dendrites have active properties, it does establish the point that an impulse spreading by passive electrotonus retains much of its amplitude and waveform and, therefore, can activate the mitral-to-granule synapses in much the same way a propagating impulse would. Thus, the IPSP produced by feedback from the granule cells would be expected to be distributed throughout the dendritic tree, as indicated in Fig. 5.12.

GRANULE CELL

Because the granule cell lacks an axon, study of its dendritic properties is obviously crucial to understanding all of its input–output functions.

In the investigation of the recurrent dendrodendritic pathway by Rall and Shepherd (1968), a model for the granule cell was developed and explored. The branching tree within the EPL was represented by an equivalent cylinder; branch diameters of 0.2–0.8 μm were assumed, and an electrotonic length of about 0.4 was estimated. The shaft diameter of the granule cell is of the order of 1 μm; for an average shaft length of 600 μm, an electrotonic length of 1.7 was estimated for the model of the combined tree and shaft. The model was used to simulate synaptic depolarization of the granule spines in the EPL. The model demonstrated that this synaptic depolarization gives rise to the extracellular potentials generated just after an antidromic impulse invades the mitral cell dendrites (Phillips et al., 1963). When the mitral cell model and the granule cell model were joined in sequence, it could be postulated that the EPSP in the granule dendritic spines is due to a dendrodendritic input from the mitral secondary dendrites. As described in the previous section, the localization and timing indicated that the spine EPSP activates inhibitory synapses onto the same secondary dendrites, to produce the long-lasting IPSP recorded in the physiological experiments (Phillips et al., 1963).

Dendritic spines. What can be said about the properties of the granule dendritic spines within the EPL, spines that have both synaptic input and output? The properties of the spines must be critical in controlling the relative effectiveness of recurrent inhibition from a single spine and lateral inhibition mediated by spread of activity to neighboring spines (see Fig. 5.8, above). This question has been addressed directly by making precise measurements of dendrites and spines in reconstructions from serial electronmicrographs (Woolf et al., 1988) and in material observed in the high-voltage electron microscope (Greer, 1988). Computational models of these measured structures have then been constructed to explore the spread of activity.

An example of two common arrangements of spines is shown in Fig. 5.13. In the top panel, spines (A–D) are arranged in a linear fashion along the dendritic branch (E). Excitatory input (from a mitral/tufted cell dendrite) to spine C produces a large EPSP in C, which undergoes electrotonic decrement in spreading into the branch E. However, there is very little decrement in spreading in the other direction, from the branch into the other spines (A,B,D), because of the impedance mismatch between dendrite and spines (see Appendix). A different arrangement is shown in the bottom panel, where several spines (B–D) arise from a common stem, a so-called complex spine. Spread of activity from C to the neighboring spines now occurs in distinct steps.

These results confirm that, although the stems of the spines may be narrow

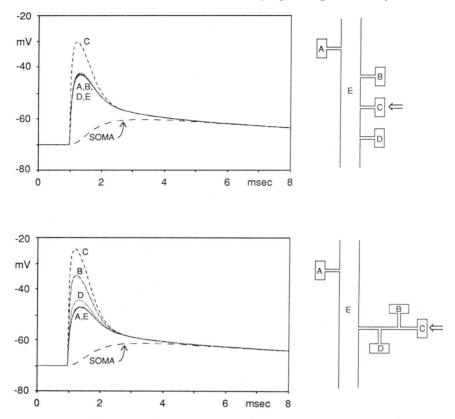

FIG. 5.13. Electrotonic models of granule cell dendrites and spines. The models, shown at right, are based on reconstructions of serial electron micrographs. An excitatory synaptic conductance change was simulated in spine C, and the synaptic potentials were measured in each spine (A–D) and in the dendritic branch. *Upper panels:* Linear arrangement of spines. *Lower panels:* A single and a complex spine. For these simulations, $R_m = 4000 \, \Omega\text{cm}^2$. (From Greer, 1988; Woolf, T.B., Greer, C.A., Shepherd, G.M., unpublished observations.)

(i.e., 0.2 μm), reasonable assumptions about the electrical properties of the membranes lead to the conclusion that electrotonic spread from spine to spine is considerable over the short distances involved. It appears therefore that lateral inhibition can be mediated by passive spread alone through the dendritic tree in the EPL. The inhibition is spatially graded according to the electrotonic decrement of the potentials in the tree. Depending on the spatial arrangements, individual spines may act as subunits in the manner discussed in Chap. 1.

As in the case of the mitral cell dendrites, these considerations do not rule out the possibility of active properties of the granule dendritic membrane, in addition to electrotonic properties. Might the spines themselves have active properties, so that spread into the branches is more effective? And might the branches have active properties? The original model of the granule cell did not rule out active spine properties, but indicated that active properties must be limited and not lead to propagation from the branches in the EPL into the main dendritic shaft (Rall and Shepherd, 1968). Subsequently, Jahr and Nicoll (1982) found that recurrent inhibition persists despite the presence of TTX, implying that voltage-gated sodium channels do not contribute to the potentials in granule dendritic spines.

PERIGLOMERULAR CELL

The PG cell has already been discussed as an example of the short-axon cell, a type we encounter in many parts of the brain. Being small, the PG cell has been relatively inaccessible to experimental studies, and there is yet no biophysical model as in the cases of the mitral and granule cells. For a first approximation to its overall electrotonic properties, the dendritic tuft may be regarded as a smaller version of the mitral dendritic tuft described above. Taking into account both the smaller diameters and the shorter lengths of the branches, it seems reasonable to conclude that an equivalent cylinder for the PG cell tuft would be similar to that for the mitral tuft, that is, $L = 0.5-1.0$. This is, again, an expression of the scaling principle for dendritic trees of different size.

PG cell functions appear to be exquisitely dependent on levels of input activity. Electrophysiological studies have suggested that at threshold there is mainly straight-through excitation of mitral cells by receptor axons; long-lasting facilitation is sometimes detectable (Getchell and Shepherd, 1975a,b). As input activity increases, the activation of PG cells, both by direct olfactory axon input and by indirect dendrodendritic synapses, begins to bring about inhibitory feedback from the PG cell dendrites. Small EPSPs probably mediate only local input– output paths through the dendrites; moderate EPSPs will lead to more extensive inhibitory actions within a glomerulus; large EPSPs, by spreading to the axon hillock, set up impulses that mediate inhibition of dendrites arising from neighboring glomeruli.

Studies of responses to natural stimulation with odors have demonstrated even more complex interactions, including the presence of an initial brief hyperpolarization at weak levels of stimulation (see next section). It thus appears that several levels of interaction can be identified within the glomerular layer, governed by the amplitudes of EPSPs and their electrotonic spread within the PG cell dendrites. Functionally, these interactions enhance transmission at detection thresholds, and provide the lateral inhibition necessary for discrimination between odors at higher odor concentrations. The PG cell dendritic tree thus provides for *multiple state-dependent input–output functions*. Similar properties have been postulated for thalamic short-axon cells, and may apply to other types of cells.

FUNCTIONAL CIRCUITS

We are now in a position to consider how the basic circuit of the olfactory bulb is involved in processing information about odor molecules.

PROCESSING OF ODOR STIMULI

We have seen that the response of a mitral or tufted cell to a single volley of impulses in the olfactory nerves is a complex sequence of excitatory and inhibitory potentials. Analysis of this sequence provided some insights into how it might arise within the basic circuit, through interactions between the mitral/tufted cells and their two sets of intrinsic neurons: PG and granule cells. Do these interactions play roles in the processing of natural stimulation with odors?

The first step in the analysis of this question is to obtain precise control over the stimulus, so that very brief step-pulses of odor can be used as stimuli, in the same manner as brief flashes of light or sound tones. When mitral cell responses are recorded to such stimuli, three main response patterns emerge (Kauer, 1974; Kauer and Shepherd, 1977). In one type, the neuron responds only with suppression of impulse activity, at all odor concentrations; this is the S (suppression) type. In another, there is usually a slower spike discharge at threshold odor concentrations, changing to an early brief spike burst followed by suppression at higher concentrations; this is termed an E (excitatory) type. Finally, there may be no detectable response to a given odor, the N (no response) type. These types have many minor variations; they are seen in their simplest form in fish and salamander, and in more complex forms in mammals.

Intracellular recordings (Hamilton and Kauer, 1985; Wellis and Scott, 1987) have revealed the synaptic potentials underlying these response types. In the experiment illustrated in Fig. 5.14, the responses to orthodromic (trace a) and antidromic (trace b) shocks were first recorded; these showed a similar sequence of impulse activation followed by a large and then a smaller long-lasting hyperpolarizing potential, as in turtles (cf. Fig. 5.7). Sensory stimulation with a brief pulse of cineole (trace c) produced an E-type response, consisting of an initial brief burst of impulses followed by a large (H2) and then a slower (H3) hyperpolarizing potential. The similarity between the responses to an artificial synchronous volley (a) and a natural stimulus pulse (c) is obvious. It is tempting therefore to suggest that activation of the synaptic circuits is similar—that is, that the natural stimulus pulse caused a near-synchronous EPSP in the glomerular tuft, leading to the impulse burst, followed by activation of dendrodendritic synaptic inhibition. The large amplitude of the hyperpolarizing potentials suggests that IPSPs are generated at or near the site of recording in the mitral cell soma, where granule cell inhibition predominates (as in the antidromically activated case of trace b). It thus appears that strong synchronous artificial and natural stimulation activate the basic circuit in a similar fashion.

At intermediate concentrations of natural stimuli (trace d), the pattern is similar, albeit later and slower, but at low concentrations some interesting differences emerge. As can be seen in trace e, the initial response is now a clear, sharp hyperpolarization (H1), followed by a slow burst and the sequence (H1, H2) of hyperpolarizing potentials. The initial hyperpolarization has not previously been seen in responses to shock stimuli, and its basis in the response to natural stimuli is still unclear. Its large amplitude points to granule cell inhibition, but, as can be appreciated by referring to the basic circuit of Fig. 5.6, the mitral cells are the route for activation of granule cells; how then can one have granule cell inhibition of mitral cells first?

One possibility is that tufted cells have the lowest thresholds for activation; because they are smaller than mitral cells, this would be an expression of the size

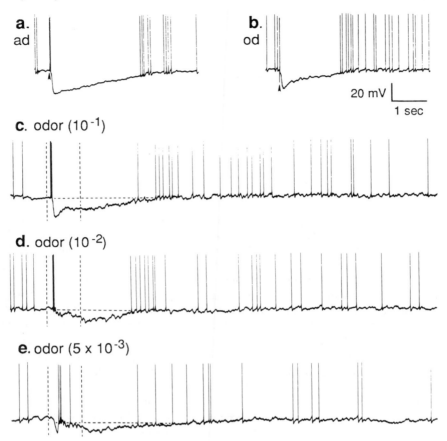

FIG. 5.14. Intracellular responses of a mitral/tufted cell in the salamander to activation by single volleys and by odor. **a.** Antidromic activation by single-shock stimulation (arrow) of the LOT (cf. Fig. 5.7A). **b.** Orthodromic activation by single-shock stimulation of the olfactory epithelium by a single-step pulse of odor (indicated by vertical dashed lines) at high concentration (10^{-1} flow dilution from vapor saturation). **c.** Odor stimulation at high concentration. **d.** Odor stimulation at moderate concentration. **e.** Odor stimulation at low concentration. See text. (Modified from Hamilton and Kauer, 1985.)

principle (see Chap. 4). One would then postulate that the tufted cells set up EPSPs in the subtype of granule cells that also connects to mitral cells (see G_I in Figs. 5.2 and 5.4). The cell types in the salamander are not as clearly differentiated as in mammals, so this postulate awaits further testing. Another possibility is that the hyperpolarization is due to a large IPSP in the glomerular tuft, mediated by dendrodendritic connections through PG cells. This would require sensory input to PG cell dendrites. These results thus provide clues to how synaptic circuits at the levels of both input processing (glomeruli) and output control (granule cells) may make powerful contributions to the processing of olfactory information.

6

RETINA *

PETER STERLING

The retina, more than any other part of the brain, is where we begin to grasp the relations between the real world (a visual scene), its *physical image* projected onto a receptor array, and the first *neural images*. The retina is a thin sheet of neural tissue lining the posterior hemisphere of the eyeball. It is actually part of the brain itself, evaginating from the lateral wall of the neural tube during embryonic development. In the embryo, as the optic stalk grows from the brain toward the overlying ectoderm, it induces the formation of the optical apparatus (cornea, pupil, lens)—the structures that cooperate to project an optical image of the world onto the retina's surface. The retina's task is to convert this optical image into a "neural image" for transmission down the optic nerve to a multitude of centers for further analysis. This is a complex task, for reasons to be explained, and this complexity is reflected in the synaptic organization.

In all vertebrate retinas the transformation from optical to neural image involves three stages: (1) phototransduction by a layer of receptor neurons; (2) transmission of their signals by chemical synapses to a layer of "bipolar" neurons; and (3) transmission of these signals, also by chemical synapses, to output neurons, termed *ganglion cells*. At both synaptic stages (receptor → bipolar and bipolar → ganglion cell) there are specialized laterally connecting neurons called, respectively, *horizontal* and *amacrine* cells, whose task is to modify the transmission across the synaptic layers. There are also centrifugal elements, *interplexiform* cells, to convey signals from the inner synaptic layer back to the outer one. These elements are shown schematically in Fig. 6.1.

A closer look at this apparently simple design, three interconnected layers and five broad classes of neuron, reveals considerable complexity (Fig. 6.2). Within a given species each class of neuron is represented by more than one and often many specific types. Each cell type is distinguished from others in its class by having a characteristic morphology, function, neurochemical properties, and connections (Rodieck and Brening, 1983; Sterling, 1983). This diversity, amounting to some 60 cellular types in the cat (Kolb et al., 1981; Sterling, 1983), was puzzling when first discovered, but the broad explanation has gradually emerged: It is impossible to encode all the information contained in the optical

---

*This chapter is dedicated to the memory of Professor Henricus J. G. M. Kuypers.

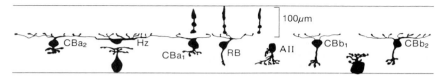

FIG. 6.1. Neuronal elements of the mammalian retina (same scale as diagrams for other regions). Symbols are defined in Fig. 6.2 legend.

image by using a single neural image. Therefore, the retina creates multiple neural images, using different cell types, for simultaneous transmission to the brain over parallel circuits. Similarly, there must be different circuits, involving different cell types, for different levels of luminance during daylight, twilight, and starlight (Sterling et al., 1987, 1988).

In this chapter, we will describe the important cell classes in the retina and some of their specific types and show how they provide, through their functional properties and synaptic connections, for parallel circuits that process different aspects of the visual world. We will then suggest how the flow of visual informa-

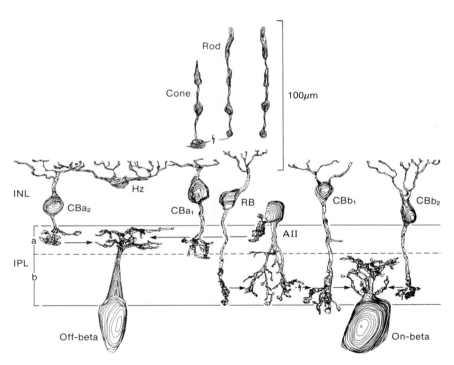

FIG. 6.2. Neuronal elements, scale enlarged in the vertical axis, to show more clearly the cell morphology and layers. Inputs: rod and cone photoreceptors. Principal neuron: ganglion cells. Intrinsic neurons: bipolar, horizontal, and amacrine cells. INL, inner nuclear layer; IPL, inner plexiform layer. RB, rod bipolar; CB, cone bipolar (subscripts a_1, b_1, etc. designate specific *types* of cone bipolar; see text); Hz, horizontal cell. AII designates a specific type of amacrine cell.

tion shifts among different circuits specialized for different levels of light intensity.

SPECIES DIFFERENCES IN RETINAL STRUCTURE AND VISUAL FUNCTION

Given the basic three-layered plan, the variation among species in retinal structure and synaptic organization is quite spectacular (Cajal, 1893; Walls, 1942). Some variants relate in obvious ways to life-style: Nocturnal animals have mainly receptors for dim light (rods); diurnal animals, mainly photoreceptors for bright light (cones). If an animal is brightly colored (usually for purposes of sexual display), it is likely to have elaborate retinal circuitry for color vision. An animal that hunts visually at a distance where the prey subtends a tiny angle (a falcon seeking a rodent, or a chameleon seeking an insect) has a region of retina specialized for high acuity (*fovea*), where the cones are extremely thin and densely packed (Reymond, 1985). An animal that hunts visually at a closer distance where the prey subtends a larger angle (a frog, whose concern for visual detail stops pretty much at the tip of its tongue) requires no such specialization.

Another critical determinant of retinal structure and synaptic organization is the amount of brain available to a particular species for further analysis of the neural image. Ganglion cells in a retina connected to a relatively small brain respond selectively to particular stimulus features that represent important aspects of the environment such as "prey" or "predator." Thus one type of ganglion cell in the frog responds vigorously to small, dark, moving stimuli but not much to anything else. It is natural, since the frog snaps at such stimuli, to call these ganglion cells "fly detectors." These and other ganglion cells with different, equally complex, responses, such as selectivity for direction and velocity of motion, seem tuned for particular "trigger features" (Barlow, 1953; Maturana et al., 1960; Barlow et al., 1964). Retinas in small-brained animals like frogs tend to have many cell types, a thick inner synaptic layer, with several sublayers, a high density of synapses, and a high proportion of "serial" synapses (Dubin, 1976).

Animals with larger brains (mammals) tend to postpone the extraction of particular features in the image until higher levels, so their retinas tend to be simpler and more easily analyzed. The cell types and connections in the mammalian retina are therefore better known, and so are the optics of the eye. Consequently, this chapter will focus on the mammalian retina, and particularly on the cat, where considerable detail is known regarding the functional architecture of circuits feeding into several specific types of ganglion cell. All measurements of neuronal size refer to the cat except where otherwise specified. Human vision is naturally of special interest. We differ from the cat in having about a tenfold greater spatial acuity (Pasternak and Merigan, 1981), which requires a fovea, and sensitivity to color. Our vision closely resembles that of the macaque monkey (Rodieck, 1988), so where the chapter refers to the functional architecture of the fovea, it is to the macaque retina, but the reader may assume that the human is similar.

NEURONAL ELEMENTS

In some regions of the brain, all members of a cell class have the same soma size, dendritic field diameter, axon diameter, cell spacing, etc. throughout the tissue; an example is the Purkinje cell in the cerebellum (Chap. 7). This is not so for the retina. As we shall see, the size, branching pattern, and distribution density of most cell types change dramatically as one goes from the central retina toward the periphery. The reasons will be discussed later; here we note that in this chapter when specific measurements are presented, they refer to the central area of the cat retina.

INPUT ELEMENTS: RECEPTORS

There are two main classes of photoreceptor: rods and cones. As shown in Fig. 6.2, both cell types are elongated, consisting of an outer segment, inner segment, cell body, axon, and synaptic terminal. The distal tip of the outer segment points toward the back of the eye, away from the light, and the tip is embedded in folds of melanotic epithelium. The outer segment contains membranous disks, stacked like saucers perpendicular to the cell's long axis, which bear the photopigment. The inner segment is similarly jam-packed, but with mitochrondria. The axon is generally short, 50 μm or less, except for the fovea, where dense packing of the cones leaves no room for connections to the correspondingly numerous, second- and third-order neurons. At the fovea, therefore, cone axons can be as long as 500 μm (Figs. 6.3 and 6.4; Rodieck, 1988; Schein, 1988).

Corresponding to their functional specializations—rods for night vision and cones for daylight—the two types of receptor differ structurally. The rod outer segment is long (25–50 μm) and thin (1–1.5 μm), permitting the rods to be packed densely in the extreme, on the order of 500,000/mm^2. This high density increases the chance of catching every available photon. The cone outer segment is short (6–8 μm) and thick (3–5 μm). Cones tend to be less densely distributed, about 5–10% of the rod density, except at the fovea, where their form and packing resemble those of rods (Steinberg et al., 1973; Rodieck, 1988). The rod axon is thin (0.25 μm), consistent with the cell's slow response to light, and the synaptic ending is small (Fig. 6.4). The cone axon is thicker (1.5 μm), consistent with the cone's faster response to light, and the synaptic ending is larger (Fig. 6.4).

In mammals the three cone types, to be discussed below, are morphologically indistinguishable, but the short-wavelength, "blue," cones stain specifically with the fluorescent dye, Procion Yellow (de Monasterio et al., 1981). This technique reveals blue cones to have a regular distribution and, as predicted from psychophysical experiments, to form a small fraction of the whole cone population. Each of the three cone types connects differently to postsynaptic cells.

In some species, such as turtles and fish, while there may be only three cone pigments, there are up to seven types of cone. Each photopigment in the outer segment can be associated with an oil droplet in the inner segment of specific color and absorption characteristics. These apparently serve as bandpass filters to

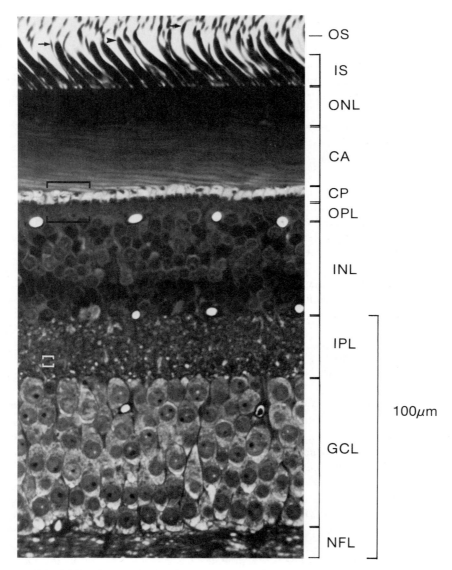

OS

IS

ONL

CA

CP
OPL

INL

IPL

100μm

GCL

NFL

FIG. 6.3. Light micrograph of radial section through monkey retina. This region, about 4° (about 1 mm) from the fovea, contains multiple ranks of ganglion cells and receptor axons and terminals that belong exclusively to cones. The outer segments that belong to these cones are in the fovea; thus the circuitry in this region serves the fovea. OS, outer segments; IS, inner segments; ONL, outer nuclear layer; CA, cone axons; CP, cone pedicles; OPL, outer plexiform layer; INL, inner nuclear layer; IPL, inner plexiform layer; GCL, ganglion cell layer; NFL, nerve fiber layer. Upper brackets indicate region shown at higher magnification in Fig. 6.4. Lower brackets indicate region shown at higher magnification in Fig. 6.11. (Photo courtesy of Y. Tsukamoto.)

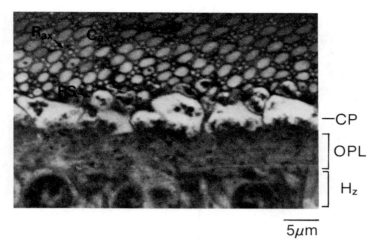

-CP

] OPL

] H_z

5µm

FIG. 6.4. Light micrograph of radial section through outer monkey retina. This region, several millimeters beyond the fovea, contains rods as well as cones. Each cone axon (C_{ax}) is surrounded by a halo of fine rod axons (R_{ax}), and the synaptic terminals of both types of receptor (CP, RS) are present at the edge of the outer plexiform layer (OPL). H_z, a row of horizontal cell somas. (Photo courtesy of Y. Tsukamoto.)

limit the spectral composition of the light entering a particular outer segment (Ohtsuka, 1985).

OUTPUT ELEMENTS: GANGLION CELLS

Ganglion cells send axons into the optic nerve and thence to the brain. Most ganglion cell somata form the innermost cellular layer of the retina, which is therefore called the ganglion cell layer (GCL) (Fig. 6.3); some "displaced" ganglion cells are found at the inner margins of the INL. Ganglion cells restrict their dendrites to the IPL. The only exception known so far is a "biplexiform" ganglion cell in monkey retina that reaches into the OPL to collect synapses directly from rods (Mariani, 1982).

Whereas there are only 3–4 fundamental types of receptor (in cat), there are close to 20 morphological types of ganglion cell (Kolb et al., 1981). Several of the morphological types have been proved (by recording and injection with Lucifer Yellow or horseradish peroxidase, HRP) to have specific physiological properties (Fukuda et al., 1984; Saito, 1983a,b; Stanford and Sherman, 1984). Thus, there is confidence that to delineate a distinctive morphology for a ganglion cell is to predict for it a distinctive physiology. As we shall relate in more detail, both rods and cones have connecting pathways to the major ganglion cell types, and each type forms a mosaic covering the whole retina. This implies that an important step in coding within the retina is to subdivide the information leaving the receptor mosaic into multiple, parallel, neural images.

The best known ganglion cells in cat retina are termed *beta* and *alpha* (Fig. 6.5; Boycott and Wässle, 1974). The beta cell body is medium sized (about 15

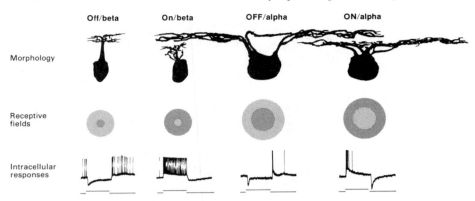

FIG. 6.5. Form and function of four types of cat ganglion cell (radial view). The beta cells have a narrow dendritic field, a narrow concentric, antagonistic receptive field, and a response with transient and sustained components. The alpha cells have a broad dendritic field, a broad concentric, antagonistic receptive field, and a response with mainly a transient component. On-center alpha and beta cells arborize deep in the inner plexiform layer, whereas off-center alpha and beta cells arborize more superficially. (Intracellular recordings from Saito, 1983a.)

μm in diameter), with medium axon and a narrow (about 30 μm), bushy dendritic tree. Beta cells are of two types: One arborizes through the full depth of the outer third of the IPL (sublamina *a*); the other arborizes throughout the full depth of the inner two-thirds of the IPL (sublamina *b*) (Fig. 6.2; Famiglietti and Kolb, 1976; Wässle et al., 1981a,b). The alpha cell body is large (25–30 μm diameter), with thick axon and a wide (180 μm), radiating, and distinctly planar dendritic tree. Alpha cells are also of two types, one arborizing in sublamina *a*, the other in sublamina *b* (Famiglietti and Kolb, 1975; Wässle et al., 1981a,b). The ganglion cells arborizing in sublamina *a* fire action potentials at light "off," and those arborizing in sublamina *b* fire at light "on" (Fig. 6.6A; Nelson et al., 1978). Henceforth, we shall call these *off*-beta, *on*-beta, and so on.

Additional forms of ganglion cell have been assigned other Greek letters— delta, gamma, epsilon—or have been numbered G_1–G_{21} according to another scheme (Kolb et al., 1981). Most of these forms have small somata, thin axons, and fine, widely ramifying dendrites. Some of these cells have complex, "feature-detecting" properties as discussed above for the retinas of lower vertebrates. In the cat, these ganglion cell types project to brainstem structures such as the superior colliculus (SC), nucleus of the optic tract, and suprachiasmatic nucleus (SCN) (reviewed by Rodieck, 1979). These areas use visual information for various specific purposes, such as centering the eyes and head on objects of visual attention (SC), setting circadian rhythms, (SCN) and so on. Almost nothing is known about the intraretinal circuitry of these ganglion cells.

INTRINSIC ELEMENTS

Bipolar neurons. Bipolar cells are the intrinsic neurons that connect receptors, directly or indirectly, to the ganglion cells. In lower vertebrates, rods and cones sometimes converge on the same bipolar neuron, but in mammals separate bi-polars are used for the two types of photoreceptor (Cajal, 1893; Boycott and Kolb, 1973).

The *rod bipolar* in cat is of one type (Fig. 6.2). It has a small soma (7 μm) located high in the ONL and a fairly narrow, candelabra-like dendritic arbor. The rod bipolar axon descends without branching to terminate at the deepest level of sublamina *b*. The axon provides output only in sublamina *b* and generally does not contact ganglion cells directly (Kolb and Nelson, 1983; McGuire et al., 1984; Freed et al., 1987).

Cone bipolar neurons are of *ten* types (Kolb et al., 1981; McGuire et al., 1984; Pourcho and Goebel, 1987; Cohen and Sterling, 1990a). Their morphological differences are fairly subtle (Figs. 6.2 and 6.6). The somata are all about 7–10 μm, the dendritic arbors have a fairly narrow field and are modestly branched. The somata of each type are located at a characteristic depth in the INL, but there is considerable variation. The similarities in dendritic form probably reflect the fact that all but one type collect from essentially the same set of cones (Cohen and Sterling, 1990a). Each bipolar collects from all the overlying cone pedicles within its reach). An exception to this rule is a type termed CBb_5 (Fig. 6.6A). Dendrites from this cell spread widely in the OPL and ignore most of the cone pedicles centered over the cell body (Cohen and Sterling, 1990a). Possibly these dendrites specifically seek blue cones which are known to be widely spaced. Bipolar cells selective for blue cones are known in the primate (Mariani, 1984).

The various types of cone bipolar cell differ most strikingly in their axonal arborizations. Five types send axons to sublamina *a*, and the other five types, including the wide-field cell (putative blue cone bipolar) send axons to sub-lamina *b* (Fig. 6.6A). Within each sublamina the five types of axon exhibit clear differences in arborization. Thus, CBb_1 has a thick stalk with clawlike branching throughout the full depth of the sublamina; the CBb_3 and CBb_4 axons both form planar arbors at about the middle of the sublamina. They differ in the delicacy of their processes, but subtly enough that one would hesitate to distinguish them on this basis alone. However they also differ distinctly in their patterns of connec-tion and in their function (Fig. 6.6B; McGuire et al., 1986; Nelson and Kolb, 1983; Cohen and Sterling, 1990a).

Horizontal cells. Horizontal cells connect almost exclusively within the OPL. Their main task is to collect widely from receptors and provide them with feedback. Fish horizontal cells are subject to centrifugal regulation via the in-terplexiform cell (Dowling, 1986); also, horizontal cells in some fish send axons to the IPL (Sakai and Naka, 1985), but this is not the general plan.

Mammalian retinas contain two types of horizontal cell (Fisher and Boycott, 1974; Kolb, 1974; Boycott et al., 1978). The somata form the upper tier of the inner nuclear layer and send processes distally into the outer plexiform layer (Fig. 6.4). In cat, the two types are termed (imaginatively) A and B (Fig. 6.7). Type A has a fairly large soma, about 12×15 μm, and stout dendrites that radiate tangentially through the OPL with little taper over a diameter of about 80 μm. Dendrites of adjacent type A cells cross each other and at these nodes form extensive electrical synapses (gap junctions) with each other (Kolb, 1977). Type

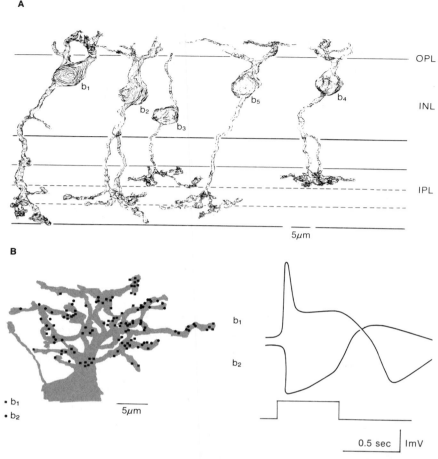

FIG. 6.6. **A.** Five types of cone bipolar cell in cat retina that innervate sublamina *b* of the inner plexiform layer (where the *on* ganglion cells arborize). These types account for all of the cone bipolar cells to this sublayer. Each cone apparently contacts at least one member of the types b_1–b_4. Furthermore, each of these bipolar types collects from all the cones immediately above it without skipping any. The type b_5 ignores the cones immediately overlying it and reaches widely to collect from outlying cones. **B.** "Push–pull" hypothesis of cone bipolar–ganglion cell function. *Left:* distribution of synapses to *on*-beta cell from b_1 and b_2 bipolar axons. *Right:* Intracellular recordings from the two types of bipolar cell showing opposite responses to stimulus. See text. (A from Cohen and Sterling, 1990a,b; B, left, from McGuire et al., 1986; Cohen and Sterling, 1990a,b; B, right, modified from Nelson and Kolb, 1983.)

B has a somewhat smaller soma and finer dendrites that radiate with taper over a narrower field (about 60 μm). Type B cells do not form electrical synapses. Both types at semiregular intervals emit extremely fine (0.2 μm) processes that ascend to the level of the cone pedicles and invaginate them.

The type B horizontal cell sends a fine axon to meander tangentially in the

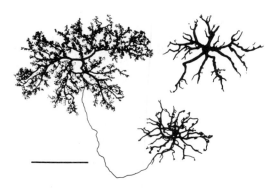

FIG. 6.7. Two types of horizontal cell from cat retina as seen in tangential view by Golgi impregnation. Dendrites of types A and B both connect with cones; axon arborization of type B connects with about 3000 rods. (From Fisher and Boycott, 1974.)

OPL for about 400 μm. It terminates in a magnificent arborization spanning several hundred micrometers (Fig. 6.7). Numerous fine processes from this arbor ascend past the layer of cone pedicles to reach the overlying tiers of rod spherules. This axonal arborization connects to as many as 3000 rods and is thought to be electrically isolated from its soma (Nelson, 1977). Thus, the axonal arborization itself is, in effect, a third type of horizontal cell, one that connects exclusively with rods.

Amacrine cells. Amacrine cells connect exclusively within the IPL (Fig. 6.2). The somata form the innermost layer of the INL; they are also numerous in the GCL, where they are called "displaced" amacrines. Amacrine processes collect input from bipolar axons and provide output to ganglion cells, other amacrines, and bipolar axons. Sometimes an amacrine process returns a synapse to the particular bipolar axon from which it collects—that is, provides a reciprocal, feedback synapse (see below). In other instances, the amacrine contacts a bipolar axon different in type from that which provided its input; that is, the amacrine feeds-forward to the bipolar. This synapse can be either chemical or electrical.

The term "amacrine" was coined by Cajal for these intrinsic neurons of the IPL because he recognized them to lack an axon. In an amacrine cell, inputs and outputs are not morphologically segregated from each other. Cajal was forced to recognize this cell class, along with the granule cell of the olfactory bulb (Chap. 5), as exceptions to his "law of dynamic polarization" (see Piccolino, 1988). Amacrine processes are extremely fine (0.5 μm or less) and interrupted at close intervals by swellings. The electron microscope shows these varicosities to be sites of chemical synaptic input and output. There are interesting exceptions to this rule: Recently a class of amacrines was found in which conventional amacrine processes spread about 100 μm in the tangential plane (Dacey, 1988). These bear spines which are probably sites of synaptic input. The distal ends of these conventional processes emit extremely fine extensions that travel for millimeters away from the soma and apparently provide synaptic outputs. In short, this amacrine cell seems to provide *multiple axons!*

Morphological diversity among the amacrine cells, even in the relatively simple mammalian retina, is considerable (Fig. 6.8). About 20 forms have been described so far (Cajal, 1893; Kolb et al., 1981), and there may be more to come. Some types arborize broadly in sublamina *a* or *b;* others form a planar arbor in a particular substratum of *a* or *b;* still others spread diffusely throughout the whole IPL. The various forms also differ in spatial extent; some are narrow field (30 μm), and others are intermediate (100 μm) or wide (1000 μm) field. When the Golgi method was the only way to visualize and classify neurons, it was impossible to be confident in sorting out distinct types of amacrine cell since the population, when viewed as a whole, varies continuously in stratification and field size. However, current methods for sorting cells by additional criteria, such as chemical staining pattern or synaptic connections, establish that, as for ganglion cells, distinctive morphology correlates strongly with many other features and is a powerful guide to classification.

Interplexiform cells. Interplexiform cells were illustrated by Cajal (1893). They have somata in the amacrine layer but send processes to the outer plexiform as well as the inner plexiform layer. The processes in the OPL do not reach the receptor terminals, a fact which distinguishes interplexiform cells from bipolars. Electron microscopy shows that the interplexiform cell receives synapses on its IPL processes and provides synaptic outputs in both the IPL and OPL. In the cat OPL, it contacts both rod and cone bipolars (Kolb and West, 1977; Nakamura et al., 1980; McGuire et al., 1984); in fish it contacts horizontal cells (Dowling, 1986). Obviously, it is well suited to convey information centrifugally. Some authors refer to the interplexiform as a sixth cell class; alternatively, one may retain the five classical types of retinal neuron and consider the interplexiform cell as an amacrine cell with centrifugal processes to the OPL, as we do here.

Centrifugal fibers. A discussion of neuronal elements in the retina would be incomplete without mentioning centrifugal fibers from the brain. In certain vertebrate orders, including fish, amphibia, reptiles, and birds, the optic nerve contains efferent axons from specific regions of the brain that terminate in the inner retina, apparently on the interplexiform cells (Dowling, 1986, 1987). Because the latter contact neurons in the outer retina, there is a structural basis for central signals to modify transmission of the neural image right at the first synaptic stage. The bird brain has a substantial central structure, the *isthmo-optic nucleus,* with a definite retinotopic organization, projecting centrifugally to the inner retina (Dowling and Cowan, 1966), and the fish brain sends fibers containing the peptide luteinizing hormone-releasing hormone (LHRH) to the retina from the olfactory bulb (Zucker and Dowling, 1987)! Thus, the idea is not farfetched that an odor, a memory, or a feeling could modify the construction of a visual image in the retina, that beauty could be literally in the eye of the beholder. However, little is known regarding the function of these centrifugal pathways beyond what

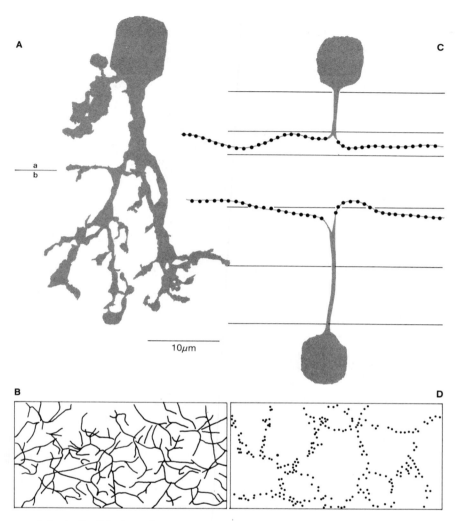

FIG. 6.8. **A.** Radial view of AII amacrine cell. The arborization in sublamina *b* of the inner plexiform layer, narrow and bushy, collects an average of 6 synaptic contacts from 30 rod bipolar axons. The output from this arbor is solely from electrical synapses. The lobular appendages in sublamina *a* provide chemical synapses (possibly glycinergic) to dendrites of *off*-alpha and -beta ganglion cells (see Fig. 6.2). **B.** Tangential view of the AII amacrine cell array, showing the close spacing and overlap of the arbors in sublamina *b*. Note that where dendrites cross each other, they do not associate; compare to Fig. 6.8D. **C.** Radial view of *on* and *off* starburst amacrine cells, showing strict lamination of their arborizations. The dendrites of *on*- and *off*-alpha ganglion cells arborize in the same strata and associate intimately with the starburst processes (see Fig. 6.18). **D.** Tangential view of *on* starburst arborizations. The starburst processes, in contrast to those of the AII, associate to form bundles. Thus, although the starburst cell has a coverage factor of 60, compared to 3 for the AII cell, the starburst network leaves huge open spaces in the IPL. (Schematics B–C courtesy of N. Vardi.)

181

has just been noted. Therefore, this chapter considers only how the beauty in the eye is transmitted to the brain and not vice versa.

CELL POPULATIONS

Each cell type at every point on the retina has a characteristic density and spacing (Fig. 6.9A; Wässle and Riemann, 1978). Table 6.1 lists as an example the densities of more than a dozen cell types in the central area of cat retina. This is the only region in any retina where the densities of so many types are known. For cell types that are interconnected (see Pathways, below), the ratios of their densities determine their ratios of convergence and divergence:

$$\frac{\text{Density of A}}{\text{Density of B}} = \frac{\text{Convergence of A} \rightarrow \text{B}}{\text{Divergence of A} \rightarrow \text{B}} \qquad (6.1)$$

This equation (Freed et al., 1987) is useful in estimating the quantitative patterns of connection because the density of a retinal cell type is relatively easy to establish. Under some circumstances, the convergence or the divergence can

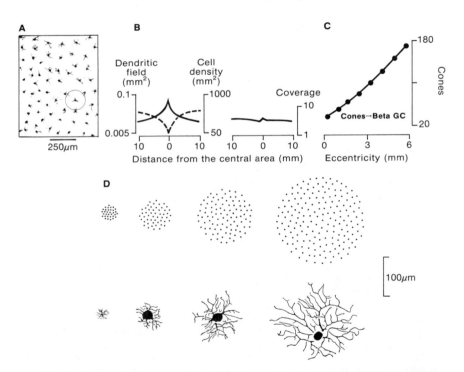

FIG. 6.9. **A.** Array of *on*-beta ganglion cells (tangential view), showing their semiregular spacing and extent of dendritic overlap. **B.** Distribution density of type A horizontal cells peaks at the center of the retina and declines toward the periphery while dendritic field area behaves reciprocally. Thus, the coverage factor is constant. **C.** Number of cones per beta ganglion cell dendritic field increases linearly from center to periphery. **D.** Increase in beta cell dendritic area and decrease in cone density. (A from Wässle et al., 1981a; B from Wässle and Riemann, 1978; C,D from Tsukamoto et al., 1990.)

be determined, but not both. Thus, it is easy to establish that one rod contacts two rod bipolar cells, but hard to determine by direct means how many rods converge on a rod bipolar. The answer can be obtained by measuring the densities of both cell types (Table 6.1) and applying the equation. The value of knowing the convergence and divergence will be illustrated later.

CIRCUITS

Most cell populations are densest in the central retina and sparser toward the periphery. A notable exception to this rule is the rod population, which reaches its maximum somewhere outside the central area, presumably to allow room for the cone population to reach *its* maximum within the central area (Rodieck, 1988). Where many cells of one type crowd together, their dendritic fields become quite narrow. This limits the convergence of signals from earlier stages and determines, in effect, the "pixel size" of the neural image transmitted by the array of that cell type. To transmit a *spatially* fine-grained neural image requires an array of closely packed, narrow-field neurons. On the other hand, to transmit a *temporally* fine-grained image requires an array of wide-field neurons with a short time constant of response. To "cover" or "tile" completely a given region of retina with a given cell type requires many cells if the dendritic field is narrow but few if the dendritic field is wide (Fig. 6.9). These considerations explain quite simply why narrow-field beta ganglion cells contribute roughly 50% of the axons to the optic nerve, and the wide-field alpha ganglion cells contribute less than 5% (Wässle et al. 1981a,b).

As the density of a given neuronal array declines from the central retina toward

Table 6.1. Densities of Specific Cell Types in Cat Area Centralis

Cell type	Density ($10^3/mm^2$)
Rod	450
Cone	27
Horizontal (A)	0.8
Horizontal (B)	2
Cone bipolar (b_1)	6
Cone bipolar (b_2)	4
Cone bipolar (b_3)	4
Cone bipolar (b_4)	4
Rod bipolar	30
Amacrine (AII)	4
Amacrine (*on* starburst)	1.4
Amacrine (*off* starburst)	0.6
Amacrine (A18)	0.02
Ganglion cell (*on*-beta)	4
Ganglion cell (*off*-beta)	4
Ganglion cell (*on*-alpha)	0.08
Ganglion cell (*off*-alpha)	0.08

the periphery, the widths of the dendritic fields increase. Thus, a beta dendritic field at the periphery can be as large as an alpha dendritic field at the center (Fig. 6.9D; Boycott and Wässle, 1974). For a given cell type, the increased dendritic area generally matches the decreased cell density so that its *coverage factor—* that is, the number of cells covering a given point on the retina—is constant (Fig. 6.9A; Boycott et al., 1978). The coverage factor for most feedforward cells (bipolar and ganglion cells) is always greater than one, but not much greater, and is far less than ten (Sterling et al., 1988). One probable reason why coverages are modestly greater than 1 is to protect against *aliasing,* the formation of moiré patterns on fine-grained arrays by fine-grained images (Hughes, 1981; Williams, 1986).

There is another kind of match between cone density and ganglion cell field area. As cone density declines from center to periphery and dendritic area rises, the number of cones converging on a ganglion cell increases (Fig. 6.9C,D). This may reflect a mechanism to cover the peripheral field with relatively few ganglion cells while maintaining the same signal/noise ratio (S/N) for all cells.

Why does the retina systematically reduce, from center to periphery, the number of ganglion cell channels and thus the amount of information transmitted to the brain per degree of visual angle? Apparently, the reason is that a considerable volume of cortex is required to analyze the information carried by each retinal channel. Brain size is limited by many factors, including skull size and metabolic demand. (The human brain consumes about 20% of the body's oxygen budget; thus, it is almost, but not quite, as demanding as Defense in the federal budget.) A compromise has evolved that devotes a large volume of cortex to analyzing a tiny patch of central retina and mounts the eye on a stepping motor (the extraocular muscles) so that the small, highly analyzed patch can be trained on any object of interest.

SYNAPTIC CONNECTIONS

Synaptic connections in the retina are confined to two distinct zones, the outer plexiform layer (OPL) and inner plexiform layer (IPL) (Fig. 6.3). Certain connections within these layers are known in detail, far more so than for any other brain region. An important advantage in the retina compared to other CNS tissues is that the distances within and between these synaptic layers are short. This allows one to trace sequences of intercellular connection across as many as five cells in the chain.

OUTER PLEXIFORM LAYER

This layer contains the axon terminals of rods and cones (Fig. 6.4), the dendritic processes of bipolar neurons, and the processes of horizontal cells. The rod synaptic terminal is round (hence *spherule*), about 3 μm in diameter, and contains a few small mitochondria and two synaptic "ribbons." The ribbon is a proteinaceous structure found in the synaptic terminals of all photoreceptors (Fig. 6.10) even among the invertebrates, as well as in bipolar axons in the retina (Fig. 6.11) and in axon terminals of vestibular hair cells. The ribbon in the rod spherule perches just above the basal surface and appears to provide a focus for

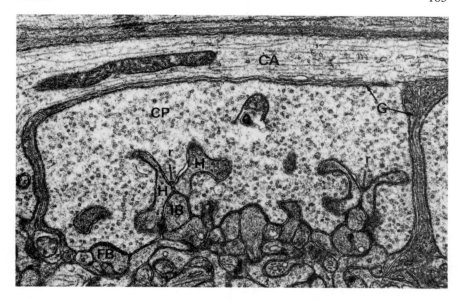

FIG. 6.10. Electron micrograph of radial section through a cone pedicle in monkey retina. Note the "triadic" synapse and the wrapping of the pedicle by glia. CA, cone axon; CP, cone pedicle; G, glia; H, horizontal cell process; FB, flat bipolar dendrite; IB, invaginating bipolar dendrite; r, synaptic ribbon. (Photo courtesy of Y. Tsukamoto.)

the synaptic "triad" (Dowling and Boycott, 1966). A single, fingerlike dendritic filament from a rod bipolar cell invaginates the spherule to terminate just beneath a ribbon, thus forming the central element of the triad. Two or three filaments from the axon terminal of the type B horizontal cell also invaginate the spherule, approach the ribbon, and form the lateral elements of the triad. The two triads in a rod spherule usually involve different rod bipolars. Consequently, although a rod bipolar neuron collects synapses from about 15 to 20 rods, each rod contributes only a single synapse (Boycott and Kolb, 1973; Sterling et al., 1988). Glial

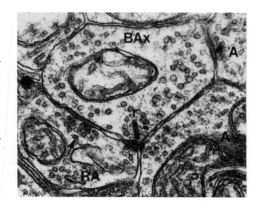

FIG. 6.11. Electron micrograph of "dyadic" synapse in the IPL of monkey retina. The bipolar axon (BA$_x$), marked by a synaptic ribbon, contacts two postsynaptic processes. These contain synaptic vesicles, indicating that they are amacrine processes, and one of them provides a feedback (reciprocal) synapse to the bipolar axon. A, amacrine; RA, reciprocal amacrine. (Photo courtesy of Y. Tsukamoto.)

membranes invest the rod spherule, but they part at the very lips of the spherule's invagination to allow access by basal processes of cone axons which provide the spherule with small electrical synapses. Through these, the rod and cone signals can mix (Kolb, 1977; Nelson, 1977; Smith et al., 1986; Freed et al., 1987).

The cone axon terminal is much larger, 5–8 μm in diameter, and flattened (footlike, hence *pedicle*). The basal processes noted above to contact rod spherules also connect electrically to adjacent cone pedicles. Thus cone signals can spread between pedicles. The pedicle contains 4–5 large mitochondria and also many synaptic ribbons (10–12 in the cat, and up to 25 in the primate). It is also filled with synaptic vesicles. In cones, as in rods, synaptic triads form deep to the synaptic ribbons (Fig. 6.10). The central triadic element in mammals is always the dendritic filament of a cone bipolar cell, and the lateral elements are always the dendritic processes from the different types of horizontal cell (Boycott and Kolb, 1973). Each cone pedicle provides multiple synapses to a cone bipolar; therefore, many of the central elements at the triads of a single cone belong to the same bipolar cell. In the cat, three types of cone bipolar form multiple central triadic elements at a pedicle (Nelson and Kolb, 1983; McGuire et al., 1984; Cohen and Sterling, 1990a). Perhaps for this reason some of the ribbons are elongated parallel to the pedicle's basal surface and accommodate several triads.

Once the cone pedicle has provided multiple synapses to three types of "invaginating" bipolar, its task is by no means complete. In addition, there are several types of cone bipolar cell whose dendritic filaments penetrate partway into the pedicle to collect multiple, "semi-invaginating" synapses (Nelson and Kolb, 1983; Cohen and Sterling, 1990a). Finally, additional types of cone bipolar extend fine dendritic tips that abut the pedicle's inferior surface where they receive multiple "basal" or "superficial" contacts (Boycott and Kolb, 1973; McGuire et al., 1984). Overall, each cone (except for the blue cones) makes multiple contacts with about nine types of cone bipolar and may produce more than 100 synapses. Little wonder that the pedicle is one of the largest axon terminals in the brain and so rich in mitochondria! (For another large terminal, see the mossy terminal of the cerebellum; Chap. 7).

The type B horizontal cell axon spreads widely and connects to rods, but to no other cell types. As noted above, it is apparently isolated electrically from its dendrites. The type A horizontal cells connect with each other electrically, but otherwise neither type A nor type B dendrites, both of which also spread widely and connect to cones, synapse with any other cell type. Such architecture certainly suggests that the task of these cells is to collect signals from many receptors and feed-back some average to the individual receptors.

INNER PLEXIFORM LAYER

This layer contains the axon terminals of bipolar neurons, the processes of amacrine neurons which are pre- and postsynaptic, and the dendrites of ganglion cells (Fig. 6.2 and 6.3). All three elements are ultrastructurally fairly distinctive. Bipolar axon terminals, as mentioned above, contain synaptic ribbons. As in receptor terminals, a ribbon perches close to the plasma membrane to form the organizing focus for a synaptic complex. At the bipolar ribbon the complex involves only two postsynaptic elements, so it is termed a "dyad" (Fig. 6.12; Dowling and Boycott, 1966). The postsynaptic elements at a dyad can both be ganglion cell dendrites, or both be amacrine processes, or be one of each. Often, the amacrine process feeds-back a chemical synapse onto the bipolar axon, in which case it is called a "reciprocal" synapse.

Although all these possibilities occur in the IPL, each cell type has its own specific pattern. For example, the rod bipolar dyad always includes one process from a type AII amacrine and one process from a small subset of other amacrine types. The AII *never* gives a reciprocal synapse at the rod bipolar dyad, but the other amacrine types *always* do. Beta ganglion cell dendrites are *never* found at the rod bipolar dyad, but alpha cell dendrites are present at about one dyad out of 30 (McGuire et al., 1984; Sterling and Lampson, 1986; Freed et al., 1987; Freed and Sterling, 1988). A bipolar axon contains 25–100 ribbons. Again, however, the number of ribbons per axon is cell-type specific. Thus, three adjacent rod bipolar axons contained, respectively, 27, 28, and 32 ribbons (McGuire et al., 1984), whereas two adjacent CBb_1 axons contained each 105 ribbons (Cohen and Sterling, 1990a).

Also present in the IPL is the common type of chemical synapse found throughout the brain. This is termed (quite reasonably) a "conventional" synapse and arises from amacrine processes (Dowling and Boycott, 1966). Synaptic vesicle size and shape do not vary much according to transmitter as in other parts of the brain; nor is Gray's scheme of classifying synapses as type 1 or 2 (Chaps. 1 and 2) of any use. Despite their known differences in function, all conventional synapses in retina look pretty much alike. Commonly in small-brained animals conventional synapses are arranged serially in long sequences (Dubin, 1976). Presumably, these sequences perform complex, local computations. However, the underlying circuits have not yet been worked out, so their actual function is unknown.

The IPL also contains electrical synapses. Some of these connect particular

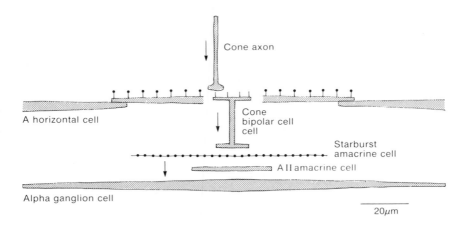

FIG. 6.12. Cable structure of various retinal cell types. See text.

types of cone bipolar cell, others connect amacrines, and still others connect particular types of amacrine and bipolar cells. The largest electrical synapses in cat IPL couple AII amacrine dendrites to the axon terminals of the CBb_1 bipolar (Sterling et al., 1988). This connection is apparently critical for vision under starlight, as will be discussed below.

In fundamental design, the ganglion cells differ from "output" cells described in other chapters of this book. For example, the ganglion cell collects a relatively modest number of synapses. A beta cell gets about 200 synapses (Sterling et al., 1988), and an alpha ganglion cell about 2000 (Freed and Sterling, 1988). In contrast, an alpha motoneuron in spinal cord may collect about 10,000 synapses, and a Purkinje cell in cerebellum may collect 100,000 synapses from parallel fibers alone! The ganglion cell is also fairly modest in soma size and dendritic extent. The beta ganglion cell is about 15 μm in diameter, and the alpha about 25 μm, compared to 50–80 μm for the motoneuron and Purkinje cell. The ganglion cell dendrites extend for about 20–200 μm, compared to 1000 μm for motoneurons. Another difference is that synapses to the ganglion cell are directed exclusively toward the dendrites with none to the soma. The motoneuron, of course, is encrusted with synapses over its entire surface. The very design of the Purkinje cell seems to hinge on an antagonism between inputs to the dendritic tree (excitatory, from parallel and climbing fibers) and inputs to the soma (inhibitory, from basket cells).

DEVELOPMENT AND PLASTICITY

Ganglion cells send their axons to the brain fairly early in embryonic development. The number is initially quite large, about 450,000 in cat, and declines gradually due to cell death, so that by birth almost the adult number has been achieved. Some additional pruning continues for a few weeks postnatally. The ganglion cell axons, upon reaching the brain, find their appropriate central targets: lateral geniculate, superior colliculus, etc. At early prenatal stages the ganglion cells fire action potentials that can drive geniculate neurons (Shatz and Kirkwood, 1984). Further, the spontaneous firing of ganglion cells seems to shape development of the geniculate laminae since blocking the action potentials with intraocular tetrodotoxin blocks segregation of the geniculate into ocular dominance layers (Shatz and Stryker, 1988).

At birth the main types of ganglion cell (*on-* and *off-*, alpha and beta) are easily recognized (Dann et al., 1988; Ramoa et al., 1988), but they are not yet connected to their retinal synaptic network because the photoreceptors are still proliferating and the bipolar axons are just descending toward the IPL (Vogel, 1978). At about the time of eye opening (6–10 days postnatal in cat), synaptogenesis enters high gear and is almost complete by four weeks postnatally (Vogel, 1978). Though each cell type probably has its own programed period and rate of synaptogenesis, there is no simple way to characterize the sequence. Thus, it does not proceed strictly centrifugally from the ganglion cell toward the receptor, nor vice versa (McArdle et al., 1977; cf. Nishimura and Rakic, 1987). Further, despite the coincidence with eye opening, the genesis of retinal wiring appears to follow a genetic program and not to be affected much by light or patterned stimulation. Neither dark rearing, nor occlusion by lid-suture (which prevents patterned stimulation, but not all light), has any obvious effect on adult retinal morphology or physiology, even though such procedures profoundly alter the structure and function of the visual cortex (Hubel and Wiesel, 1977).

The adult mammalian retina does not appear to be very plastic, in contrast to the retinas of lower vertebrates. The retina in fish continues to grow throughout life, adding nerve cells to all three layers in concentric rings at the periphery. The optic tectum, which in fish is the main target of the optic nerve, also continues to add neurons, but in concentric crescents rather than in complete rings. This creates a topological mismatch between the retina and its map on the tectum. The map therefore requires continuous readjustment, which is accomplished by the continuous retraction of old retinotectal synapses, growth of the optic axons across the tectum, and the formation of new synapses (Easter and Stuermer, 1984; Reh and Constantine-Paton, 1984). If the neural retina is totally removed from the eye of a fish or amphibian, cells from the pigment epithelium dedifferentiate and divide, regenerating a whole new neural retina whose ganglion cell axons find their way to the brain and reconnect properly.

Another sort of plasticity to consider is the shift in retinal wiring that accompanies changes in the life-style of certain organisms. For example, the adult frog eats flies and has a type of ganglion cell tuned by specific "circuits" (patterns of connection between specific cell types) to "flylike" stimuli (Barlow, 1953). But the tadpole eats algae and lacks this cell type. The fly-detecting ganglion cell and its neural circuitry develop as part of the many complex changes that accompany metamorphosis (Frank and Hollyfield, 1987). It seems fairly certain, however, that its emergence is directed by a genetic program that operates whether or not the frog is ever confronted with a fly.

There is at least one sense in which the retina does exhibit phenomena that could be described as plastic. To achieve the best possible performance over a huge dynamic range, it apparently adjusts the relative strengths of different circuits, and even the coupling strength of electrical synapses and the gains of chemical synapses (Smith et al., 1986; Sterling et al., 1987; Yang and Wu, 1989b).

BASIC PATHWAYS

At this point in other chapters of this book, the section head would read "Basic Circuits." In order to treat the complexity of the retina, we will introduce a distinction: "Pathway" will refer to a *linear* (one-dimensional) sequence of synaptically connected neurons, whereas "circuit" will refer to the same sequence of neurons when their synaptic connections are considered in terms of populations in three dimensions. To appreciate the difference at a glance, compare Fig. 6.2 to Figs. 6.14 and 6.20. The distinction reminds us that whereas certain insights emerge from knowledge of a pathway, others require knowledge of the whole circuit. With a pathway identified (physiological responses of the neurons as well as their connections), one discovers the hierarchical transformations that occur across its successive levels. However, this provides no insight into the *mechanism* of the transformation at a given stage, because this is embodied in the full pattern of connections in two dimensions. Therefore, the *purpose* of the whole sequence of transformations emerges only when the two-dimensional connections at all stages along the pathway are added to the linear sequence to give the full circuit in three dimensions (Sterling et al., 1988).

DIRECT PATHWAYS

Some of the main feedforward pathways discovered so far in the cat retina are shown in Fig. 6.2. Pathways from cones are short, involving only three neurons: cone → cone bipolar → ganglion cell. They also exhibit a striking *divergence* into parallel pathways at the very first synapse: Each cone contacts two pairs of bipolars. One pair (a_1, a_2) sends axons exclusively to sublamina *a* of the IPL, and the other pair (b_1, b_2) sends axons exclusively to sublamina *b*. Types a_1 and a_2 converge on the *off*-beta cell; b_1 and b_2 converge on the *on*-beta cell (McGuire et al., 1984, 1986). The b_1 and b_2 pathways lead also to the *on*-alpha ganglion cell (Freed and Sterling, 1988); probably the a_1 and a_2 pathways lead also to the *off*-alpha ganglion cell, but this is unproven.

Two different pathways lead from rods (Fig. 6.2). Both are longer than the cone pathways, and both involve an electrical synapse. One begins with the electrical synapses already noted from rod spherule to cone pedicle. The rod signal entering the cone pedicle has access to all the parallel cone bipolar pathways described above (Kolb, 1977; Smith et al., 1986). The other rod pathway begins with chemical synapses from the rod spherule to the rod bipolar cell. The rod bipolar contacts by means of a chemical synapse the amacrine type called "AII," and this cell provides two strikingly different outputs. In the *on* sublamina of the IPL it forms large electrical synapses with the axon terminals of the b_1 cone bipolar, and in the *off* sublamina it forms chemical synapses onto ganglion cell dendrites (both alpha and beta) and onto cone bipolar axon terminals (Famiglietti and Kolb, 1975; Freed et al., 1987; Sterling et al., 1988).

Several aspects of function can be inferred from this diagram. The rod–cone electrical synapse implies that rod signals should be present in cones and cone bipolar cells, and Nelson (1977) has observed this. The convergence of rod and cone signals implies that the ganglion cells, both *on*- and *off*-, alpha and beta, should operate over the full range of luminance covered by rod and cone systems. This also has been observed (Barlow et al., 1957). These pathway diagrams also raise certain questions. Broadly, what is the point of such extensive parallelism? Why are there parallel *on* and *off* pathways, and why are there parallel bipolar pathways to each type of ganglion cell? Further, because the same types of bipolar contact both alpha and beta cells, why should there be two classes of ganglion cell? Finally, because rods converge on the cone bipolar pathways, why is there a separate rod bipolar pathway? We shall return to these points.

INDIRECT PATHWAYS

Many synaptic contacts onto ganglion cells arise from amacrine cells. About 30% of the synapses to the *on*-beta cell are amacrine (Kolb, 1979; Sterling et al., 1988), and so are about 85% of the synapses to the *on*-alpha cell (Freed and

Sterling, 1988). Because visual signals to the amacrines arise from the bipolar cells, and because there are 20 or more types of amacrine cell, there must be a rather rich set of indirect pathways of the general form: bipolar → amacrine → ganglion cell. Though undoubtedly important, especially to the alpha cell, the facts about which *types* of bipolar and amacrine interconnect with each other and with ganglion cells are virtually unknown.

VISUAL TRANSDUCTION

We turn now to consider the functional properties of the synaptic circuits. These properties are finely tuned to the specific characteristics of the inputs from the photoreceptors, which in turn are set by the mechanisms of transduction of the visual stimuli.

Rods and cones share the same basic mechanism for transduction. The plasma membrane of the outer segment bears numerous cation channels permeable to Na^+ and Ca^{2+}. Each channel opens when it binds two molecules of cGMP. In the dark, the intracellular concentration of cGMP is high, so many channels flicker open for several milliseconds at a time. At any instant about 10^4 are open, which causes a continuous inward current due mainly to Na^+, and depolarizes the receptors (to about -40 mV) in the dark. The Na^+ entering the outer segment is extruded continuously at the inner segment, which explains the large energy demand at this region and thus the need for its densely packed mitochrondria (Attwell, 1986).

The optical image focuses on the photoreceptor array at the level of the inner segments. The photons forming this image pass through the plasma membrane and are trapped: The dense packing of mitochondria gives a high refractive index to the inner segment and thus the properties of a "wave guide" (Enoch, 1981). Photons cannot escape but funnel toward the outer segment where, in their passage through the stacked membrane disks, each strikes and is absorbed (with high probability) by a molecule of photopigment. The photon's energy isomerizes the pigment molecule, leading through an amplifying cascade of reactions to a brief fall in the concentration of cGMP. This closes the membrane Na^+ channels, which reduces the "dark current" and hyperpolarizes the outer segment. The voltage change descends through the inner segment, where voltage-gated channels shape it temporally and boost its amplitude (reviewed by Attwell, 1986). However, the photon-induced voltage at the axon terminal is not all-or-nothing, but graded with light intensity.

The rod and cone differ intrinsically in their degree of amplification. The fall in cGMP concentration caused in a rod by a quantal absorption is enough to suppress about 4% of the dark current. The same event in a cone has about 70-fold less amplification and suppresses only about 0.06% of the dark current. Consequently the rod gives a modest but detectable signal to a single photon and saturates to greater than 100 absorptions within its integration time (Baylor et al., 1984). The cone's response to a single photon is not detectable, but its response amplitude does not saturate completely even in the brightest light.

Animals with significant color vision (macaque monkey, but not cat) have at least three types of cone (Rodieck, 1988). The main difference between cone types is in the primary structure of the "opsin" protein that, linked covalently to Vitamin A, forms the photopigment. The amino acid sequences of opsins in the three cone types differ slightly, just enough to shift the peaks of their broad absorption spectra slightly toward longer or shorter wavelengths (Nathans et al., 1986). All three cone types respond to visible light of all wavelengths, but each type absorbs somewhat more efficiently than the others over a certain range.

DENDRITIC PROPERTIES

COMPARISONS BETWEEN AXONS AND DENDRITES

Classically "dendritic" has implied passive membrane and current flow in a centripetal direction, usually toward the soma or axon hillock. "Axonal" has implied active membrane and regenerative propagation away from the soma toward the presynaptic terminal. As discussed in Chaps. 1 and 2, one of the main insights gained from the study of synaptic organization is the realization that these structure–function relations are not simple. Some of the earliest and clearest evidence has come from the retina, where the classical definitions of "dendrite" and "axon" tend to dissolve. The membranes of most retinal processes, both axons and dendrites, appear to be passive. There are exceptions. For example, several types of voltage-sensitive channel are present in the photoreceptor inner segment (see above), and even without special manipulations of the ionic environment it is often possible to evoke a spike in a cone (Piccolino and Gerschenfeld, 1980). Furthermore, certain types of amacrine cell normally respond to light with action potentials (Werblin and Dowling, 1969). However, voltage-sensitive membranes apparently do not contribute in a major way to most retinal computations. Only at the output of the retina, where the *results* of the computations must be conveyed rapidly over a long distance, do action potentials assume importance for coding.

Another sense in which the dendrite–axon distinction fades is in the very notion of neuronal polarity. Although retinal neurons along the major feedforward pathways are conventionally polarized, neurons responsible for lateral connections, those that collect and disperse signals widely, exhibit no obvious polarity. These neurons, the horizontal cells of the OPL and the amacrine cells of the IPL, were a problem to Cajal's formulation and defense of the "law of dynamic polarization of nerve cells." Cajal was well aware of this difficulty; for his resolution of it, see Piccolino (1988).

FEEDFORWARD ELEMENTS

Assuming that retinal processes are passive, several points can be inferred from Fig. 6.12. The main dendrites of the feedforward neurons have a fairly narrow lateral spread and branches of medium caliber. Thus, the collecting region for dendrites of most cone bipolar types (in the area centralis) is no more than about 15 μm wide, and the axonal spread is similar. Dendrites and axon are both about 1 μm in diameter. The AII amacrine and beta ganglion cell dendritic trees are both about 30 μm wide, with individual dendrites about 1 μm, tapering to 0.5

μm in diameter. The ganglion cell with the broadest lateral spread is the alpha cell; it has a dendritic field about 180 μm wide and correspondingly thicker dendritic branches. One effect of this design is that all synapses are well within one space constant of the soma. Synapses even on distal branches are little attenuated. (For possible exceptions, see Koch et al., 1982a.)

LATERAL ELEMENTS

The lateral elements illustrated in Fig. 6.12 differ architecturally from the feed-forward elements; furthermore, OPL design differs from that of the IPL. In the OPL, the long, thick processes of horizontal cells interconnected by low-resistance junctions seem designed to spread signals over long distances. The input from each cone is delivered to a horizontal cell spine which connects over a high-resistance segment (spine neck) to the low-resistance horizontal cell dendritic network. Consequently, a cone signal suffers considerable attenuation, and the horizontal cell responds weakly to a spatially restricted stimulus that activates only a few cones. On the other hand, having suffered the large voltage drop across the spine neck, subsequent attenuation is minor. Consequently, cone signals injected far from the soma are nearly equal to those injected at the soma. Conversely, the large, low-resistance network acts as a constant current source, supplying near and distant spine heads with relatively similar amounts of current. The overall effect of the electrotonic network of type A horizontal cells is to average cone signals over a broad region and feed-back the spatially averaged signal onto every cone. This electrotonic system, due to its large expanse of membrane, has a relatively large capacitance and thus a long time constant. This serves to delay the feed-back slightly, which is important for reasons to be discussed.

In the IPL, the filamentous processes of amacrine cells seem designed to limit sharply the spatial spread of signals (Fig. 6.12; Miller and Bloomfield, 1983). A synaptic contact to an amacrine varicosity will have a larger effect on intracellular voltage and thus on transmitter release at that site and the immediate neighborhood. However, owing to the high internal resistance of the thin process, the voltage will decay sharply with distance. The short segment of dendritic branch, owing to its limited expanse of membrane and its high conductance during a postsynaptic potential (PSP), must have a relatively short time constant (see Appendix). This would make it useful for facilitating higher temporal frequencies of a signal from the bipolar inputs to ganglion cells (Vardi et al., 1989).

NEUROTRANSMITTERS AND SYNAPTIC ACTIONS

Essentially all of the transmitters identified elsewhere in the brain exist also in the retina. Transmitters can be assigned to certain cell types with confidence and to others only with more circumspection. The level of confidence depends mainly on the number of criteria that have been satisfied in each case. Multiple tests are needed because the result of any given test can be falsely positive or falsely negative. Considerable progress over the last decade suggests certain broad principles (Daw et al., 1989).

PHOTORECEPTORS

The main feedforward pathways through receptor → bipolar → ganglion cell involve the excitatory amino acid glutamate. The evidence is strongest for the photoreceptors, which contain glutamate and, when depolarized, release it onto neurons that bear various types of glutamate receptor (Miller and Slaughter, 1986; Massey and Redburn, 1987; Nawy and Copenhagen, 1987). This simple conclusion represents untold person-years of difficult but indirect effort. Two recent experiments, technical *tours de force,* are both direct and apparently conclusive. In the first (Copenhagen and Jahr, 1989), a rod from turtle retina was sucked by its outer segment into a micropipette through which it could be electrically stimulated. The tip of a second pipette, bearing a patch of neuronal membrane ripped from a cultured hippocampal neuron, was moved close to the rod axon terminal. The membrane patch, which was in the "outside-out" configuration (see Chap. 2), contained the N-methyl-D-aspartate (NMDA) type of glutamate receptor typical of hippocampal neurons. Electrically depolarized, the rod released a transmitter that opened ion channels linked to glutamate receptors in the membrane patch of the "sniffer" pipette. The other experiment (Ayoub et al., 1989) was to suck the rod axon terminal into a pipette containing glutamate dehydrogenase plus nicotinamide adenine dinucleotide (NAD) and directly measure the release of glutamate by an increase in fluorescence due to the formation of $NADH_2$.

At least as important as the transmitter is the nature of the postsynaptic receptor. Different receptors may be coupled (positively or negatively) to different ion channels, and this allows a single presynaptic element to exert quite different effects, even opposite ones, on several postsynaptic elements. Consider the action of glutamate released by the cone pedicle onto bipolar cells (Fig. 6.6B). One class of bipolar, exemplified by CBb_1, bears receptors activated by 2-amino-4-phosphonobutyrate (APB), and these *depolarize* to light onset. Another class of bipolars, exemplified by CBb_2, bears receptors antagonized by *cis*-2,3-piperidinecarboxylic acid (PDA), and these *hyperpolarize* to light onset (Slaughter and Miller, 1981, 1983). The functional significance of leading the presynaptic signal into parallel bipolar pathways of different polarity is discussed below.

Another reason for coupling the same presynaptic transmitter to different receptors is to create postsynaptic elements with different temporal responses. It is obvious in Fig. 6.6B that the signal in the b_1 bipolar cell has both a transient and a sustained component, whereas the signal in the b_2 bipolar cell has only the sustained component. Similar observations of different temporal filtering have been made on bipolar and horizontal cells in fish and turtle (Ashmore and Copenhagen, 1980; Copenhagen et al., 1983; Nawy and Copenhagen, 1987).

BIPOLAR CELLS

Evidence that the bipolar neurons release glutamate as a transmitter is slender by comparison. It rests mainly on the responsiveness of ganglion and amacrine cells to glutamate applied ionophoretically and its various agonists and antagonists (Massey and Miller, 1988). As noted above, there are many types of bipolar neuron, and there are reasons to believe that some may use other transmitters. For example, certain bipolar neurons accumulate exogenous [^{3}H]glycine and contain endogenous glycine, as shown immunocytochemically (Cohen and Sterling, 1986; Pourcho and Goebel, 1987). Furthermore, some ganglion cells bear glycine receptors, and respond to ionophoresis of glycine with an increased conductance to Cl$^-$ which is blocked by the glycine antagonist, strychnine (Bolz et al., 1985).

The suggestion that certain bipolar neurons would inhibit ganglion cells is appealing because it would explain cases where bipolar types of opposite response polarity contact the same ganglion cell. In the example shown (Fig. 6.6B), the depolarizing bipolar cell would excite the *on*-beta ganglion cell by increasing release of glutamate. Simultaneously, the hyperpolarizing bipolar cell would disinhibit the ganglion cell by decreasing release of inhibitory transmitter. Such a mechanism has been termed "push–pull" (McGuire et al., 1986), in analogy to a certain type of electronic amplifier. The possible function of such a mechanism is discussed below (Functional Circuits). However, it is difficult to *prove* that the glycine present in a bipolar cell such as CBb$_2$ is actually the transmitter. Although the ganglion cell may respond to ionophoresed glycine, the substance might normally be delivered by glycinergic amacrine cells. Furthermore, the presence of glycine in a bipolar axon might merely reflect its coupling by gap junctions to a glycine-accumulating amacrine cell (Cohen and Sterling, 1986).

HORIZONTAL CELLS

Many of the lateral circuits in retina use inhibitory amino acids (Daw et al., 1989). Evidence is strong that horizontal cells use GABA. In the cat, horizontal cells neither accumulate exogenous GABA nor stain with antisera to glutamic acid decarboxylase (GAD), the GABA-synthetic enzyme. However, the cells *do* stain with antiserum to GABA itself (Chun and Wässle, 1988; Wässle and Chun, 1988), and they also show in situ hybridization to mRNA for the glutamic acid decarboxylase. Furthermore, the axon terminals of photoreceptors hyperpolarize to ionophoresis of GABA, as they should if it is truly the horizontal cell's transmitter (Tachibana and Kaneko, 1984). Recent evidence suggests that release of GABA by horizontal cells does not occur by the conventional, Ca^{2+}-dependent, quantal release mechanism, but rather by a Ca^{2+}-independent "carrier" mechanism (Schwartz, 1987).

AMACRINE CELLS

Although quite a few different transmitters are associated with amacrine cells, there are more cell types than transmitters. Thus, in the cat, about five types of amacrine cell synthesize, accumulate, and store GABA (Freed et al., 1983). Several types accumulate glycine (Sterling, 1983; Pourcho and Goebel, 1985); two amacrine types, termed "starburst" (Famiglietti, 1983), synthesize acetylcholine (Masland and Tauchi, 1986) (see Chap. 1). There are also several amacrine types that accumulate indoleamine (e.g., Sandell and Masland 1986), and at least one that by several criteria is dopaminergic (Pourcho, 1982; Voigt and Wässle, 1987). In fact, when all the members of each type that accumulate or synthesize a particular transmitter are summed, it is clear that there are not enough amacrines to go around. Some have to be counted twice. This implies colocalization of two transmitters in the same cell type, and this has been clearly demonstrated in several instances (reviewed by Miller, 1988). It turns out that GABA is pretty ubiquitous, being associated with the amacrine types that contain acetylcholine, dopamine, and indoleamine. The functional significance of having two transmitters in the same neuron remains to be determined.

The transmitters in the IPL, like glutamate in the OPL, also have multiple types of receptor. Thus, for acetylcholine there are several types of nicotinic receptor as well as a muscarinic receptor, and for GABA there are type A and B receptors. For dopamine there are D_1 and D_2 receptors for dopamine, and for indoleamines that are also several types (reviewed by Daw et al., 1989; Massey and Redburn, 1987).

Certain cell types, with a fairly stereotyped morphology across species, also have the same transmitter. Thus, the "starburst" amacrine, first described in the rabbit (Famiglietti, 1983), is recognized in the rat, cat, monkey, and bird. Invariably it is represented by both *on* and *off* types, and invariably these contain acetylcholine (Masland and Tauchi, 1986).

PEPTIDES

Many of the peptides present elsewhere in the brain are present also in retina (reviewed by Massey and Redburn, 1987). The list varies somewhat between species, but includes such compounds as enkephalin, endorphin, somatostatin, thyroid hormone-releasing hormone (TRH), vasoactive intestinal peptide (VIP), substance P, calcitonin gene-related peptide (CGRP), and adrenocorticotropic hormone (ACTH). The effects of these compounds on various cell classes in retina are just beginning to be studied, and their contributions to visual performance are completely obscure. Perhaps the most interesting observation so far is that TRH suppresses the maintained activity and light response of *on* ganglion cells and enhances the responses in *off* ganglion cells—but only in relatively bright light. In dim light, TRH has no effect (Bolz and Thier, 1985). Clearly, to link the peptide-containing amacrines with known feedforward circuitry will be a critical advance.

FUNCTIONAL CIRCUITS

GENERAL ASPECTS OF GANGLION CELL FUNCTION

Receptive fields. A ganglion cell's functional connections to the two-dimensional receptor mosaic determine what is called its "receptive field." A receptive field is mapped by exploring the receptor sheet with a small spot of light while recording action potentials from the ganglion cell. Using this method, Barlow (1953) showed in the frog, and Kuffler (1953) in the cat, that a common type of receptive field has two distinct regions, arranged concentrically and antagonistically (Fig.

6.5). Illuminating the receptors in the region centered immediately over the ganglion cell's dendritic field elicits a sharp change in firing rate. The *on* cells increase their firing, and the *off* cells decrease their firing to the onset of light. The response to the offset of light is the reverse. Illuminating the receptors over a much broader region surrounding the center (4–5 times the center diameter) antagonizes the response to the center. Thus, an *on* center ganglion cell has an *off* surround, and vice versa (Kuffler, 1953).

The distribution of sensitivity across the center of a ganglion cell receptive field is not uniform but dome-shaped, peaking at the middle and declining toward the edge. This is also true for sensitivity in the surround. These mutually antagonistic, domelike distributions of sensitivity are reasonably fitted by Gaussian functions. Therefore, the overall sensitivity profile of a ganglion cell has come to be described as the "difference of Gaussians," one narrow and tall, the other broad and shallow (Fig. 6.13A) (Rodieck, 1965; Enroth-Cugell and Robson, 1966). This arrangement ensures that the ganglion cell will respond, not to absolute luminance but to the difference in luminance between the center and the surround. In short, it responds to the *contrast* in a stimulus, and this is probably a foundation for the constancy of our perception of black and white, to be discussed below (Shapley and Enroth-Cugell, 1984).

This basic receptive field structure is constant from the brightest to the dimmest daylight (luminance range of 3–4 log units). Nor does it change in twilight

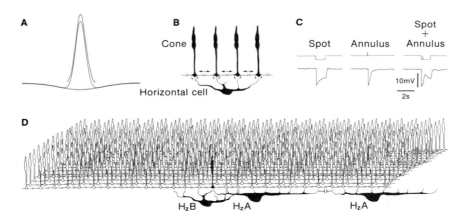

FIG. 6.13. **A.** Difference-of-Gaussians sensitivity profile (one-dimensional), the classical quantitative model of a ganglion-receptive field. **B.** One-dimensional view of cone–horizontal cell pathways responsible for establishing a center–surround receptive field at the level of the cone synaptic terminal. **C.** Intracellular recordings from turtle cone demonstrating its center–surround receptive field: A spot hyperpolarizes the cone on which it is centered; an annulus causes a brief hyperpolarization by scattering light onto the center; the annulus presented together with the spot demonstrates the depolarizing effect of the surround. **D.** Two-dimensional view of the cone–horizontal cell circuit. Each cone pedicle integrates the signals from about 2500 surrounding cones before its signal is fed-forward to the bipolar cells. (A–C from Gerschenfeld et al., 1980; D from Smith and Sterling, 1990b.)

or even moonlight, over an additional decline in luminance of about 4 log units (Enroth-Cugell et al., 1977; Sterling et al., 1987). However, when the luminance level of starlight is reached, about the last log unit above absolute threshold, the receptive field structure changes abruptly (Barlow et al., 1957). The center apparently expands so that a fully dark-adapted beta cell in the area centralis will connect to about 1500 rods (Sterling et al., 1988). Accompanying this change is a marked weakening (virtual loss) of the antagonistic surround (Barlow et al., 1957; Derrington and Lennie, 1982). Sensitivity in the center is so heightened that a single quantal event—that is, the isomerization of a single rhodopsin molecule—causes 2–3 spikes in the ganglion cell (Barlow et al., 1971; Mastronarde, 1983). The ganglion cell sums linearly the signals from the quantal events in different receptors so that under starlight its maintained discharge to a steady background is proportional to the luminance, rather than as before, to the contrast (Barlow and Levick, 1969).

Temporal properties. At the onset of light a ganglion cell fires action potentials at high frequency for a brief period (transient component). The firing rate then decays to a lower frequency (sustained component). The beta cell gives both a strong transient and a strong sustained component in bright light, but the alpha cell gives primarily a transient and much less sustained component. Thus, these cell types were initially called by some investigators "sustained" and "transient" cells (Fig. 6.5; Cleland et al., 1971). However, it must be realized that in dimmer light photoreceptors respond more slowly. Consequently, the transient response must weaken in all ganglion cells as the light dims, so there comes a point where the beta and alpha cells can no longer be distinguished by this criterion.

Circuits for the ganglion cell receptive fields. Once the ganglion cells' spatial and temporal properties were known, it was of great interest to determine the properties of the intrinsic retinal neurons (bipolars, amacrines, and horizontal cells), for this would reveal how ganglion cell responses arise. To record from these intrinsic cells is technically challenging. Whereas a ganglion cell's behavior can be monitored by extracellular recording of action potentials, the intrinsic retinal neurons, lacking all-or-nothing spikes, can be monitored only by intracellular recording, which is far more difficult. The first intracellular recordings from morphologically identified intrinsic neurons were achieved by Werblin and Dowling (1969) in the mudpuppy, and by Kaneko (1970) in the goldfish.

They found bipolar cells to have concentric, antagonistic receptive fields, both *on* and *off* center, and horizontal cells to have broad, uniform fields. It was natural to suggest that direct connections [receptor → bipolar cell] form the bipolar cell's receptive field center, whereas indirect connections [receptor → horizontal cell → bipolar cell] form its surround. However, Baylor and his colleagues (1971) soon showed that the cone itself has an excitatory receptive field much larger than its own aperture, and a much broader antagonistic surround (Fig. 6.13). Apparently, the basic center–surround structure is established at the cone terminal—that is, *before* the first feedforward synapse.

With this background, we are in a position to present specific anatomical circuits by which cone receptive fields carry forward to specific types of ganglion cell. These provide parallel circuits that may be tuned for optimal processing of visual information under different conditions such as bright daylight, twilight, and starlight.

CIRCUIT FOR DAYLIGHT: THE *ON*-BETA GANGLION CELL

The *on*-beta ganglion cell *dendritic* field encompasses roughly 36 cones which connect to it via nine b_1 cone bipolar axons (Fig. 6.14A). These contribute a total of about 100 chemical synapses, but they do so unequally: Axons at the middle of the beta cell dendritic field contribute many synapses, axons at the edge contribute only a few. Consequently the cones innervating bipolar cells at the middle of the field will affect the beta cell strongly, whereas cones innervating bipolars at the edge will effect the beta cell only weakly. The resulting "synaptic weighting function" is shown in Fig. 6.14B (Cohen and Sterling, 1990b). The *on*-beta *receptive* field is considerably wider than the dendritic field. The recep-

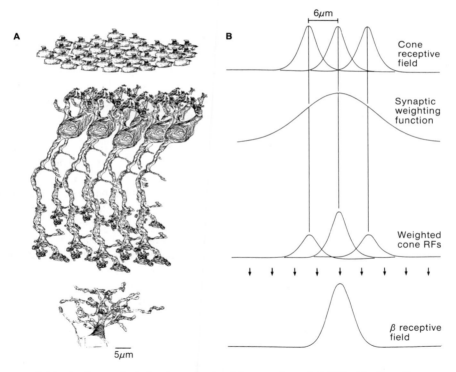

FIG. 6.14. **A.** Anatomical circuit connecting 36 cones through 9 CBb$_1$ bipolar cells to an *on*-beta ganglion cell. The central bipolars contribute many synapses to the beta cell, whereas the peripheral bipolars contribute only a few. This distribution produces the synaptic weighting function shown in B. **B.** Beta ganglion cell receptive field generated by "convolution" of cone receptive fields with bipolar synaptic weighting function. (A from Cohen and Sterling, 1990a,b; B from Smith and Sterling, 1990b.)

tive field *center* encompasses about 100 cones, roughly three times the number in the dendritic field, and the surround is even wider, and includes about 2500 cones (Cleland et al., 1979; Linsenmeier et al., 1982). How do cones beyond the beta *dendritic* field contribute to the beta cell *receptive* field? Probably through their contributions to cones that *do* connect directly to the beta cell. If many cones beyond the dendritic field contribute through a few cones within the dendritic field, then the cone receptive field itself must be rather wide (Smith and Sterling, 1990a).

How wide can be estimated in several ways: One is to calculate what the cone receptive field should be if the array of 36 cones (6 μm spacing) sums linearly according to the bipolar cell synaptic weighting function (Fig. 6.14B) to produce the beta cell receptive field (Smith and Sterling, 1990a). Another method is to simulate the cone receptive field using a large-scale compartmental model of the cone-horizontal cell network (Fig. 6.13D) (Smith and Sterling, 1990b). Both methods suggest that the receptive field center of the cone is about 90% as wide as that of the beta cell. The receptive field surround of the cone is also computed to be nearly as wide as that of the beta cell surround, but somewhat shallower.

According to the compartmental model, the cone center, although almost as wide as the beta cell center, is not dome-shaped. Rather, it is quasi-exponential, with the rapid-then-slow decay characteristic of long electrotonic cables (Fig. 6.14B). This is to be expected if the cone receptive field arises by electrical coupling to its neighbors. How then is the dome-shaped sensitivity profile established ultimately in the beta cell? Apparently, it is established by convolution of the quasi-exponential cone receptive field profile with the broad bipolar → ganglion cell synaptic weighting function (Fig. 6.14B).

Circuit architecture versus beta cell performance. How does the architecture of this circuit relate to the quality of the beta cell's performance? The answer requires us to consider what basic task the beta cell performs in the natural environment. A natural scene is full of fine detail that must be projected as an optical image onto the photoreceptors and converted into a neural image for projection by the ganglion cells to the brain. The finest-grained neural image that can be sent to the brain must belong to the arrays of *on-* or *off*-beta cells because each has the narrowest sampling aperture (smallest receptive field center) of all the ganglion cell arrays. The fine detail present in a natural scene is nothing like that in an eye chart or a newspaper where the contrast is great (black/white = 1/10). Contrasts in a natural scene tend to be extremely low, more like 90/100 or even 99/100 (Srinivasan et al., 1982). Such are the contrasts provided by a rabbit in an amber field, a frog in the grass, a toad on the sand, etc. Under the best circumstances, both humans and cats can detect such low contrast, and it is probably fair to say that the beta cell's most important function is to create a fine-grained neural image using the lowest contrasts that it possibly can.

To represent a low-contrast scene as an optical image requires alot of light. One can verify this simply by viewing an image containing fine detail at a distance where it begins to blur. If one decreases the intensity of the image, either by turning down the light or by viewing it through a dark glass, the detail will be utterly lost. If one increases the intensity, the image will be clearer unless the light is already very bright. This can hardly be news, for who is unaware that visual acuity deteriorates after sunset? But why? Consider the explanation by Albert Rose, a pioneer in video engineering.

Rose (1973) likens the retina to a black canvas on which particles of light (photons) are used to paint a scene in the style of the pointillists. To render one picture element (pixel) black and the others white (high contrast) requires at least $N - 1$ photons, where N = the number of pixels in the array. However, to render this single pixel *gray*—say, a luminance 99% of the surrounding white pixels—requires the gray pixel to receive 99 photons while the others get 100. Thus, the less the contrast in a scene, the more light is needed to represent it in an image (Fig. 6.15A).

There is an additional fact of physics to consider—that the arrival of a photon at a given point in space is random in time. This temporal fluctuation in photon arrival, termed *photon noise,* is proportional to the square root of the average number of photons. Consequently, to paint the pale-gray pixel noted above (1% contrast) using random white dots requires, not 100 photons, as with a determinate paintbrush, but the square of 100—that is, 10,000 photons. To represent, for an instant, in gray the finest detail that human optics can project on the retinal canvas would require a single cone to register about 10,000 photons, which is about what is available in strong daylight. In short, to see fine detail at low contrast, every possible photon must be counted to minimize photon noise.

The noise in the optical image from photon fluctuations carries forward into the neural image where there are additional sources of noise. Each critical step in

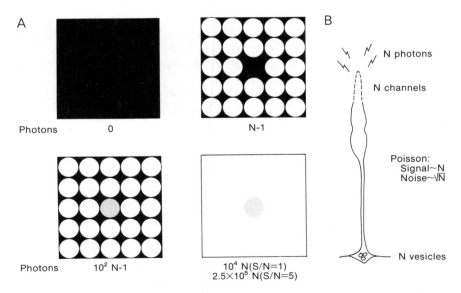

FIG. 6.15. **A.** Explanation, after Rose (1973), of why so many photons are needed to represent a low-contrast optical image. See text. **B.** Multiple sources of noise at several stages before the first feedforward synapse. See text.

the creation and transmission of the neural image depends on indeterminate processes (Fig. 6.15B). These, like photons, are governed by Poisson statistics so their noise level also follows the "square root law." For example, the photochannels in the receptor outer segment flicker open and shut as they bind and release molecules of cGMP. The closing of a given channel at any instant does not represent a fall in the concentration of cGMP (any more than a single bump of a gas molecule on the wall of a container represents pressure) and therefore does not represent a transduction event. Only a fall in the *average* number of open channels signifies transduction, so again, the S/N ratio is $N/\sqrt{N}$. To represent at the first neural stage a 1% difference in the optical image will require 10,000 photosuppressible channels, which is about what is available at any instant in one receptor (Attwell, 1986). It may be noted that the Poisson statistics governing these properties also govern synaptic vesicle release at the axon terminal (see Chap. 1).

Im summary, every stage of processing in the synaptic circuits of the retina repeats the critical problem of building a low-contrast image by means of indeterminate processes. What requires 10,000 photons will require 10,000 photochannels, 10,000 vesicles at the first synapse, and so on. Retinal circuitry seems to have been shaped by this ceaseless struggle: to maximize the signal and minimize the noise. With this in mind, let us return to the beta cell circuit. Several questions can now be addressed regarding the functional architecture:

1. Why should the center of the smallest beta cell receptive field collect signals from 100 cones?
2. Why should the weighting of these signals at the level of the beta cell be dome-shaped?
3. Why should the center–surround receptive field be established at the cone axon terminal—that is, *before* the first feedforward synapse?
4. Why should there be parallel *on* and *off* arrays, and why should they be established at the first feedforward synapse?

Questions such as these have been reviewed in regard to the physiological optics, photoreceptor mechanisms, and ganglion cell transmission by Barlow (1981). Here we review them in regard to the neural circuitry. All of these features apparently reflect the beta cell's primary task: to signal fine detail at low contrast. The reason why the beta cell center collects signals from 100 cones is that these signals are partially correlated due to local correlations in natural scenes. Collecting them reduces noise arising from Poisson fluctuation of photons, transduction cascades, membrane channels, and quanta of synaptic transmitter. One reason why the weighting function for summing cone signals is dome-shaped is that this is optimal for improving the S/N ratio when the correlations between the cone signals are less than 1. This suggests that the Gaussian center of the ganglion cell, and the circuitry that produces it, might have evolved to maximize the S/N ratio in the ganglion cell and thus improve the cell's contrast sensitivity (Tsukamoto et al., 1990).

We noted earlier (Cell Populations) that with increasing distance from the central area the beta cell dendritic field diameter increases and the cone density falls. The result is a net linear increase in the number of cones converging on the beta cell (Fig. 6.9). The increased *number* of cones tends to improve the S/N ratio according to the square-root law. However, this tendency is weakened by the increased cone spacing which tends to decrease the strength of their correlations. For beta cells these two factors balance, which may ensure that beta cells all across the retina have the same S/N ratio (Tsukamoto et al., 1990).

The reason why the center–surround receptive field is established at the level of the cone axon terminal is to protect the signal against corruption by Poisson noise added during feedforward transmission. The way to protect a signal from noise is to amplify it so that the signal becomes large compared to any added noise. The range over which the cone's membrane potential can modulate transmitter release is only tens of millivolts, but the range of luminance in the optical image may vary by several log units. If the cone had to devote its limited voltage range to coding the full luminance range, there could be very little amplification (Fig. 6.16).

The solution is to measure the average luminance over a fairly broad region and use it to predict what the luminance at the center of that broad region "should" be. This prediction can then be subtracted from the actual luminance at the center, because the mean luminance, being common to all the elements, is redundant. With the signal representing mean luminance removed, the cone terminal can devote its modest dynamic range to amplifying the *difference* between the predicted and actual. This difference, which represents local contrast, carries most of the important information. Thus, the "predictive coding" accomplished by surround antagonism at the level of the cone terminal serves to reduce the level of redundancy in the signal (Fig. 6.16). This effective compression of the signal allows greatest amplification and thus the best protection against noise (Barlow, 1981; Srinivasan et al., 1982; Buchsbaum and Gottschalk, 1983; Laughlin, 1987). Clearly, were this amplification deferred until *after* the first synapse, the signal would be irreparably corrupted by noise at the first stage of transmission.

The reason why there should be parallel *on* and *off* arrays at the first feedforward synapse is to increase the cone terminal's effective dynamic range. The voltage swing of a cone moves the membrane potentials of two bipolar cells in opposite directions (Fig. 6.6B). Thus, to a given contrast, the cone terminal will increase the transmitter output of one bipolar and decrease the transmitter output of another—in effect, doubling the dynamic range of the cone's synaptic output. The outputs of the bipolar pair can be arranged in several ways. Both members can connect to the same ganglion cell, with one being excitatory and the other possibly inhibitory. This is the "push–pull" mechanism mentioned above (Fig. 6.6B). Alternatively, both members of the bipolar pair can be excitatory, each connecting to a different ganglion cell type. This creates the independent arrays, also described above, of *on* and *off* ganglion cells (Fig. 6.5). Thus, the retina may exploit both possibilities.

The push–pull arrangement would serve increased amplification, which in turn would both protect against noise and also improve the time constant of the cell's response. The independent *on* and *off* arrays allow ganglion cells to signal

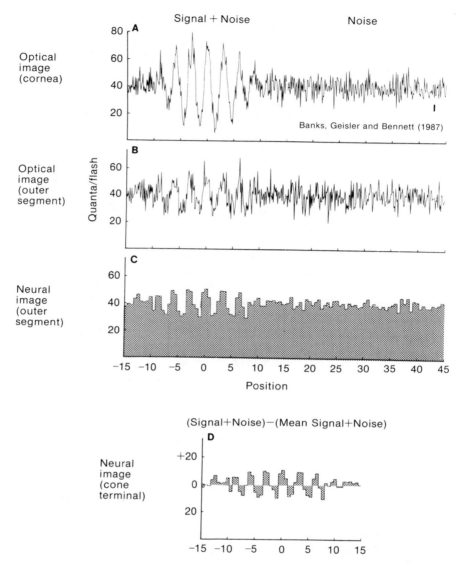

FIG. 6.16. A fine, low-contrast grating is flashed briefly on a small patch of a CRT. Abscissa, minutes of arc. **A,B.** The spatial distribution of quanta at the cornea and cone outer segments. **C.** The spatial distribution of the neural response at the level of outer segment. At this level, the signal is not far above the noise and is noticeably in danger of corruption at the next stage (synaptic transmission). If the signal were amplified at this stage to protect it against noise, most of the dynamic range would be occupied by the DC level (calculated according to an "ideal observer" model by Banks et al., 1987). **D.** The signal to be amplified after application of the "predictive coding" scheme (Srinivasan et al., 1982). See text.

both increments and decrements in luminance by increased firing rate. This protects against Poisson noise that arises during spike generation. Such noise will decrease with the square root of the number of spikes. If there were only *on* ganglion cells, the *S/N* ratio would be worse for decrements in luminance than for increments, and visual thresholds would be worse for decrements This has recently been demonstrated in primates by testing sensitivity to luminance increments and decrements in the presence and absence of the drug APB which blocks transmission to the *on* bipolar cells (Sandell and Schiller, 1988).

A PARALLEL CIRCUIT FOR DAYLIGHT: THE *ON*-ALPHA GANGLION CELL

The circuit of the *on*-alpha cell resembles that of the *on*-beta cell in that the alpha cell collects synapses from the same array of b_1 cone bipolar cells. However, whereas the beta cell collects from about nine members of the b_1 array, the alpha cell collects from roughly 150 members, each of which contributes 1–7 synapses. More synapses are collected near the center of the alpha dendritic field than toward its edge—principally, it seems, because that is where there is more alpha cell dendritic membrane. This creates for the alpha cell a broad, domelike synaptic weighting function, which convolved with the cone receptive fields, contributes to the alpha cell's broad, difference-of-Gaussians receptive field (Fig. 6.17; Freed and Sterling, 1988).

The alpha cell differs from the beta cell in having a more transient response (Fig. 6.5). One source of this difference arises simply from the different geome-

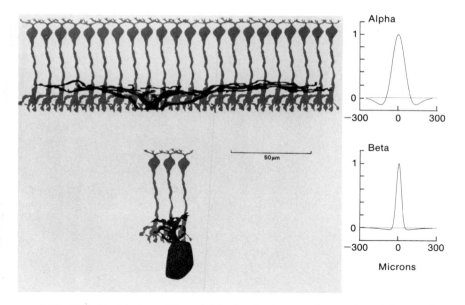

FIG. 6.17. *Left:* Radial view of b_1 cone bipolar circuits to *on*-beta and *on*-alpha ganglion cells. *Right:* Difference-of-Gaussians receptive field profiles for the same ganglion cells. See text. (Left, from Freed and Sterling, 1988; right, from Smith and Sterling, 1989a.)

try of convergence from the b_1 cone bipolar array (Fig. 6.17). The center of the b_1 bipolar receptive field is wide (about 50 μm) compared to the spacing within the array (about 10 μm). Consequently, when only a few b_1's converge (as on a beta cell), their centers will tend to be concentric and thus mutually reinforcing, and so will their surrounds. However, when a large array of b_1 bipolars converges (as on an alpha cell), any point in the middle of the field will be overlapped by several b_1 bipolar centers and many b_1 bipolar surrounds. This causes a much deeper surround Gaussian for the alpha than for the beta cell (Fig. 6.17). Consequently, a small maintained stimulus to the receptive field center of an alpha cell will deliver excitation from a few b_1 bipolar centers, and with some delay, withdrawal of excitation due to stimulation of many b_1 bipolar surrounds. This wiring will tend to emphasize the transient and suppress the sustained component in the alpha's response (Freed and Sterling, 1988).

Eighty-five percent of the synapses to the *on*-alpha cell arise from amacrine varicosities (Freed and Sterling, 1988). Some of these may derive from *starburst* amacrine cells since these arborize in the same plane as the alpha dendrites and course in intimate association with them over considerable distances (Fig. 6.18; Vardi et al., 1990). At light *on*, a starburst process releases a pulse of acetylcholine (Masland et al., 1984), and this transmitter is known to excite the alpha cell (Schmidt et al., 1987). Thus, the main feedforward circuit [cone $\rightarrow$ b_1 $\rightarrow$ alpha] may be supplemented by a parallel loop [cone $\rightarrow$ b_1 $\rightarrow$ starburst $\rightarrow$ alpha]. This loop may have the "highpass filter" properties suggested for amacrine processes (see Dendritic Properties, above) and therefore might further enhance the transient portion of the alpha ganglion cell's response.

It is fair to ask why there should be an alpha cell array in addition to a beta cell array, particularly since the large aperture of the alpha cell renders it poor for signaling fine spatial detail. One reason is that the very receptive field structure that renders the beta cell sensitive to fine detail at low contrast (narrow center,

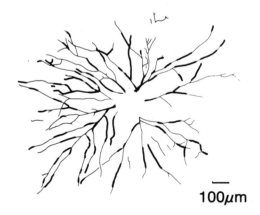

100μm

FIG. 6.18. Alpha ganglion cell dendritic tree (tangential view) showing intimate association over about 25% of its total length with bundles of starburst amacrine processes. (From Vardi et al., 1990.)

broad surround) renders it *in*sensitive to coarser details. Thus, the beta cell's contrast sensitivity declines at low spatial frequencies (Derrington and Lennie, 1982). The alpha cell fills this gap, because its large center renders it sensitive to lower spatial frequencies. It is as though the two cell types view the world through screens of different mesh.

Another advantage of the alpha cell is that it can do for temporal correlations in the visual scene what the beta cell does for spatial correlations. A low-contrast spot moving rapidly across the cone mosaic will add to each cone only modest numbers of extra photons. It would be impossible, by examining the output of any single cone, to distinguish this signal from photon fluctuation. However, the S/N ratio could be improved by summing the temporally correlated signals from a sequence of cones. In this case, the most valuable information in the signal is that which is most sharply demarcated in time—that is, the transient. Furthermore, the larger the region that can be devoted to temporal averaging, the greater the sensitivity to high velocity. Thus, both major features of the alpha cell, its large receptive field center and its transient response, suit it to extend the range of motion detection beyond what the beta cell can do.

CIRCUIT FOR TWILIGHT: THE *ON*-BETA GANGLION CELL

As the sun sets and twilight falls, our visual perception of the world changes. Colors gradually shift in hue and then desaturate, though spatial resolution survives to some extent. Thus, in a garden the red roses turn purple and then black, but the structure of the bush remains distinct. Only later, as the stars appear, do the details of the foliage dissolve into shadow. These perceptual changes seem to depend on changes in the circuits that drive individual ganglion cells. How might this happen?

The shift in hue and subsequent loss of color reflect a gradual decline in the amplitude of cone signals (and therefore a loss of trichromacy) and concomitant shift to rods. Rods saturate at luminances that cause more than about 100 photoisomerizations per integration time (about 200 msec); thus rods are useless in daylight. However, in twilight at luminances where cone responses are declining, rod responses are rising. Rods are most sensitive to somewhat shorter wavelengths (toward the blue) than are cones; thus the rose shifts in hue (Sterling et al., 1987).

The schematic circuit for twilight is shown in Fig. 6.19. Each cone in the area centralis is surrounded by about 40–50 rods, and its axon manages to contact all of these by means of electrical synapses. Therefore, as the photosignal from the cone outer segment declines in twilight, compensatory photosignals enter the pedicle from all the surrounding rods. Now, all the circuits driven by input to the cone pedicle (cone–cone electrical synapses, horizontal cell feedback synapses, and feedforward synapses onto cone bipolar cells) continue to work as in daylight. No change in the basic center–surround receptive field would be expected, and indeed none is found at the level of the ganglion cell (Enroth-Cugell et al., 1977). On the other hand, one may well expect a decline in temporal sensitivity due to the slower responses of rods, and a decline of contrast sensitivity due to the increase in photon noise that inescapably accompanies the fall in light intensity. Thus, the initially puzzling finding of a robust rod signal in the cone (Nelson, 1977) is explained: The rod signal is robust in the cone because it represents the summation of signals from many rods. This rod signal apparently serves to drive in failing light the same cone circuits that drive the ganglion cell in daylight (Smith et al., 1986).

As ambient luminance declines further, there comes a point where photons are so thinly spread that every rod absorbs only one photon within its integration time. Now the signal from the rod outer segment is tiny (<1 pA) and represents a single quantal event, the isomerization of one molecule of rhodopsin (Baylor et al., 1984). As luminance falls still further, toward the range of moonlight, the number of quantal signals in the *pool* of rods converging on a cone declines. Thus, the number of rods effectively converging onto the cone falls drastically and this reduces the ability of rods to drive (through electrical synapses) the cone circuits. Another problem at such low luminance is that the rod–cone electrical network, which is isopotential when all rods are active, becomes progressively less so. Therefore, a quantal signal arriving at a rod axon terminal will tend to dissipate into the low-resistance rod–cone electrical network. Little photovoltage will be transmitted to either the cone or the rod axon terminals. This point is illustrated in Fig. 6.19. One is led to conclude that at such low ambient luminance, the electrical synapses between rods and cones must *uncouple* in order to protect transmission of the quantal photovoltage to the rod terminal. There it can modulate a chemical synapse to the rod bipolar cell (Smith et al., 1986). Direct evidence for decreased electrical coupling between rods and cones in dim light has recently been obtained in salamander by Yang and Wu (1989b).

CIRCUIT FOR STARLIGHT: THE *ON*-BETA GANGLION CELL

The circuit responsible for beta cell function in starlight is thought to involve the rod bipolar $\rightarrow$ AII amacrine pathway (Fig. 6.20). The architecture of this circuit seems well suited for its proposed function, namely to convey each quantal signal at maximum gain to the ganglion cell (Sterling et al., 1988). Consider the sequence shown on the right in Fig. 6.20. One rod *diverges* at successive stages to two, five, and finally eight interneurons. Thus, the quantal signal arising in one rod would expand to eight copies at the level of the b_1 bipolar axon, and this would mobilize *several hundred* synapses onto beta ganglion cells. This would amplify the quantal signal and thus protect it from certain corruption by noise at the level of the ganglion cell spike generator. The reconvergence occurring at the last stage of this circuit (from eight copies in the b_1 bipolar to two in the *on*-beta cell) tends to remove accumulated synaptic noise by signal averaging: The eight copies of the signal are time-locked and thus mutually reinforcing when they converge. However, the synaptic noise accompanying each copy of the signal is temporally random and thus not reinforced by the convergence. At the last stage in this circuit there is *di*vergence of the quantal signal to at least two *on*-beta cells (Fig. 6.20). This matches the finding by Mastronarde (1983) that adjacent *on*-beta cells show coupled firing to the same quantal event.

The *convergence* of about 1500 rods onto one *on*-beta cell is shown on the left in Fig. 6.20. This number of rods accounts for the spatial extent of the beta cell's expanded, pure, center receptive field under starlight (see above). This circuit

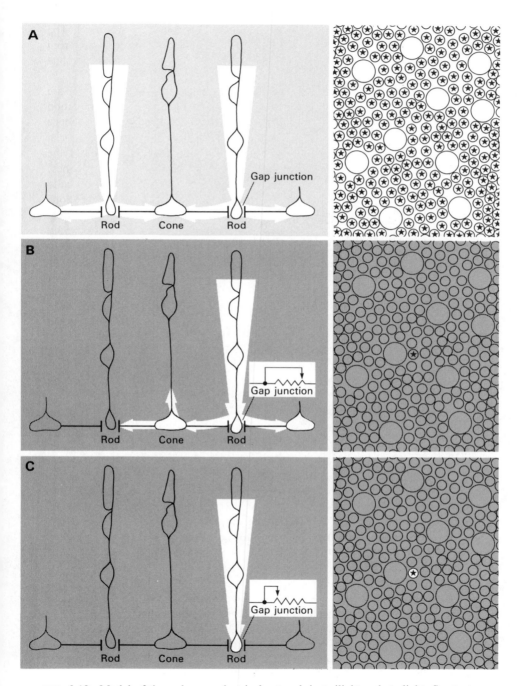

FIG. 6.19. Model of the rod–cone electrical network in twilight and starlight. See text. (From Smith et al., 1986.)

Convergence   Divergence

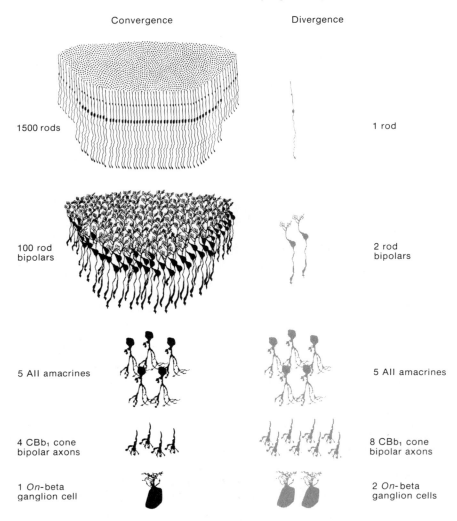

1500 rods   1 rod

100 rod
bipolars   2 rod
                                            bipolars

5 AII amacrines   5 AII amacrines

4 CBb₁ cone
bipolar axons   8 CBb₁ cone
                                            bipolar axons

1 *On*-beta
ganglion cell   2 *On*-beta
                                            ganglion cells

FIG. 6.20. Convergence and divergence in the rod bipolar circuit to the *on*-beta ganglion cell. This circuit apparently conveys single quantal events to the beta cell in starlight. See text. (From Sterling et al., 1988.)

also accounts for the ganglion cell's maintained discharge in the dark of about 15 spikes per second. This discharge has been attributed to quantal "dark events" (thought to arise from thermal isomerization of single rhodopsin molecules), which would be transmitted through the circuit with the same gain of 2–3 spikes per event as though they were caused by photon absorption. The number of these events in the receptive field of a ganglion cell has been estimated at about 6 per second (Barlow et al., 1971; Mastronarde, 1983). Fifteen hundred rods would produce this many events if there were about 0.004 dark events/rod/sec, and this is fairly close to what was measured by Baylor et al. (1984) in single monkey rods.

These considerations may also explain why the number of rods converging on the beta cell is 1500 and not 150 or 15,000. If there were only 150 rods, quantal events would be rare in the ganglion cell (less than one per second) and the cell would be silent for considerable periods. This would violate an important principle of information theory, which is to *use* all the available channels. If there were as many as 15,000 rods, the maintained discharge due to quantal dark events would be on the order of 150 spikes per second. Against this high background discharge, the incremental 2–3 spikes caused by a real photon absorption would be obscured. Thus, 1500 rods is the minimum needed to maintain transmission in a beta channel at the lowest natural luminances; it may also be the maximum that can be usefully converged at high gain, given the finite thermal stability of rhodopsin at 37°C (Sterling et al., 1988).

Wiring of the cone bipolar versus rod bipolar circuits. The wiring patterns of the cone bipolar and rod bipolar circuits differ strikingly. The cone bipolar circuit has only three stages (Fig. 6.14). Most of the spatial convergence for the beta cell receptive field, both center and surround, is accomplished at the level of the cone terminal (via lateral connections in outer plexiform layer) *before* the first feedforward chemical synapse: 2500 cones → 1 cone (Fig. 6.13D). Thereafter, the cone bipolars convey in parallel to the beta cell what amounts to multiple copies of virtually the same local contrast signal (Smith and Sterling, 1990a,b). The rod bipolar circuit has five stages (Fig. 6.20). Spatial convergence is accomplished stepwise, starting *after* the first feedforward synapse: 1500 rods → 100 rod bipolars (15 : 1) → 5 AII amacrines (20 : 1) → 4 CBb$_1$ (1.25 : 1) → 1 *on*-beta (4 : 1). These architectural differences apparently relate to the different problems faced by the two circuits in establishing the best possible S/N ratio under different ambient luminance.

In daylight and twilight (cone bipolar circuit), when photons are relatively (but not overwhelmingly) abundant, the major source of noise is photon fluctuation (Banks et al., 1987). Under this circumstance the most efficient coding procedure is to compress the signal so that it can be maximally amplified at the first chemical (noisy) synapse. The surround mechanism (predictive coding) subtracts redundant components from the signal by negatively feeding-back the average signal. The center mechanism (collective coding) pools the strongly correlated signal components via electrical (nonnoisy) synapses. This may ensure optimal use of the cone → cone bipolar chemical synapses. By spreading out the signals from the center cones to their neighbors, and vice versa, presynaptic voltages are averaged. This tends to hold voltages in all the terminals to the steepest region of the input/output curve for transmitter release where gain is maximized (Chaps. 1 and 2; Laughlin, 1987).

In starlight (rod bipolar circuit), photons are so scarce that their fluctuations are huge. There is essentially no redundancy in the signal and thus no need for the sort of surround antagonism employed by the cone bipolar circuit. The best that a ganglion cell can do under these conditions is to count every photon. The critical problem in attempting to do so is that the S/N ratio for a quantal event in a rod is at best only 4–5 (Baylor et al., 1984). Were 1500 rods to connect electrically as do the cones, they would pool mainly their *noise*. The noise in the one rod containing the quantal signal would increase as the square root of the number of rods converging—in this example, by a factor of almost 40. Obviously the quantal signal would be utterly lost.

The reason to converge rods via chemical rather than electrical synapses may be that a chemical synapse has a *threshold,* which can be used to filter or "clip" the noise and pass the signal (Baylor et al., 1984; Freed et al., 1987). For the ganglion cell to signal a quantal event reliably, the S/N ratio at the ganglion cell can be no worse than about 5. Otherwise there would be too many false positives (Rose, 1973; Barlow, 1982). Because the convergence of 1500 rods would increase noise by nearly 40 fold, the degree of noise suppression between the rod and the beta cell can be no less than this. It is easy to imagine that such a degree of noise suppression might require several stages, and this may explain why convergence in the rod bipolar circuit is accomplished stepwise. Once the major convergence steps are accomplished via chemical synapses (by the level of the AII amacrine), the signal can pass at the next stage via electrical synapse where there is actually *divergence* (5 AII $\rightarrow$ 8 CBb$_1$) leading, as described above, to a final stage of major amplification and signal averaging (Sterling et al., 1988).

This account of circuits in the central area of the cat retina is by no means exhaustive. Nothing has been said about ganglion cell types other than alpha and beta even though they contribute 50% of the axons in the optic nerve. Neither has much been said regarding amacrine circuits except for the AII and starburst types, despite the fact that 30% of the synapses to the *on*-beta cell and 85% of the synapses to the *on*-alpha are amacrine. These omissions have less to do with the length of this chapter than with the fact that on these topics almost nothing is known.

A great puzzle still to be solved is what circuits underlie the directionally selective responses of certain ganglion cells. Such cells have a characteristic dendritic morphology, easily recognized in rabbit and turtle retina (Amthor et al., 1984; Jensen and Devoe, 1983). Directional selectivity is abolished by GABA antagonists (Caldwell et al., 1978), and the starburst amacrine (which contains GABA as well as acetylcholine) is thought to contribute to the mechanism, but *how?* Hypothetical circuits for directional selectivity are described in Chap. 1.

CONCLUDING REMARKS

IS THERE A DIFFERENCE BETWEEN BLACK AND WHITE? (LAO-TSE 600 B.C.E.)

To this ancient question there is a now definite scientific answer. Black-and-white is our perception when the ratio of light reflected from a surface onto adjacent regions of the retina is about 10. This perception is independent, over a wide range, of the *intensity* of light. If one is indoors, the white on this page reflects less light than would the black print if one were reading outdoors in the sun. On the other hand, at nightfall, as twilight fades and stars appear, the discernible difference between print and empty regions on the page diminishes, gradually rendering the print illegible, even though the ratio of light reflected from the print and the page is constant. In starlight, one can still see large forms and shadows, but the experience of black and white is gone. This experience, its constancy over a wide range of luminance, and its ultimate decay depend on the nature of light itself and on the functional architecture of the synaptic circuits in the retina.

In this chapter, we have distinguished the circuits that serve different luminance levels (daylight, twilight, and starlight). This raises the next great puzzle: How does the retina switch from one circuit to the next and back again? The answers are crucial for understanding the neural basis for dark and light adaptation. Further, since each of these circuits must operate over several log units of light intensity, how are their gains continuously adjusted to match the ambient illumination (Shapley and Enroth-Cugell, 1984)? Are there special circuits for gain control, and do they differ for the cone bipolar and rod bipolar pathways? Intuitively, the amacrine cells with the slower-acting transmitters and modulators, such as catecholamines, indoleamines, and peptides, should be involved. On the other hand, hardly any detail has accumulated regarding the pathways of the neurons containing these substances, let alone their three-dimensional circuits. Nor are there theories or computer simulations to suggest the best strategies for gain control. For example, should it be done in the OPL or the IPL?

These are puzzles for the future. However, with the rapid growth in understanding retinal synaptic organization, and the power of current methods, these puzzles may soon be solved.

7

CEREBELLUM

RODOLFO R. LLINÁS AND KERRY D. WALTON

The cerebellum, a very distinct region of the brain, derives its name as a diminutive of the word *cerebrum*. To the ancient anatomists this was a second, smaller brain in its own right. This is particularly explicit in the German language, where *Kleinhirn* ("cerebellum") translates literally to "small brain." It occupies, in all vertebrates, a position immediately behind the tectal plate and straddles the midline as a bridge over the fourth ventricle. In addition, it is the only region of the nervous system to span the midline without interruption.

The cerebellum has undergone an enormous elaboration through evolution, in fact, more so than any other region of the central nervous system including the cerebrum. Indeed, in *Homo sapiens* the cerebellum has increased in size four fold in the past ten million years, as opposed to the entire brain, which has increased in mass three fold (Jansen, 1969; Romer, 1969). On the other hand, the cerebellum has maintained its initial neuronal structure, almost invariant, throughout vertebrate evolution. Thus, its size but not its wiring has changed in evolution. As an example, the cerebellar cortex in a frog has an area approximately 12 mm^2—that is, 4 mm wide and 3 mm in a rostrocaudal direction. In humans, the cerebellar cortex is a single continuous sheet having an area of 50,000 cm^2 (1000 mm wide in the rostrocaudal direction and an average of 50 mm wide in the mediolateral direction). This is 4×10^3 times more extensive than that of a frog (Braitenberg and Atwood, 1958). This cortex folds into very deep folia (Fig. 7.1), allowing this enormous surface to be packed into a 6 cm × 5 cm × 10 cm volume. Because the cerebellar cortex is very long in the rostrocaudal direction, most of the foldings occur in that direction.

The basic functional design of the cerebellum is that of an interaction between two sets of quite different neuronal elements, those of the *cortex* and those in the centrally located *cerebellar nuclei*. The cerebellar cortex receives two types of afferents, the climbing and the mossy fibers, and generates a single output system, the axons of Purkinje cells (Ramón y Cajal, 1904). The cerebellar nuclei receive collaterals from the climbing and mossy fibers (Bloedel and Courville, 1981) and are the main targets for the Purkinje cell axons. The cerebellum as a whole is connected to the rest of the central nervous system by three large fiber bundles, the cerebellar peduncles.

The function of the cerebellum must be considered within the context of the

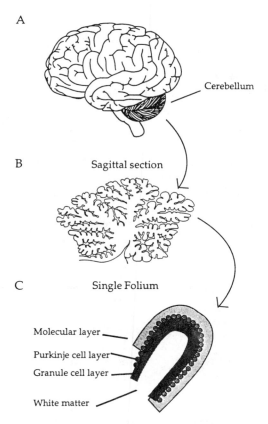

A

Cerebellum

B   Sagittal section

C   Single Folium

Molecular layer

Purkinje cell layer

Granule cell layer

White matter

FIG. 7.1. **A.** Drawing of the lateral view of the human brain showing the cerebellum. **B.** Midsagittal section of the cerebellum. **C.** Drawing of a single folium, showing the three layers of cerebellar cortex and the white matter.

rest of the nervous system since it is not a primary way station for sensory or motor function; that is, its destruction does not produce sensory deficits or paralysis. Nevertheless, lesions of the cerebellum are accompanied by well-defined and often devastating changes in the ability of the rest of the nervous system to generate even the simplest motor sequences used to attain motor goals. Indeed, the cerebellum is essential to the execution of specific movements, as well as to placing motor sequences in the context of the total motor state of the individual at a given instant. Such a function is called *motor coordination* and relates to many different levels of brain function. It is not surprising then that the cerebellum should have a complex neuronal organization and that it should be vigorously connected with the rest of the brain. The enormous Purkinje cells are the sole link between the cerebellar cortex and the cerebellar nuclei. These neurons are the largest neuronal elements in the brain, with respect to the number of synapses they receive, and probably also with regard to complexity of their integrative properties. In this chapter we will show how the role of the cerebellum in motor coordination arises from the interplay between the intrinsic excitability of the Purkinje cell and of the cerebellar nuclear cell membrane, and from the crystallike organization of the synaptic connectivity in the cerebellar cortex.

NEURONAL ELEMENTS

The cerebellar cortex is one of the least variable of central nervous system structures with respect to its neuronal elements (Ramón y Cajal, 1904; Palay and Chan-Palay, 1974). In fact, a basic circuit present in all vertebrates is now well recognized as being composed of the Purkinje cell, the single output system of the cortex, and two inputs: (1) a monosynaptic input to the Purkinje cell, the climbing fiber; and (2) a disynaptic input, the mossy fiber–granule cell–Purkinje cell system.

Because the Purkinje cell bodies are arranged in a single sheet, the Purkinje cell layer, the cerebellar cortex is divided into two main strata: (1) the level peripheral to the Purkinje cell layer known as the molecular layer; and (2) the layer deep to the Purkinje cells (i.e., toward the white matter), the granular layer. Central to the granular layer is the white matter formed by the input and output nerve fiber systems of this cortex (Fig. 7.1B,C).

INPUT ELEMENTS

Climbing fibers. The two types of cerebellar afferents, the climbing fibers and the mossy fibers, represent opposite extremes among the afferents in the central nervous system. The climbing fibers originate from only one brainstem nucleus, the inferior olive. The main inputs to the olive originate in the spinal cord, the brainstem, the cerebellar nuclei, and the motor cortex. Olivary axons are long, fine (1–3 μm), and myelinated. They cross the brainstem at the level of the inferior olive, after which they course rostrally to enter the cerebellum via the inferior cerebellar peduncle. In entering the cerebellar mass, they give off collaterals to the cerebellar nuclei and proceed toward the cerebellar cortex after branching into several fine fibers. The fibers lose their myelin as they penetrate through the granular layer before meeting with the Purkinje cell dendrites in the molecular layer (Fig. 7.2A). Each fiber branches repeatedly to "climb" along the entire Purkinje cell dendritic tree, for which reason they were named "climbing fibers" by Ramón y Cajal. Each Purkinje cell receives only one climbing fiber. However, a given inferior olivary cell axon branches to form several climbing fibers. There may be as many as ten climbing fibers generated by a single inferior olivary cell.

Mossy fiber–parallel fiber pathway. The other cerebellar afferents, the mossy fibers, originate from many central nervous system (CNS) regions. Chief among them are the vestibular nerve and nuclei, the spinal cord, the reticular formation, the cerebellar nuclei, and the cortex via the pontocerebellar pathway, perhaps one of the most massive pathways in the brain. These fibers enter through the middle

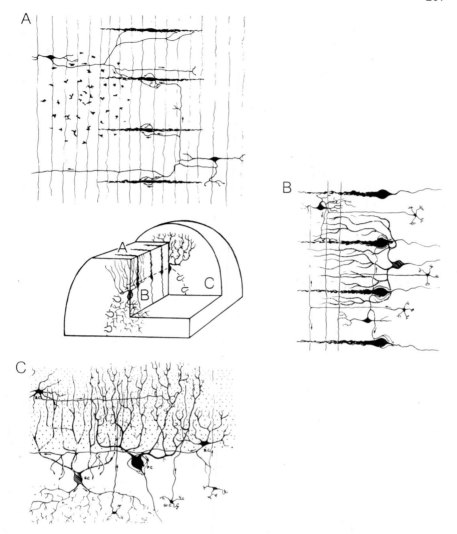

FIG. 7.2. Geometric organization of the neuronal elements of the cerebellar cortex. Three planes of section through a cerebellar folium: **A.** Tangential plane (looking down on the cortical surface). **B.** Transverse (medial to lateral) plane. **C.** Sagittal (anterior to posterior) plane.

and rostral cerebellar peduncles and send collaterals to the deep cerebellar nuclei before branching in the white matter and synapsing on the granule cells (Chan-Palay, 1977). Thus, unlike the climbing fibers, mossy fibers do not synapse directly on Purkinje cells, but on the small granule cells lying directly below them (Fig. 7.2B). This connectivity increases the number of Purkinje cells ultimately stimulated by one mossy fiber axon. Also, because mossy fibers branch profusely in the white matter, a given mossy fiber innervates several folia. The synapses between mossy fibers and granule cells occur as the fine branches of the mossy fibers twine through the granule layer axons. The contacts are made as the mossy fiber enlarges and generates tight knottings along its length. These portions of contact are called "mossy fiber rosettes." One mossy fiber may have 20–30 rosettes (Fig. 7.4A).

An integral part of the mossy fiber input pathway is the granule cell axon, which completes the disynaptic input connection to the Purkinje cells. The axon of the granule cell, usually nonmyelinated, projects upward, past the Purkinje cell layer, into the molecular layer. On its way, it may form synapses with the dendritic trunk of Purkinje cells. In the molecular layer, the axon splits into two branches which take diametrically opposite directions forming the shape of an uppercase T (Fig. 7.2B). Fibers forming the horizontal part of the T are found in all depths of the molecular layer. Because these fibers are precisely arrayed parallel to each other along the length of a folia, they have been named parallel fibers. These are perpendicular to the plane of the Purkinje cell dendrites (Fig. 7.3A), so that each Purkinje cell dendritic tree in humans may be intersected by as many as 200,000 parallel fibers (Braitenberg and Atwood, 1958).

OUTPUT ELEMENTS

Purkinje cells. As stated above, the Purkinje cell is the only output element of the cerebellar cortex. These cells, which reach numbers as high as 15×10^6 in humans, were among the first neurons recognized in the nervous system (Purkinje, 1837) (Fig. 7.4E). Each cell has a large and extensive dendritic arborization, a single primary dendrite, a spherelike soma (20–40 μm), and a long, slender axon that is myelinated when it leaves the granule cell layer. The dendrites of a typical human Purkinje cell may form as many as 200,000 synapses with afferent fibers—more than any other cell in the CNS.

The Purkinje cell dendrites extend densely above the Purkinje cell layer through the molecular layer toward the boundary of the cortex. The unusual arrangement of the Purkinje cell dendrites makes them at once the most conspicuous structural element in the cerebellar cortex and provides an important clue to its functional organization. The entire mass of tangled, repeatedly bifurcating branches is confined to a single plane, very much like a pressed leaf. Moreover, the planes of all the Purkinje cell dendrites in a given region are parallel, so that the dendritic arrays of the cells stack up in neat ranks; adjacent cells in a single plane form equally neat, but overlapping, files (Fig. 7.2A). To a large extent, this orderly array determines the nature and number of contacts made with other kinds of cells. Thus, parallel fibers running perpendicular to the plane of the dendrites intersect a great many Purkinje cells, by the very manner in which these elements are organized.

The Purkinje cell is not merely a transmitter or repeater of information originating elsewhere. As we shall see, its output is determined by its synaptic interactions with other neurons, by their interactions with one another, and by its quite complex intrinsic membrane properties.

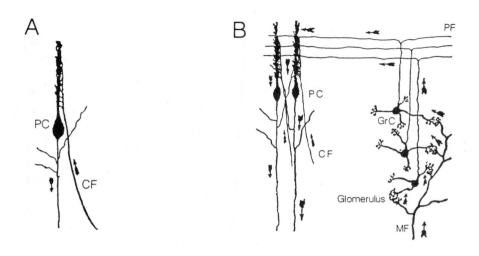

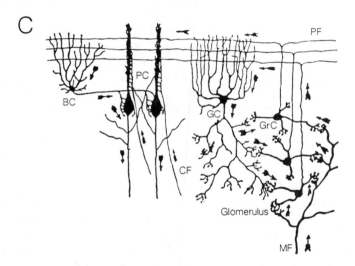

FIG. 7.3. Drawing of the cerebellar afferent circuits and intrinsic neurons. **A.** The climbing fiber–Purkinje cell circuit. A fine branch of an axon from the inferior olivary nucleus (CF) climbs over the extensive arborization of the Purkinje cell (PC) dendritic tree; note the axon collaterals of the Purkinje cell axon. The Purkinje cell is viewed in profile here since it is drawn from a coronal section of the cerebellar cortex. **B.** In the glomeruli, activity in the mossy fibers (MF) excites granule cells (GrC), whose axons project toward the surface of the cortex where they bifurcate to form parallel fibers (PF); these in turn pass through many Purkinje cell dendrites with which they form excitatory synapses. **C.** In this drawing, the two afferent systems shown in A and B are combined and the two main types of intrinsic neurons are depicted: (1) the Golgi cells (GC), with cell bodies just below the Purkinje cell layer; and (2) the basket cells (BC), with cell bodies in the molecular layer. (Modified from Cajal, 1904.)

INTRINSIC ELEMENTS

The basic circuit common to all cerebella contains only one excitatory intrinsic neuron, the granule cell. This basic circuit is augmented by three types of inhibitory interneurons: the Golgi cells of the granule cell layer (Fig. 7.2C, GrC), and the basket (Fig. 7.2C, BC) and stellate cells of the molecular layer which are elaborated progressively in evolution. We will begin with the granule cells.

Granule cells. These are the smallest cells in the cerebellum, with an oval or round soma 5–8 μm in diameter. They are densely packed in the granule cell layer, which occupies about one third of the cerebellar mass. In fact, these cells are the most numerous in the CNS; there are about 5×10^{10} cerebellar granule cells in the human brain. Each cell has four or five short dendrites (each less than 30 μm long) that end in an expansion. Their thin (0.1–0.2 μm in diameter) ascending axon has varicosities, where synapses are formed, before it bifurcates to form the parallel fibers (see above). After bifurcating, the parallel fiber may run for 6 mm (3 mm on each side) before coming to an end.

Golgi cells. There are two sizes of Golgi cells: (1) large ones (somata 9–16 μm in diameter), which are found mainly in the upper part of the granular cell layer; and (2) smaller ones (somata 6–11 μm in diameter), which are found in the lower half of the granular layer. They have extensive radial dendritic trees (Fig. 7.3A) that extend through all layers of the cortex (Fig. 7.3B). They receive input from the parallel fibers in the molecular layer and from climbing and mossy fiber collaterals in the granular layer. Their axons branch repeatedly in the granular layer, where they terminate on granular cell dendrites in the cerebellar glomeruli (see below). There are approximately as many Golgi cells as Purkinje cells.

Basket and stellate cells. These are both found in the molecular layer, receive input from parallel fibers, and may be considered to be members of a single class. The processes of both cell types are oriented transversely to the long axis of the folia (see Fig. 7.2C, BC).

The stellate cells are generally found in the outer two thirds of the molecular layer. The smallest stellate cells, in the most superficial regions of the molecular layer, have 5 to 9-μm-diameter somata, a few radial dendrites, and a short axon. Deeper stellate cells are larger, have more elaborate dendritic arborizations that radiate in all directions, and varicose axons which can extend parallel to the Purkinje cell dendritic plane as far as 450 μm. There are about 16 times as many small stellate cells as Purkinje cells.

Basket cells are found in the deep parts of the molecular layer, near the Purkinje cell layer. Their dendrites ascend into the molecular layer, in some instances as far as 300 μm. Their axon extends along the Purkinje cell layer at right angels to the direction of the parallel fibers. They may spread over a distance equal to 20 Purkinje cell widths and may contact as many as 150 Purkinje cell bodies. During its course, the horizontal segment of a basket cell axon sends off groups of collaterals that descend and embrace the Purkinje cell soma and initial segment (Fig. 7.2B; see also below). As many as 20–30 different basket cells are believed to wrap their axon terminals around each Purkinje cell soma, forming a basketlike meshwork resembling that on a Chianti bottle (Hámori and Szentágothai, 1966). Basket cell axons also ascend to contact the Purkinje cell dendritic tree. There are about six times as many basket cells as Purkinje cells.

CEREBELLAR NUCLEI

The Purkinje cell axons proceed through the granular cell layer, where they are myelinated, and course through the white matter to the cerebellar nuclei. Here they make inhibitory synapses on the projecting cells of the nuclei (Ito et al., 1964). There are three cerebellar nuclei on each side of the midline; each receives input from a region of the cortex directly above it and projects to specific brain regions. The most medial nuclei, the fastigial, receive input from the midline region of the cerebellar cortex, the vermis. They project caudally to the pons, medulla, vestibular nuclei, and spinal cord and rostrally to the ventral thalamic nuclei. Lateral to the vermis are the newer parts of the cerebellar cortex, the paravermis (projecting to the interposed nuclei) and the hemispheres (projecting to the convoluted dentate nuclei). These two deep cerebellar nuclei project rostrally to the red nucleus and ventral thalamic nuclei. Fibers also project to the pons, cervical spinal cord, reticular formation, and inferior olive. There is a pattern of innervation of the cerebellar nuclei within this broad radial organization whereby the rostrocaudal and mediolateral groups of Purkinje cell axons parcel each cerebellar nucleus into well-defined territories (Voogd and Bigaré, 1980).

These cells are not uniform in size: Cells of small, medium, and even large diameter (≈ 35 μm) are found. The large cells have $10-12$ long dendrites (about 400 μm long) that radiate to encompass a sphere. There are a few small cells with short axons, but the majority have long axons that leave the nuclei. The cerebellar nuclei are quite complex; there are two distinguishable cell populations in the fastigial and interpositus nuclei, and three in the dentate nucleus in cat (Palkovits et al., 1977).

The cerebellar nuclei are not simply "throughput" stations; rather the synaptic integration that takes place here is a fulcrum for cerebellum function. Indeed, it is here that information from the cerebellar cortex is integrated with direct input from the mossy and climbing fibers. (This will be discussed in the section on Basic Circuits). As in Purkinje cells, the intrinsic properties of the nuclear neurons are very important to their function (see Intrinsic Membrane Properties).

SYNAPTIC CONNECTIONS

Over 100 years ago, Ramón y Cajal (1888) published his description of the cerebellum. In this study of Golgi-stained material, the synaptic connections were already indicated, as shown in Fig. 7.2, as were the directions of flow of impulses in this cortex. Recent electron microscopic studies, which have provided additional information about the type of synaptic connections and their fine structure (cf. Palay and Chan-Palay, 1974), have confirmed Cajal's initial description. The synaptic connections among the elements in the cerebellum will be discussed by layer, not by cell type, in order to highlight the local circuits at each level of the cerebellum. We will begin with the granular cell layer.

GRANULAR CELL LAYER

Two cell types receive input here, the granule cells and the Golgi cells. The synapses onto granule cells take place in the cerebellar "glomeruli." The rosettes, which occur along the fine branches at the terminals of mossy fibers, form the core of each glomerulus. Excitatory synapses (Gray's type 1) are made between the rosettes and the interdigitating dendrites from as many as 20 granule cells. This can be seen in the electron micrograph in Fig. 7.4A, where a large mossy fiber presynaptic terminal (mf) is seen to be surrounded by several granular cell dendrites (g). The presynaptic terminal contains spherical presynaptic vesicles about 450 Å in diameter. Golgi cell axon terminals surround the rosettes, where they make inhibitory (Gray's type 2) synapses onto the granule cell dendrites. All are encapsulated by a glial lamella which marks the border of each glomerulus.

In the granular layer, Golgi cells receive excitatory (type 1) input from the mossy fibers. These synapses are formed on the Golgi cell dendrites and somata. Thus, mossy fiber volleys excite Golgi and granule cells. Climbing fibers also contact Golgi cells in the granular cell layer. Finally, Purkinje cell recurrent axon collateral varicosities and terminals make inhibitory (type 2) synapses on Golgi cell dendritic trunks and primary branches.

PURKINJE CELL LAYER

The synapse formed in this region is between the basket cell axon terminal and the Purkinje cell soma and initial segment (Fig. 7.4B–D). As many as 50 basket cell axon branches make intricate arborizations surrounding the somata, which form many axosomatic synapses, as shown in the electron micrograph in Fig. 7.4D. Even though basket cell terminals cover both the soma and the axon hillock of the Purkinje cells, only a few synapses with the typical structure of Gray's type 2 (see Chap. 1) have been observed at the axon hillock level; however, a rather impressive morphological structure known as the *pinso terminale* may be found at this level (Ramón y Cajal, 1888). This terminal portion is not a chemical synapse, but similar to the electrical inhibitory synapse in Mauthner cells. These synapses very effectively shut down the output of the cortex.

MOLECULAR LAYER

Climbing fiber–Purkinje cell connection. Among the afferent systems to central neurons, none is more remarkable in extent and power than the climbing fiber junction with Purkinje cells. This junction is unusual not only for its large coverage of a considerable portion of the Purkinje cell dendritic tree, but also because, as we have seen, only one climbing fiber afferent contacts each Purkinje cell. The synapses are made between varicosities (2 μm across) on the climbing fiber and stubby spines on the soma and main dendrites of the Purkinje cell; as

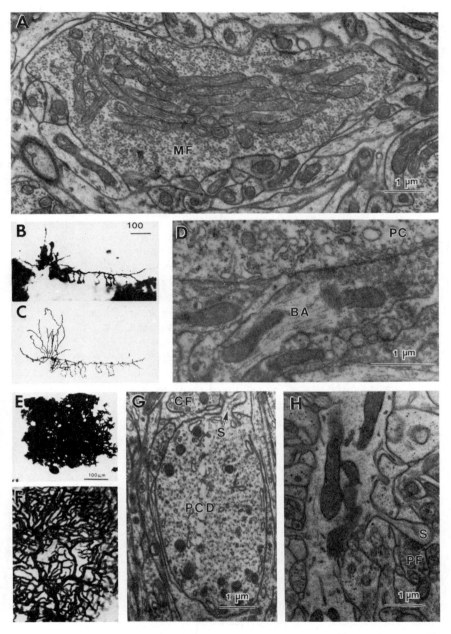

FIG. 7.4. Morphology of some synaptic connections: **A.** Mossy fiber (mf) rosette. **B.** Golgi-stained basket cell. **C.** Computer reconstruction of basket cell. **D.** Purkinje cell soma (PC) and basket cell axon (BA). **E.** Golgi-stained mammalian Purkinje cell. **F.** Detail of smooth and spiny branchlets of Purkinje cell. **G.** Purkinje cell dendrite (PCD) and climbing fiber terminal (CF) contracting a spine. **H.** Spiny branchlet ultrastructure showing spines (S) and a parallel fiber synapse (PF) on a spine. Calibration in B: 100 μm. (Micrographs courtesy of Dr. Dean Hillman.)

223

many as 200 contacts may be made between a climbing fiber and its Purkinje cell. Each contact synapses on 1–6 spines. A climbing fiber terminal contacting a Purkinje cell spine is shown in Fig. 7.4G, where the preterminal enlargement is indicated by arrows. The presynaptic vesicles are round and 440–590 Å in diameter. Morphologically the presence of a climbing fiber synapse seems to exclude nearby parallel fiber–Purkinje cell contacts. The Purkinje cell dendrites can thus be divided into a central area covered by the climbing fibers and the more peripheral spiny dendritic portion that is contacted by parallel fibers.

Parallel fiber–Purkinje cell connection. In contrast to the climbing fibers which contact mainly Purkinje cell dendrites, the parallel fibers terminate on the dendrites of all the neuronal elements in the cerebellar cortex, the only exception being the granule cells. Thus, parallel fibers contact the dendrites of Purkinje cells, basket cells, stellate cells, and Golgi cells. On the Purkinje cells, parallel fibers synapse with the spines on the terminal regions of the Purkinje cell dendrites, in regions called *spiny branchlets.* The synaptic junction is formed between the head of a spine and a globular expansion of the parallel fiber; the spine penetrates the swollen part of the fiber. The electron micrograph in Fig. 7.4H illustrates a Purkinje cell spiny branchlet with three spines (marked by arrows). A synapse with a parallel fiber is clearly seen on the upper right-hand spine. The synaptic vesicles are spherical and 260–440 Å in diameter. A parallel fiber forms synapses with one out of the 3–5 Purkinje cells that it traverses. Thus, most of the parallel fibers passing a dendritic tree will not form synapses. There is such a large number of parallel fibers that as many as 200,000 synapses on one Purkinje cell dendrite may be formed in man, by far the largest number of synaptic inputs to any central neuron. The ascending portion of the granule cell axon has varicosities that are presynaptic to spines on the lower dendrites of Purkinje cells.

Plasticity of Purkinje cell connectivity. Thus, in the Purkinje cell dendritic tree, the climbing fiber input is normally proximal to the parallel fiber input (Fox et al., 1967). When damage to the climbing fibers occurs in the adult animal, however, spines proliferate in large numbers on Purkinje cell smooth dendrites. These are promptly invaded by newly formed parallel fiber contacts (Sotelo et al., 1975), indicating a tug of war or a territoriality between the two systems. Also, destruction of the parallel fibers promotes multiple climbing-fiber innervation (Mariani et al., 1977), indicating that a true competition for a Purkinje cell dendritic tree exists between parallel and climbing fiber afferents and even between climbing fiber afferents themselves. It also indicates that a single climbing fiber cannot provide all the necessary input, since Purkinje cells do become innervated by climbing fibers after parallel fiber damage.

Recently quantitative studies have been made of the changes in the parallel fiber–Purkinje cell synapse localized after lesioning of the parallel fiber input. In one set of experiments, the parallel fibers were sectioned and the molecular layer was undercut (to destroy the granule cells) (Hillman and Chen, 1984). The number, size, and average contact area of the parallel fiber–Purkinje cell synapses were evaluated 2–3 weeks after the lesion and compared to control values from unlesioned animals. It was found that the number of parallel fibers contacting a Purkinje cell was reduced in relation to the extent of the lesion, but that the area of synaptic contact of the surviving synapses was proportionately increased. Thus, there was a change in the position and size of the synapses in response to perturbations, but the total area of synaptic contact remained stable. Change in the size of the presynaptic boutons was not accompanied by a change in the presynaptic grid densities or the number of synaptic vesicles (Hillman and Chen, 1985a). This suggests that as the size of the boutons increased, there was a parallel increase in the morphological correlates of the neurotransmitter release machinery. Stabilization of the total synaptic area has also been seen in other areas of the CNS (see Hillman and Chen, 1985b).

Other connectivity in the molecular layer. In addition to Purkinje cells, the dendrites of stellate, basket, and Golgi cells receive inputs in the molecular layer (see Fig. 7.2C). The stellate cells in turn make inhibitory synapses into Purkinje cell dendritic shafts. Parallel fiber swellings make excitatory synapses onto stellate cell dendritic spines. The basket cells receive excitatory synaptic connections from climbing fibers and parallel fibers and are inhibited by Purkinje cell axon collaterals. Parallel and climbing fibers make the same *en passant* synapses with basket cell dendrites as with Purkinje cell dendrites. Finally, Golgi cell dendrites receive excitatory synapses from the parallel fibers. These axodendritic synapses are by far the largest number of synapses onto Golgi cells.

CEREBELLAR NUCLEI

Five different types of synaptic terminals have been distinguished on the basis of the characteristics of their membrane attachment and shape of synaptic vesicles. Both axosomatic and axodendritic synapses are found. The presynaptic terminals are made by mossy fibers, climbing fibers, axon collaterals from projecting nuclear cells are from Purkinje cells (Palkovits et al., 1977). Purkinje cell axons have 2–3 branches which arborize extensively in the nucleus, describing a narrow cone. Synapses are formed at the terminals and at *en passant* thickenings along the length of the axon. Synapses are usually formed with dendritic thorns or spines of nuclear cells, although some synapses are axosomatic. The thickenings and terminals have dispersed ovoid vesicles, usually found where they contact the dendrites of nuclear neurons. It has been calculated that a Purkinje cell may contact as many as 35 nuclear cells, but the majority of contacts are made with 3–6 Purkinje cells. In addition to this divergence, there is convergence since there are about 26 nuclear cells for each Purkinje cell. There are about 860 Purkinje cell axons for each nuclear cell.

Complex synaptic combinations such as serial and triadic synapses are found in the cerebellar nuclei (Hámori and Mezey, 1977), as seen in the retina (Chap. 6) and thalamic nuclei (Chap. 8). These imply a quite complex interaction between the afferents and nuclear cells. In these synapses the first presynaptic element may be a Purkinje cell axon terminal, a brainstem afferent terminal (climbing or mossy fibers), or an axon terminal collateral. The last probably involve projecting nuclear cell axon collaterals. The second terminal in the sequence, which is both pre- and postsynaptic, derives from the cerebellar nuclear cells themselves; they are either axon collaterals of projecting neurons or Golgi type II interneurons. Although such synapses are a regular feature of the nuclei, they do not form as large a percentage of synapses as in some sensory systems.

BASIC CIRCUIT ORGANIZATION

There are three main circuits in the cerebellum, two circuits in the cortex, which include afferent fibers as shown in Fig. 7.2 and one circuit in the deep nuclei. They are diagramed in Fig. 7.5.

MOSSY FIBER CIRCUIT

The sequence of events that follows the stimulation of mossy fibers was first suggested by János Szentágothai of the Semmelweis University School of Medicine in Budapest: The stimulation of a small number of mossy fibers activates, through the granule cells and their parallel fibers, an extensive array of Purkinje cells and all three types of inhibitory interneurons. Subsequent interactions of the neurons tend to limit the extent and duration of the response. The activation of Purkinje cells through the parallel fibers is soon inhibited by the basket cells and the stellate cells, which are activated by the same parallel fibers. Because the axons of the basket and stellate cells run at right angles to the parallel fibers, the inhibition is not confined to the activated Purkinje cells; those on each side of the beam or column of stimulated Purkinje cells are also subject to strong inhibition. The effect of the inhibitory neurons is therefore to sharpen the boundary and increase the contrast between those cells that have been activated and those that have not.

At the same time, the parallel fibers and the mossy fibers activate the Golgi cells in the granular layer. The Golgi cells exert their inhibitory effect on the granule cells and thereby quench any further activity in the parallel fibers. This mechanism is one of negative feedback: through the Golgi cells the parallel fiber extinguishes its own stimulus (Fig. 7.5A). The net result of these interactions is the brief firing of a relatively large but sharply defined population of Purkinje cells.

CLIMBING FIBER CIRCUIT

Stimulation of a group of climbing fibers produces a powerful excitation of Purkinje cells. The stimulus elicits prolonged bursts of high-frequency action potentials. The climbing fibers also activate Golgi cells, which inhibit the input through the mossy fibers (Fig. 7.5B). Thus, when climbing fibers fire, their Purkinje cells are dominated by this input. The climbing fiber input to basket and stellate cells sharpens the area of activated Purkinje cells.

CEREBELLAR CORTEX—DEEP NUCLEI CIRCUIT

Electrical activation of mossy fiber inputs to the cerebellar system generates an early excitation in the cerebellar nuclei since the collaterals terminate directly on

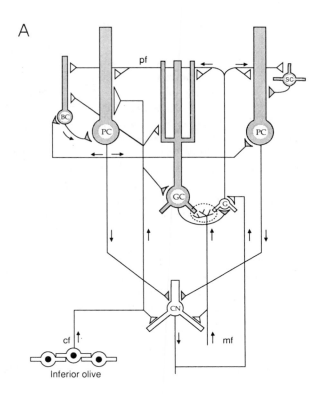

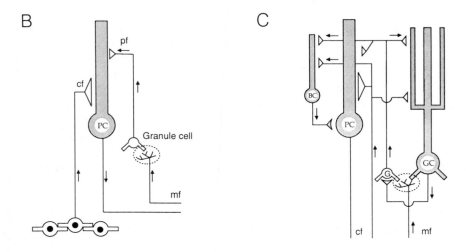

FIG. 7.5. Diagram of the basic circuit in the mammalian cerebellum. **A.** This circuit includes all the elements making specific synaptic connections in the cerebellar cortex and nuclei. **B,C.** Simplified diagrams of cerebellar cortex showing the afferent circuits (B) and the intrinsic neurons (C). BC, basket cell; cf, climbing fiber; CN, cerebellar nuclear cell; G, granule cell; GC, Golgi cell; mf, mossy fiber; PC, Purkinje cell; pf, parallel fiber; SC, stellate cell.

the cerebellar nuclear cells (Fig. 7.5C). The same information then proceeds to the cerebellar cortex, which in turn produces an early excitation of Purkinje cells to be translated into inhibition at the cerebellar nucleus. This inhibition is followed by a prolonged increase in excitability of the cerebellar nuclear cells. The increased excitability is the result of two actions: (1) disinhibition due to reduced Purkinje cell activity, which in turn is due to the inhibitory action of basket and stellate cells after the initial activation of Purkinje cells; and (2) CN cell intrinsic properties (see below). The Purkinje cell inhibition is also due indirectly to the inhibitory action of the Golgi interneuron, which, by preventing the mossy fiber input from reaching the molecular layer, reduces the excitatory drive to Purkinje cells. The CN projecting neurons themselves send axon collaterals to cortical inhibitory interneurons including basket cells, which thus provide recurrent inhibition of the CN neurons as seen in spinal motoneurons (Chap. 4).

INTRINSIC MEMBRANE PROPERTIES

In Chap. 2, it was emphasized that the functional characteristics of a neuron are the outcome of a complex interplay between its intrinsic membrane properties and its synaptic interactions. In no part of the brain is this exemplified more vividly than in the cerebellum. Indeed, as already mentioned in Chap. 2, the Purkinje cell is one of the best known models for demonstrating these properties. Because of this importance, we will consider the intrinsic membrane properties separately in this section before addressing the synaptic actions of the system.

PURKINJE CELLS

The intrinsic membrane properties of cells may be considered as independent of synaptic input, although interaction of synaptic potentials with intrinsic membrane properties shapes the activity of the cell. Intrinsic properties are usually studied by determining the response to direct activation, that is, to depolarizing or hyperpolarizing current injected into the cell, usually into the soma. Purkinje cell electrical activity may be recorded under in vivo or in vitro conditions; however, since the most reliable recordings are obtained in vitro, our understanding of the electrical properties of the mammalian Purkinje cell membrane has come mainly from studies of cerebellar slices (Llinás and Sugimori, 1978, 1980a,b). Antidromic activation of a Purkinje cell is characterized by a large spike having an initial segment/soma dendritic (IS–SD) break that is in many ways similar to that obtained in vivo from motoneurons and other central neurons.

Direct stimulation of Purkinje soma via the recording microelectrode demonstrates that these cells fire in a way that is quite different from that seen in other neurons. Indeed, square current pulses lasting about 1 sec (Fig. 7.6A) produce, at just threshold depolarization, a repetitive activation of the cell. That is to say, with long pulses the neuron fires, but a single isolated spike cannot be generated by this type of stimulation. This burst of activity is produced by a low-threshold, sodium-dependent conductance that does not inactivate within several seconds and serves to trigger the fast action potentials. This sodium conductance is different from that responsible for the fast action potentials seen in virtually all

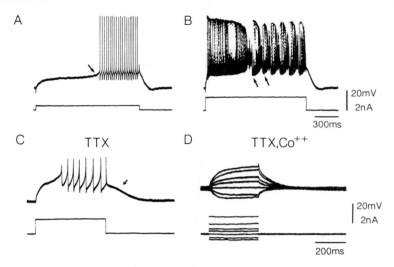

FIG. 7.6. Intrinsic properties of mammalian Purkinje cells recorded in vitro. **A.** A prolonged, threshold current pulse injected into the soma of a Purkinje cell elicits a train of action potentials after an initial local response (arrow). **B.** Increased current strength elicits high-frequency firing and oscillatory behavior (arrows). **C.** After addition of TTX to the bath, the fast action potentials are blocked, and the slowly rising action potentials underlying the oscillations seen in B are revealed. A slow after depolarization may also be seen. **D.** Addition of cobalt chloride (Co^{2+}) to the TTX perfusate removes all electroresponsiveness, indicating that the slow action potentials in C were calcium dependent. (Modified from Llinás and Sugimori, 1980a.)

nerve cells: It is activated at a lower voltage and does not inactivate. With increased stimulation the onset of the repetitive firing moves earlier. Also, at the end of the initial pulse of firing, a reduction in the amplitude of the spikes is followed by a rhythmic bursting, as marked by arrows in Fig. 7.6B.

Pharmacological studies in cerebellar slices have shown that the fast action potentials and the bursting have different ionic mechanisms. Removal of extra-cellular sodium or the application of tetrodotoxin (TTX, a sodium conductance blocker) to the bath causes a complete abolition of the fast action potentials seen in Fig 7.6A and B but leaves a late, slow-rising burst potential intact, as shown in Fig. 7.6C. This slow bursting of Purkinje cells has been found to be generated by a voltage-activated calcium conductance followed by a calcium-dependent potassium conductance increase. We know the spikes are calcium dependent because they are seen in the absence of sodium, and because they are blocked by the removal of calcium from the extracellular medium or by ions that block the slow calcium conductance (cobalt, cadmium, manganese), as shown in Fig. 7.6D. When the calcium in the bathing solution is replaced by barium, the after hyperpolarization is reduced and the bursting response is converted into a prolonged single action potential. This demonstrates the presence of a calcium-activated potassium conductance, since it is known that barium does not activate the calcium-activated potassium conductance. All electroresponsivenes is gone following calcium and sodium blockade, as shown by the application of both tetrodotoxin (TTX) and cobalt to the extracellular medium (see Fig. 7.6D). Thus, at the somatic level, Purkinje cells have not one, but three main mechanisms for spike generation: (1) a sodium-dependent spike similar to that seen in other cells, which is blocked by the absence of extracellular sodium or by the application of TTX; (2) a low-threshold, noninactivating sodium spike; and (3) a calcium-dependent action potential, which has a slow rising time and a rather rapid return to baseline.

The distribution and properties of voltage-gated channels in the dendrites will be discussed below (Dendritic Properties).

CEREBELLAR NUCLEAR CELLS

The electrical properties of cerebellar nuclear (CN) neurons were first studied in detail in in vitro preparations (Jahnsen, 1986a–b; Llinás and Mühlethaler, 1988b). Like Purkinje cells, CN cells have a collection of ionic conductances that give them complex firing abilities. CN cells have a noninactivating sodium conductance similar to that described in Purkinje cells, in addition to the usual sodium- and potassium-dependent conductances which generate fast action potentials. The firing of CN cells depends on their resting potential. If a cell is depolarized with a current pulse from the resting potential as in Fig. 7.7A, the cell fires a train of action potentials. However, if the same current pulse is injected when the cell is held hyperpolarized from the resting potential, all-or-nothing bursts are seen, as shown in Fig. 7.7B and D. Also, if a hyperpolarizing current pulse is injected into a CN neuron, a burst of action potentials is seen at the end of the current injection (Fig. 7.7C).

This "rebound response" following hyperpolarization is important in CN cell function. This is easily understood since Purkinje cells are inhibitory and generate inhibitory postsynaptic potentials (IPSPs) in CN cells. The ionic basis for these burst responses was determined by pharmacological studies. Thus, after eliminating the fast sodium conductance by application of TTX, although the fast action potentials seen in Fig. 7.7D are blocked, a slowly rising spike is elicited from the hyperpolarized membrane potential (Fig. 7.7E). The threshold for these spikes is lower than that for the fast sodium-dependent action potentials; they are therefore called low-threshold spikes (LTS). They are calcium dependent since they are blocked after addition of cobalt or removal of calcium from the bath, and are insensitive to TTX. The presence of a LTS is probably of major functional significance in these neurons because climbing fiber activation of Purkinje cells following such bursts can easily be elicited following the powerful IPSPs produced by this input (see Fig. 7.13) (Llinás and Mühlethaler, 1988b).

SYNAPTIC ACTIONS

CLIMBING FIBER ACTION ON PURKINJE CELLS

Although, for the most part, a one-to-one relationship exists between a climbing fiber and a given Purkinje cell (each Purkinje cell receives one climbing fiber), the cell of origin of this afferent (the inferior olivary neuron) is probably capable

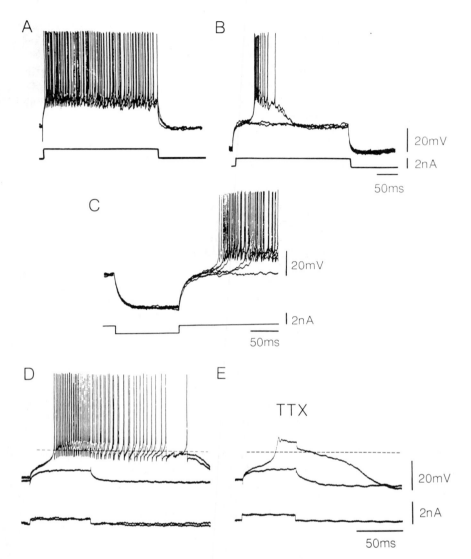

FIG. 7.7. Intrinsic properties of cerebellar nuclear neurons. **A.** A depolarizing current injection from resting potential elicits tonic firing. **B.** When the same strength current pulse is delivered from a hyperpolarized membrane level, an all-or-none burst response is elicited. **C.** Hyperpolarizing current injection from the resting potential elicits a strong rebound burst of action potentials from a slow depolarization. **D.** Response to current injection from a hyperpolarized level (resting potential marked by broken line). **E.** Addition of TTX to the perfusate blocked the fast action potentials, revealing a slowly rising, prolonged depolarization and after depolarization; these responses were then blocked by addition of Co^{2+} to the bath. (Modified from Llinás and Mühlethaler, 1988b.)

of producing more than one climbing fiber afferent and probably as many as ten. One of the most powerful synaptic junctions in the CNS is that between the climbing fiber afferent and the dendrite of a Purkinje cell. This has been called a "distributed" synapse since a single presynaptic fiber makes contact with the postsynaptic cell at many points throughout the Purkinje cell dendritic tree, and the synapse is distributed over a large surface area. This is in contrast to more typical synapses, as between a 1a terminal and motoneuron, where there are only a few, relatively localized points of contact (see Chap. 4). Eccles et al. (1966a) demonstrated electrophysiologically that stimulation of the inferior olive produces a powerful activation of the Purkinje cell. This synaptic excitation is characterized by an all-or-nothing burst of spikes that shows little variability from one activation to the next. These are called complex spikes. Several complex spikes recorded from an isolated preparation are superimposed in Fig. 7.8A and B. It is now known that the spikes on the broad EPSP are produced at the dendrite by a voltage-activated calcium conductance (see later) and at the somatic and axonic levels by the usual Hodgkin–Huxley sodium–potassium spikes (Llinás and Sugimori, 1978, 1980b).

Climbing fiber responses in Purkinje cells may be elicited by placing a stimulating electrode in the white matter near the midline. This "juxtafastigial" stimulation activates inferior olivary axons in the white matter. Following a juxtafastigial stimulus, climbing fiber synapses are activated simultaneously and produce a very large unitary EPSP in the postsynaptic dendrite. This unitary synaptic potential usually has an amplitude of 40 mV and lasts 20 msec. The all-or-nothing character of the climbing fiber burst actually represents the all-or-nothing character of the presynaptic spike in the climbing fiber. If the Purkinje cells are hyperpolarized far enough to prevent the cell from spiking, the all-or-nothing character of the EPSPs may be seen (Fig. 7.7C).

Under conditions in which the sodium- and calcium-dependent spikes are prevented, the chemical nature of the synapse may be studied in detail and its distributed character clearly demonstrated. Depolarization of the soma or dendrite can produce a reduction in amplitude and an actual reversal of the sign of the climbing fiber EPSP, as shown in Fig. 7.8D. A large increase in the EPSP amplitude is seen when the membrane potential is moved in the hyperpolarizing direction (lower traces in D). The reversal in sign (shown in Fig. 7.8D, 22.1 nA) is then necessary and sufficient evidence to indicate that a synaptic junction is chemical in nature (see Chap. 2).

The fact that different parts of the EPSP (the peak and falling phase) reverse at different levels of depolarization (see biphasic reversal at 22.1 nA in Fig. 7.8D) indicates that the synapse occurs at multiple sites having different distances from the site of recording in the soma (Llinás and Nicholson, 1976). Because a current point source, a microelectrode, is utilized to change the membrane potential, the potential change along the dendrite is maximum near the site of impalement and decreases with distance. Because the synapses closest to the site of recording generate most of the rising phase of the recorded EPSP, this component is the first to reverse. Those synapses located at a distance generate the slowest components (owing to the cable properties of the dendrites) and are less affected by the

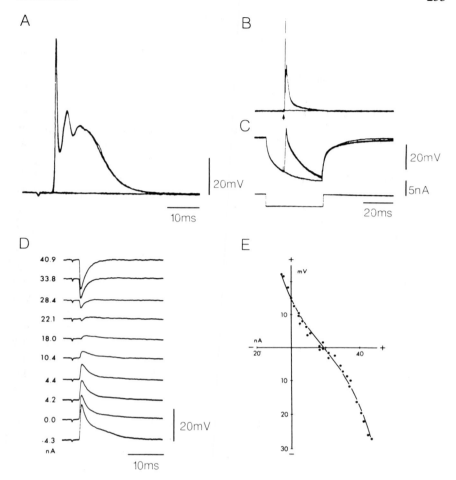

FIG. 7.8. Climbing fiber activation of mammalian Purkinje cells in vitro. **A.** All-or-none complex spikes in a Purkinje cell evoked by white matter stimulation are superimposed. **B.** In another Purkinje cell, five threshold white-matter stimuli (arrow) evoke very uniform complex spikes on four occasions. **C.** If threshold stimuli are delivered when the cell is hyperpolarized (to prevent action potential firing), the all-or-none climbing fiber EPSP may be seen. **D.** Reversal of climbing fiber EPSP. Notice that, as expected in a distributed synapse, the reversal is biphasic, with the early portion of the potential reversing at lower levels of injected current than the late part; this may be seen at 18, 22, and 28 nA. **E.** Plot of the voltage–current relation for the EPSP reversal shown in D. (Modified from Llinás and Nicholson, 1976, and Llinás and Mühlethaler, 1988a.)

current injection. Recordings similar to those obtained in vitro can also be obtained in vivo.

Activation of the climbing fiber afferents generates a burst of action potentials at the Purkinje cell axon. The frequency of this response is high, generally 500/sec. Indeed, it is higher normally than that seen after parallel fiber stimulation, suggesting that one of the possible functions of the climbing fiber system is to produce a discharge of distinct bursts of action potentials. As discussed later, climbing fiber activation also produces very sharp IPSPs in the target neurons of Purkinje cells.

PARALLEL FIBER ACTION ON PURKINJE CELLS

As discussed fully above, mossy fiber inputs activate Purkinje cells via the parallel fiber–Purkinje cell synapse (Eccles et al., 1966b). Early investigators named these responses "simple spikes." Purkinje cell responses to spontaneous activity in the parallel fibers are illustrated in Fig. 7.9A; notice that during this recording period two complex climbing fiber spikes were also recorded (marked by arrows). This circuit can be activated by direct parallel fiber stimulation of the cerebellar surface or via the mossy fiber–granule cell–parallel fiber pathway from white matter stimulation. Both types of stimulation generate short-latency EPSPs in Purkinje cells.

This postsynaptic potential differs from that generated by the climbing fiber in two ways: First, it is graded as shown by the response to juxtafastigial stimuli of increasing intensity (Fig. 7.9B,C). Second, it is generally followed by an IPSP (see trace, Fig. 7.9B). The IPSP is generated by activation of the inhibitory interneurons of the molecular layer. The parallel fiber synaptic depolarization can generate action potentials at the somatic level as well as dendritic calcium spikes if the stimulus is large enough (see below). Because parallel fiber activation of Purkinje cells is followed by a disynaptic inhibition, this synaptic sequence is reviewed in detail in conjunction with the inhibitory systems in the next subsection of this chapter.

INHIBITORY SYNAPSES IN THE CORTEX

Inhibitory neurons are organized in the cerebellar cortex into two main categories: Those that reside in the molecular layer are the basket and stellate cells, whereas those that reside in the granular layer are the Golgi cells.

Granular layer. In the granular layer the main inhibitory system is the Golgi cell axonic plexus. This plexus releases GABA, inhibiting granule cell dendrites within the granule cell glomerulus. Indeed, while mossy fibers activate the terminal dendritic claws of the granule cells, the Golgi cell axons also distribute their contacts on the dendrites of the granule cells and may block the synaptic action of the mossy fibers by the release of GABA. The inhibition that ensues is so powerful as to totally block the parallel fiber input to the cerebellar cortex (Eccles et al., 1966d).

Molecular layer. In the molecular layer, inputs from a climbing fiber and from the parallel fibers represent the two types of excitatory afferents terminating on a Purkinje cell. The Purkinje cell, however, receives input from three inhibitory systems as well: one subserved by the basket cell, the second by the stellate cell, and the third by the catecholamine system arising from the locus coeruleus (Bloom et al., 1971; Pikel et al., 1974). Activation of the *basket cells* generates a

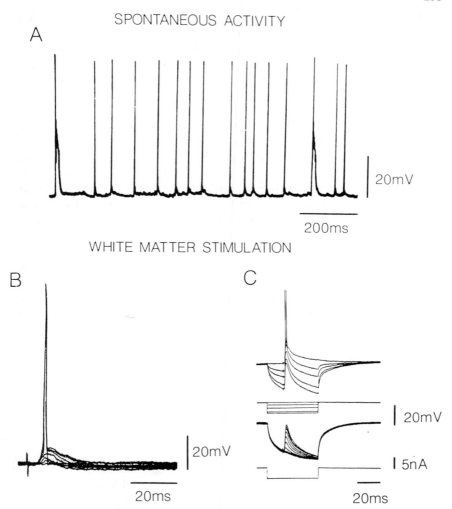

SPONTANEOUS ACTIVITY

A

20mV

200ms

WHITE MATTER STIMULATION

B

C

20mV

20mV

20ms

5nA

20ms

FIG. 7.9. Mossy fiber activation of mammalian Purkinje cells in vitro. **A.** Spontaneous activity in the mossy fiber–parallel fiber system gives rise to fast, simple spikes in Purkinje cells which are in contrast to the two broad, climbing fiber-evoked complex spikes in the trace. **B.** White matter stimulation of increasing strength evoked graded EPSP–IPSP sequences due to mossy fiber–parallel fiber activation. **C.** When such stimulation is delivered during hyperpolarizing pulses of increasing amplitude (middle trace), the parallel fiber-mediated EPSP may be seen (top trace); the bottom trace illustrates the graded nature of the synaptic potential. (Modified from Llinás and Mühlethaler, 1988a.)

graded inhibition at each side of the activated bundle of parallel fibers (Andersen et al., 1964; Eccles et al., 1966b,c). This can be seen clearly when recordings are made lateral to the beam of stimulated parallel fibers. In this case, at low stimulus intensity, only the IPSP is seen; however, if the stimulus intensity is increased, more Purkinje cells are excited by the parallel fibers and an EPSP–

IPSP sequence is seen (Fig. 7.10A). The basket cell IPSP is generated by a membrane conductance increase to chloride, most probably by the release of gamma-aminobutyric acid (GABA, see below). The second inhibitory system is that represented by the *stellate cells* which synapse mainly on Purkinje cell dendrites. Electrophysiologically they have the same pattern of inhibition as that of basket cells.

Monoaminergic inhibition. The third inhibitory system in the molecular layer is that of the locus coeruleus; its catecholamine-mediated inhibition generates a large, prolonged hyperpolarization in Purkinje cells (Hoffer et al., 1973). Although intriguing questions arise about the function of this system, it is possible (due to its rather widespread character) that it is related to the general state of wakefulness of the animal rather than to specific cerebellar functions. Indeed, morphologically the system consists of rather thin filamentous afferents that reach the cerebellar cortex and bifurcate widely to cover not only the neuronal elements but probably also the vascular system (Bloom et al., 1971).

PURKINJE CELL ACTION ON CEREBELLAR NUCLEAR CELLS

Perhaps one of the most surprising findings in the physiology of the cerebellum is the fact that the only output of the cerebellar cortex, the Purkinje cells, exercises

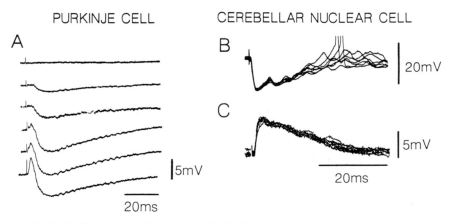

PURKINJE CELL   CEREBELLAR NUCLEAR CELL

FIG. 7.10. Inhibitory synaptic potentials in Purkinje cells and cerebellar nuclear cells. **A.** In this case, the stimulating electrode was placed on the cerebellar surface lateral to the recorded Purkinje cell because, under such conditions, powerful IPSPs could be recorded in the Purkinje cell. As the stimulus intensity was increased (lower traces), the band of activated parallel fibers became wider, and finally the parallel fibers synapsing on the recorded Purkinje cell were themselves activated; thus an EPSP preceded the IPSP. **B.** IPSPs recorded in a cerebellar nuclear cell. Stimulation of the white matter between the cerebellar cortex and nuclei may elicit graded EPSPs and IPSPs. For particular locations and amplitudes of stimulation, IPSPs may be elicited in the absence of an early EPSP, as shown here. These IPSPs are very regular, often triggering rebound firing of the cell, as seen here. **C.** That these large potentials are synaptic potentials is shown by their reversal upon injection of a hyperpolarizing current.

an inhibitory input onto the cerebellar nuclear neurons (Ito et al., 1964). This finding indicates that the cerebellar cortex is the most sophisticated inhibitory system in the brain, not only because of its refinement of connectivity and the integrative ability of these neurons, but also because of the extent of information reaching the cerebellar cortex. Indeed, there are as many neurons in the cortex (5 $\times$ 10^{10}) as there are neurons in the rest of the brain. The powerful GABA inhibition of the Purkinje cells on the cerebellar nuclei also demonstrates the rich biochemistry of the system (Obata et al., 1967). The Purkinje cells project in a radial pattern onto the nuclei as discussed previously (Neuronal Elements).

Electrical stimulation of the cerebellar white matter can elicit quite complex sequences of EPSPs and IPSPs in CN cells. Here we will consider the simplest case—where white matter stimulation is limited to the Purkinje cell axons. In this case, only IPSPs are recorded. The records shown in Fig. 7.10B were made from a CN cell in a cerebellum–brainstem preparation isolated from adult guinea pig (Llinás and Mühlethaler, 1988b). In the example shown in Fig. 7.10C, several IPSPs were elicited; it can be seen that their onset and amplitude are very reliable (four traces are superimposed) and that they can be easily reversed in sign by current injection, as in this example. The response of CN cells to white matter stimulation is not always so straightforward, as will be discussed below (Functional Circuits).

NEUROTRANSMITTERS

GLUTAMATE

Several indirect lines of evidence indicate that glutamate is the neurotransmitter in granule cells. It depolarizes Purkinje cells when applied ionophoretically to the dendrites (Krnjević and Phillis, 1963; Curtis and Johnston, 1974; Sugimori and Llinás, 1981) and the dendrites are more sensitive than the soma to the glutamate. Further, naturally occurring L-glutamate is more potent than the D-glutamate isomer (Chujo et al., 1975; Crepel et al., 1982). In frog Purkinje cells, the reversal potential of the glutamine-elicited EPSP is close to that for parallel-fiber evoked EPSPs (Hackett et al., 1979).

Neurochemical studies have shown that the glutamate content is lower than normal in cerebella in which the number of granular cells has been reduced by X-irradiation (Valcana et al., 1972; McBride et al., 1976), by virus infection (Young et al., 1974), or by mutation (Hudson et al., 1976; Roffler-Tarlov and Turey, 1982). Also, compared to control values glutamate uptake is reduced in synaptosomal preparations from cerebella in which the granule cell number has been reduced (Young et al., 1974; Rohde et al., 1979). Quantitative localization of glutamate immunoreactivity has shown that parallel and mossy fiber terminals have significantly higher levels of glutamate than do Golgi or Purkinje cells (Somogyi et al., 1986).

Finally, glutamate release from synaptosomal preparations of rat cerebellum is reduced in synaptosomes prepared from X-irradiated agranular cerebella. The release is dependent on calcium, antagonized by increased magnesium, and stimulated by membrane depolarization caused by elevated levels of potassium.

These characteristics mimic those essential for neurotransmitter release (Sandoval and Cotman, 1978).

GABA

The most likely candidate for the neurotransmitter liberated from basket cells and Purkinje cells is GABA.

Basket cell inhibition or Purkinje cell electrical activity is blocked by application of agents known to block GABA receptors, such as bicuculline or picrotoxin. This effect has been demonstrated in several ways, involving a reduction in the ability of basket cell activation to (1) depress Purkinje cell spontaneous firing (Curtis and Felix, 1971), (2) depress Purkinje cell antidromic field potentials (Bisti et al., 1971), or (3) elicit IPSPs in Purkinje cells (Dupont et al., 1979). Also, ionophoretic application of GABA inhibits Purkinje cell spontaneous activity (Kawamura and Provini, 1970; Okamoto et al., 1976; Okamoto and Sakai, 1981). Finally, immunocytochemical studies have demonstrated the presence of the GABA-synthesizing enzyme glutamic acid dehydrogenase (GAD) in basket cell terminals around Purkinje cell somata (McLaughlin et al., 1974; Chan-Palay et al., 1979; Oertel et al., 1981). Basket cells also take up radioactive GABA (Hökfelt and Ljungdahl, 1972; Sotelo et al., 1972).

Although the inhibitory transmitter of stellate cells has not been decisively established, it may be taurine (Frederickson et al., 1978).

The inhibitory nature of Purkinje cells was first demonstrated in Deiters' nucleus. Ionophoretic application of GABA hyperpolarizes Deiters' neurons (Obata et al., 1967), a target of Purkinje cell axons. IPSPs following Purkinje cell activation, as well as GABA-induced potentials, reverse near the same membrane potential and are mediated by an increased conductance to chlorine (Obata and Shinozaki, 1970; ten Bruggencate and Engberg, 1971). Picrotoxin blocks both Purkinje cell IPSPs and GABA potentials in Deiters' neurons. A reduction in GAD activity in the interpositus cerebellar nucleus is associated with destruction of the cerebellar hemisphere of the same side. Immunocytochemical studies have associated GAD activity with Purkinje cell axon terminals (Fonnum et al., 1970). In fact, GAD activity in Purkinje cell axon terminals is very high; 350–1000 mM GABA can be synthesized per hour (Fonnum and Walberg, 1973a).

MONOAMINERGIC AFFERENTS

In addition to the inhibition produced by local circuit neurons, elements of the cerebellar cortex (in particular, the Purkinje cells) may be inhibited by release of norepinephrine following activation of the locus coeruleus (see Foote et al., 1983). This form of inhibition, first demonstrated by Bloom and his collaborators (Siggins et al., 1971), suggests that Purkinje cell excitability may be depressed for protracted periods by the release of norepinephrine from terminals arising from the brainstem neurons. The terminals, rather than synapsing at specific points, seem widespread within the cortex. Their activation apparently produces a widespread release of catecholamines that hyperpolarize the Purkinje cells. Such hyperpolarization seems to be mimicked by application of cyclic adenosine-3′,5′-monophosphate (AMP) (Siggins et al., 1971), and norepinephrine may function by the activation of an electrogenic sodium pump similar to those in other central neurons (Phillis and Wu, 1981). Indeed, the possibility that an electrogenic sodium pump may be activated by norepinephrine is indicated, since the hyperpolarization is accompanied by a decreased ionic conductance change (Siggins et al., 1971).

There is also evidence for dopaminergic cerebellar afferents to the cerebellar nuclei, and to the Purkinje and granular cell layers of the cortex (Simon et al., 1979). The raphe nuclei, which synthesize and release serotonin, project fibers to all parts of the cerebellar nuclei and cortex (Takeuchi et al., 1982). These terminate at mossy fiber rosettes diffusely throughout the granular layer; in the molecular layer they bifurcate like parallel fibers and synapse with the intrinsic neurons (Chan-Palay, 1977). In the molecular and granular layers, beaded fibers with fine varicosities have been labeled with serotonin-specific antibodies (Takeuchi et al., 1982).

MOSSY FIBERS

No neurotransmitter candidates have been clearly identified to be liberated from mossy fibers. There is some indirect evidence for several candidates, however. For example, among the peptides, somatostatin-immunoreactive fibers have been shown to enter the cerebellum (Inagaki et al., 1982); these probably end as mossy fibers. Acetylcholine (ACh) is present in some mossy fiber terminals isolated as synaptosomes (Israël and Whittaker, 1965), and some mossy fibers contain acetylcholinesterase (Phillis, 1968). Choline acetyltransferase, the enzyme for ACh synthesis, has been demonstrated on mossy fibers and glomeruli by using immunocytochemistry (Kan et al., 1978, 1980). However, a role for ACh as a transmitter has not been supported by pharmacological or physiological studies.

DENDRITIC PROPERTIES

MICROELECTRODE RECORDINGS

That dendrites are capable of electroresponsive activity and are not simple passive cables was first shown in Purkinje cells. The earliest recordings indicating that the dendrites are active were made from alligator cerebellum. Here intradendritic recordings revealed large dendritic spikes in response to parallel fiber stimulation (Llinás and Nicholson, 1971). Injection of hyperpolarizing current allowed these spikes to be dissected into several all-or-none components. From these early studies it was proposed that there are several "hot spots" in the dendrites that are capable of spike generation, and that dendritic spikes travel toward the soma in a discontinuous manner. Subsequent intradendritic recordings from pigeon Purkinje cells showed that dendritic spikes are calcium dependent (Llinás and Hess, 1976). It was not until cerebellar slice preparation was used, however, that the dendritic properties of Purkinje cells were revealed in all their complexity.

The types of spontaneous action potentials that may be seen at different levels in a mammalian Purkinje cell soma and dendrites are illustrated for an in vitro experiment in Fig. 7.11. A typical bursting is seen at the somatic level (B). Recordings obtained at different levels in the dendritic tree are shown in C, D, and E. The decrease in amplitude of the fast spike which occurs as recordings are made further from the soma indicates clearly that the fast sodium action potentials seen at the soma do not actively invade the dendrites. Rather, they are electrotonically conducted and can be detected only to about middendritic level, their amplitude decrementing rather quickly with distance.

The bursting calcium-dependent spike, on the other hand, can be seen to be large and rather prominent in the upper dendrites, indicating a differential distribution for sodium and calcium conductances. Furthermore, direct stimulation of dendrites after application of TTX, as shown in Fig. 7.12A, produces two types of calcium-dependent electroreponsiveness. A small stimulus can generate a plateaulike response and a burst of action potentials. Because both responses can be blocked by cobalt, cadmium, or D600 (see Fig. 7.12B), it must be concluded that the dendrites of the Purkinje cell are capable of generating

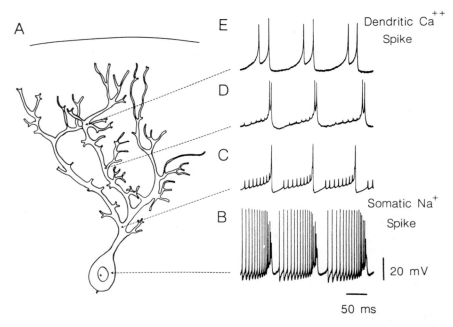

FIG. 7.11. Composite illustration of recordings made from different regions of a Purkinje cell in vitro. **A.** Drawing of typical mammalian Purkinje cell. **B.** Fast action potentials dominate this recording, with slower membrane oscillations. **C–E.** As the electrode moves away from the soma, (1) the amplitude of the fast, Na-dependent action potentials progressively decreases until they are not seen in the most distal branches; and (2) the slow, prolonged, calcium-dependent action potentials increase in amplitude and become distinct in the distal branches. Although the dendritic spikes are discontinually propagated toward the soma, the somatic spikes do not actively invade the dendrites. (Modified from Llinás and Sugimori, 1980a.)

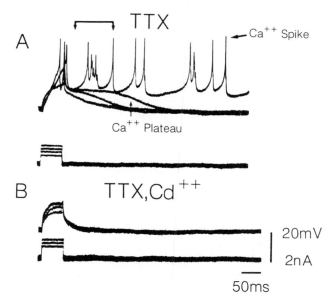

FIG. 7.12. Purkinje cell dendritic recordings in vitro. **A.** Intradendritic recording in the presence of TTX. Short depolarizing pulses elicit Ca-dependent plateau potentials and Ca spikes. As the current amplitude is increased, the plateau responses increase in duration, and full spike bursts are generated. **B.** The calcium dependence of both the plateau and the spike bursts is demonstrated by their complete abolition after Cd^{2+} has been added to the TTX bathing solution.

calcium-dependent spikes, which may be either of a prolonged plateau form or clear all-or-nothing action potentials.

The Purkinje cells thus demonstrate the following set of voltage-dependent ionic conductances. As discussed earlier, in the soma, there are (1) a rapid, inactivating Hodgkin–Huxley sodium current that generates a fast spike; (2) a fast voltage-activated potassium current that generates the after hyperpolarization following a fast spike; and (3) a calcium-activated potassium conductance. In addition, the somatic membrane displays a noninactivating voltage-activated sodium conductance capable of generating repetitive firing of the Purkinje cell following prolonged depolarization. At the dendritic level, on the other hand, excitability seems to be due mainly to a voltage-activated calcium conductance increase. This conductance may generate a low plateau potential or calcium spikes (Fig. 7.12A), and the spikes may be followed by potassium activation due to an increase of both voltage-activated and calcium-activated conductances to potassium.

It is therefore clear that the complex electrical responses observed in these cells after direct stimulation or activation of climbing or parallel fibers are largely due to the electroresponsive properties of the Purkinje cells themselves.

OPTICAL RECORDING

Recently, optical probes have been used to mark the spatial distribution of voltage-sensitive ionic channels in Purkinje cells. The sodium conductance is restricted to the soma and axon as visualized by using fluorescently labeled TTX (Sugimori et al., 1986).

Mapping of the distribution of an increase in intracellular calcium concentration ($[Ca^{2+}]_i$) during spontaneous and evoked Purkinje cell activity allows visualization of the probable location of calcium channels in the somadendritic membrane. This has been done in experiments using Arsenazo III absorption (Ross and Werman, 1986), and Fura-2 as a calcium indicator. Experiments using the fluorescent Ca^{2+} indicator Fura-2 have shown that during spontaneous bursting, the $[Ca^{2+}]_i$ increases first in the fine dendritic branches, where the increase is also the largest (Tank and Llinás, 1988). The $[Ca^{2+}]_i$ is later seen to increase in the dendritic trunk, and by this time it has begun to subside in the fine dendrites. The $[Ca^{2+}]_i$ in the soma increases very little. This temporal sequence of increased calcium activity, first in the distal and then in the proximal dendrites, supports the electrophysiological description of the two calcium conductances— the low-threshold plateau and all-or-none calcium-dependent dendritic spikes (see Fig. 7.11). The presence of voltage-activated calcium channels in the spiny branchlets provides a mechanism whereby parallel fiber EPSPs can be enhanced by slow local increases in calcium conductance. In contrast, when the synaptic activity is in the larger dendritic branches, full calcium-dependent dendritic spikes can be generated in the main dendritic tree. Climbing fiber synapses tend to depolarize the main dendritic tree, producing full dendritic spikes. Thus, the distribution of calcium channels over the dendritic tree is a critical element in the fine tuning of the electrophysiological sophistication of this most remarkable cell.

If a cell loaded with Fura-2 is depolarized by somatic current injection, the increased $[Ca^{2+}]_i$ in the dendrites is not uniform. Rather, there are well-localized areas of marked increases, supporting the earlier hypothesis of "hot spots" of calcium influx (Llinás and Nicholson, 1971).

FUNCTIONAL CIRCUITS

We have seen that there are two main types of afferents to Purkinje cells: (1) the climbing fiber–Purkinje cell system (Fig. 7.2A), which is organized into groups of specifically and synchronously activated Purkinje cell having quite different spatial locations (Armstrong, 1974); and (2) the mossy fiber–granule cell–parallel fiber–Purkinje cell system (Fig. 7.2B), in which Purkinje cells are activated at given loci in very specific geometrical patterns owing to the particular spatial relationship between the parallel fibers and the Purkinje cells dendrites. In the latter case, instead of the one-to-one relationship seen between the Purkinje cell and its climbing fiber afferent, a many-to-many relationship is present. Moreover, the directionality of the parallel fibers has been shown to be all important in determining the peculiar orthogonal organization of these fibers with respect to the lateral spread of the Purkinje cell dendritic tree. Thus, in the climbing fiber mode of activation, specific Purkinje cells may be activated, whereas with mossy and parallel fiber input, large numbers of Purkinje cells are activated in rows.

In considering the functional circuit of the cerebellum, the cerebellar nuclear cells must be included, for what the cortex ultimately does is to help determine the firing of these cells. CN cells are regulated in three ways:

1. By excitatory input from collaterals of the cerebellar afferent systems.
2. By inhibitory inputs from Purkinje cells activated over the mossy or climbing fiber pathways. Mossy fiber activation should generate a *tonic inhibition,* due to simple spikes in Purkinje cells firing synchronously at 30–40 Hz (Bell and Grimm, 1969), as well as a *transient inhibition* following a specific mossy fiber volley.
3. By the climbing fiber system, which originates in the inferior olivary nucleus, where the neurons are known to have powerful pacemaker properties and to be electrotonically coupled (Llinás and Yarom, 1986). This system generates rhythmic and synchronous activation of Purkinje cells, and thus not surprisingly *rhythmic inhibition* of CN neurons.

Let us consider more closely this last circuit, which includes the inferior olive, cerebellar nuclei, and Purkinje cells (Figs. 7.13 and 7.14). If the activity of the circuit is taken to start at the inferior olive, the axons of these cells may be followed to the cortex as climbing fibers or to the CN as collaterals. As recorded from the soma of a CN cell, activation of this pathway by white matter stimulation generates the complex response shown in Fig. 7.13B. This response may be considered as having five parts as shown in the figure: (1) antidromic excitation of the CN cell (1 in Fig. 7.13B); (2) Direct excitation of Purkinje cells, which is seen in the CN as a small IPSP (2 in Fig. 7.12); (3 and 4) a second EPSP–IPSP sequence (3 and 4 in Fig. 7.13B) with a latency of 3–4.5 msec. The EPSP is due to climbing fiber collateral activation of the CN cells, and the IPSP is generated following synaptic activation of the Purkinje cells. Finally (5), a rebound response is recorded, which is due to the intrinsic membrane properties of the CN cells themselves (Fig. 7.13B). Thus, the response in Fig. 7.13B is a combination of the properties of synaptic circuit and the intrinsic properties of the Purkinje cells and CN cells.

In fact, in this circuit the membrane properties of CN neurons are particularly important because the punctate and rather powerful synaptic EPSP–IPSP sequences are often followed by a rebound spike burst, as is seen in Fig. 7.13B. This means, then, that if a sufficient number of inferior olivary neurons, having a common rhythmicity, are activated synchronously at any particular time, a large and equally synchronous activation of Purkinje cells will occur. In fact, this is what occurs when harmaline, a tremorgenic agent known to act directly on the inferior olive (de Montigny and Lamarre, 1974; Llinás and Volkind, 1973; Llinás and Yarom, 1986), is administered. Such activation would be reflected in the CN as inhibition followed by a rebound activation. The activity in the circuit in the presence of harmaline is shown in the diagram in Fig. 7.14. This activity has been shown in vitro and is probably also the case in vivo, as indicated by

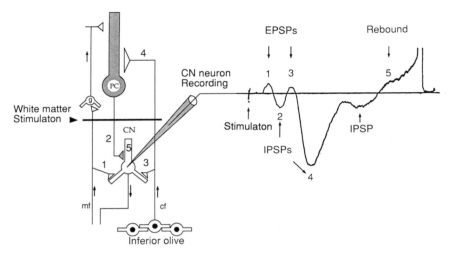

FIG. 7.13. Response of cerebellar nuclear cells to white matter stimulation. *Left:* Drawing of elements activated after white matter stimulation. *Right:* Response of cerebellar nuclear cell to white matter stimulation. White matter stimulation activates mossy fibers, climbing fibers, and Purkinje cell (PC) axons. The first response (1), a graded EPSP, is due to activation of the mossy fiber collaterals; the second (2), a small IPSP, is due to direct stimulation of Purkinje cell axons. The third response (3), a graded EPSP, is due to activation of climbing fiber collaterals. Finally (4), the powerful IPSP and smaller IPSPs follow climbing fiber activation of Purkinje cells. Although the cell is at the resting potential, the hyperpolarization is often sufficient to elicit a rebound response in the cerebellar nuclear cell (5).

multiple-electrode recording studies of Purkinje cell activity in the rat cortex (Llinás, 1985). The climbing fiber system generates rostrocaudal bands of synchronous Purkinje cell activation, and Purkinje cells within such bands (which are about 200 μm across) fire with the same frequency, and at intervals as close as 0.5–1.0 msec, indicating that the olivocerebellar system is quite tightly organized.

Spontaneous background inhibitory potentials recorded from CN neurons indicate that Purkinje cells also fire with some degree of synchronicity in the absence of harmaline. The powerful rebound excitation following these IPSPs is in turn transmitted to the rest of the CNS as an excitatory input and to the inferior olive as an inhibitory input (via the GABAergic nucleoolivary pathway) (Mugnaini and Oertel, 1981; Sotelo et al., 1986). This pathway is important since it seems to have a decoupling effect on the electrotonic junction in the inferior olive (Sasaki and Llinás, 1985; Sotelo, et al., 1986). In short, the olivocerebellar system would serve as an oscillatory circuit capable of generating timing sequences such as observed in tremor and in the organization of coordinated movements (Llinás, 1985). The mossy fiber–parallel fiber system provides a continuous and very delicate regulation of the excitability of the cerebellar nuclei, brought about by the tonic activation of simple spikes in Purkinje cells, that ultimately generates the fine control of movement known as motor coordination.

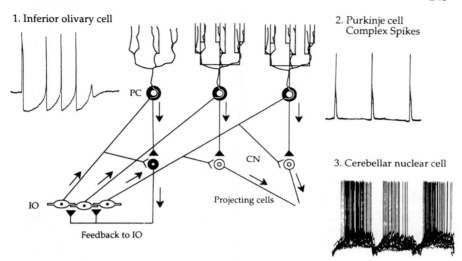

1. Inferior olivary cell

PC

IO

Feedback to IO

CN

Projecting cells

2. Purkinje cell
   Complex Spikes

3. Cerebellar nuclear cell

FIG. 7.14. Diagram of the circuit involved in the production of rhythmic activity in the olivocerebellar system: (1) Rhythmic activity in the inferior olivary neurons is transmitted to the Purkinje cells, where it is transformed to complex spikes (2) to the cerebellar nuclear projecting cells (white somata) and inhibitory cells (filled soma) eliciting EPSPs. Complex spikes trigger high-frequency firing of Purkinje cell axons that impinge on the cerebellar nuclear cells with powerful IPSPs and rebound firing (3). Thus bursts of spikes are transmitted to the rest of the nervous system including the cerebellum (as mossy fibers). The cerebellar nuclear cells projecting to the inferior olive (IO) are inhibitory and synapse in the glomeruli. (Filled synaptic terminals are inhibitory and open synaptic terminals are excitatory.)

The fact that the mossy fibers inform the cerebellar cortex of both ascending and descending messages to and from the motor centers in the spinal cord and brainstem gives an idea of the ultimate role of the mossy fiber system; it informs the cortex of the place and rate of movement of limbs and puts the motor intentions generated by the brain into the context of the status of the body at the time the movement is to be executed.

Because the Purkinje cell is itself an inhibitory neuron, the entire output of the elaborate cerebellar cortical neuronal network in the cerebellar cortex is therefore the organized, large-scale inhibition of neurons in the cerebellar nuclei. It should be recalled that of all the cells residing in the cortex only the granule cells are excitatory; all the rest are inhibitory. As illustrated in Figs. 7.12 and 7.13, an understanding of their membrane properties and synaptic interactions provides fundamental insight into the functioning of the cortex.

8

THALAMUS

S. MURRAY SHERMAN AND CHRISTOF KOCH

The thalamus is the gateway to the neocortex, and as such these two main components of the vertebrate telencephalon have evolved in close relation to each other. Virtually all routes to the cortex are relayed by the thalamus, although a few diffuse and poorly understood pathways from other brainstem sites do exist. Our conscious perception of the world around us depends on information reaching the cortex and being analyzed there, and thus the thalamus represents a key link in this process.

As we shall see in this chapter, however, the thalamus does much more than act merely as a passive and machinelike relay of information to the cortex. Instead, the ability to pass through this gateway is determined by specialized neuronal circuitry. The gate can be: (1) completely open, which results in the relay of all information to cortex; (2) completely closed, which cuts off cortex from the outside world; or (3) partially open, which permits certain information to reach cortical levels. Thus the thalamus filters the flow of information to cortex and as such is an important neuronal substrate for many forms of attention (Singer, 1977; Sherman and Koch, 1986).

OVERALL ORGANIZATION

The thalamus is most highly developed in mammals and especially so in primates. All sensory systems pass through the thalamus on their way to the neocortex. This includes somatosensory information from the muscles, deep tissues, and skin; visual information from the eyes; auditory information from the ears; gustatory information from the taste buds; and olfactory information from the nose, after relaying through the olfactory cortex (which is the paleocortex, rather than neocortex). Each part of the thalamus, in turn, receives fibers from the area of the cortex to which it projects (E. G. Jones, 1985).

The thalamus can be divided on the basis of connectivity and embryological origin into three main divisions: dorsal thalamus, ventral thalamus, and epithalamus. The dorsal thalamus, which is the largest division, has massive reciprocal connections with cerebral cortex and striatum; in fact, virtually the whole cortex receives a projection from the dorsal thalamus. Authors often mean "dorsal thalamus" when they refer simply to "thalamus." The ventral thalamus does not innervate cortex. However, it does receive innervation from cortex, and most of its subnuclei, one of which is the reticular nucleus of the thalamus (RNT; also known as the nucleus reticularis thalami or the thalamic reticular nucleus), have reciprocal connections with specific nuclei in the dorsal thalamus (E. G. Jones, 1985; Ohara and Lieberman, 1985). The epithalamus lacks direct afferent or efferent connections with the cortex and is actually more closely associated with the hypothalamus; it will not be considered further here.

The dorsal thalamus can be divided into a number of discrete nuclei (Fig. 8.1). We now recognize that many, and perhaps all, of these nuclei have unique functional correlates, with specific input and output routes. These routes are limited to the same hemisphere, since no contralateral connections involving any thalamic nucleus have been found. An exhaustive survey of all dorsal thalamic nuclei is beyond the scope of this chapter (see E. G. Jones, 1985, for a more thorough account), but examples of the best studied nuclei follow. The *lateral geniculate nucleus* (LGN) relays input from the retina to visual cortex. There are actually two LGN divisions: The dorsal division, which is part of the dorsal thalamus, projects to cortex, and, unless otherwise specified, is what we mean by "LGN"; the ventral division, which is part of the ventral thalamus, also receives retinal input but projects only subcortically, mostly to the midbrain. The *medial geniculate nucleus* (MGN) receives auditory input from the inferior colliculus and projects to auditory cortex. The *ventral posterolateral nucleus* (VPL)

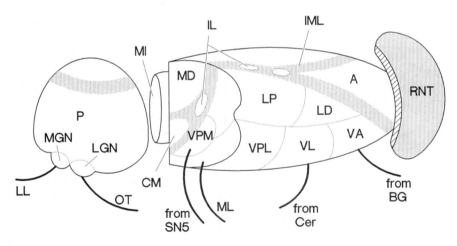

FIG. 8.1. Schematic three-dimensional view of right thalamus with many of its major nuclei. A cut is placed in the posterior part to reveal a representative cross section. Some of the important ascending afferents are also shown. To prevent obscuring of the dorsal thalamus, only the rostral tip of the reticular nucleus of the thalamus (RNT) is shown. Other abbreviations: A, anterior; BG, basal ganglia; Cer, cerebellum; CM, centromedian; IL, intralaminar nuclei; IML, internal medullary lamina; LD, lateral dorsal; LGN, lateral geniculate nucleus; LL, lateral lemniscus; LP, lateral posterior; MD, mediodorsal; MGN, medial geniculate nucleus; MI, midline nuclei; ML, medial lemniscus; OT, optic tract; P, pulvinar; SN5, main sensory and spinal nuclei of the 5th nerve; ST, spinothalamic; VA, ventral anterior; VPL, ventral posterolateral; VPM, ventral posteromedial. See E. G. Jones (1985) for details of connectivity of these nuclei. (Redrawn from Brodal, 1981.)

transmits somatosensory input from the body, providing the cortex with information about touch, pressure, joint position, temperature, and pain; its contiguous companion is the *ventral posteromedial nucleus* (VPM), which transmits somatosensory information from the head. The VPL receives ascending input from the spinal cord and dorsal column nuclei in the medulla, whereas the VPM receives input from the 5th cranial nerve via the main sensory and spinal nuclei of this nerve. The *basal ventral medial nucleus* receives gustatory input from the parabrachial nucleus of the pons and projects to the primary somatosensory cortex. The pulvinar is innervated by the superior colliculus and the pretectum. The *ventral lateral nucleus* receives the majority of its input from the deep cerebellar nuclei and projects to the primary motor cortex. The *ventral anterior nucleus* is innervated by the basal ganglia and projects to both the motor cortex and the basal ganglia.

THE LGN AS THE PROTOTYPICAL THALAMIC NUCLEUS

At the level of synaptic circuitry, more is known about the LGN than about any other thalamic structure, and this nucleus has been more thoroughly studied in the cat than in any other species. It seems likely that many of the organizational principles of the cat's LGN apply generally to other dorsal thalamic nuclei across mammals, although our present knowledge of most other such nuclei is too sparse to be truly comfortable with this generalization. Nonetheless, many of the specific examples for the functional organization of the thalamus derive from the cat's LGN, and most of the discussion of thalamus below refers to the LGN. It is thus worth briefly introducing this nucleus to the reader.

Figures 8.2 and 8.3 illustrate the laminar patterns of the cat's LGN (see Sherman and Spear, 1982; Sherman, 1985; Sherman and Koch, 1986). It is comprised of several laminae, most of which are innervated by one or the other retina. In addition to this segregation based on ocular origin, axons from neighboring retinal loci innervate neighboring geniculate zones, thereby setting up an orderly point-to-point map of visual space within the LGN. This is known as a retinotopic map, and analogous maps exist within other thalamic nuclei, such as the VPL, VPM, and MGN (E. G. Jones, 1985). Most is known about the A-laminae (laminae A and A1) of the LGN, which form a reasonably matched pair, with lamina A innervated by the contralateral retina and lamina A1 innervated by the ipsilateral retina. The other main geniculate zones are the C-laminae, and the medial interlaminar nucleus which, despite its name, is really just a part of the LGN.

NEURONAL ELEMENTS

The neuronal elements of the thalamus can be divided into three components: the extrinsic afferent inputs to the nucleus, the relay cells (or principal neurons) that project to cortex, and the interneurons (or intrinsic neurons).

INPUTS

Figure 8.4 schematically illustrates the major afferents for a typical dorsal thalamic nucleus. Seen in this perspective, the retinal or lemniscal afferents to

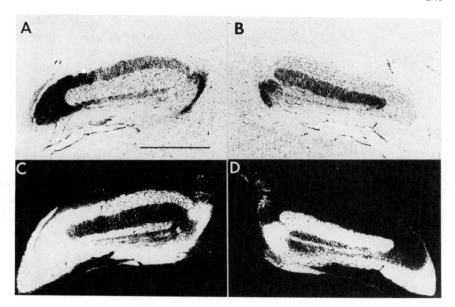

FIG. 8.2. Photomicrographs of the cat's LGN, showing its lamination as seen in coronal view near the middle of the nucleus. The sections were treated for autoradiography after retinogeniculate terminals from the right eye were labeled by injection of that eye with tritiated proline. The labeled terminals are dark in the bright-field views and bright in the dark-field views. Although not labeled, the perigeniculate nucleus is the thin band of neurons lying just above lamina A. See Fig. 8.3a for labeling plus an interpretation of these photomicrographs. **A.** Bright-field view of left LGN. **B.** Bright-field view of right LGN. **C.** Dark-field view of left LGN. **D.** Dark-field view of right LGN. (From Sherman, 1985.)

relay cells are one class among several and thereby are a minority. The other afferents include long pathways from the cortex and brainstem reticular formation plus local inputs from RNT cells and interneurons.

Retinal or lemniscal afferents. The best characterized input to a dorsal thalamic nucleus is that which conveys the main sensory message to be relayed to cortex. For the LGN, this input arises from the ganglion cells of the retina, whose axons form the optic nerve and tract. The number of retinogeniculate axons from each retina varies with species; it is slightly under 100,000 in cats and is roughly 1 million in monkeys and humans (Rakic and Riley, 1983; Williams et al., 1983). Comparable input to the VPL and MGN derives, respectively, from the medial lemniscus and lateral lemniscus. For simplicity, we shall refer to these afferents as the retinal (or retinogeniculate) and lemniscal (or lemniscothalamic) afferents; generic terms for other afferents are thus nonretinal and nonlemniscal.

In cats, retinal ganglion cells can be divided into at least three physiologically and morphologically distinct classes: X cells, Y cells, and the remainder, which we shall refer here to as W cells. Their main features are summarized in Table 8.1 (see also Sherman and Spear, 1982; Sherman, 1985; and Chap. 6). Other

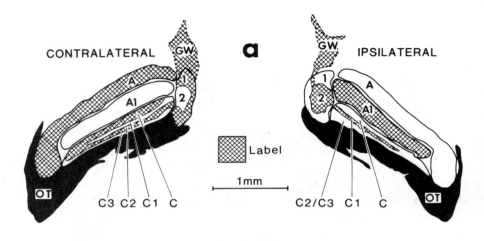

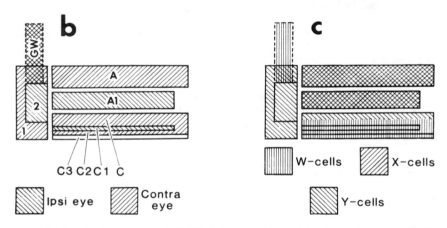

FIG. 8.3. Lamination of the cat's LGN. **a.** Interpretation of the photomicrographs in Fig. 8.2. For the left (*contralateral*) LGN, the right nasal retina projects to laminae A, C, and C2, lamina 1 of the medial interlaminar nucleus (MIN), and the geniculate wing (GW); for the right (*ipsilateral*) LGN, the right temporal retina innervates laminae A1 and C1, lamina 2 of the MIN, and the GW. Not shown is MIN lamina 3, which would appear more rostrally in the LGN; this lamina is innervated by axons from the contralateral temporal retina. Neither retina innervates lamina C3 (which is innervated by the midbrain), and both retinas innervate the GW, which is the only geniculate region binocularly innervated. **b.** Schematic view of ocular input to geniculate laminae shown in *a*. **c.** Schematic view of distribution of W, X, and Y cells by lamina. W cells are limited nearly exclusively to the C-laminae and GW. X cells are limited nearly exclusively to the A-laminae. Y cells are found in the A-laminae, the MIN, and the top tier of lamina C. OT, optic tract. (From Sherman, 1985.)

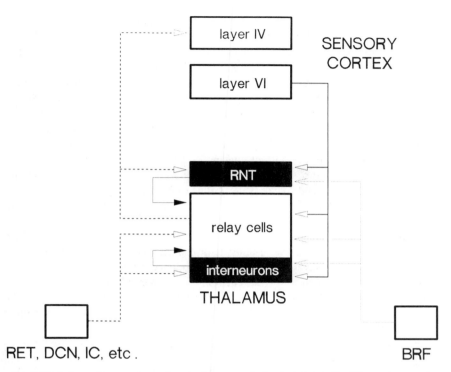

SENSORY
CORTEX

THALAMUS

RET, DCN, IC, etc .   BRF

FIG. 8.4. Schematic summary of main inputs to thalamic relay cells. The main sensory input arrives from retina (RET) via the optic tract, the dorsal column nuclei (DCN) via the medial lemniscus, and the inferior colliculus (IC) via the lateral lemniscus. Local GABAergic inhibitory input is provided by interneurons and cells of the reticular nucleus of the thalamus (RNT). Other main inputs include axons from layer 6 of sensory cortex and cholinergic, noradrenergic, or serotonergic axons from the brainstem reticular formation (BRF). (Redrawn from Sherman and Koch, 1986.)

Table 8.1. Main Features of W, X, and Y in Retina and LGN Cells

Property	W cells[a]	X cells	Y cells
Receptive field size	Large	Small	Medium
Contrast sensitivity	Poor	Fair	Good
Spatial resolution	Poor	Good	Fair
Temporal resolution	Poor	Fair	Good
Axonal conduction velocity	Slow	Medium	Fast
Retinal ratio	10–20%	75–80%	5–7%
LGN relay cell ratio	10%	40–50%	40–50%

[a]Here we refer only to the subset of W cells that seem to be involved in retino-geniculate innervation; see Sherman (1985) for details.

251

mammals, including primates, have comparable retinal ganglion cell classes (Stone, 1983; Rodieck and Brening, 1973; Sherman, 1985). Every X and Y cell, but only a minority of W cells, innervates the LGN. The A-laminae in the cat's LGN include only X and Y cells (Fig. 8.3c). We know most about X and Y cells there, but, unfortunately, very little is yet known about W cells. Thus most of our comments are restricted to the X and Y pathways through the geniculate A-laminae. Each of these retinal cell types innervates a unique geniculate cell type, thereby establishing W, X, and Y classes of geniculate cell and parallel, functionally distinct W, X, and Y pathways. This organization of W, X, and Y pathways has led to the important concept of *parallel processing,* whereby each of the pathways analyzes somewhat different aspects of the visual scene (for details, see Sherman and Spear, 1982; Stone, 1983; Shapley and Lennie, 1985; Sherman, 1985). Parallel processing seems to be a feature of all sensory systems (Dykes, 1983; E. G. Jones, 1985).

Retinogeniculate axons appear to be excitatory and glutamatergic, which means they use the amino acid glutamate or a similar compound as a neurotransmitter (Kemp and Sillito, 1982). Recent evidence suggests that part of this excitatory input is mediated via receptors for N-methyl-D-aspartate (NMDA; Moody and Sillito, 1988; Lo and Sherman, 1989). Likewise, at least some of the lemniscal input to VPL uses a glutamate-like substance and NMDA receptors (Salt, 1988).

Within the appropriate geniculate lamina, each retinal axon arborizes repeatedly to form many short branches, each of which ends in a knoblike terminal (Bowling and Michael, 1984; Sur et al., 1987). It is at these terminals that synapses are formed. The terminal arborizations are relatively restricted, of the order of 100–300 μm in diameter, although X arbors are consistently smaller than are Y arbors. The restriction of these terminal arbors is necessary for maintaining the retinotopic map. Retinal X axons outnumber Y axons by roughly 10 to 1, but the ratio of X to Y geniculate cells postsynaptic to these retinal axons is only about 2 to 1 (Sherman, 1985; see also below). This change in X/Y ratios between retina and LGN seems to result largely from the fact that retinogeniculate Y arbors are more extensive than are the X arbors. Most geniculate cells receive their retinal input from a single axon, or a very small number. There is thus little convergence in retinogeniculate circuitry, although considerable divergence exists, because each retinal axon innervates roughly 5–10 (for X axons) or 30–50 (for Y axons) geniculate neurons. The limited convergence maintains small receptive fields among geniculate neurons and helps to preserve spatial acuity in the visual system.

The nature of lemniscal input to other thalamic nuclei is not known to the same detail as is the case for retinogeniculate axons. However, the basic features seem similar for all thalamic nuclei (E. G. Jones, 1985). Thus lemniscal afferents to the VPL and MGN exhibit the same general morphology as do retinogeniculate axons, and there seems to be little convergence among lemniscothalamic connections.

Cortical afferents. The most numerous inputs to thalamus originate among layer six pyramidal cells of the cortex (see Fig. 8.4). Like retinal or lemniscal axons, these cortical axons are excitatory and appear to be glutamatergic (Giuffrida and Rustioni, 1988). Strong reciprocity exists in thalamocortical connections, because the cortical input for each thalamic nucleus generally originates from the same cortical area that is innervated by the thalamic nucleus in question. Thus for the LGN, this cortical pathway emanates from visual cortex (mostly areas 17, 18, and 19), and roughly half of these layer 6 cells contribute to the corticogeniculate pathway. Likewise, somatosensory and auditory cortex projects back, respectively, to the VPL and MGN.

The anatomically dominant input to thalamus arises from cortex. In fact, there seems to be at least an order of magnitude more corticothalamic axons than thalamocortical ones. Thus roughly four million axons from visual cortex innervate the geniculate relay cells of the A-laminae (Sherman and Koch, 1986). Each cortical axon innervates many thalamic neurons, thereby establishing considerable divergence and convergence in the corticothalamic pathway. Nonetheless, the corticothalamic pathway faithfully adheres to the map established in the thalamic nucleus (see above); for instance, the corticogeniculate pathway conforms to the retinotopic map in the LGN. Corticogeniculate neurons seem to be heterogeneous and probably represent several functional classes (Tsumoto and Suda, 1980), although they have not yet been properly classified and it is not clear to what extent other corticothalamic pathways contain functional subsets of axons.

Brainstem afferents. Other inputs to the thalamus emanate from various brainstem sources, and these have not yet been thoroughly studied. Afferents from the brainstem reticular formation in the pons and midbrain (see Fig. 8.4) include cholinergic neurons (i.e., using acetylcholine as a neurotransmitter) of the pedunculopontine tegmental nucleus, noradrenergic neurons (i.e., using norepinephrine) of the locus coeruleus, and serotonergic neurons (i.e., using serotonin) of the raphe nucleus. These inputs can either excite or inhibit thalamic neurons (see Chap. 2 and below).

The LGN receives additional although sparse brainstem inputs that may be unique to the visual pathways, and these are thus omitted from Fig. 8.4. These include afferents from the superior colliculus and parabigeminal nucleus of the midbrain and from the pretectal nucleus of the optic tract (NOT). The parabigeminal input is cholinergic, but the neurotransmitters for the collicular and pretectal inputs are not known. They will not be further considered in this chapter, and the reader is instead referred to the discussion of this in several recent papers (Harting et al., 1986; Fitzpatrick et al., 1988).

Inputs from the RNT. A final extrinsic source of innervation to each dorsal thalamic nucleus derives from the RNT (Jones, 1985; Ohara and Lieberman, 1985; Sherman and Koch, 1986). The RNT should not be confused with the brainstem reticular formation. The RNT forms a shell anteriorly and dorsally around the dorsal thalamus (see Fig. 8.1). Generally, each dorsal thalamic nucleus (e.g., the LGN, VPL, MGN, etc.) has a subnucleus of the RNT associated with it, and reciprocal connections are formed between them (E. G. Jones, 1985; Ohara and Lieberman, 1985; Sherman and Koch, 1986). That is, relay cell axons en route to cortex pass through the appropriate RNT zone, where they emit collateral terminals, and the RNT cells in turn project axons back into the dorsal thalamic nucleus. It is worth noting that corticothalamic axons also pass through the appropriate RNT zone en route to their thalamic destination, and they also provide collateral innervation to these RNT cells. Finally, the RNT is also innervated by the same regions of brainstem that innervate the dorsal thalamus. The RNT cells are GABAergic (i.e., they use gamma-aminobutyric acid, or GABA, as a neurotransmitter) and inhibit their dorsal thalamic targets.

For the cat's LGN, the related RNT zone is known as the perigeniculate nucleus, and it lies just dorsal to lamina A. Among purists, there is some controversy as to whether or not the cat's perigeniculate nucleus is a part of the RNT or whether it is a special GABAergic cell group unique to the LGN. Because all of its connections seem completely analogous to other RNT regions, and because no other RNT zone for the LGN has yet been defined, we shall consider it to be a part of the RNT.

RELAY NEURONS

Relay (or projection) neurons, which represent roughly 75% of the cells in most thalamic nuclei (but see below), are the only output of a dorsal thalamic nucleus. They project to cortex with a collateral innervation of the RNT en route. Roughly 300,000 relay cells reside in each of the A-laminae of the cat's LGN (Sanderson, 1971), and Fig. 8.5A,B illustrates relay X and Y cells there. These are fairly representative of thalamic relay cells in other nuclei.

Interesting morphological differences between these geniculate relay cell types have been documented (Sherman, 1985). Compared to the Y cells, the X cells are slightly smaller with thinner dendrites and thinner axons; the dendritic arbors of Y cells tend to be spherical, whereas those of X cells tend to be oriented perpendicularly to laminar borders and thus along projection lines; and the Y cells tend to have smooth dendrites, whereas many appendages (i.e., collections of knobs, thorns, or spines) exist on X cell dendrites, particularly near proximal branch points. These appendages mark the postsynaptic sites of retinal input and synaptic glomeruli (see below).

The projection of relay cells concentrates in layer 4 of the cortical target area, with a smaller terminal zone in layer 6. In the cat, geniculate cells project to both striate cortex (area 17 or V1) and extrastriate cortex (mostly area 18, but also area 19 and the lateral suprasylvian cortex). The X cells project exclusively to area 17, whereas the Y and W cells project to striate and extrastriate cortices. However, in primates, nearly all geniculate neurons project only to area 17. Similar relationships hold for other thalamic nuclei, since multiple projections from VPL to somatosensory cortex and MGN to auditory cortex have been described.

INTERNEURONS

Roughly 25% of the cells in most thalamic nuclei are local interneurons. However, as an example of the bewildering variation in relative numbers of relay cells and interneurons, the cat's LGN and VPL plus the rat's LGN have roughly a 3 : 1 relay cell to interneuron ratio, but the rat's VPL has practically no interneurons

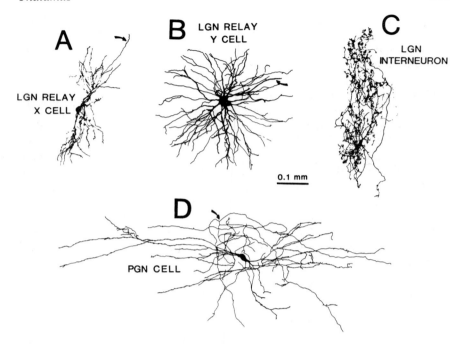

FIG. 8.5. Tracing of four representative neurons from the cat's LGN and perigeniculate nucleus, which is the equivalent of the RNT for the A-laminae of the LGN. Each of the cells was first studied physiologically and then labeled intracellularly with horseradish peroxidase. Where obvious, the axon is indicated by an arrow. **A.** Relay X cell. **B.** Relay Y cell. **C.** Interneuron. **D.** Perigeniculate neuron. (Redrawn from Sherman and Koch, 1986.)

(Ohara et al., 1983; Ralston, 1983; Spreafico et al., 1983; Fitzpatrick et al., 1984). Thus analogous nuclei in the same animal (e.g., the rat's LGN and VPL) can vary in this regard, as can homologous nuclei across species (e.g., the VPL of cats and rats).

Interneurons have been best described for the LGN, but they seem basically similar in other thalamic nuclei. Geniculate interneurons have small cell bodies with long, thin, sinuous dendrites (Fig. 8.5C). The dendrites are notable for giving rise to bulbous appendages connected to the stem dendrite by long (10 μm or more), thin (usually less than 0.1 μm in diameter) processes; these appendages usually occur in clusters. Overall, the dendrites with their bulbous appendages look like the terminal arbor of an axon, and thus Guillery (1966) referred to these dendrites as "axoniform" in appearance. In fact, these bulbous appendages represent a major synaptic output of the cell, since they are synaptic terminals that are both presynaptic and postsynaptic to other elements in the geniculate neuropil (Ralston, 1971; Famiglietti and Peters, 1972; Hamos et al., 1985; Ralston et al., 1988). Many of the synapses from interneurons are thus dendritic in origin.

Most or all of these interneurons also have a conventional axon that arborizes locally, typically within the dendritic arbor (Hamos et al., 1985; Montero, 1987). However, recent evidence suggests that at least some interneurons in the cat's VPL nucleus may lack an axon (Ralston et al., 1988). In the cat's LGN, interneurons seem to be associated mostly with the X pathway, since they receive retinal input only from X axons and contact mostly only relay X cells (Sherman and Friedlander, 1988). All interneurons are GABAergic, and both their dendritic and axonal outputs inhibit their postsynaptic targets.

We have previously described RNT cells as a source of nonretinal or nonlemniscal afferents to the dorsal thalamus. Although this is correct, RNT cells do not project beyond the thalamus, instead providing local, GABAergic, inhibitory input to thalamic relay cells. They are thus functionally in many ways similar to interneurons, and many investigators group them with interneurons as local inhibitory cells. For this reason, we illustrate the morphology of a perigeniculate neuron (Fig. 8.5D) along with a relay X and Y cell and an interneuron in Fig. 8.5. It is not yet clear what, if any, fundamentally different role in retinogeniculate and lemniscothalamic transmission is played by the RNT cells and interneurons.

SYNAPTIC CONNECTIONS

TYPES OF SYNAPTIC TERMINAL

The synaptology of both relay cells and interneurons has been described on the basis of electron microscopic studies. Most of these studies have concentrated on the LGN and VPL with rather similar results (Guillery, 1971; Wilson et al., 1984; Hamos et al., 1985; E. G. Jones, 1985; Montero, 1987; Ralston et al., 1988). The following description derives from the LGN.

Four major types of synaptic terminal exist in the LGN (Guillery, 1971), and their origins are largely defined. RLP terminals (*r*ound vesicles, *l*arge profiles, and *p*ale mitochondria) form asymmetrical synapses and comprise 10–20% of all synaptic profiles. They derive from retina. RSD terminals (*r*ound vesicles, *s*mall profiles, and *d*ark mitochondria) also form asymmetrical synapses and are the most numerous, comprising roughly half of all terminals. Most RSD terminals derive from cortex, although some derive from brainstem sources. *F* terminals (*f*lattened vesicles) form symmetrical synapses and represent a little more than one quarter of the terminals in the LGN. Two subtypes, F1 and F2, have been recognized. Although a constellation of features can distinguish them, the most salient are that F1 terminals derive from axons and are strictly presynaptic, whereas F2 terminals are dendritic in origin and are both presynaptic and postsynaptic. F1 terminals mostly arise from axons of RNT cells and interneurons, although some brainstem axons may also form F1 terminals; F2 terminals derive from dendrites of interneurons.

INPUTS TO RELAY CELLS

Reconstructions at the electron microscopic level reveal that geniculate relay cells in the cat receive roughly 4000 synapses, nearly all onto their dendrites with rare contacts on their somata (Wilson et al., 1984). Figure 8.6 schematically

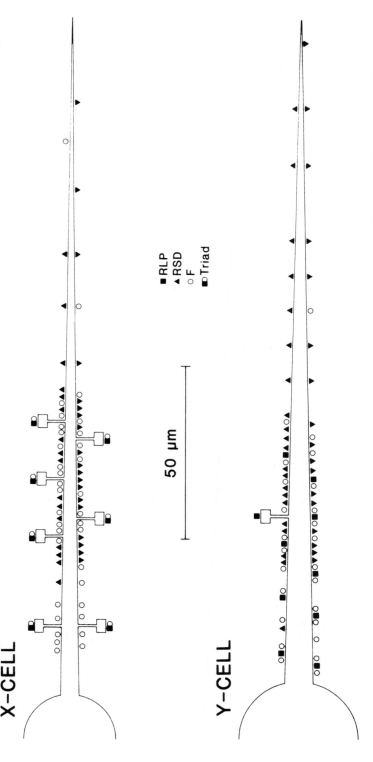

FIG. 8.6. Schematic representation of the distribution of synaptic terminals onto a typical dendrite of a relay X and Y cell. For simplicity, only a single, unbranched dendrite is shown for each neuron. Each type of synaptic terminal (RLP, RSD, and F), and also synaptic triads are indicated by separate symbols. The density (synapses per micrometer of dendritic length) of each terminal type is also represented by the relative number of synaptic terminals. Dendritic appendages are denoted by the T-shaped attachments to dendrites. Although all of the small squares represent F terminals, these can be divided into F1 and F2 types. F2 terminals, which derive from interneuronal dendrites, are largely limited to X cells as parts of the triadic synaptic arrangements located in glomeruli; F1 terminals, which derive from axons of interneurons, RNT cells, and brainstem cells, are found on both X and Y cells and do not participate in triads. The RSD terminals on the more peripheral dendrites derive nearly exclusively from corticogeniculate axons, whereas those more proximally located include many from brainstem axons. See text for details. (From Wilson et al., 1984.)

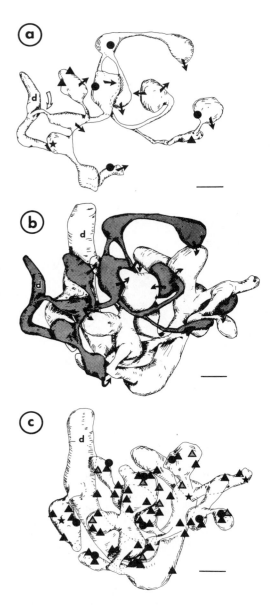

FIG. 8.7. Reconstruction of a glomerular zone in the geniculate A-laminae of the cat's LGN, showing the F2 terminals from an intracellularly labeled interneuron, the postsynaptic cluster of appendages from a relay X cell, and the location of synaptic contacts; each scale bar represents 1.0 μm. **a.** Labeled processes from the interneuron. A thin stem dendrite (d) emits an extremely fine process (open arrow) that arborizes into 12 F2 terminals connected by extremely fine processes. These terminals are postsynaptic to retinal or RLP terminals (circles), unlabeled F terminals (triangles; most or all of these may be F1 terminals, but they were not sufficiently reconstructed to be certain), and an RSD terminal (star). The labeled F2 terminals also form synaptic outputs (solid arrows). **b.** Combined reconstruction of the labeled interneuron's processes from *a* (stippled area) and unlabeled postsynaptic processes from *c* (clear area). The synapses from the F2 terminals onto the relay X cell's appendages are illustrated (solid arrows; these represent

258

summarizes the distribution of various types of synaptic input on the dendritic arbors of these relay X and Y cells. Relay cells in other thalamic nuclei probably have a comparable pattern of synaptic inputs. For both relay X and Y cells, inputs from retinal and F terminals concentrate in the proximal region of the dendritic arbor, whereas RSD input (i.e., mostly of cortical origin) dominates distal dendrites. However, major differences between X and Y cells exist in the types of F terminal present and in the detailed nature of the retinal input. To explain these differences first requires a description of the glomerulus and the synaptic triad.

A *glomerulus* is a complex synaptic structure (Fig. 8.7). Glomeruli seem to be related to interneurons, and it is interesting that the rat's VPL, which lacks interneurons (see above), also lacks glomeruli (Ralston, 1983). For the A-laminae of the cat's LGN, glomeruli include a major set of inputs to proximal dendrites of relay X cells, but they do not seem to be associated to any degree with the Y pathway (Sherman, 1988). Glomeruli are common in other thalamic nuclei, but the pattern of specificity for functional types outside of the LGN is presently unknown (Jones, 1985; Ralston et al., 1988). Glomeruli have terminals of all four types noted above, and these terminals complexly interrelate with each other. A retinal terminal is typically located at or near the glomerular center and is surrounded by a number of other terminals. This retinal terminal contacts two different postsynaptic elements: an F2 terminal that derives from dendritic appendages of interneurons; and a dendritic appendage, or less frequently a dendritic shaft, of a relay X cell. The retinal terminal usually contacts several F2 terminals within a glomerulus, and all of the synapses formed by the retinal terminal are asymmetrical. The interneuron's F2 terminals, in turn, make symmetrical synaptic contacts onto the *same* postsynaptic element of the relay X cell, be it dendritic appendage or shaft, contacted by the retinal terminal.

Because three terminal types are involved, this special neuronal circuit within the glomerulus is known as a triad (see Koch, 1985, for a detailed hypothesis concerning the role of these triadic circuits). Figure 8.8 illustrates a triad involving RLP and F2 terminals and a dendritic appendage of a relay X cell. F1 and RSD terminals are also present in the glomerulus, and these may contact both F2 terminals and the relay X cell in triadic arrangements. A retinal terminal is usually the common presynaptic element in the triad, but occasionally other terminals, such as those from the brainstem, can serve this function. Both a retinal terminal and brainstem axon can share the same F2 terminal and postsynaptic relay X cell process in triadic circuitry within a glomerulus.

The vast majority of retinal input to relay X cells is filtered through this complicated circuitry of the glomerulus. Retinal input to Y cells is simpler and

---

the same solid arrows as in *a*). **c.** Unlabeled postsynaptic dendrite (d) from a relay X cell with eight appendages that receive all of the neuron's synaptic input in the reconstructed zone. These include nine synapses from RSD or retinal terminals (circles), nine from F2 terminals of the labeled interneuron (stippled triangles; these correspond to the solid arrows in *a* and *b*), 40 from unlabeled F terminals (solid triangles), and three from RSD terminals (stars). The 16 triadic synaptic arrangements are illustrated by overlapping pairs of symbols for synapses from RLP and F terminals. (From Hamos et al., 1985.)

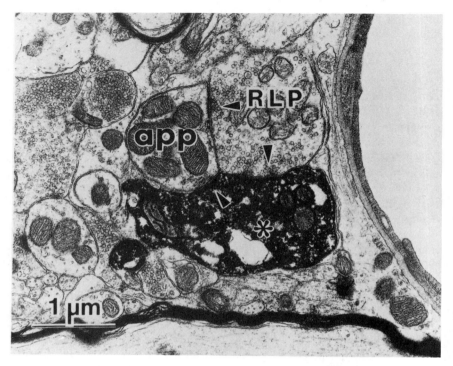

FIG. 8.8. Electron micrograph of a triadic synaptic relationship from the A-laminae of the cat's LGN. An interneuron was labeled with horseradish peroxidase, which created an electron-dense reaction product; its labeled F2 terminal is dark in this micrograph. A retinal terminal (RLP) contacts both the labeled F2 terminal and the dendritic appendage (app) of a relay X cell. The labeled F2 contacts the same appendage. Synaptic contacts are indicated by arrowheads. Scale: 1.0 μm.

involves conventional asymmetrical synapses onto proximal dendritic shafts (Wilson et al., 1984; Sherman, 1988). F2 terminals are nearly always limited to glomeruli, and the lack of glomeruli associated with the Y pathway results in very few such terminals contacting Y cells. More than 90% of the F input to these cells is of the F1 variety, whereas roughly two thirds of F input onto X cells is of the F2 variety.

INPUTS TO INTERNEURONS

Input to interneurons has been worked out in less detail (Hamos et al., 1985). As with our previous examples, most of our detailed knowledge of interneurons stems from studies of the LGN, but comparable studies in other thalamic nuclei, especially the VPL, reveal basically similar properties for thalamic interneurons (Ralston et al., 1988). In the LGN, many retinal, RSD, and F1 terminals contact interneurons. Much of this input is focused onto their dendritic appendages, which are the presynaptic F2 terminals. Input is also formed onto dendritic shafts and somata, and these are the only geniculate neurons that seem to receive significant retinal input onto their somata.

DENDRITIC CABLE PROPERTIES

RELAY CELLS

Both X and Y classes of relay cell are electrically rather compact, with dendritic arbors extending for roughly one length constant (Bloomfield et al., 1987; Bloomfield and Sherman, 1989). In practice, this means that even the most distally located synaptic input can have significant effects on the soma and axon, with attenuation of postsynaptic potentials never exceeding one third to one half (see Fig. 8.9). The values of neuronal input resistance for these cells are 15–25 MΩ (megohms), and their passive membrane time-constants are 8–11 msec.

One of the reasons for the electrotonically restricted dendritic arbors of X and Y relay cells is the nature of their dendritic branches. These branches closely adhere to Rall's "3/2 branching rule" (Bloomfield et al., 1987), whereby the diameters of the daughter dendrites raised to the 3/2 power and summed equals the diameter of the parent dendrite raised to the 3/2 power (Rall, 1977). Such branching matches impedance on both sides of the branch point and permits efficient current flow across these branches in *both* directions (see Appendix). This maximizes the transmission of distal postsynaptic potentials to the soma. This also implies that a potential generated anywhere in the dendritic arbor or at the soma will be efficiently transmitted throughout the dendritic arbor. Among other things, this means that the discharge of an action potential will depolarize the entire dendritic arbor by tens of millivolts, and this could have significant effects on voltage-dependent processes in the dendrites (see below).

INTERNEURONS

Unlike relay cells, interneurons are not electrotonically compact (Bloomfield and Sherman, 1989), partly because their dendrites are thinner and longer than those of relay cells. More importantly, the dendritic branch points of interneurons violate the "3/2 branching rule," because daughter branches tend to be too thin. This leads to poor current flow across these branch points. As a result, much of the synaptic circuitry in distal dendrites, including that involving the F2 terminals, is functionally isolated from the soma and axon (Fig. 8.9). Ralston (1971) proposed some time ago that synaptic input onto the F2 terminals of interneurons in the cat's VPL would also be isolated from the soma.

Recent computational modeling based on these observations suggests an interesting mode of operation for these interneurons (Sherman, 1988; Bloomfield and Sherman, 1989), which is schematically depicted by Fig. 8.10. Clusters of dendritic appendages, which are major sites of input and output, represent local circuits whose computations are largely independent of activity in other clusters and in the soma; the axonal output is controlled instead by input to the soma and proximal dendrites. This output appears to be mediated by conventional action potentials (Sherman and Friedlander, 1988). Also, while the dendritic F2 outputs

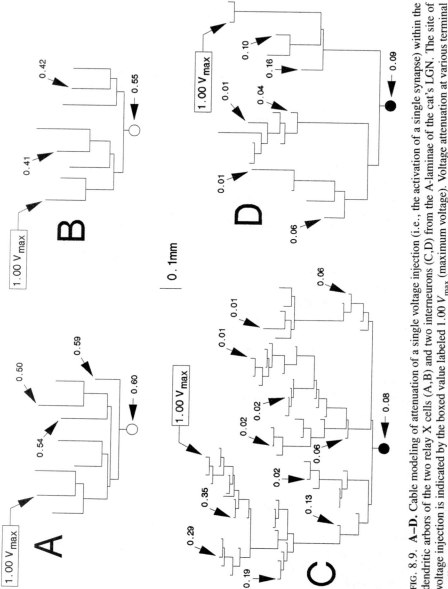

FIG. 8.9. **A–D.** Cable modeling of attenuation of a single voltage injection (i.e., the activation of a single synapse) within the dendritic arbors of the two relay X cells (A,B) and two interneurons (C,D) from the A-laminae of the cat's LGN. The site of voltage injection is indicated by the boxed value labeled $1.00 \ V_{max}$ (maximum voltage). Voltage attenuation at various terminal endings within the arbor and soma is indicated by arrows and given as fractions of V_{max}. Note that voltage never falls below 0.5 of its maximum value anywhere within the dendritic arbor or soma for either relay cell. However, considerable voltage attenuation is evident for the interneurons so that very little of the synaptic current will reach the soma. (From Bloomfield and Sherman, 1989.)

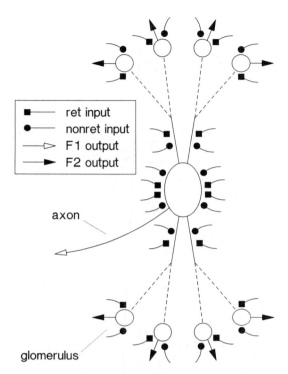

FIG. 8.10. Schematic view of hypothesis for functioning of interneurons in the cat's LGN. Retinal and nonretinal inputs are shown to the glomeruli as well as to the proximal dendrites and soma. The glomerular inputs are acted upon and lead to F2 outputs from the dendrites, whereas the inputs to the proximal dendrites and soma led to F1 outputs from the axon. The dashed lines indicate the electrotonic isolation between glomeruli and the proximal dendrites plus soma. This isolation suggests that the two sets of synaptic computations, peripheral for the glomerular F2 outputs and proximal for the axonal F1 outputs, transpire in parallel and independently of each other. Most glomeruli are also functionally isolated from one another.

innervate relay X cells through glomeruli, the axon forms F1 terminals that innervate dendritic shafts of unknown origin outside of glomeruli (Hamos et al., 1985; Montero, 1987; Sherman, 1988). This suggests that the interneuron simultaneously does double duty: Integration of the axonal F1 outputs via action potentials depends on one set of proximal inputs and involves one type of postsynaptic target, whereas integration of the dendritic F2 outputs depends on local inputs and involves a different postsynaptic target.

BASIC CIRCUIT

Enough is known about the cat's LGN to provide a schematic circuit diagram, including a fair estimate of the numbers of neuronal elements present. Of course, many of the specific features of this diagram still depend on a certain amount of guesswork, but the broad outlines seem clear. It seems likely that these broad outlines apply as well to other thalamic nuclei.

Each of the A-laminae of the cat's LGN contains roughly 400,000 neurons (Sanderson, 1971). Of these, perhaps 300,000 are relay cells and 100,000 are interneurons. The interneurons seem exclusively part of the X pathway. Slightly more relay X cells (150,000–200,000) than relay Y cells (100,000–150,000) seem to exist (Sherman, 1985). These geniculate neurons are innervated by slightly fewer than 100,000 retinogeniculate axons; however, the X:Y ratio among these axons is much higher—at roughly 10:1—than is the case for geniculate relay cells (Sherman, 1985). The reason is that retinogeniculate Y axons innervate many more geniculate neurons than do X axons. The geniculate cells in the A-laminae are also innervated by more than four million corticogeniculate axons (Sherman and Koch, 1986), although the details of how these axons innervate relay X and Y cells plus interneurons are not yet clear. We also still lack estimates for the number of afferent axons from the RNT and brainstem reticular formation.

The basic organization of the cat's LGN is summarized schematically in Fig. 8.11. Many of the details of this circuit, including the differences between the X and Y pathways, have been described above. These relay cells also receive input from cortex and from the brainstem reticular formation. Major inhibitory input derives from local GABAergic cells, which are the interneurons and RNT cells. Both of these GABAergic cells are innerated by cortex and by the brainstem reticular formation. In addition, RNT cells are innervated by axon collaterals from the relay cells, and interneurons receive input from retinal X axons.

Although much of the circuitry outlined in Fig. 8.11 is sketchy, the following conclusions can be tentatively drawn. Much of this repeats earlier points, but it is offered here as a concise summary. Relay cells receive retinal or lemniscal input onto proximal dendrites in close association with GABAergic input. The GABAergic input derives from RNT cells and interneurons. Also innervating proximal dendrites are inputs from the brainstem reticular formation. Distal dendrites are dominated by cortical input, but the electrotonic compactness of relay cells implies that even these distal inputs are quite important functionally.

Figure 8.11 also summarizes some differences between the X and Y pathways, and perhaps this can be taken as a reflection of the kinds of variation present throughout thalamic circuitry. Three main differences exist: the nature of retinal or lemniscal input, the presence of glomeruli, and the role of interneurons. Retinal input to relay Y cells is fairly straightforward, innervating proximal dendritic shafts in simple contact zones. Retinal input to relay X cells is much more elaborate, because it involves complicated triadic relationships that include dendritic terminals of interneurons. Glomeruli are also a major feature of X but not Y circuitry, and the glomerulus may be viewed as a major filter of reti-

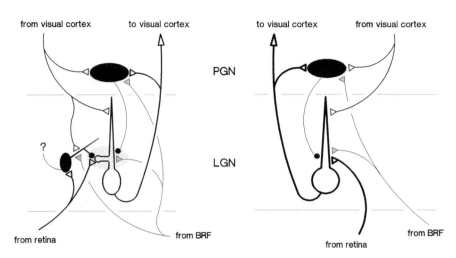

FIG. 8.11. Schematic view of X and Y circuits for the A-laminae of the cat's LGN. **A.** X pathway. Much of the input to X relay cells (open), including inputs from retina, from dendrites of interneurons (filled), from the perigeniculate nucleus (PGN, a part of the RNT), and from the brainstem reticular formation (BRF), is filtered through glomeruli (stippled region). Retinal terminals engage in triadic relationships with terminals from the interneuron's dendrites and dendritic appendages on the relay cell. Cortical input dominates the peripheral dendrites of the relay cell. The interneuron is also innervated from retina, cortex, and the BRF; the target of the interneuron's axon remains unknown, except that it is extraglomerular. The PGN cell is innervated from geniculocortical axons, corticogeniculate axons, and BRF axons. **B.** Y pathway. This diagram is much simpler, because interneurons do not appear to be present. The retinal axon contacts the relay cell (open) on proximal dendritic shafts among axon terminals, from cortex, PGN, and brainstem. Cortical and brainstem inputs to the relay and PGN cells are similar to that shown in A. Although not illustrated, it seems that at least some PGN axons can innervate both relay X and Y cells.

nogeniculate transmission (see above). Finally, interneurons also seem to be intimately related to X but not Y circuitry. They are innervated by retinal X axons, and their dendritic outputs nearly exclusively innervate relay X cells. The axonal targets of interneurons largely remain a mystery; however, the axons use F1 terminals and contact extraglomerular dendritic shafts, whereas the dendritic outputs use F2 terminals and contact dendritic appendages in glomeruli.

It should be emphasized that the circuit schematically represented by Fig. 8.11 is preliminary and greatly simplified. Many questions still remain. For one example, what is the interrelated pattern of innervation involving single cortical axons, RNT cells (or interneurons), and relay cells? The implication of this last question is illustrated in Fig. 8.12 which shows two very different functional circuits that adhere to our superficial knowledge of interconnections among these cell populations. Figure 8.12A shows a true feedback inhibitory circuit in which

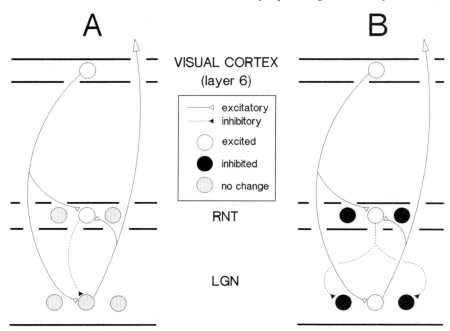

FIG. 8.12. Different circuits involving the RNT. **A.** Simple feedback inhibition for which thalamic relay cells and RNT cells are reciprocally connected. Activity of relay cell leads to its own inhibition after two synaptic delays. Likewise, activation of a corticothalamic axon leads to monosynaptic activation of the relay cell followed by disynaptic inhibition. These circuits thus act through negative feedback to keep the firing rates of thalamic relay cells from changing very much. **B.** A more complex circuit for which relay cells and RNT cells are connected with a lateral displacement. Activity of a relay cell now will disynaptically inhibit its neighbors. Because these neighbors normally activate RNT cells that inhibit the relay cell in question (not shown), this leads to disinhibition of that relay cell. Furthermore, a similar displacement of corticothalamic connections will both excite the relay cell and disinhibit it.

a relay cell's axon collateral excites an RNT cell that, in turn, inhibits this same relay cell; also, the cortical axon that monosynaptically excites the relay cell disynaptically inhibits it through the RNT cell. Figure 8.12B depicts a very different picture: RNT cells excited by the relay cell inhibit only its neighbors, which would have the effect of *disinhibiting* that relay cell by preventing its neighbors from inhibiting it through the RNT. Also in Fig. 8.12B, the cortical axon excites the relay cell directly and disinhibits it indirectly by disynaptically inhibiting that relay cell's neighbors. Indirect evidence for the cat's LGN favors the circuit shown in Fig. 8.12B (see Sherman and Koch, 1986). However, this conclusion is tenuous, and our main point here is that, even though we can draw detailed circuit diagrams such as that shown by Fig. 8.11, we still cannot answer many functional questions about this circuit until we know more about single cell connections.

MEMBRANE PROPERTIES

SYNAPTIC RESPONSES

Figure 8.13A,B illustrates the responses of a geniculate relay cell to a synchronous volley of impulses arriving via its retinal afferents. Such responses typically consist of a monosynaptic EPSP followed by a disynaptic or multisynaptic IPSP. The IPSP is generated from interneurons and/or RNT cells (cf. Figs. 8.6, 8.8, 8.11, and 8.12). This excitatory/inhibitory sequence is exhibited by most thalamic cells in response to a lemniscal volley. Because of the monosynaptic nature of the retinal or lemniscal input, a relay cell responds to such input with a relatively fixed latency (Fig. 8.13C).

The responses shown in Fig. 8.13 were obtained under artificial experimental conditions from a preparation physiologically distorted by various drugs and anesthetics. During natural activity, these thalamic responses would be heavily modulated by nonretinal or nonlemniscal inputs. In order to comprehend how nonretinal pathways can control responses of thalamic neurons to retinal inputs, and similarly how nonlemniscal pathways control responses of VPL and MGN neurons to their lemniscal inputs, we must first understand the intrinsic electrophysiological properties of these thalamic neurons.

INTRINSIC CONDUCTANCES

The intrinsic electrophysiological properties of neurons play a great role in determining their integrative characteristics (see Chap. 2 of this book). We can no longer view a thalamic cell as being a simple response element that linearly sums its synaptic inputs to determine its axonal output. Instead, these cells have a variety of active membrane conductances, many controlled in a conventional manner by ligand binding of neurotransmitters, but some controlled by membrane voltage and others controlled by concentration levels of certain ions, such as Ca^{2+}.

Both in vitro and in vivo experiments directed at different thalamic nuclei across several mammalian species have revealed a surprising plethora of intrinsic membrane conductances present in *all* thalamic neurons, both in the dorsal thalamus nuclei and within RNT neurons (McCormick and Prince, 1988; Steriade and Llinás, 1988). These conductances all lead to currents that alter the membrane potential. The number of active conductances described for thalamic neurons continues to grow, and the present number is at least six. Which conductances are active can greatly affect the nature of the thalamic neuron's relay of its input to cortex. Conductances found in thalamic neurons are generally found in many other brain cells as well, and for the most part these have been described in detail in Chap. 2.

Na$^+$ conductances. Two voltage-dependent Na$^+$ conductances have been described. The fast, inactivating Na$^+$ conductance described by Hodgkin and Huxley is voltage dependent and subserves the conventional action potentials. The other Na$^+$ conductance is persistent and noninactivating. This creates a

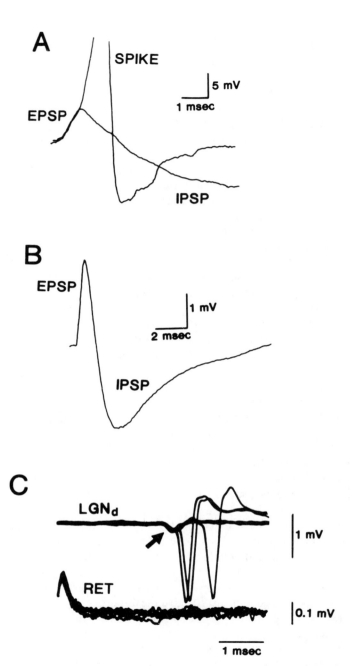

FIG. 8.13. Examples of retinogeniculate transmission in cats. **A.** Intracellular records from a geniculate relay cell showing response to electrical stimulation of the optic chiasm. Responses to two different levels of stimulation are superimposed. At the lower level, a monosynaptic EPSP is followed by a disynaptic IPSP. At the higher level, threshold is reached for an action potential (SPIKE). **B.** Computer average of 100 responses at the lower level of stimulation for the same cell as in **A.** This shows the EPSP and IPSP more clearly. **C.** Extracellular records from a geniculate neuron (LGN) and its simultaneously

plateau depolarization and partially regulates the generation of rhythmic firing in thalamic neurons.

Ca²⁺ conductances. Likewise, there are at least two voltage-dependent Ca^{2+} conductances. One has a high threshold and is most likely located in the dendrites; rather little is known about this conductance. The other has a lower threshold and plays a dramatic role in retinogeniculate and lemniscothalamic transmission described more fully below; it is often known as the *low-threshold* *Ca²⁺ spike* or *conductance.* The currents associated with these Ca^{2+} conductances have been called, respectively, I_L for the high-threshold one and I_T for the low-threshold one (see Chap. 2). The low-threshold spike occurs both in RNT cells and in thalamic relay neurons, but preliminary evidence suggests that, at least for the LGN, this conductance plays little role in interneurons (McCormiock and Pape, 1988).

K⁺ conductances. A number of voltage- and/or Ca^{2+}-dependent K^+ conductances exist that give rise to various membrane currents (see Chap. 2). The best known is the *delayed rectifier* (I_K), which is part of the action potential and repolarizes the neuron following the Na^+ conductance. Several others (I_A, I_C, and possibly I_{AHP}) hyperpolarize the neuron for varying lengths of time following a conventional action potential. The amount of this hyperpolarization determines the cell's relative refractory period, which limits its maximum firing rate.

IMPORTANCE OF THE LOW-THRESHOLD SPIKE

The low-threshold spike is especially important for the control of retinogeniculate and lemniscothalamic transmission (see Chap. 2). It is activated by a small depolarization from rest (thus the low threshold) that opens membrane channels permeable to Ca^{2+}, thereby increasing a Ca^{2+} conductance (Steriade and Llinás, 1988). The ensuing, all-or-nothing Ca^{2+} spike is relatively slow and has a triangular wave form (Fig. 2.4D). Superimposed onto this spike are 2–7 fast, conventional Na^+ action potentials. This entry of Ca^{2+} into the cell activates one or more voltage- and Ca^{2+}-dependent K^+ conductances that hyperpolarize the neuron for 50–200 msec. This hyperpolarization is sufficiently strong to create a relative refractory period during which the cell is fairly inexcitable and can thus no longer relay retinal or lemniscal information to cortex. Interestingly, if the membrane potential is always held more positive than approximately −60 mV, the low-threshold Ca^{2+} conductance is *inactivated.* This inactivation can be

---

recorded retinal input (RET). The trace showing the geniculate cell's responses is triggered from the preceding action potential of the retinal afferent. For the geniculate trace, at a fixed delay after each retinal spike, a brief potential occurs which is known as the *s-potential* (arrow). The s-potential is thought to represent the extracellularly recorded EPSP. In many, but not all, traces, the s-potential is followed soon after by an action potential. Thus each geniculate spike is preceded at a fairly fixed time by a retinal spike, although not every retinal spike evokes one in the postsynaptic cell. (A,B redrawn from Bloomfield and Sherman, 1988; C redrawn from Cleland et al., 1971.)

reversed, or *deinactivated*, if the membrane potential is held below approximately -65 mV for at least 50–100 msec. When deinactivated, the low-threshold spike can be triggered by a small depolarization.

Any change in activity that hyperpolarizes the cell sufficiently long to deinactivate this Ca^{2+} conductance primes the cell for the initiation of the low-threshold spike. Physiologically, such deinactivation can result either from active hyperpolarization through GABAergic or other inputs, or from reduction of a tonic depolarizing input, such as the corticothalamic pathway. Any sufficiently large depolarization can then *activate* the low-threshold spike. This can be done either via an EPSP (e.g., from a retinal or lemniscal input) or via the relative depolarization that ensues from removing the deinactivating source (e.g., cessation of the hyperpolarizing input or restoration of the tonic depolarizing one). In fact, often the hyperpolarization due to the series of K^+ conductances following the low-threshold spike is sufficient in time and amplitude to deinactivate this spike, and the passive repolarization following these K^+ conductances activates it (Steriade and Llinás, 1988). This can be repeated for many cycles until some other input to the cell breaks through. The result is a neuron that enters a phase of cyclic, rhythmic activity at 6–10 Hz.

THALAMIC GATING FUNCTIONS

RELAY VERSUS BURST RESPONSE MODES

Thalamic relay cells therefore exhibit two distinct response modes: a *relay* mode and a non-relay *burst* mode (Sherman and Koch, 1986). If the cell is at rest or slightly depolarized so that its membrane is above roughly -60 mV, it is in the relay mode. Now a suprathreshold EPSP will induce a train of normal action potentials (see Fig. 2.5D). Within physiological limits, the frequency of action potentials monotonically rises with increasing EPSP amplitude. Under these conditions, transmission through the thalamic relay neuron to the cortex will lead to a pattern of input to the cortex that faithfully reflects the pattern of retinal or lemniscal input to the thalamus. However, once a thalamic neuron becomes hyperpolarized sufficiently long to deinactivate a low-threshold spike, it enters the burst mode (Fig. 2.5E). Now a suprathreshold EPSP will trigger a low-threshold spike, and many such spikes may ensue as the cell bursts rhythmically. Until this bursting cycle is broken, the neuron no longer relays sensory input to the cortex because its firing pattern bears no resemblance to the pattern of its retinal or lemniscal inputs.

RNT CONTROL OF RESPONSE MODES

Although thalamic neurons may switch between relay and burst modes at any time, the relay mode appears to be the state of most thalamic neurons in the awake, alert animal, whereas the burst mode dominates during less alert periods, including drowsiness and quiet or non-REM sleep (Steriade and Llinás, 1988). During such inattentive periods, the EEG in all mammals, including humans, becomes highly synchronized, and fast, rhythmic spike-like electrical phenomena known as *spindles* can be seen (see Fig. 8.14). These spindles have a frequency of 7–14 Hz.

This dominant feature of the synchronized EEG is generated in the thalamus (Steriade and Llinás, 1988). Studies of thalamic neurons have shown that all RNT cells can spontaneously generate rhythmic discharges at a rate of approximately 10 Hz. The low-threshold spike appears to be the underlying cause of this endogenous bursting behavior, and the oscillations can be generated within individual RNT cells. Dendrodendritic synapses exist among RNT neurons, and these could serve to synchronize entire RNT regions. Recent studies have demonstrated that groups of deafferented RNT neurons can generate such synchronized oscillatory activity in the absence of external input (Steriade and Llinás, 1988).

Because RNT neurons provide an inhibitory, GABAergic input to thalamic relay cells, the RNT entrains its oscillatory activity onto these relay cells. That is, the synchronized bursts of RNT activity would lead to waves of hyperpolarization among relay cells; this would deinactivate low-threshold spikes in the relay cells, and they would synchronously enter the burst mode. By themselves, neurons in the LGN or VPL do not spontaneously generate spindle rhythmicity, since disconnecting the projection cells from the RNT via surgical or chemical means abolishes the oscillations (Steriade and Llinás, 1988).

ROLES OF OTHER INTRINSIC CONDUCTANCES

The low-threshold spike is the best studied and most dramatic membrane conductance relating to thalamic cell function, but other conductances are present that play more subtle roles in gating of the transmission of retinal or lemniscal input to cortex. Because these are not yet well established or studied for thalamic neurons, only several will be briefly noted below.

Afterhyperpolarization. The afterhyperpolarization (AHP) following action potentials is important for the integrative properties of the neuron. That is, the AHP is a hyperpolarization that establishes a relative refractory period for the neuron, and thus the strength and duration of the AHP controls the extent to which the neuron adapts to long-lasting excitatory inputs. This is known as spike frequency *adaptation* or *accommodation,* and is reflected by the inability of the neuron to respond at high frequencies to a prolonged afferent input. Results from both the guinea pig and cat indicate that action potentials in thalamocortical cells are followed by a prolonged AHP with an overall duration of up to 70 msec, whereas their duration in RNT cells is much shorter (8–10 msec). The different duration of the AHP is reflected in much higher firing rates for reticular than for thalamocortical cells. The basis of the AHP is an increase in one or possibly two Ca^{2+}-dependent K^+ conductances, since removal of extracellular Ca^{2+} or intracellular injection of Ca^{2+} buffers prevents these long-lasting AHPs.

Studies of bullfrog sympathetic ganglion cells (see Chap. 3) and rodent hippocampal neurons (see Chap. 11) have shown that most of the AHP and its consequent spike frequency accommodation can be essentially suppressed by local

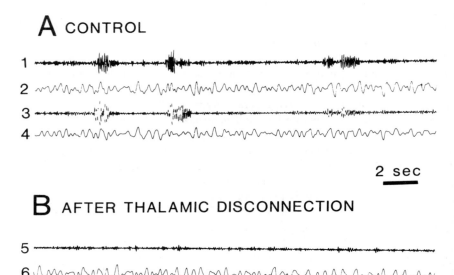

A CONTROL

1

2

3

4

2 sec

B AFTER THALAMIC DISCONNECTION

5

6

7

8

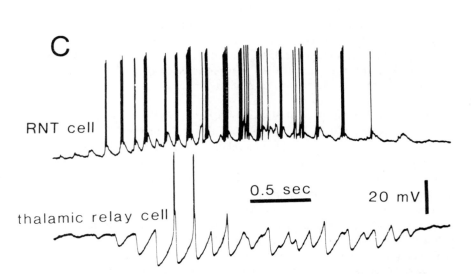

C

RNT cell

0.5 sec   **20 mV**

thalamic relay cell

FIG. 8.14. Relationship of thalamus to spindle activity in the cortical electroencephalogram (EEG) of cats. **A,B.** Effect on the EEG of rostral thalamic transections that disconnect the thalamus from the cortex. The numbering of traces is as follows: 1 and 5, higher-frequency EEG (7–14 Hz) for right hemisphere; 2 and 6, lower-frequency EEG (0.5–4.0 Hz) for right hemisphere; 3 and 7, higher-frequency EEG (7–14 Hz) for left hemisphere; 4 and 8, lower-frequency EEG (0.5–4.0 Hz) for left hemisphere. **A.** Normally (before transection), each hemisphere shows activity in both the higher (7–14 Hz) and the lower (0.5–4.0 Hz) filtered traces. **B.** After transection, the higher frequencies are

272

application of norepinephrine and/or acetylcholine (Sherman and Koch, 1986; Steriade and Llinás, 1988). This suppression involves the blockade of a slow, Ca^{2+}-dependent K^+ current known as the I_{AHP} (see Chap. 2). Because thalamic relay cells display an AHP and receive cholinergic and noradrenergic input from the brainstem, a similar mechanism may work in the thalamus to vary the gain of lemniscothalamic transmission.

Plateau potential. Thalamic neurons show a voltage-dependent Na^+ conductance, which shows very little if any inactivation. This leads to a persistent, inward current known as $I_{Na,p}$ (see Chap. 2), and it generates plateaulike potentials that can outlast the duration of the initial stimulus. It serves as a sort of counterweight to I_{AHP}; that is, it tends to counterbalance the Ca^{2+}-dependent K^+ conductance by generating a slow rebound depolarization. The $I_{Na,p}$ thus has a definite role in controlling the gain of lemniscothalamic transmission.

High-threshold Ca^{2+} spikes. Presumed intradendritic recordings in thalamic relay cells have also revealed a voltage-dependent, high-threshold Ca^{2+} conductance, similar to that observed in Purkinje cells (Steriade and Llinás, 1988; see also Chap. 7). This conductance triggers all-or-none depolarizing responses in the dendrites and soma that are followed by the activation of Ca^{2+}-dependent K^+ conductances (Steriade and Llinás, 1988). It is likely that dendritic spikes play a crucial role in relaying to the cell body postsynaptic potentials that are generated on dendrites.

SYNAPTIC TRANSMITTERS

GLUTAMATERGIC INPUTS

The major inputs to relay cells are the retinal or lemniscal inputs and the cortical inputs, and these all appear to be glutamatergic (see above). Of particular interest here is the nature of the postsynaptic receptors to glutamate-like substances. These basically fall into two major classes (for a review, see Mayer and Westbrook, 1987; see also Chap. 2).

The first class is what we shall refer to as the *non-NMDA* receptors. Based on the pharmacology of agonists and antagonists, many authors recognize two or three classes of non-NMDA receptor, but for the purposes of our treatment, we can lump them together. The non-NMDA receptor acts in a fairly straightforward way, and its activation increases the conductance to Na^+ and perhaps other cations, thereby depolarizing the cell.

---

selectively eliminated from the EEG. **C.** Activity of thalamic neurons during an EEG spindle. During the spindle, the RNT neuron (*top*) undergoes a long-lasting, slow depolarization that elevates its firing rate. In contrast, a thalamic relay cell (*bottom*) is hyperpolarized rhythmically, and the rebound from these hyperpolarizations often leads to low-threshold Ca^{2+} spikes. The elevated firing in the RNT cell seems to cause the rhythmic hyperpolarizations in the relay cell. (A,B revised from Steriade et al., 1987; C revised from Steriade and Llinás, 1988.)

The second class is the *NMDA* receptor. All cells responsive to glutamate-like substances seem to have non-NMDA receptors, and only a subset of these also have NMDA receptors. The NMDA receptor has the interesting property of being both voltage and transmitter dependent. At relatively depolarized levels of the membrane, activation of the receptor increases the conductance of Na^+ and other cations (mostly Ca^{2+} and some K^+). However, at increasing membrane hyperpolarization, Mg^{2+} ions can clog the ion channel and reduce the conductance: these Mg^{2+} ions are permitted to approach the channel by a hyperpolarized membrane but are driven away by a depolarized membrane. The range over which membrane depolarization can increase conductance of the channel associated with the NMDA receptor seems to vary across cells, but can extend from -140 mV to -40 mV. Thus, in order for an EPSP to be generated via an NMDA receptor, two events must occur simultaneously: the presynaptic presence of a glutamate-like neurotransmitter coupled with a postsynaptic depolarization sufficient to unblock the channel. In other words, the NMDA receptor complex can act as a sort of molecular AND gate.

The two receptors offer different means of gating synaptic transmission. With the conventional, non-NMDA receptor, EPSP amplitude decreases with membrane depolarization as the reversal potential for the EPSP is reached. However, with the NMDA receptor, EPSP amplitude actually increases with membrane depolarization over some physiological range, usually peaking near the resting potential, before further membrane depolarization toward the reversal potential reduces the EPSP amplitude. Evidence exists for the LGN and VPL that at least some of the relay cells use NMDA receptors for their retinal or lemniscal inputs (Moody and Sillito, 1988; Salt, 1988; Lo and Sherman, 1989). For these relay cells, any input that tonically modulates membrane potential (e.g., the cortical input) can determine the size of the retinal or lemniscal EPSP and thus offer a sensitive means of gating retinogeniculate or lemniscothalamic transmission (see Koch, 1987).

GABAERGIC INPUTS

Thalamic relay cells receive a large GABAergic input from RNT cells and interneurons. To a first approximation, these GABAergic inputs will inhibit the relay cells and thereby depress retinogeniculate and lemniscothalamic transmission. However, there are two provisos or complications to be added to this approximation.

First, two distinct GABAergic receptors, $GABA_A$ and $GABA_B$ receptors, are present in the thalamus, and both seem to exist on relay cells (Sherman and Koch, 1986; Bloomfield and Sherman, 1988). Activation of the $GABA_A$ receptor increases a Cl^- conductance, whereas activation of the $GABA_B$ receptor increases a K^+ conductance. Because the reversal potential for K^+ is much more negative (at roughly -100 mV) than that for Cl^- (at roughly -70 mV), $GABA_B$ activation results in more hyperpolarization than does $GABA_A$ activation. However, the neuronal conductance increase and thus the decrease in neuronal input resistance is much greater with $GABA_A$ than with $GABA_B$. As a result, $GABA_A$ inhibits more by clamping the membrane at a subthreshold level and thus *shunting* EPSPs, and GABA$_B$ inhibits more by hyperpolarizing the membrane. The GABA$_A$ response is thus much more nonlinear, acting more like a voltage multiplication, whereas the GABA$_B$ response is more linear, acting like simple voltage subtraction. Also, the GABA$_A$ response is somewhat faster than is the GABA$_B$ response.

Second, although both GABA$_A$ and GABA$_B$ activation counteract EPSPs and thus are inhibitory, they might also play an important role in controlling the relay versus burst response modes of relay cells. That is, both GABAergic responses may provide sufficient hyperpolarization to deinactivate the low-threshold spike. GABA$_B$ receptors would be more effective in this regard, because they provide more hyperpolarization and less shunting of an activating input.

BRAINSTEM INPUTS

As discussed above, various nuclei in the brainstem project in a diffuse manner to both the ventral and the dorsal thalamus. The best studied of these afferent inputs derive from the cholinergic, noradrenergic, and serotonergic neurons of the brainstem reticular formation, but the specific input patterns vary across species and thalamic nuclei (Singer, 1977; Sherman and Koch, 1986; Steriade and Llinás, 1988). For instance, in the cat, cholinergic inputs predominate to the LGN but not to the VPL (Fitzpatrick et al., 1989), and cholinergic inputs to the rat's LGN are relatively sparse (Levey et al., 1987). The effects on thalamic neurons of stimulating various sites in the brainstem reticular formation have proven complex and difficult to understand, although it seems clear that these brainstem sites are partially responsible for mediating arousal and alertness or drowsiness and sleep. The use of in vitro slice preparations as well as intracellular electrophysiology has permitted us to come to a much clearer understanding of the modulatory role of the brainstem reticular formation (Sherman and Koch, 1986), particularly for the cholinergic inputs.

Cholinergic inputs. The application of acetylcholine onto the mammalian thalamus leads to a complex constellation of effects. For RNT neurons, acetylcholine causes a small, slow, long-lasting *increase* in a K$^+$ conductance, and this is mediated by an M$_2$ muscarinic receptor. This increase in K$^+$ conductance leads to a long-lasting hyperpolarization, which in turn deinactivates the low-threshold spike and switches the RNT neuron into the burst mode (McCormick and Prince, 1987a,b). A similar slow and long-lasting *increase* in a K$^+$ conductance has been observed upon application of acetylcholine in GABAergic thalamic interneurons in the cat's LGN (McCormick and Pape, 1988). However, since the Ca^{2+} underlying the low-threshold spike appears to be absent in interneurons, this increase in K$^+$ conductance serves only to inhibit action potentials.

Application of acetylcholine to relay cells produces a more complex response (McCormick and Prince, 1987a,b). The most prominent response is a rapid depolarization due to an increased cation conductance that is subserved by a nicotinic receptor. This nicotinic response is often followed by a slower depolarization due to activation of a muscarinic receptor that decreases a K$^+$ conductance. These nicotinic and muscarinic depolarizations, by inactivating the low-threshold Ca^{2+} spike, keep the neuron in its relay mode. Some relay cells may also respond with a slow increase in a K^+ conductance, which is subserved by an M_2 muscarinic receptor and is much like the muscarinic response of RNT cells and interneurons.

On balance, the final impact on thalamocortical processing of this ascending brainstem cholinergic input remains something of a mystery. On the one hand, this input reinforces retinogeniculate and lemniscothalamic transmission by depolarizing the relay cells and inhibiting the interneurons. This depolarization makes the relay cells more sensitive to retinal and lemniscal input and encourages them to remain in the relay mode. On the other hand, cholinergic input seems to switch RNT cells into the burst mode, and, as noted above, this could indirectly switch relay cells into this mode as well. Perhaps the final answer is to be found in more detailed studies of thalamic circuits, because, as suggested by Fig. 8.12B, connections between the RNT and its accompanying dorsal relay nucleus may represent a sort of indirect, positive feedback circuit at the single cell level rather than a negative feedback. It may be that cholinergic axons enhance the relay mode of some relay cells directly while indirectly promoting the burst mode of others through the RNT. When such details are better understood, the role of the ascending cholinergic input might become much clearer.

Noradrenergic inputs. The noradrenergic innervation of the thalamus arises largely, if not exclusively, from neurons in the locus coeruleus. The postsynaptic effects of norepinephrine seem more straightforward than are those of acetylcholine (McCormick and Prince, 1988). In the LGN, MGN, and RNT, norepinephrine produces a decrease in K^+ conductance that leads to a slow depolarization lasting for more than 1 min. The effective resting potential in all thalamic cells is thereby shifted to more positive values, promoting the relay mode and facilitating the ability of retinal or lemniscal EPSPs to trigger conventional action potentials.

Serotonergic inputs. A rather poorly understood innervation of thalamus originates with serotonergic neurons of the dorsal raphe nucleus in the brainstem. The function of this input has been studied sparsely in the LGN, and its overall action there seems to depress retinothalamic transmission (see Sherman and Koch, 1986). However, we must know more about both the responses of thalamic neurons to serotonin as well as the distribution of this serotonergic input to other thalamic nuclei.

FUNCTIONAL SIGNIFICANCE OF THALAMIC CIRCUITS

Noted above is the observation that each geniculate cell receives the vast majority of its retinal input from one or very few retinal ganglion cells of the same type (left or right retina, on or off center, X or Y). Thus the receptive field of each geniculate cell is nearly identical to that of its retinal input: Geniculate cells display circular receptive fields organized into concentrically arranged, antagonistic centers and surrounds. Subtle differences have been described between receptive fields of geniculate cells and those of their retinal inputs, and these mostly involve greater inhibition seen postsynaptically (reviewed in Sherman and Spear, 1982; Sherman, 1985; Shapley and Lennie, 1985). Perhaps the most dramatic difference is the presence of a purely inhibitory receptive field for the nondominant eye having a homonymous position to that of the dominant eye's receptive field. Such details have yet to be established for other sensory pathways, but preliminary data suggest a resemblance in receptive field properties between MGN or VPL cells and their lemniscal inputs. For the purposes of the present discussion, we conclude that no significant receptive field transformation occurs at the level of thalamus.

This absence of a major receptive field transformation across the retinogeniculate synapse stands in stark contrast to the obvious transformations observed when progressing through the synaptic zones of retina or cortex and also across the geniculocortical synapse. Comparable transformations exist as well in other parts of the visual system, such as the superior colliculus and extrastriate visual cortex. Similar transformations also exist outside the thalamus in other sensory

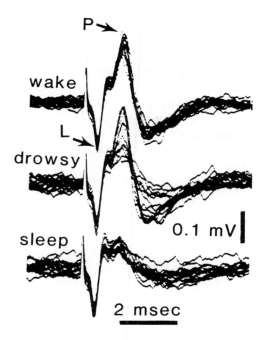

FIG. 8.15. Effects of different levels of consciousness on the ability of the thalamus to transmit synaptic information to cortex. Illustrated are field potentials recorded from the cat's ventral anterior thalamus in response to electrical stimulation of the brachium conjunctivum. Input from the brachium conjunctivum is analogous to lemniscal input for the ventral anterior nucleus. The response consists of a presynaptic afferent "lemniscal" component (L) plus a postsynaptic relayed component (P). The amplitude of the latter component is a measure of the gain of lemniscothalamic transmission. Note that, although the presynaptic component is unchanged, the postsynaptic component gets progressively smaller as the animal descends into sleep through drowsiness. (Revised from Steriade and Llinás, 1988.)

systems, such as the spinal cord and cortex for the somatosensory system, and the inferior colliculus and cortex for the auditory system. These other, extra-thalamic transformations represent obvious functional roles for these other regions of sensory systems: synaptic zones there clearly form more complex receptive field properties as the hierarchy is ascended, and this provides a basis for these sensory systems to extract information about stimuli in the world outside.

It is this absence of any clear functional change in receptive fields across the retinogeniculate or lemniscothalamic synapse that has prompted many investigators to think of specific sensory thalamic nuclei as merely passive relay stations for signals from the periphery to cortex. However, such a trivial function belies morphological data presented above for the LGN—that only a minority of the synapses (10–20%) present in the neuropil and onto relay cells is retinal in origin. The function of the vast majority of synaptic input seems invisible to conventional receptive field approaches. In fact, evidence accumulated over the past decade strongly suggests that this nonretinal input serves to gate or control the gain of retinogeniculate transmission.

This provides a unique role for the thalamus: it is not merely a passive relay, but instead actively filters the flow of information to cortex, and the nature of the filtering is dependent on the animal's state of consciousness and alertness (see Fig. 8.15). This active filtering has not been revealed by the usual receptive field studies, possibly because the anesthetics used in such studies block action of many of the nonretinal pathways, but recording in unanesthetized animals has revealed considerable state-dependent variation in responsiveness of geniculate neurons. While it seems clear what general role the LGN and other thalamic nuclei play, further work is needed to reveal how many different types of retinogeniculate and lemniscothalamic filtering exist and precisely how these filtering functions are achieved.

9

BASAL GANGLIA

CHARLES J. WILSON

The basal ganglia are a richly interconnected set of brain nuclei found in the forebrain and midbrain of mammals, birds, and reptiles. In many species, including most mammals, the forebrain nuclei of the basal ganglia are the most prominent subcortical telencephalic structures. The large size of these structures, and their similarity in structure in such a wide range of species make it likely that they contribute some very essential function to the basic organizational plan of the brain of the terrestrial vertebrates. However, the assignment of a specific functional role for the basal ganglia has been difficult, as it has for other brain structures that have no direct connections with either the sensory or motor organs.

The most widely accepted views of basal ganglia function are based on observations of humans afflicted with degenerative diseases that attack these structures. In all cases, these diseases produce severe deficits of movement. None of the movement deficits are simple, however, or easily described. In some, such as Parkinson's disease, movements are more difficult to make, as if the body were somehow made rigid and resistive to changes in position. In others, such as Huntington's disease, useless and unintended movements interfere with the execution of useful and intended ones. These clinical observations have led most investigators to view the basal ganglia as components of a widespread system that is somehow involved in the generation of voluntary movement, but in complex and subtle aspects of that process.

The anatomical connections of the basal ganglia link it to elements of the sensory, motor, cognitive, and motivational apparatus of the brain. These connections are best appreciated within the context of the arrangement of the several nuclei that make up the basal ganglia. A diagram showing the arrangement of the most prominent of these nuclei as they appear in a frontal section of the human brain is shown in Fig. 9.1. The major structures are the caudate nucleus, putamen, globus pallidus, substantia nigra, and subthalamic nucleus. Also seen in the diagram are the two largest sources of input to the basal ganglia, the cerebral cortex and the thalamus.

The connectional relationships between these structures are shown in the diagram in Fig. 9.2. In dealing with this complexity, it is helpful to distinguish between inputs and outputs. With regard to the caudate nucleus and the putamen,

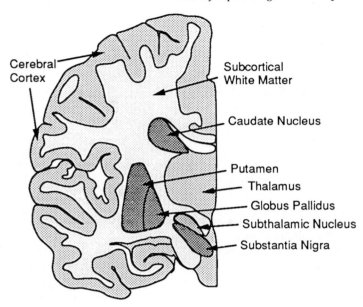

FIG. 9.1. Schematic representation of a transverse section through a human brain hemisphere showing the sizes and locations of several important components of the basal ganglia.

inputs from sensory, motor, and association cortical areas converge with inputs from the thalamic intralaminar nuclei, dopaminergic inputs from the substantia nigra (pars compacta), and serotonergic inputs from the dorsal raphe nucleas. Similar connections, but arising from limbic cortex and hippocampus, are formed in a third structure, the nucleus accumbens. These three parts of the basal ganglia (caudate nucleus, putamen, and nucleus accumbens) are very similar in their internal structure and are often referred to together as the *neostriatum*. They receive most of the fibers that enter the basal ganglia from other parts of the brain. Together they make up most of the volume of the basal ganglia as well.

The output from the neostriatum projects almost exclusively to other basal ganglia structures. The main targets of these axons are three nuclei: the external segment of the globus pallidus, the internal segment of the globus palliduas, and the pars reticulata of the substantia nigra. These three structures are very similar in their cellular organization. Two of them, the internal segment of the globus pallidus and the substantia nigra pars reticulata, project to structures outside the basal ganglia and provide the main output pathways for the results of neuronal operations performed within the nuclei. Their targets are primarily in the thalamus (mostly in the ventral tier thalamic nuclei which project to frontal cortex), lateral habenular nucleus, and in the deep layers of the superior colliculus. The external segment of the globus pallidus projects mainly to the subthalamic nucleus, a small but important component of the basal ganglia that, like the neostriatum, projects to the globus pallidus and substantia nigra. Inputs from outside the basal ganglia can enter the system at several points, including

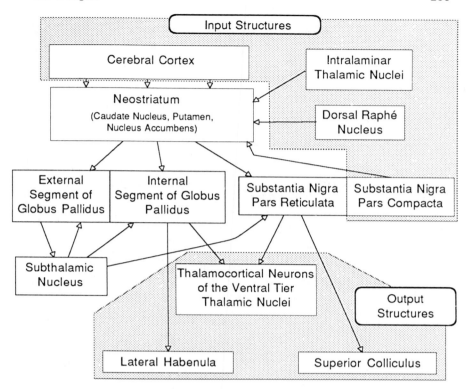

FIG. 9.2. Diagram showing some of the most important connectional relationships among the nuclei that make up the basal ganglia, their input structures, and output structures. The key role of the neostriatum as the recipient of most synaptic input from outside the basal ganglia is emphasized, as is the flow of information in one direction through the basal ganglia.

the subthalamic nucleus, substantia nigra, and globus pallidus. However, by far most inputs enter at the level of the neostriatum. Thus, a large portion of the mystery of basal ganglia organization and function is contained within the internal organization of the caudate nucleus, putamen, and nucleus accumbens. Most of this chapter will address that organization.

NEURONAL ELEMENTS

The neuronal elements of the neostriatum tend to be of modest size and number, but considerable complexity. They are represented in Fig. 9.3 in the standard scale to facilitate comparison with other brain regions, and in Fig. 9.4 at higher magnification to permit close examination of important details of the cellular elements.

INPUTS

As already indicated, input fibers to the neostriatum arise primarily from the cerebral cortex, the intralaminar nuclei of the thalamus, the dopaminergic neu-

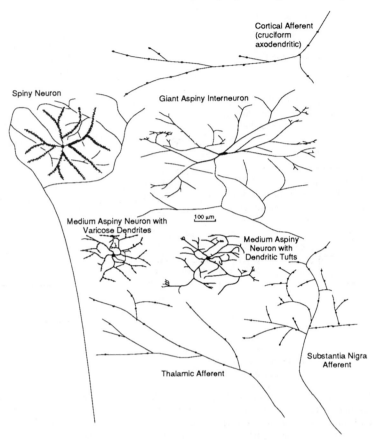

FIG. 9.3. The afferent fibers and neuron types of the neostriatum, shown at the standard scale. The axon and local axonal arborization of the spiny neuron is shown, as is a small portion of the axonal arborization of the giant aspiny interneuron. The axons have been omitted from the other two aspiny neuron types.

rons of the substantia nigra (pars compacta), the serotonergic neurons of the dorsal raphe nucleus, and the basolateral nucleus of the amygdala. Less numerous inputs also arise from the external segment of the globus pallidus and from the substantia nigra, pars reticulata (see review by Graybiel and Ragsdale, 1983).

Cortical afferents. Until recently, the arborization patterns of afferent fibers in the neostriatum were studied almost exclusively with the Golgi method. Although the Golgi method provides an excellent image of the stained fibers, the origin of a stained axon is often indeterminant because it usually cannot be followed all the way back to its cell body of origin in a distant part of the brain. Investigators using the Golgi techniques to study the neostriatum were able to identify afferent fibers by tracing them to the internal capsule, but they could not be certain which

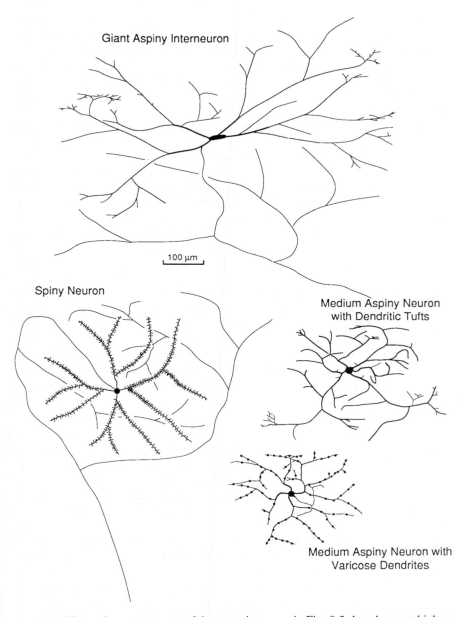

Giant Aspiny Interneuron

⊢ 100 µm ⊣

Spiny Neuron

Medium Aspiny Neuron
with Dendritic Tufts

Medium Aspiny Neuron with
Varicose Dendrites

FIG. 9.4. The major neuron types of the neostriatum, as in Fig. 9.3, but shown at higher magnification to illustrate the morphological differences between cell types.

fibers were from cerebral cortex, which were from substantia nigra, and so on. More recently, the morphology of identified axons has been studied by means of axonal staining with the lectin *Phaseolus vulgaris* leucoagglutinin (PHA-L). The lectin is injected extracellularly in an afferent structure, and stains a group of neurons around the injection site. After a period of 1–2 weeks, the lectin will have moved, either by diffusion or by axonal transport, so that the axons of some of those neurons will be completely and diffusely stained over very long distances. The selectivity and intensity of staining compare favorably with the Golgi method, and the PHA-L technique has the advantage of experimental choice of the group of neurons whose axons are to be stained.

Studies employing either technique have shown that afferent fibers to the neostriatum arborize mostly in the pattern described by Cajal as "cruciform axodendritic" (Fox et al., 1971). This pattern is illustrated for cortical and thalamic afferents in Fig. 9.3. The term *cruciform axodendritic* means that the fibers take a relatively straight course through the tissue, crossing over dendrites and making synapses with them en passant. The implication of this kind of arborization pattern is that individual fibers cross the dendritic fields of many neurons, but do not make many synapses with any given cell. Conversely, neostriatal neurons can be expected to receive inputs from a large number of afferent fibers, but not to receive many synapses from any one of them. This pattern is generally followed by fibers from all the afferent systems. The reader will recognize that these are the same rules that govern the input connections of parallel fibers to Purkinje cells (Chap. 7) and lateral olfactory tract fibers to olfactory pyramidal neurons (Chap. 10). So far, at least, there seems to be no input pathway in the neostriatum that can be compared with the climbing fibers of the cerebellar cortex.

Because convergence of many different axons seems to be the dominant pattern of axonal arborization in the neostriatum, there has been great interest in the spatial patterns formed by axonal arborizations from afferent structures, especially from the cerebral cortex. Studies using modern axonal tracing methods have established that the projections arising from even a very small portion of the cerebral cortex may extend through a large region of the neostriatum, being especially extensive in the rostrocaudal direction (e.g., Künzle, 1975; Jones et al., 1977). Although it has not yet been experimentally confirmed, most investigators presume that single afferent fibers correspondingly extend over large portions of the neostriatum. If so, one might think it to be geometrically necessary that inputs from wide areas of the cortex would have access to a common pool of neostriatal neurons, and that no cortical area could exercise exclusive control of any neostriatal neuron. This scheme suggests a sort of combinational logic circuit in the neostriatum, in which a striatal neuron may be excited only if there is convergent input from the correct combination of cortical (and perhaps other) pathways.

If this is even approximately correct, it becomes very important to learn which areas of cortex send axons to which areas of the neostriatum. One particularly wants to know if some cortical areas overlap greatly in their projections to the neostriatum, whereas fibers from other cortical areas never converge. Initial studies of the arborization patterns of selected cortical areas yielded the provocative suggestion that cortical regions interconnected by strong corticocortical connections (and therefore likely to be functionally related) project to similar, perhaps overlapping portions in the neostriatum, whereas functionally unrelated cortical areas had nonoverlapping domains in the neostriatum (Yeterian and Van Hoesen, 1978). Subsequent, more detailed examination of the axonal projection patterns, however, revealed that the axonal arborizations possess a rich internal patterning (Selemon and Goldman-Rakic, 1986; Malach and Graybiel, 1986). Within the rather large general area of neostriatum occupied by fibers from a specific cortical region, there are areas of relative concentration and rarefaction of inputs. Another cortical area projecting to this same region may exhibit either a similar or a complementary pattern of fiber arborization. In the case of the latter, there will be very little fine-grain overlap in the arborization volumes of the two cortical areas. These findings do not refute the idea that converging inputs to the neostriatum may define a restricted set of logical combinations of inputs capable of exciting neostriatal neurons, and that spatial location in the neostriatum may represent these logical combinations. They have instead enriched the possibilities for the kinds of combinations that could occur. At the same time, however, the idea has become much more complex and difficult to verify anatomically.

In the study of axonal innervation patterns, it is often forgotten that the postsynaptic neurons have dendrites that can reach across the domains created by the axonal arborizations and sample from more than one of them. This introduces still more complicated possibilities for the convergence of synaptic inputs. It is typical of brain organization that some cells restrict their dendritic fields to correspond to the geometry of axonal arborizations, whereas others create a higher-order level of organization by reaching out to receive combinations of nonoverlapping but adjacent inputs. The possible input combinations are determined by both the pattern of axonal arborizations and the patterns of dendritic branching. It is therefore important not only which axons actually overlap with others in the target zone, but also which are neighbors and which never are. Axonal domains that are not adjacent are unlikely to converge on any cells, regardless of their dendritic fields, whereas neighboring axonal arborizations are likely to converge on some cells even if they do not overlap.

In summary, the rules governing the spatial organization of cortical afferent fiber arborizations in the neostriatum are beginning to be worked out, and will continue to be an important area for future anatomical study of the neostriatum. It will be necessary to interpret this organization within the context of the dendritic fields of each of the neostriatal cell types, and in relation to the mosaic clusters of neostriatal cells (see below).

Other afferents. Many of the characteristics of corticostriatal afferent fibers are also shared by thalamic, nigral, and amygdalar afferents, although they have been studied in much less detail. Like the corticostriatal axons, inputs from these structures also project to large areas within the neostriatum, and do so in the cruciform axodendritic pattern. Also like the corticostriatal axons, their terminal arborizations are not of uniform density, but show a clustered fine-grain organization suggestive of complex patterns of convergence within larger input domains.

PRINCIPAL NEURON

Most of the neurons in the neostriatum are principal neurons. This stands in contrast to brain structures such as the retina or cerebellar cortex, where the principal neurons are few in comparison to interneurons. The neostriatal principal neurons are called "spiny neurons" because of the large numbers of dendritic spines that cover their dendrites (e.g., DiFiglia et al., 1976; Wilson and Groves, 1980). As shown in Figs. 9.3 and 9.4, the cell bodies of these cells range from 12 to 20 μm in diameter, and they give rise to a small number of dendritic trunks with diameters of 2–3 μ. The cell bodies and the initial dendritic trunks are usually free of spines. The smooth trunks divide within 10–30 μm of their origins to give rise to spiny secondary dendrites which may branch one or two more times. A spiny neuron generally has 25–30 dendritic terminal branches, which radiate in all directions from the cell body to fill a roughly spherical volume with a radius of 0.3–0.5 mm (300–500 μm). The density of dendritic spines increases rapidly from the first appearance of spines at about 20 μm from the soma, to a peak at a distance of about 80 μm from the soma. The peak spine density can be as high a 4–6 per 1 μm of dendritic length, making the neostriatal principal neuron one of the most spine-laden cells in the brain (in density, not total spine number). The spiny dendrites taper gradually from about 1.5 μm to only 0.25 μm at the tips, and the spine density likewise tapers gradually, reaching about half the peak value at the dendritic tips (Wilson et al., 1983c). The total number of spines, and the implications of spines for the function of the spiny neuron are discussed below.

The axon of the spiny cell arises from a well-defined initial segment on the soma or a proximal dendritic trunk. The main axon emits several collaterals before leaving the vicinity of the cell body, and these give rise to a local collateral arborization. The local axonal arborizations of spiny neurons are so rich that they caused many earlier investigators, using the Golgi method, to conclude that these cells must be interneurons. The spiny cells were among the first cells to be studied by intracellular injection of horseradish peroxidase (HRP) (see Kitai et al., 1976a), which revealed that their collateral axonal arborizations were even much more elaborate than visualized with the use of the Golgi method. A photomicrograph of an intracellularly stained spiny neuron from the rat is shown in Fig. 9.5. More recent studies using axonal tracing and intracellular labeling have shown that all the spiny cells project outside the neostriatum besides contributing to the axonal plexus in the region around the cell body (e.g., Preston et al., 1980; Chang et al., 1981).

A second kind of principal neuron has been described, although this cell is much less common (DiFiglia et al., 1976; Bolam et al., 1981). It is characterized by a much lower dendritic spine density, a larger soma with occasional dendritic spines, and fewer but longer and less branched dendrites. This cell has been

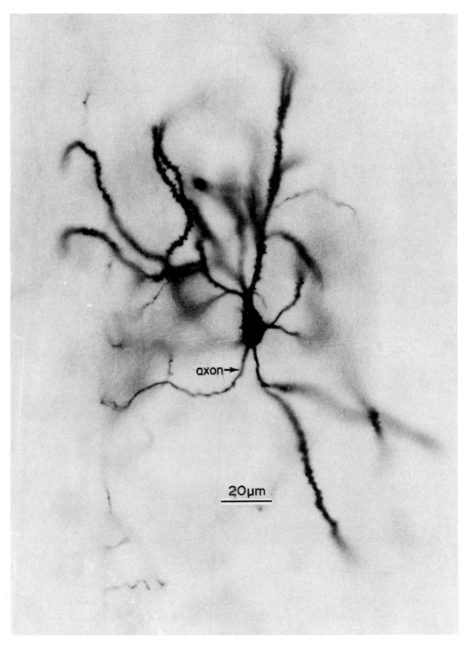

FIG. 9.5. A photomicrography of an intracellularly stained neostriatal spiny projection neuron. Dendritic spines can be seen on all but the most proximal portions of the dendrites. The axon initial segment is indicated by the arrow, and some fine branches of the local axonal arborization are visible.

shown to project to the substantia nigra, but because of its relative rarity, little more is known about it.

Golgi studies of the neostriatum have revealed a great diversity of aspiny neuron types. The number of cell types that can be described on morphological grounds may be as high as eight or nine, and in most schemes there are still neurons that cannot be categorized (e.g., Chang et al., 1982). Together these cells account for only a small proportion of the cells in the neostriatum, and it is not at all clear what the existence of so many different kinds of rare interneurons could mean. For purposes of this discussion, as for most other practical purposes at this time, it is enough to describe only three interneuron types that have been examined in sufficient detail to be characterized as functionally and structurally separate categories. These are (1) the giant aspiny interneuron (aspiny II cell of DiFiglia et al., 1976); (2) the medium-sized sparsely spined cells (aspiny III cell of DiFiglia et al., 1976); and (3) the medium sized smooth and varicose dendrite cells (aspiny I cell of DiFiglia et al., 1976). The morphological characteristics of these cells are shown in Figs. 9.3 and 9.4.

Giant aspiny interneuron. Although representing less than 2% of all the cells in the neostriatum, the largest cells in the tissue have long fascinated students of the basal ganglia. These cells were originally recognized as interneurons by Kölliker, in the late 1800s. Because they are few and large, however, many later investigators have assumed (by analogy with so many other brain structures) that they are the principal cells, and that the numerous small neurons are interneurons. In recent years, the discovery that most or all of the largest cells are cholinergic, and that acetylcholine-containing axons do not participate in the efferent projections to globus pallidus and substantia nigra, has led to a general acceptance of the interneuronal status of the giant cell of Kölliker.

These cells have been described in Golgi-stained sections by DiFiglia and Carey (1986) and Chang and Kitai (1982), and using intracellular staining by Bishop et al. (1982). They usually have an elongated cell body, up to 50 or 60 μm in length, but commonly only 15–25 μm in their shortest diameter. A few stout dendrites arise from the soma and branch in a radiating fashion, with some dendrites extending 0.5–0.75 mm (500–750 μm) from the soma. The distal dendrites exhibit some irregularly shaped appendages and varicosities. The axons of these neurons arise from dendritic trunks. They may be myelinated initially, but lose their myelin in the reductions of axonal diameter that occur in repeated bifurcations. The axon branches many times in the fashion classically associated with interneurons, that is, with daughter branches being approximately equal in size and forming approximately 120° angles with each other and with the parent branch. The resulting arborization consists of a dense plexus of extremely fine axonal branches that fill the region of the dendritic field (commonly up to 1 mm from the soma) and sometimes go beyond, but usually do not leave the nucleus. Perhaps some do project to the cortex, at least in cats (Jay-araman, 1980).

Medium-sized interneurons. The medium-sized interneurons are usually divided into two categories, although the names given to these categories are not con-sistent among authors. The most obvious difference between the two neuron types is the existence of some dendritic spinelike appendages and fine-tufted dendritic terminations on one class of cells, and usually smooth dendrites, often with prominent varicosities, on the other kind of cell. These correspond to the aspiny I and aspiny III neurons of DiFiglia et al. (1976), and the medium type III/IV neurons and the medium type V neurons of Chang et al. (1982), respec-tively. Most of the cells are medium sized, that is, with somata about the same size as the spiny neurons; some cells of both types are larger than those but smaller than the giant cell of Kölliker. Perhaps they should be called medium and large cells, but they are usually referred to simply as medium cells.

The two classes of neurons are represented in Figs 9.3 and 9.4 They both have rounded somata and smooth or sparsely spinous dendrites that fill a volume roughly equivalent to that of a spiny cell dendritic field. The axon of the varicose dendrite cell is apparent in Golgi-stained neurons and has been described by many authors. The axon of the other cell type is usually not well stained, but some local axonal arborization has been reported. The classification of both of these neuron types as interneurons is based primarily on the absence of any evidence for a projecting axon, rather than a complete visualization of their axonal arborization.

EFFERENT AXONS

The axons of spiny neostriatal neurons are gathered into small fiber fascicles that perforate the gray matter of the neostriatum, giving it the striated appearance for which it is named. Although these axons form a major fiber system of the forebrain, they are not large or heavily myelinated. In rats, the axons are near the threshold diameter for myelination (about 0.25 μm) and may or not be my-elinated. Some axons are myelinated briefly near the cell body but lose their myelin thereafter (Chang et al., 1981). In primates, the axons are larger and more consistently myelinated, although the gradual tapering in axonal diameter that occurs over the course of these axons is accompanied by a thinning and partial loss of the myelination in these species as well (Fox and Rafols, 1976). At least two different kinds of fibers can be identified in the rat, both of which arise from the spiny cells. One kind of spiny neuron gives rise to an axon that arborizes primarily in the external segment of the globus pallidus, whereas another kind projects little or not at all to this structure but instead arborizes in the internal segment of globus pallidus and in the substantia nigra (Chang et al., 1981).

In the globus pallidus and substantia nigra, the axons of neostriatal spiny neurons arborize in a very characteristic *longitudinal axodendritic* pattern. This pattern, which contrasts sharply with that of afferent fibers in the neostriatum, is characterized by individual neostriatal efferent axons running parallel to dendrites of the pallidal and nigral target neurons, making multiple synaptic contacts that almost completely ensheath the dendrites of the postsynaptic cells (e.g., DiFiglia and Rafols, 1988).

CELL POPULATIONS

According to studies of Nissl-stained sections which show all cell bodies, there are approximately 111 million neurons in the neostriatum of the human (Fox and Rafols, 1976). Of these, only about 1–2% are large, aspiny interneurons. Principal neurons make up the largest population, representing about 77% of cells in monkeys. The principal neurons are responsible for a larger proportion of the total cell population in cats, rats, and mice, being perhaps as much as 95% of all cells (Graveland and DiFiglia, 1985a). The remaining cells, about 3–4% in carnivores and rodents and perhaps as many as 23% in primates, are divided among the various types of medium-sized interneurons. There are approximately 540,000 neurons in the lateral segment of globus pallidus, and 170,000 cells in the internal segment. On the basis of these numbers alone, there must be an impressive convergence of inputs from the neostriatum onto the output cells of the globus pallidus. Not all neostriatal principal neurons project to the external segment of the globus pallidus, and the exact ratio is not known. If only half of them did, then there would be about 55,000,000 neostriatal neurons contributing to the innervation of 540,000 cells in the external segment of the globus pallidus, and a minimal convergence ratio of 100 : 1. Even higher ratios would obtain for the internal segment of the globus pallidus and the substantia nigra, pars reticulata. In view of the longitudinal axodendritic synaptic arrangement in these structures, the influence of neostriatal output on the cells of these structures must be very strong.

Because the arrangement of neostriatal cell bodies and the arborization patterns of afferent axons do not follow any simple geometrical arrangement, quantitative estimates of the degree of convergence and divergence are not easily obtained for neostriatal afferents. The evidence currently available, however, suggests that all the asymmetrical synapses formed on the heads of dendritic spines arise from afferent fibers (although not all afferent synapses are of this type; see below). This suggests that we can derive a lower limit to the number of afferent synapses formed onto individual spiny neurons by counting the number of dendritic spines per cell. Counts of spine density have been made in several species, and they are generally in agreement. Although spine density is somewhat lower in humans and monkeys than in rats and cats, the dendrites are slightly longer as well, making the number of spines per dendrite approximately the same (Graveland and DiFiglia, 1985b). The integrated spine density in the rat has been measured, and varies among cells from about 300 to 500 per dendrite (Wilson et al., 1983c). There being 25–30 dendrites per spiny neuron, this yields a count of about 7,500–12,000 spines per cell. Probably each dendritic spine gets at least one excitatory synapse, which means that each spiny neuron has at least this many excitatory inputs. It is remarkable, in view of this large excitatory innervation, that most spiny neurons are electrophysiologically silent most of the time. How this comes about requires analysis of the functional organization of the spiny neuron and its dendritic tree (see below).

SYNAPTIC CONNECTIONS

The typical synaptic connections of the neostriatum are shown in Fig. 9.6. We will describe briefly each of the main types of connections.

CORTICAL AND THALAMIC CONNECTIONS

The most noticeable feature of the neostriatum as seen in the electron microscope is the large number of small axons forming asymmetrical synapses (Gray's type 1; see Chap. 1) on the heads of dendritic spines. This kind of synapse accounts for about 85% of all synapses in the neostriatum (Kemp and Powell, 1971a). It is the characteristic synaptic type formed by afferents from cerebral cortex, and from parts of the intralaminar nuclei of the thalamus (e.g., Kemp and Powell, 1971b). The axons from cerebral cortex and thalamus are similar in morphology, both exhibiting small round synaptic vesicles. A smaller number of cortical and thalamic afferent synapses are made onto dendritic shafts and somata of neo-striatal neurons. Some of these are formed on dendritic shafts of spiny projection neurons, and others are on the somata and dendrites of the aspiny neurons. One area of the thalamus has been shown to form preferentially axodendritic rather than axospinous synapses (Dubé et al., 1988). The spiny cell that is the target of this projection is somewhat unusual in appearance, and this may represent a special subspecialization within the thalamostriatal pathway.

SUBSTANTIA NIGRA CONNECTIONS

The nigrostriatal projection has been more intensely studied, because it has presented an interesting and puzzling inconsistency. There are several ways to label a pathway so that its axons can be identified in the electron microscope. For example, there is an EM variant of the axonal tracing technique employing extracellular injections of radioactive amino acids, which are taken up by cells and converted to proteins and transported down the axons. With this method, axons arising from one part of the brain can be identified in electron micrographs of sections taken from another area, in which they terminate. Another approach is to destroy the neurons in one area, and use the characteristic changes that accompany axonal degeneration to identify the axons in a distant, otherwise unharmed brain area. Still another is to label the axons that form a particular projection by making use of a biochemical marker that is characteristic of the projection. In the case of the dopaminergic neurons of the nigrostriatal pathway, this is especially easy because they have a very large complement of the enzyme tyrosine hydroxylase (TH), which is involved in the synthesis of dopamine. Immunocytochemistry using antibodies to TH is therefore an excellent way to demonstrate which axons are from the substantia nigra. Two other methods take advantage of special properties of the dopaminergic axons. One is destruction of these axons with the selective neurotoxin, 6-hydroxydopamine, which is be-lieved to kill specifically dopaminergic axons in the neostriatum. The degenerat-

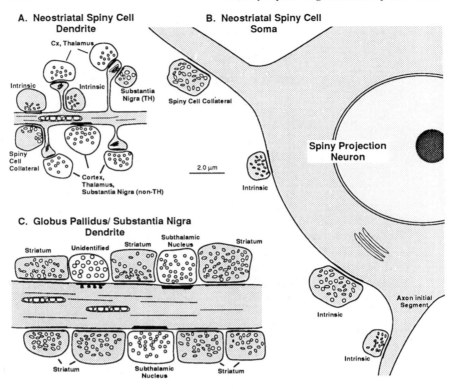

FIG. 9.6. The major synaptic types of the neostriatal spiny neuron and the neurons in the substantia nigra and globus pallidus that receive input from it. **A.** The distal spiny dendrites of the spiny neurons. Inputs with round synaptic vesicles form asymmetrical inputs primarily on dendritic spines, but occasionally on the shafts of the dendrites. These arise mostly from afferent fibers, especially from the cerebral cortex (Cx) and thalamus, but also from other structures, such as the non–TH-staining axons from the substantia nigra (TH = tyrosine hydroxylase). Spiny cell collaterals, TH-staining axons from the substantia nigra, and intrinsic intrastriatal connections (from aspiny neurons) form symmetrical synapses with pleomorphic and flattened vesicles on the stalks of dendritic spines, on the proximal part of the spine heads, and on dendritic shafts. Axodendritic synapses are overrepresented so that all the synapse types could be shown. Most of the surface of the dendritic shaft is free of synaptic input. **B.** The inputs to the proximal surface of the spiny neurons are primarily intrinsic to the neostriatum, arising from the axons of interneurons and from the collateral arborizations of the spiny neurons. They form symmetrical synapses with pleomorphic or flattened synaptic vesicles, and are present at very low density on the aspiny initial portion of the dendrites, the somata, and the axon initial segments.

ing axons can be identified by the characteristic appearance of degenerating neuronal tissue. The second is the labeling of dopaminergic neurons with another dopamine analogue, 5-hydroxydopamine. This analogue does not kill dopaminergic axons but is specifically absorbed by them, and is directly visible in the electron microscope in tissue fixed in the conventional way.

The reason for describing all of these methods is that although they are all well-proven tools in other systems, in the nigrostriatal pathway they produce different, apparently inconsistent results. For example, after autoradiographic labeling of the nigrostriatal pathway, or selective destruction of that pathway using either conventional lesions of the substantia nigra or treatment with 6-hydroxydopamine (e.g., Hattori et al., 1973), or using labeling with 5-hydroxydopamine (e.g., Groves, 1980), the axons that are labeled in the neostriatum appear similar to those of the corticostriatal or thalamostriatal projections. That is, they are of Gray's type 1 in that they contain small, round synaptic vesicles and form asymmetrical synapses on the heads of dendritic spines. If the TH method is used instead, a different kind of axon is labeled, which contains larger, more variably shaped synaptic vesicles, and forms symmetrical synapses on dendritic shafts, somata, and the stalks of dendritic spines (e.g., Freund et al., 1984). Both experiments have been repeated several times, and there is no readily available explanation, other than to reject arbitrarily one of the two sets of observations. In the diagram shown in Fig. 9.6, two nigrostriatal endings are shown, one labeled TH, and the other non-TH. It remains to be seen whether both of them are really there. One point that can be made from this is that trusted experimental techniques sometimes produce inconsistent findings. It is permissible to postpone judgment on such difficult issues while awaiting the new insight that will provide an explanation.

AXON COLLATERAL CONNECTIONS

The boutons that arise from the local collaterals of spiny neurons contain large synaptic vesicles of variable shape, and form symmetrical synapses onto the stalks of dendritic spines, dendritic shafts, somata, and initial segments of axons (Wilson and Groves, 1980; Bishop et al., 1982; Somogyi et al., 1982). These synapses are similar in appearance to those stained positively with TH. Like the TH synapses, local spiny cell collaterals that end on dendritic spines share that spine with some other input, often a cortical fiber. Some of the terminals of spiny cell axons form synapses with dendrites, somata and initial axonal segments of other spiny neurons, but some are definitely made with these portions of the aspiny cells, including the giant aspiny neuron (Bolam et al., 1986).

INTERNEURON CONNECTIONS

Synapses formed by the axons of the giant aspiny cell, and those of at least some medium interneuron types (Takagi et al., 1983; DiFiglia and Aronin, 1982; Bolam et al., 1984; Phelps et al., 1985), contribute to a third major morphological synaptic type in the neostriatum. These synapses have small flattened vesicles, and form symmetrical synapses on the stalks of dendritic spines, dendritic shafts, somata, and initial segments; they therefore fall into Gray's type 2 (see Chap. 1). They are common on spiny neurons, and axons from both giant aspiny neurons and medium interneurons have been shown to terminate in this way on identified spiny cells. Terminals of this type, and the terminals with large pleomorphic synaptic vesicles, form the primary synaptic types found on the somata and proximal dendritic surface of the spiny cell.

Inputs to the giant aspiny neuron are of the same three major morphological types, but they do not show the specific localization on the cell surface that is observed on the spiny neurons (Chang and Kitai, 1982; DiFiglia and Carey, 1986). Boutons with small, round vesicles form asymmetrical synapses on dendrites of all sizes, and even on the somata of aspiny neurons. Likewise, symmetrical synapses formed by boutons with large pleomorphic vesicles, and by boutons with small flattened synaptic vesicles, are seen on all parts of these cells. The synaptic density on the giant neuron is much lower as well, with large parts of the surface area of the cell being free of synapses. It is likely that despite its large size, the giant interneuron receives many fewer synaptic contacts than the spiny cell.

OUTPUT CONNECTIONS

The output axons of the spiny neurons form synapses in the substantia nigra and globus pallidus that are similar in morphology to the synapses formed by their local axonal terminals (Chang et al., 1981). In both the globus pallidus and substantia nigra, the dendrites of most neurons are completely encased in a quiltwork of synaptic terminals. The majority of those terminals are derived from the neostriatal efferent fibers. A second kind of terminal found on these dendrites, containing small round vesicles and forming asymmetrical contacts, is derived from the subthalamic nucleus (Kita and Kitai, 1987), whereas a third type, containing larger round vesicles and also forming symmetrical contacts, is from an unidentified source.

BASIC CIRCUIT

The known connections between elements in the neostriatum are shown in Fig. 9.7. Although this diagram is no doubt incomplete, especially in its portrayal of the connections of interneurons, it does show the essential features of the connections of the spiny cells, which represent the principal cells of the neostriatum. Two identical spiny cells are shown in the diagram, with their local axon collaterals making inhibitory synaptic connections between them. A giant aspiny interneuron and one medium aspiny interneuron are shown as intrinsic inhibitory neurons, making synapses primarily with the principal cells. Although the direct effects of interneurons on neostriatal spiny neurons have not yet been observed in experiments, these cells are shown as inhibitory on the basis of the distribution and morphology of the synapses they make on the neostriatal neurons. The neurons whose action is least understood are the giant interneurons and the medium aspiny neuron with nonvaricose dendrites. Both of these cell types make symmetrical synaptic contacts characterized by small flattened vesicles on the spine stalks, dendritic shafts, and somata of the spiny cells. This is the pattern expected for inhibitory neurons in the neostriatum. This is not definitive evidence, however, and what little is known about the action of the transmitters released by these cells (acetylcholine and somatostatin, respectively) does not fit the classical view of inhibitory transmission (see below). To indicate this uncertainty, the synapses made by the giant aspiny interneurons are marked with question marks.

The medium interneuron with varicose dendrites stains very intensely for

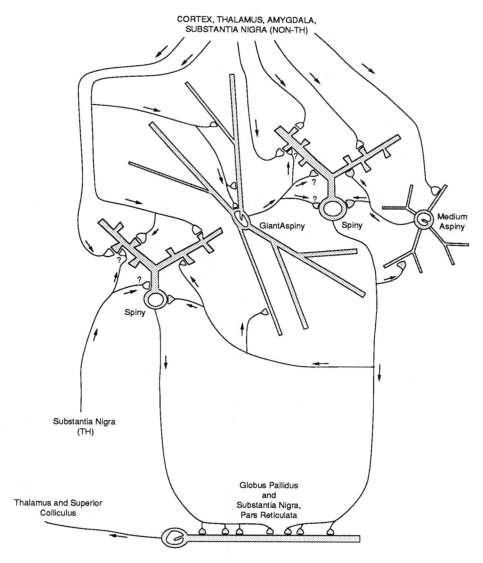

CORTEX, THALAMUS, AMYGDALA,
SUBSTANTIA NIGRA (NON-TH)

GiantAspiny

Spiny

Medium
Aspiny

Spiny

Substantia Nigra
(TH)

Thalamus and Superior
Colliculus

Globus Pallidus
and
Substantia Nigra,
Pars Reticulata

FIG. 9.7. Simplified basic circuit of the neostriatum and its outputs. Excitatory afferent input from several sources is received by all cell types. Because of their large numbers and heavy afferent innervation, most of these are formed on spiny neurons. Spiny cells contribute inhibitory input to all cell types. The interneurons regulate activity through the direct pathway consisting of the afferent fibers and the spiny projection neurons. The effect of their synaptic input to the spiny neuron is unknown, as indicated by the question marks. The output of the spiny neurons converges on the dendrites of target cells in the substantia nigra and globus pallidus, which themselves project to the thalamus and superior colliculus.

GABA, and is likely to behave in a more familiar inhibitory manner. The medium aspiny interneuron in Fig. 9.7 represents this cell.

Afferent inputs go to all of the neostriatal cell types, where most make excitatory synapses predominantly on the dendritic spines of the spiny neurons. Included in this category are inputs from cerebral cortex, the thalamic nuclei, amygdala, and the TH-negative inputs from the substantia nigra. The TH-containing axons are shown separately in Fig. 9.7. These axons end upon spine stalks and dendritic shafts, rather than spine heads, and do not make the classical Gray's type 1 synapse formed by other neostriatal afferent fibers. Because the physiological action of these fibers does not allow for a simple classification as excitatory or inhibitory, these terminals are also marked with question marks.

Regardless of the uncertainty about the action of interneurons, the neostriatum is unusual in that the principal neurons, as well as many of the interneurons, are inhibitory. In this regard, it is similar to the cerebellum (Chap. 7). The next neuron in the output pathway, the principal neuron of the globus pallidus and substantia nigra, is also shown in Fig. 9.7, and it is also inhibitory. Excitatory influences in the basal ganglia arise mostly from incoming fibers.

Appreciation of the operation of this mostly inhibitory circuit requires a knowledge of the firing patterns of the neurons involved. This will be described in a later section, but for the present purposes it is important to understand that neostriatal spiny cells fire very rarely, and in episodes that last for only about 0.1–3.0 sec. Most interneurons likewise do not fire impulses rapidly. Thus, tonic intrinsic inhibitory interactions between the neurons are probably not important contributors to the membrane potential of the spiny neuron under resting conditions, but rather limit firing during episodes of excitation that arise because of activity in excitatory inputs. The cells in globus pallidus and substantia nigra, on the other hand, fire tonically at very high rates. Their tonic firing produces a constant inhibition of neurons in the thalamus and superior colliculus. The firing of spiny neostriatal neurons can cause a transient pause in that tonic inhibition, releasing thalamic and superior colliculus neurons to respond to excitatory inputs that would otherwise be subthreshold. Thus the neostriatum acts to disinhibit neurons in thalamus and superior colliculus in response to excitation in its afferent fibers, and the interneurons of the neostriatum help to regulate the duration, strength, and spatial patttern of that disinhibition.

NEUROTRANSMITTERS

The neurotransmitters in the neostriatum will be summarized in relation to the simplified basic circuit of Fig. 9.8.

SPINY NEURON

The neurotransmitter of the spiny neuron has long been believed to be GABA (e.g., Precht and Yoshida, 1971). In recent years, several peptides have also been localized in spiny neurons, especially dynorphin, substance P, and enkephalin. These are not all present in every spiny neuron, but instead seem to mark particular subsets of cells (e.g., Penny et al., 1986a,b). Enkephalin, for example, is present in very large quantities in the spiny neurons' axon terminals in the

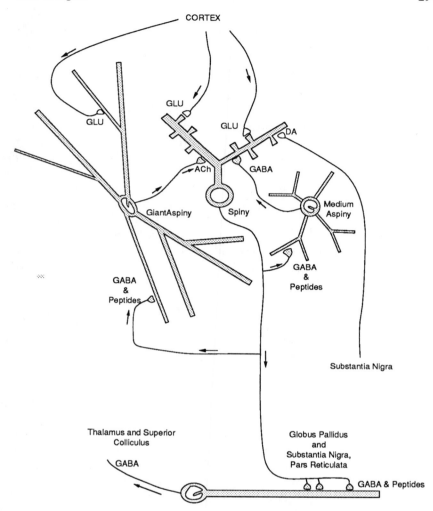

FIG. 9.8. Neurotransmitters that are present at some of the synaptic connections represented in Fig. 9.7. The spiny neurons contain GABA and the peptides enkephalin, substance P, and dynorphin (but not all of these in every cell). The medium aspiny neuron with varicose dendrites (called "medium aspiny" here) contains GABA, whereas the giant aspiny neuron synthesizes the neurotransmitter acetylcholine. Other intrinsic cell types containing somatostatin and other peptides are not shown. The only afferent pathways shown are the cortical input, which releases glutamate, and the dopaminergic pathway from the substantia nigra.

globus pallidus, but is relatively sparse in the substantia nigra. Substance P and dynorphin both show the opposite pattern, being heavily concentrated in axon terminals in the substantia nigra, but much less so in the globus pallidus. The cell bodies of spiny neurons likewise contain either enkephalin or substance P, but usually not both. GABA, on the other hand, coexists with either transmitter.

These observations suggest that spiny neuron subpopulations with differing axonal projections probably also differ in their content of peptide cotransmitters. The functions of the peptides in the striatal efferent projection are not known.

INTERNEURONS

The neurotransmitter of the giant aspiny interneuron is acetylcholine (e.g., Bolam et al., 1984; Phelps et al., 1985). Although these neurons are few in number, their axonal arborizations are very large and dense, and they provide a very rich cholinergic innervation to the neostriatum. This is a good example of the error that is committed if we judge the importance of a cell type purely by the number of cells. The cholinergic neurons of the neostriatum, and the system of cholinergic synapses that they give rise to in the neostriatal neuropil are known to exert a powerful influence on the firing of the spiny neurons and the final output of the neostriatum. Pharmacological treatments for many human disorders, including Huntington's disease, Parkinson's disease, and even schizophrenia often rely upon manipulation of transmission at cholinergic synapses in the neostriatum. It is surprising, given the importance of those synapses, that the action of acetylcholine on neostriatal spiny neurons has not received more attention from neuropharmacologists.

One class of medium aspiny interneurons is GABAergic (Bolam et al., 1983; Oertel and Mugnaini, 1984). These neurons are easily distinguished from the spiny cells by their especially high concentration of GABA, which causes them to stand out from other GABAergic neurons in immunocytochemical studies, and their content of the calcium-buffering protein parvalbumin. Parvalbumin is present in some, but not all, classes of GABAergic neurons in various parts of the brain, including the cerebral cortex, hippocampus, and thalamus (Heizman, 1984; Celio, 1986; Kosaka et al., 1987; Gerfen et al., 1985). In the basal ganglia, it is found in the GABAergic principal cells of the substantia nigra and globus pallidus, as well as in the strongly GABA-positive aspiny cells in the neostriatum. In all cases, it appears to be concentrated in cells that are capable of maintaining high rates of firing. The parvalbumin-containing neurons of the neostriatum are morphologically the same as the fast firing interneurons, and although no direct demonstration of their identity has yet been made, it is very likely that these will prove to be the same cell type.

Another medium-sized interneuron that does not contain GABA contains a very high concentration of the neuropeptide somatostatin (DiFiglia and Aronin, 1982; Takagi et al., 1983). Except for their size, the morphology of these cells is very similar to the giant interneuron. There are no reports of the physiological properties of these cells.

INPUT FIBERS

A variety of evidence indicates that the neurotransmitter in the corticostriatal pathway is glutamate (e.g., McGeer et al., 1977). The thalamostriatal pathway is less clear. The strong similarities between the corticostriatal and thalamostriatal pathways, in both morphology and physiology, apparently do not extend as far as a common neurotransmitter. Whatever the transmitter of the thalamic projection neurons is, however, it must be similar to glutamate in its postsynaptic action. The nigrostriatal projection is almost entirely dopaminergic, and the input from the raphe contains serotonin.

MOSAIC ORGANIZATION OF THE NEOSTRIATUM

In the human neostriatum, and in the neostriatum of some primates, the neurons can be seen to be clustered into groups of high cell density separated by areas of lower cell density. No such internal organization can be seen in most other animals. Despite the apparent cell clusters of primates, the neostriatum has long been believed to be structurally homogeneous. This stood in contrast to the cerebral and cerebellar cortices, olfactory bulb, and other brain structures that possess a very prominent layered organization.

In the past several years, it has become evident that the clusters of cells in the primate neostriatum, called *cell islands,* represent a fundamental feature of the organization of this structure, and are present in a less visible, perhaps less differentiated, form in all mammals. Unlike layered structures, however, this mosaic organizational plan does not separate the tissue into many compartments, but rather only two. In primates, these are clear enough, and are called cell islands and *matrix* (Goldman-Rakic, 1982). The matrix is the somewhat less cell dense neostriatal tissue between the cell islands. A very important finding for the interpretation of this organization was the observation that afferent fibers, particularly fibers of cortical origin, respect these tissue compartment boundaries (e.g., Goldman and Nauta, 1977; Ragsdale and Graybiel, 1981). Certain cortical areas project preferentially to the matrix and mostly avoid the islands. This observation encourages a search for traces of cell islands and matrix in other animals, like the cat or the rat, which have no visible clustering of the neurons to indicate the boundaries of the mosaic. When this is done, cortical axons that should project to the matrix are seen to fill a space in the neostriatum that is perforated with many small irregularly shaped spaces that seem to correspond to the islands seen in primates. It is not yet certain that these are actually the same, and thus they are given a slightly different name: *striosomes,* or sometimes just *patches* (Graybiel et al., 1981; Herkenham and Pert, 1981).

In the rat, some areas of the cerebral cortex have been shown to project preferentially to the striosomes as well (Gerfen, 1984; Donoghue and Herkenham, 1986). Unfortunately, the functional properties of the cortical areas that have been shown to project into the striosomes are not yet clear; therefore, it is not possible to make any generalizations about the function of the striosomes versus the matrix. Other afferent pathways also observe the boundaries between striosomes and matrix. Although no thalamic input to the striosomes has yet been identified, some intralaminar nuclei have been shown to project preferentially to the matrix. It is expected that there are thalamic projections to the striosomes, even though they have not yet been seen, because neostriatal spiny neurons recorded everywhere in the nucleus have been shown to exhibit monosynaptic EPSPs in response to thalamic stimulation (Kocsis et al., 1976). The dopaminergic inputs from the substantia nigra also observe the boundaries between striosomes and matrix. In the rat, there are two different sets of nigrostriatal dopaminergic neurons, one that innervates the patch (striosome) compartment, and one that innervates the matrix (Gerfen, 1985).

The observation that afferent axons respect the boundaries of internal neostriatal tissue compartments has no immediate functional implications. Because most afferent fibers make synapses on the dendrites of neostriatal neurons, afferents that are confined to the matrix could easily be making synapses with neurons of the striosomes, if those cells send their dendrites into the matrix (see Neuronal Elements, above). A similar uncertainty holds for the internal interconnections among neostriatal neurons. It is important, therefore, to ascertain whether the dendritic and axonal fields observe the boundaries, and to determine this for each of the cell types. This work has not been completed. It seems clear from the present data that all the cell types already described for the neostriatum are represented in both compartments. Studies of the spiny neurons in rat have suggested that these cells strictly obey the compartment boundaries (Herkenham et al., 1984; Penny et al., 1988), with cells in the striosomes keeping their dendritic fields restricted to the striosomes, and cells in the matrix having their dendritic fields contained within the matrix (see Bolam et al., 1988, however, for an alternative point of view). Similarly, the local axonal arborizations of the spiny neurons are apparently restricted to the compartment containing the cell body (Penny et al., 1988). This observation, together with the precise segregation of afferent fibers, means that the most direct pathway through the neostriatum, going from afferent fibers to the principal neurons and out to the target structures, consists of two parallel and independent pathways. The shapes and proportions of the striosomes in comparison to the neurons (as they appear in rats) are illustrated in Fig. 9.9.

Interneurons may form associational pathways between the striosomes and matrix. An especially likely candidate for this role is the giant interneuron. This cell, with its long radiating dendrites and large intrastriatal axonal field, cannot easily fit within a striosome, at least in the rat (in which the striosomes are particularly small). A few observations (Penny et al., 1988) suggest that this cell ignores striosomal boundaries, sending its dendrites, and probably its axon, into both compartments. These relationships are shown in Fig. 9.9. In this figure, the medium aspiny neurons are also shown to cross striosomal boundaries. This is based upon less direct observations using immunocytochemical methods, which are described below.

Given that the striosomes and matrix receive different synaptic inputs, it is natural to wonder whether they might have different output targets. Early results by Graybiel et al. (1979) showed that neurons projecting to the cat globus pallidus were preferentially located in the matrix. More recent studies have shown that projections to substantia nigra are also segregated between the compartments, with striosomes projecting preferentially to the substantia nigra pars compacta, where the dopaminergic nigrostriatal neurons are located, and the matrix projecting to pars reticulata, where nondopaminergic neurons projecting to the thalamus and superior colliculus are found (Gerfen, 1985).

A particularly exciting development has been the discovery that neostriatal compartments can be distinguished on the basis of their cytochemical charac-

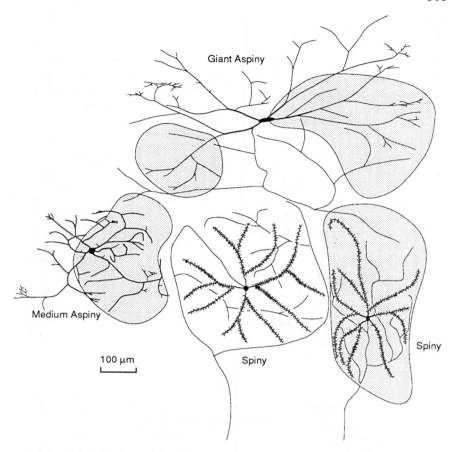

FIG. 9.9. The mosaic internal organization of the neostriatum in relationship to the dendritic fields of the major cell types, drawn at the same scale. The striosomal compartment of the neostriatum is shown stippled, whereas the matrix is white. Two different populations of spiny neurons, one associated with the striosomes and one with the matrix, are represented by the two spiny cells. Their dendrites and local axonal fields are mostly confined to the compartment containing the cell body. The aspiny interneurons, however, have dendritic fields that cross compartment boundaries.

teristics (e.g. Graybiel et al., 1981). Cells in the striosomes and matrix, although similar in many respects, are sufficiently different biochemically to allow demonstration of the compartments by using many standard cytochemical methods. For example, in many species the concentration of acetylcholinesterase is detectably higher in the matrix, allowing visualization of the striosomes as acetylcholinesterase-poor zones in sections stained for the presence of this enzyme (e.g., Graybiel et al., 1981). Another useful biochemical marker is the mu opiate receptor, which is present in much higher concentration in the neuropil of the striosomes (e.g., Herkenham and Pert, 1981). The peptides met-enkephalin and substance P, both very abundant in the neostriatum, are also differentially localized, but in a more complex way. In some parts of the neostriatum, the striosomes are richer in these substances, whereas in other parts these substances are preferentially expressed in the matrix. The functional meaning of these differences is not known. Their existence proves, however, that we cannot consider the mosaic organization of the neostriatum to be a simple topographical relationship, representing nothing more than the arrangement of afferents and efferents. The differences between the two compartments run deeper than that.

Some of the immunocytochemical studies of the neostriatal mosaic have immediate implications for the synaptic organization of the neostriatum. An example is the observation that cholinergic synaptic boutons, as visualized by using immunocytochemistry for cholineacetyltransferase, are present in higher density in the matrix than in the striosomes (Graybiel et al., 1986). This difference is apparently not accompanied by a corresponding difference in the distribution of the cholinergic cell bodies. From this we must conclude that the cholinergic neurons have a preference for axonal arborization in the matrix. Similarly, somatostatin, which is contained in a class of medium-sized interneurons, is preferentially located in the matrix, despite the existence of somatostatin neurons in both compartments (Cheeselet and Graybiel, 1985; Gerfen, 1985). Immunocytochemical staining of somatostatin neurons is sufficient to show that the dendrites of these cells can cross the boundaries between striosome and matrix, and this is also indicated in Fig. 9.9.

SYNAPTIC ACTIONS

Because the neostriatum is a central structure, several synaptic relays removed from the direct sensory input pathways, it is not possible to analyze its synaptic actions by natural stimulation, as in so many other regions considered in this book. Analysis of synaptic actions therefore must rely on activation of neostriatal input and output pathways by electrical stimulation. We shall summarize the evidence for the main types of excitatory and inhibitory actions that control the activity of the striatal neurons as revealed by this experimental approach.

ACTIONS OF CORTICAL AND THALAMIC INPUTS

Stimulation of the cerebral cortex in intact animals sets up a synchronous volley of impulses in a subset of the corticostriatal axons, which evokes a large-amplitude EPSP in the spiny neurons of the neostriatum (e.g., Wilson et al., 1982, 1983b; Wilson, 1986a). The latency of this EPSP matches that of the fastest corticostriatal axons, and shows a constant latency despite changes in stimulus intensity or frequency, as expected for a monosynaptic EPSP. The behavior of this EPSP suggests that it represents the sum of the action of many synapses, each of which has only a very weak effect at the soma. Its amplitude is finely graded with stimulus intensity. That is, very small changes in stimulus intensity produce correspondingly small changes in EPSP amplitude. Likewise, the EPSP shows no minimal threshold amplitude. Thus the synaptic potential components contributed by individual axons are too small to detect in a conventional intracellular recording, and the EPSP that is recorded must be composed of many such small EPSPs.

This initial EPSP is followed by a long-lasting hyperpolarization, and upon its termination, by a period of depolarization and increased synaptic noise. These components of the response are illustrated in Fig. 9.10Aa. Among these late components of the response, we should find the effects of recurrent collaterals of the spiny neurons that fired action potentials in response to the initial EPSP. The complexity of the response observed in the in vivo preparation suggests that it would be useful to carry out further analysis in a simplified preparation, such as slices of neostriatum. Such experiments have not usually been performed using cortical stimulation because it is difficult to get any significant piece of the corticostriatal pathway intact in a slice. They have instead used intrastriatal stimulation or stimulation of the subcortical white matter. Because of the nature of current spread from stimulation sites, these two kinds of stimulation are probably the same thing. They stimulate not only the corticostriatal fibers but all afferent fibers, the axons of spiny neurons, and striatal interneurons. Nonetheless, the response to this kind of stimulation is very informative. As shown in Fig. 9.10Ab, excitatory inputs in vitro never produce the pronounced long-lasting inhibition or rebound excitation seen in the intact animal. Instead, the EPSP, which looks similar to that observed in vivo, is followed by only a small and short-lasting IPSP. That IPSP may be due to recurrent axon collaterals of the spiny neurons, or to inhibitory interneurons excited by the stimulus, or even to an inhibitory afferent pathway (Kita et al., 1985a,b).

Stimulation of the thalamus in intact animals produces effects that are very similar to those of cortical stimulation (e.g., Wilson et al., 1983a). It is important to note in such experiments that when stimulating one area we are often engaging the activity of many structures indirectly. This is illustrated for the thalamocorticostriatal system in Fig. 9.10B. We cannot stimulate thalamostriatal cells without also stimulating thalamic neurons that project to cortex, and stimulation of these thalamocortical fibers has profound effects on cortical activity. Also, some cortical neurons projecting to the thalamus are stimulated antidromically, and their recurrent collaterals in the cortex are activated. It is important to attempt to separate these polysynaptic effects from the direct response to stimulation by experimentally interrupting the polysynaptic pathways. When such experiments are performed, the cortical and thalamic responses are greatly simplified, and become almost identical EPSPs (Fig. 9.10Ac). Although it is usually assumed that the corticostriatal input is more important than the thalamic one, there is no indication of this in a comparison of the sizes of the responses that can be produced by stimulation of the two structures. Their maximal EPSP amplitudes are approximately equal.

The results of the experiments described above indicate that the late components of the responses seen in intact animals are not due to the action of intrastriatal circuits. If the long-lasting hyperpolarization and late depolarization components of the response are not due to intrastriatal circuitry, then what could they be?

Measurement of the effect of hyperpolarizing and depolarizing currents, and of

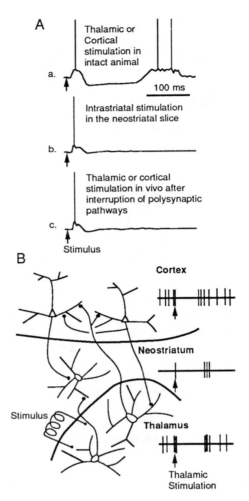

FIG. 9.10. Synaptic potentials evoked in spiny neurons by stimulation of the cerebral cortex and thalamus, and their interpretation. **A.** Synaptic potentials in three different preparations. Trace *a* shows the response evoked by stimulation of the cortex or thalamus (the response is the same for either site) in an intact animal. Initially there is a large EPSP, which can trigger an action potential, followed by a long period of membrane hyperpolarization and inhibition of synaptic responses. This period is followed by a period of rebound excitation. Trace *b* represents the response evoked by local stimulation in neostriatal slices, part of which is probably due to stimulation of cortical and thalamic afferents. Trace *c* shows the response to cortical or thalamic stimulation in an animal in which the polysynaptic pathways shown in B have been interrupted experimentally. **B.** Diagram showing some of the synaptic connections among the cerebral cortex, thalamus, and neostriatum that contribute to the long-lasting inhibition and rebound excitation observed in spiny neostriatal neurons after cortical or thalamic stimulation. Typical responses of a cortical, striatal, and thalamic neuron upon stimulation of the thalamus are illustrated on the right. Stimulation of the cortex produces similar responses, due to activation of the same circuits.

conductance changes that accompany the membrane potential change, can help to provide an answer. The long-lasting hyperpolarization is slightly increased in amplitude when the cell is hyperpolarized, and decreased with membrane depolarization. The effect is approximately equal in magnitude to the effect of the same manipulation on the EPSP. This suggests that the ionic currents responsible for the long-lasting hyperpolarization resemble those responsible for the EPSP. Secondly, measurement of whole-cell input resistance (R_N) during the long-lasting hyperpolarization shows that the cell has a net increase in resistance during this time. This measurement is difficult because of the large anomalous rectification of spiny neurons in this voltage range (see below) which has to be compensated out of the analysis. When this is done, the long-lasting hyperpolarization is seen to be accompanied by a decrease in conductance (increase in the resistance) of the cell. This is what would be expected if the long-lasting hyperpolarization were a *disfacilitation*—that is, the removal of an excitatory input.

This interpretation is strengthened by examination of the behavior of neurons in the cerebral cortex and thalamus after stimulation of either structure. It has long been known that stimulation of the thalamus produces an initial excitatory effect in the cerebral cortex, which is followed by a long-lasting inhibitory period. The inhibitory period is itself followed by a "rebound" excitation (Thalamus, Chap. 8; and Neocortex, Chap. 12). A similar pattern of excitation and inhibition is observed in both cortex and thalamus after application of a stimulus to the cortex. This pattern is precisely what would be required if the long-lasting hyperpolarization and subsequent excitatory period in the neostriatum were due to a removal and subsequent reassertion of tonic excitatory influences from the cortex and thalamus. This scheme for the generation of the long-lasting effects of cortical or thalamic stimulation is illustrated in Fig. 9.10B.

ACTIONS OF SUBSTANTIA NIGRA INPUTS

Responses of spiny neurons to stimulation of the nigrostriatal afferents have been difficult to characterize. The nigrostriatal fibers are notoriously resistant to electrical stimulation, and they lie very close to other fiber systems that can produce large excitatory responses in the neostriatum (e.g., cortical efferent fibers in the cerebral peduncle and internal capsule). Simple stimulation of the nigrostriatal pathway therefore produces results that are difficult to interpret.

Early studies employing stimulation of substantia yielded mixed results. Using extracellular recordings, both excitatory and inhibitory responses were observed. It was difficult to prove in these studies that the responses observed in extracellular recording experiments were really due to the direct effect of the nigrostriatal axons, rather than indirect polysynaptic pathways within the neostriatum and between the neostriatum and related structures. Intracellular recordings were more consistent in showing excitation (see review by Kitai, 1981). Intracellular recordings provide a more direct measurement of the effect of stimulation because they allow visualization of subthreshold responses and a more direct view of the response to synaptic activation. The latency of the responses is also more accurately measured, because the synaptic responses are directly visible, rather than inferred from the latency of action potentials that may follow by a variable additional delay. In these ways, the results of the intracellular recording experiments were more reliable, but when the antidromic conduction time of nigrostriatal neurons was measured, it was discovered to be much longer than the latency of the EPSP observed in striatal neurons after stimulation of the substantia nigra. Further analysis of the EPSPs observed in the intracellular recordings of nigrostriatal responses revealed that a large proportion of these were actually due to activation of fibers of passage (Wilson et al., 1982). A response due purely to the nigrostriatal pathway is difficult to demonstrate in an unambiguous way, even after removal of the confounding fibers of passage.

In parallel with these efforts to understand the nigrostriatal input, other investigators were performing experiments employing ionophoretic application of dopamine and dopamine agonists to neostriatal neurons in vivo. These likewise produced mixed results. In most experiments, changes in firing rate due to dopamine application were recorded extracellularly, and the effect of dopamine on the cell was inferred from the firing rate changes. In these studies, some neurons increased their firing rate, whereas others decreased it. In a small number of experiments, the effects of dopamine were observed more directly by intracellular recording. These experiments (Kitai et al., 1976b; Herrling and Hull, 1980) usually revealed depolarizations, but often these were accompanied by a decrease in neuronal excitability and a decrease in firing rate. Understanding such unusual responses requires an analysis of the ionic mechanisms underlying them, and this analysis cannot be performed in the in vivo preparation.

Partly in response to these technical problems, studies of the effects of the nigrostriatal input onto neostriatal neurons have recently concentrated on the direct application of dopamine onto striatal neurons in vitro. Biochemical and pharmacological studies have established the existence of two different dopamine receptors in the neostriatum. The physiological effects of one of them, called the D2 receptor, are well characterized from these studies. In substantia nigra neurons, where it is concentrated, this receptor causes a fast IPSP, due to an increase in the conductance of the membrane to potassium (Lacey et al., 1987).

The effect of the other receptor, called D1, is less clear. It appears that the D1 receptor may be the most important for the direct effect of the nigrostriatal afferents on spiny neurons, whereas the D2 receptor is localized primarily on presynaptic dopaminergic terminals. Application of dopamine to neostriatal neurons in vitro does not have a profound effect on their membrane potentials or input resistances. Some authors have reported a small depolarization, sometimes associated with a decrease in firing rate (e.g., Calabresi et al., 1987; Akaike et al., 1987). The decrease in firing rate appears to be related to a change in firing threshold, and also a decrease in the anomalous rectification of the membrane in the depolarizing direction. This effect on anomalous rectification could be of great importance in regulating the responsiveness of the cells to other inputs (see below). In any case, it seems most likely that the effect of D1 dopamine receptor activation will be found to be a modification of the physiological properties of the spiny cell membrane, rather than a classical fast synaptic potential. Although the problem of dopamine's effect on neostriatal neurons has been a particularly difficult one, recent improvements in the understanding of neuromodulators provide good reason to hope that a definitive answer will soon be forthcoming.

ACTIONS OF INTERNEURONS

The action of the giant cholinergic interneurons on the spiny cells of the neostriatum are known mostly through studies of the direct application of ACh to neurons in vitro. Again, there are multiple receptors known to exist in the neostriatum, and the relative contributions of each to synaptic transmission between the neurons are unknown. Nicotinic cholinergic receptors have been reported to contribute to the fast EPSP evoked in neostriatal neurons by local stimulation in slices. Muscarinic receptors have been reported to underly slow depolarizations or hyperpolarizations, the first probably due to a decrease in potassium conductance, and in either case simultaneously causing a decrease in the amplitude of fast EPSPs elicited by intrastriatal stimulation (Kita et al., 1985c; Dodt and Misgeld, 1986). As in the case of the D1 dopamine receptor, the muscarinic effects of ACh do not indicate a classical neurotransmitter function, but rather a way of altering the membrane properties of the spiny neuron, and thereby its responsiveness to other inputs.

The responses of the giant aspiny interneuron to cortical and thalamic inputs have also been described (Wilson et al., 1990). These cells exhibit less complicated EPSPs, with fast rise times and little or no late response components. The maximal amplitudes of the EPSPs are much smaller than they are for the spiny neurons. It is possible, however, to detect discrete components in the EPSPs in these neurons, especially with low-intensity stimuli. Thus the EPSPs evoked in the aspiny neurons appear to consist of the action of fewer axons than those in the spiny neurons, with each synaptic contact creating a larger EPSP as observed from the soma. These properties are consistent with there being relatively fewer excitatory afferent synapses on these neurons, and their placement on relatively proximal dendritic shafts and even somata.

DENDRITIC PROPERTIES

The properties of dendrites constitute a theme running through most of the chapters of this book, and perhaps nowhere are they more important than in the spiny neurons of the neostriatum. As noted in the Appendix, in the analysis of the linear properties of dendrites, three electrophysiological parameters are particularly important: the input resistance (R_N), the membrane time constant (tau, τ), and the electrotonic length of the dendritic tree (L). We will discuss how the spiny neuron has served as a case study for the experimental determination of these properties. This will give the student a greater appreciation for the crucial importance of the dendrites in determining the input–output characteristics of the spiny neuron.

For many neurons, one observes much higher input resistances when the cells are recorded in tissue slices than when recorded in intact animals. Two arguments are usually presented to explain this fact. First, tonic synaptic activity, which is generally lost in the preparation of the slices, will act to lower apparent input resistance by the opening of synaptic ionic conductance channels. Second, the damage done by the microelectrode is probably somewhat less under the mechanically more stable conditions of the in vitro recording chamber.

In the case of neostriatal spiny neurons, the input resistance recorded in slices is 20–60 MΩ, which in fact matches very well with that obtained in vivo (Sugimori et al., 1978; Kita et al., 1984; Bargas et al., 1988). The reason is probably that the most important influence on input resistance in spiny neurons is neither damage done by the electrode nor synaptic activity; instead, the biggest determinant is the action of a fast anomalous rectification. This rectification can be seen by passage of current pulses of different amplitudes through the spiny neuron (Fig. 9.11). If the cell were to show no rectification, the *input resistance* (the ratio between the size of the voltage shift and the amplitude of the injected current that causes it) would remain constant regardless of the size of the current pulse (dashed line in the graph on the right in Fig. 9.11): That is, increasing the size of the current pulse in constant steps would produce voltage deflections that also increase by constant steps. In contrast, the usual behavior of spiny neurons is illustrated by the curve shown on the right in Fig. 9.11. These cells show a marked anomalous rectification over their entire natural range of membrane potentials. In anomalous rectification, input resistance decreases with increasing membrane polarization. Contributing to the anomalous rectification in the depolarizing direction is a slowly developing voltage-sensitive depolarization, which appears as a slowly developing depolarization (arrow in Fig. 9.11) super-

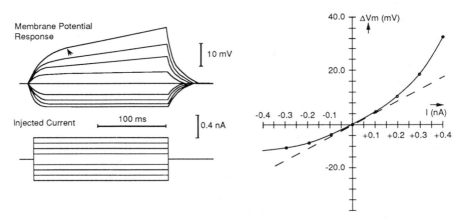

FIG. 9.11. The response of the neostriatal neuron to application of transmembrane current pulses through an intracellular microelectrode. *Left:* The membrane potential response and injected current waveforms for current pulses of eight different but equally spaced intensities. The arrow in the largest response indicates the onset of the slow depolarizing membrane response, which is superimposed upon the linear portion of the response to depolarizing currents. *Right:* The steady-state current–voltage relationship is shown. The dashed line indicates the behavior that would be expected of a linear neuron. The deviations from this line in both depolarizing and hyperpolarizing directions, are typical of that seen in neostriatal spiny neurons.

imposed upon the linear charging curve to depolarizing currents. The onset of anomalous rectification in neostriatal spiny neurons is fast relative to the time constant of the neuron, so it affects the entire time course of the charging curve. (Anomalous rectification is not really an anomaly. It is actually quite common. It is called "anomalous" because it was not expected on the basis of the original biophysical theory of membranes.)

It is difficult to apply the linear cable theory, discussed in the Appendix, to such a nonlinear cell. Certainly, the time constant of the cell cannot be a constant if the membrane resistivity is altered by any shift in membrane potential. It is therefore perhaps not surprising that there has been some disagreement about the time constants of the cells. Recent experiments in tissue slices using small current pulses (to minimize the effects of anomalous rectification) have yielded values near 5 msec for the time constant under these circumstances (see discussion of the significance of the membrane time constant in the Appendix). It is quite possible to increase the time constant to 10 or 15 msec by depolarizing the cells slightly, however, and longer time constants have been obtained with small changes in the extracellular potassium concentration (Bargas et al., 1988).

Again by means of small current pulses in cells recorded in vitro, the electrotonic length of the equivalent cylinder of the dendrites has been measured to be about 1.5–1.8 length constants. This electrotonic length seems very long, and it is especially surprising in view of the relatively short dendrites of these cells (about 200 μm). This result is predicted by cable theory, however, because of the spiny nature of the dendrites. The dendritic spines create a very large surface area. Measurement of this area is made difficult by the small dimensions of the spines. These structures cannot be accurately measured in the light microscope (remember that the resolution of the light microscope allows measurements to the nearest 0.25 μm or thereabouts, whereas the stalks of dendritic spines can be as small as 0.1 μm).

Estimation of the surface areas of neostriatal spiny neurons has been accomplished by using a combination of reconstruction of serial sections from electron micrographs and high-voltage electron microscopy (HVEM) of thick (5 μm) sections (Wilson, 1986b). These measurements have shown that the dendritic field of spiny neurons cannot be approximated as a cylinder, or even a tapered cylinder. Its surface area is dominated by the dendritic spines, being greatest at about 80 μm from the soma, where the dendritic spine density is highest, and tapering off slowly to the tips of the dendrites. This distribution of surface area is shown in Fig. 9.12. Because the dendritic diameter does not change in proportion to the surface area, the leakage of current from the dendrites of the spiny neuron, and the low-pass filtering effect of the dendrites are both increased dramatically by the presence of spines. This effect is shown in Fig. 9.13 for a spiny dendrite with different spine densities. It is clear that the addition of dendritic spines has a very large effect on the effective electrotonic length of the dendrites, even in neurons with very high membrane resistivities. When this is taken into account for the neostriatal spiny neurons, the measurements of time constant (τ), electrotonic length (L), and input resistance (R_N) of the cells can be

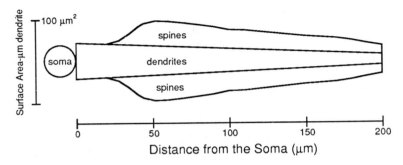

FIG. 9.12. The distribution of membrane on the neostriatal spiny neuron as a function of distance from the soma. The total somatic membrane is represented by the diameter of the circle labeled "soma." The remaining membrane area is shown as a density (μm^2/μm of linear distance from the soma), and the contributions of the spines and dendrites are indicated. (From Wilson, 1986b.)

seen to agree on a membrane resistivity (R_m) ranging from about 5,000 to 15,000 Ω-cm^2, depending upon the state of the anomalous rectification (which itself depends upon membrane potential).

We are now in a position to pull together these considerations of dendritic properties and understand how they contribute to the input–output operations of the spiny neuron. The small amplitudes of individual synaptic responses on spiny neurons are probably due to their placement on dendritic spines, and on the electrotonically long dendrites. Assuming that the dendritic spines act passively under the conditions of these experiments, simulations of neostriatal dendritic spines predict that individual (unitary) EPSPs as recorded from the soma are indeed very weak (Wilson, 1984). The large maximal amplitude of the EPSPs evoked by afferent stimulation can be explained by the large number of inputs converging on each cell, and by the nonlinear behavior of the neostriatal spiny neuron when depolarized by large inputs. As already described, the input resistance of spiny neurons is very dependent upon membrane potential, due to the action of a fast anomalous rectification. When large numbers of synapses depolarize the neuron, its membrane resistance increases. The effect of the increased resistance will be to shorten the dendrites electrotonically, and make the cell more sensitive to subsequent inputs. In addition, large depolarizations can be expected to trigger slow depolarizing potentials in the dendrites. These slow potentials are due to calcium currents (I_L; see Chap. 2) that can only be activated in neostriatal neurons by large depolarizations. Whereas EPSPs recorded in the somata may not seem large enough to trigger these currents, EPSP amplitudes are much greater in the dendrites (due to the dendritic location of the synapses responsible for the EPSPs). Thus the largest EPSPs observed may in fact represent a combination of synaptic and voltage-sensitive components (Kita et al., 1985a; Cherubini and Lanfumey, 1987). These largest EPSPs probably also reflect a contribution due to voltage-sensitive glutamate receptors (NMDA recep-

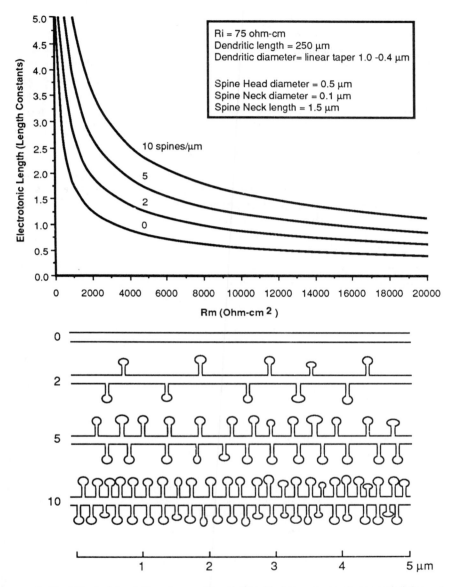

FIG. 9.13. Effect of dendritic spines on the relationship between membrane resistivity and electrotonic length for dendrites. Spine densities for each of the curves are indicated below. (Modified from Wilson, 1988.)

311

tors) that are not activated by small EPSPs which occur on a background of hyperpolarization (Cherubini et al., 1988).

When acting together, these features of the spiny neuron can be expected to make it relatively insensitive to excitatory inputs, and resistant to depolarization when the membrane is hyperpolarized. When the level of excitation in a portion of the dendritic tree becomes somewhat higher, however, the dendritic properties of the spiny neuron can be expected to change in a way that makes it much more sensitive to excitatory synaptic excitation.

FUNCTIONAL OPERATIONS

It is clear that progress is being made toward a complete analysis of the synaptic connections, membrane properties, and cell morphology of the neostriatal spiny neuron. We turn finally to consider the insight this information provides into the firing patterns that underly the normal functions of the neostriatum.

NATURAL FIRING PATTERNS

Neostriatal spiny neurons exhibit a very characteristic firing pattern. Even in unanesthetized and behaving animals, the cells are usually silent. Occasionally, the cells fire a train of several action potentials that lasts from 0.1 to 2.0 sec, and then become silent once again (see Fig. 9.14, top). The train of action potentials is not really stereotyped enough to be called a burst. The discharge rate during the episode of firing usually does not become greater than 40 per second, and the cells usually do not fire rhythmically during the episode (suggesting that firing rate is not limited by spike afterpotentials). In behaving animals, these episodes of firing are sometimes locked to the onset of movements (e.g., DeLong, 1973; Kimura et al., 1984).

Why are the cells silent so much of the time? The anatomical arrangement of the neostriatal spiny neuron and of the neostriatum itself indicates that there are thousands of excitatory synapses on each spiny neuron. The results of intracellular recording experiments described above suggest that many of these afferents are active under ordinary conditions, and the cells are normally depolarized by 5–20 mV above the resting potential. Otherwise, removal of afferent activity could not cause a hyperpolarization of this size. Why are these synapses not sufficient to cause the cell to fire? One reason is simply the high threshold of spiny neurons. Other influences apparently acting to reduce the effects of the synaptic activation that is present are the low-input resistance of the neuron (due mostly to the large amount of anomalous rectification exhibited by these cells in the usual range of membrane potentials), the electrotonically extended nature of its dendrites (as discussed above), and perhaps the action of inhibitory synaptic input.

Intracellular recording experiments show that when spiny neurons do fire, the train of action potentials arises from a prolonged episode of depolarization accompanied by an increased synaptic noise (Wilson and Groves, 1981). The depolarizing episodes also show several other revealing characteristics. They cannot be triggered by depolarizing current pulses; they cannot be terminated by hyperpolarizing current pulses; and they do not disappear when the cell is hyper-

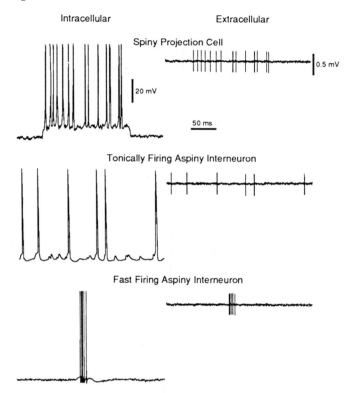

FIG. 9.14. Representative firing patterns of three kinds of neostriatal neurons. The patterns of activity as seen in extracellular recording are illustrated on the right, and the intracellular membrane potentials that lead to these firing patterns are illustrated on the left.

polarized below spike threshold by passage of current into the soma. In these ways, the depolarizing episodes are reminiscent of the period of depolarization that follows the long-lasting hyperpolarization after stimulation of neostriatal afferent pathways (see Fig. 9.10). This similarity suggests the possibility that the naturally occurring episodes of depolarization are caused by an increase in synaptic input to the neuron. When excited electrically, large numbers of afferent fibers are excited synchronously. Under more natural conditions, such large-scale synchrony in the corticostriatal or thalamostriatal pathway is unlikely. In anesthetized animals, and perhaps also in waking animals, there may, however, be local corticostriatal and thalamostriatal synchrony—that is, correlations in firing of small converging subsets of corticostriatal and thalamostriatal neurons that may be important for shaping the firing patterns of the cells (Katayama et al., 1980). But the sharp onsets and offsets of the depolarizing episodes, and their relatively constant amplitudes, are difficult to explain by synaptic synchronization alone. These features probably require an interaction between synaptic activity and the membrane properties of the spiny neuron. For example, anomalous rectification, and the other membrane properties described above, should act to sharpen the onset and offset of the depolarizing episodes by increasing the sensitivity of a neuron to inputs as it becomes more depolarized, and by decreasing the sensitivity to inputs as the cell becomes polarized again.

Most neostriatal neurons are spiny cells, and nearly all recordings of single neurons that are obtained from the neostriatum are from spiny cells. It is therefore not necessary to obtain anatomical verification of the cell type associated with responses that are common to most neurons. On the other hand, identification of firing patterns and synaptic responses of aspiny neurons requires intracellular staining for the determination of cell type. Some recordings of identified *giant aspiny neurons* have been reported (Wilson et al., 1990). The firing pattern observed for these cells is illustrated in Fig. 9.14 (middle). They are tonically active cells, firing irregularly with average rates less than 20 Hz. Action potentials from these cells arise from brief depolarizations that resemble unitary synaptic events. These vary in size, but are often as large as 5 mV. Action potentials arise from the largest of these, but often are triggered by the summation of two or three. At least some of these depolarizations are probably synaptic potentials evoked by afferent fibers, because stimulation of the cortex or thalamus can briefly increase the rate of their occurrence, whereas the rate falls during the period of long-lasting inhibition that is seen in spiny neurons. In extracellular recordings from behaving monkeys, neurons with firing patterns like that of the giant aspiny neuron have been reported to show a unique kind of response during execution of learned movements. Unlike the spiny neurons, which fire in relation to the movement itself, tonically firing neurons fire in relation to the sensory cue that triggers the movement (Kimura et al., 1984). This is not to say that the tonically active neurons are sensory cells. They do not respond to the same stimulus when it is not a signal for initiation of a movement.

Even less complete observations of identified medium-sized aspiny neurons are available. Some recordings of intracellularly stained neurons with the appearance of the varicose dendrite interneuron have been made, and they show that it fires in short bursts with high spike frequency (Fig. 9.14, middle) (Galarraga, personal communication). Cells firing in this pattern can be seen also in extracellular recordings from the neostriatum, although their activity in relation to movement is unknown.

The end result of activity in the neostriatum must be expressed as a change in the activity of those neurons in the globus pallidus and substantia nigra that receive an input from the neostriatum. Although the spiny neurons contain many peptides and other substances, physiological studies indicate that the primary fast effect of activity in spiny neurons is a GABAergic inhibition of the target cells. Those cells in globus pallidus and in substantia nigra pars reticulata that are the target of this inhibition have very high rates of tonic activity. It is not clear whether excitatory input is required for these cells to fire as they do, but in any case they usually fire rhythmically at a rate determined primarily by their own membrane characteristics (Nakanishi et al., 1987). Inhibition exerted by (usually silent) neostriatal spiny neurons during their brief episodes of firing causes a momentary decrease in the rate of this tonic firing. The fast firing cells of the globus pallidus and substantia nigra are themselves GABAergic neurons which exert a tonic inhibition on their target cells in the thalamus and superior colliculus (e.g., Ueki, 1983; Deniau and Chevalier, 1985). Thus excitation of neostriatal spiny neurons leads to a disinhibition of the otherwise suppressed activity of neurons in the thalamus and superior colliculus.

COMPLEX INTEGRATIVE TASKS

Pauses in the firing of neurons in the globus pallidus and substantia nigra are observed in behaving animals, and the behavioral conditions that lead to their occurrence offer important clues to the function of the basal ganglia.

An example from an experiment by Hikosaka and Wurtz (1983) is shown in Fig. 9.15. In this experiment, the firing of substantia nigra neurons was recorded simultaneously with eye movement in awake, behaving monkeys. Responses to eye movements are shown under two conditions. In one, shown in Fig. 9.15A, the monkey has been trained to gaze at a point of light (fixation point) until the light is turned off. The timing of the fixation light is indicated by line F in Fig. 9.15A. After the fixation light is turned off, the monkey is to make a sudden eye movement (saccade) to a point at a new position, the target position. The target position must be remembered, because the target light is not turned on at the time that the eye movement is to be made. The target light is turned on later (at the

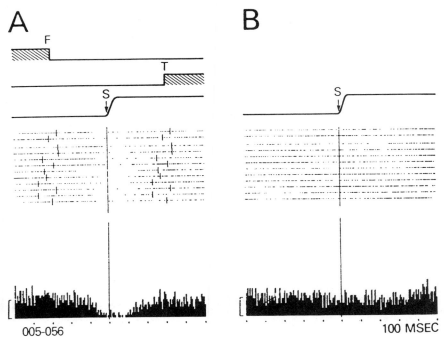

FIG. 9.15. The firing of a substantia nigra neuron during an eye movement to a remembered position as part of a learned response (**A**), and during a comparable movement made spontaneously (**B**). (From Hikosaka and Wurtz, 1983.)

time marked T in Fig. 9.15A), and the eye movement is made in the dark. In the other condition, there are no lights at all, and the monkey makes spontaneous movements. The experimenters selected trials in which the spontaneous movement was similar to the movements that occurred under the more controlled conditions. Responses under these conditions are shown in Fig. 9.15B. The firing of the neuron is indicated in two ways. In one, it is shown as a series of horizontal raster lines, in which each line represents a single trial, and each point is a single action potential. All responses are aligned to the movement, with small vertical lines indicating the exact timing of the fixation and target lights for each trial. Below these is a histogram, in which the number of action potentials summed for all trials is plotted against time relative to the eye movement.

In both cases, eye movements occur in the dark, with no immediate sensory stimulus to trigger or guide them. The motor details of the eye movement are also the same in Fig. 9.15A and B. The responses of the neuron are dramatically different, however. In Fig. 9.15A, when the eye movement is to a remembered target and the performance of the eye movement is motivated by a reward, the substantia nigra neuron shows a profound pause in firing, almost certainly due to increased activity in the striatonigral pathway. In Fig 9.15B, when the eye movement is unrelated to the task and unrewarded, the substantia nigra neuron shows no change in firing.

Although not all neurons responded in the way shown in Fig. 9.15, none of the substantia nigra responses studied by Hikosawa and Wurtz (1983) were readily categorized as strictly sensory or motor. Instead, the responses of substantia nigra neurons during eye movements reflected more global circumstances of the experiment, including elements of what are usually considered cognitive and motivational, as well as sensory and motor, factors. These observations, and similar ones obtained in other nuclei of the basal ganglia, seem to give eloquent testimony to the convergence of afferent input from diverse cortical and thalamic regions, and processing of that information by the intrinsic circuits of the neostriatum.

10

OLFACTORY CORTEX

LEWIS B. HABERLY

We turn now to the synaptic organization of the cerebral cortex. Although the neocortex, which forms the convoluted mantle of the human brain, is often considered to be synonymous with the cerebral cortex, there are two additional types of phylogenetically old cerebral cortex that have attracted a great deal of experimental interest in recent years. These are the olfactory cortex (paleocortex) and hippocampal formation (archicortex). These areas, which are predominantly three-layered in mammals, display a parallel phylogenetic development. They can be identified with scrutiny in sharks, are clearly recognizable but structurally primitive in amphibians, and in reptiles display a striking similarity to the corresponding regions in mammals in cytoarchitecture, cell types, and axonal connections (reviewed in Haberly, 1990). By contrast, the six-layered neocortex is present only in mammals.

An often posed question is the nature of the evolutionary relationship between the phylogenetically old cortices and the neocortex. Unfortunately, there are no extant descendants from the reptile-to-mammal evolutionary series that can provide a direct answer to this question, and consequently, arguments concerning this relationship are indirect and speculative (see Haberly, 1990). Regardless of the nature of the evolutionary link, however, what is important for our purposes is that the three types of cortex display many common features. Over the following chapters it will become apparent that paleocortex, archicortex, and neocortex show striking similarities in cellular morphology and physiology, neurochemistry, synaptic relationships, and local circuitry. These similarities have allowed the phylogenetically old types of cortex with their easily analyzed, precise laminar organizations to serve as model systems for cerebral cortical function. As described at the end of this chapter, the olfactory cortex is particularly well-suited for addressing questions concerning associative memory function. Increasing use is also being made of the olfactory cortex for study of the neural basis of epilepsy because of its great susceptibility to seizure activity (see Piredda and Gale, 1985; McIntyre and Wong, 1986; Racine et al., 1988).

Olfactory cortex is usually defined as those areas that receive direct synaptic input from the olfactory bulb (see Chap. 5). As in the neocortex and hippocampal formation, many different olfactory cortical areas can be distinguished on the basis of differences in cytoarchitecture and connectivity (Fig. 10.1) (see reviews

A. Olfactory areas

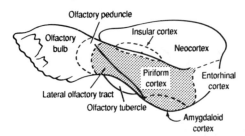

D. Insular cortex, Thalamus, Hypothalamus

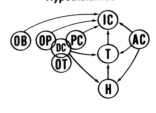

B. Olfactory bulb input

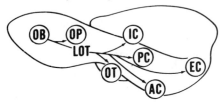

E. Hippocampal formation

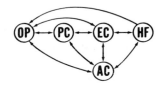

C. Associational connections

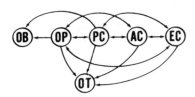

F. Cholinergic, Monoaminergic

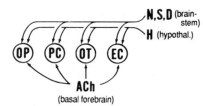

FIG. 10.1. Olfactory cortical areas and connections. **A.** Lateral aspect of opossum brain. The stippled area is the piriform cortex, the largest olfactory cortical area. **B.** Projection areas of the olfactory bulb (OB) via the lateral olfactory tract (LOT). AC, amygdaloid cortex; EC, entorhinal cortex; IC, insular cortex; OP, olfactory peduncle; OT, olfactory tubercle; PC, piriform cortex. **C.** Associational connections of olfactory cortical areas and projections to the olfactory bulb. **D.** Inputs to the insular cortex (IC), thalamus (T), and hypothalamus (H) from olfactory areas. DC, deep cells in PC, OT, and OP. **E.** Connections between the hippocampal formation (HF) and olfactory cortical areas. **F.** Monoaminergic and cholinergic inputs to olfactory cortex. N, S, D, noradrenergic, serotonergic, and dopaminergic cells in the brainstem; H, histaminergic cells in the hypothalamus; ACh, cholinergic cells in the basal forebrain.

by Switzer et al., 1985; Price, 1987). The largest olfactory area, on which this chapter will focus, is the piriform cortex (also termed the pyriform or prepyriform cortex). Other cortical regions that receive direct olfactory bulb input include small areas within the olfactory peduncle (anterior olfactory "nucleus," tenia tecta, and dorsal peduncular cortex), the olfactory tubercle, entorhinal cortex, insular cortex, and cortical areas associated with the amygdala (Fig. 10.1A,B) see Switzer et al., 1985; Price, 1987). Even though the insular cortex and entorhinal cortex can be considered as olfactory cortex by the traditional definition in terms of olfactory bulb input, these areas cannot be considered to be paleocortex. The insular area is clearly neocortical in its laminar structure, and the entorhinal cortex has a unique structure comprising four or five laminae that precludes categorization as paleocortex, archicortex, or neocortex. In some schemes it is referred to as juxtoallocortex, between allocortex (three-layered cortex) and isocortex (six-layered cortex) (see Pribram and Kruger, 1954).

NEURONAL ELEMENTS

The cytoarchitecture and major neuronal elements of the piriform cortex are illustrated in Fig. 10.2. This diagram is based on classical anatomical methods and recent studies employing intracellular dye injections.

CYTOARCHITECTURE

The piriform cortex is usually described in terms of three layers (Fig. 10.2). Layer I is the superficial plexiform layer that contains dendrites, fiber systems, and a small number of neurons. It has been divided into a superficial part, layer Ia, that contains afferent fibers from the olfactory bulb, and a deep part, layer Ib, that contains association fibers from other parts of the piriform cortex and other olfactory cortical areas (Price, 1973).

Layer II is a compact layer of cell bodies. It can be divided into a superficial part, layer IIa, in which semilunar cells are concentrated, and a more densely

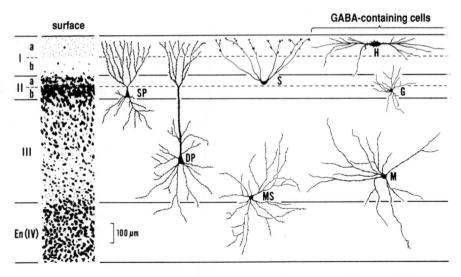

FIG. 10.2. Cytoarchitecture and major cell types of piriform cortex. SP, superficial pyramidal cell; DP, deep pyramidal cell; S, semilunar (pyramidal-type) cell; MS, multipolar cell with spiny dendrites; M, multipolar cell with smooth dendrites; H, superficial horizontal cell; G, small globular soma cell. SP, DP, and S cells are confined to the layers indicated; MS and M cells are found also in the endopiriform nucleus (En); G cells are found in all layers. See Figs. 10.5 and 10.6 for afferent and intrinsic fiber systems.

packed deep part, layer IIb, which is dominated by pyramidal cell somata (Haberly and Price, 1978a).

Layer III displays a marked gradient in structure from superficial to deep (Cajal, 1955; Valverde, 1965; Haberly, 1983). Its superficial part contains a moderately high density of pyramidal cells and large numbers of basal dendrites descending from pyramidal cells in layer II. With increasing depth, the number of cells decreases, the density of myelinated and unmyelinated axons increases, and multipolar cells become the predominant population.

Deep to layer III is the endopiriform nucleus, which is extensively interconnected with the overlying cortex. This nucleus, either alone or together with the deep part of layer III, has also been termed layer IV of the piriform cortex (O'Leary, 1937; Valverde, 1965). The endopiriform nucleus contains multipolar cells that closely resemble those in the deep part of layer III, but are more densely packed (Tseng and Haberly, 1989a).

INPUTS

The mitral cells in the olfactory bulb are the major source of afferent input to the piriform cortex and other olfactory cortical areas. The olfactory peduncle, olfactory tubercle, and adjoining portion of the anterior piriform cortex also receive a projection from tufted cells (Haberly and Price, 1977; Skeen and Hall, 1977). Axons of mitral and tufted cells reach the piriform cortex and other olfactory areas by way of the lateral olfactory tract (LOT), which extends caudally to approximately the middle of the piriform cortex (Fig. 10.1A). Axons in this tract are predominantly myelinated but small in diameter (mean of 1.3 μm in the rat; Price and Sprich, 1975).

LOT axons give rise to large numbers of thin myelinated and unmyelinated collaterals. These leave the LOT throughout its length and spread obliquely across the entire surface of the piriform cortex and other olfactory areas (Devor, 1976). Synaptic terminals of these afferent axons are found exclusively in layer Ia, which has a sharply defined, deep boundary with layer Ib (Price, 1973; Schwob and Price, 1984). This mode of afferent input via tangentially spreading superficial fibers is shared by portions of the hippocampal formation, but contrasts with the neocortex where input fibers enter from the deep white matter. It is interesting to note, however, that in primitive mammals certain neocortical areas have afferent input from the thalamus that is predominantly to the superficial part of layer I (Ebner, 1969). Also, in the dorsal cortex of reptiles, which some believe to be homologous to the neocortex of mammals, afferent input fibers spread tangentially in a superficial subzone, just as in the olfactory cortex (Hall and Ebner, 1970).

CENTRAL INPUTS

In addition to the afferent system from the olfactory bulb, the olfactory cortex receives inputs from the basal forebrain, brainstem, thalamus, and hypothalamus (Haberly and Price, 1978a). Projections from areas that contain cholinergic, noradrenergic, serotonergic, dopaminergic, and histaminergic cells are described below under Neurotransmitters (see also Fig. 10.1F).

As in the other two types of cerebral cortex, many olfactory cortical areas are reciprocally connected with each other (Fig. 10.1C) (Haberly and Price, 1978a,b; Luskin and Price, 1983a). The piriform cortex, olfactory peduncle, and amygdaloid cortex also send fibers to the olfactory bulb (Luskin and Price, 1983a). The olfactory tubercle is unique since it receives inputs from, but does not project to, other olfactory cortical areas. Extensive crossed projections also link many olfactory cortical areas, but in contrast to most ipsilateral associational projections these connections are not reciprocal (Haberly and Price, 1978a,b).

OUTPUTS

Output pathways from the olfactory cortex have been described to the neocortex, thalamus, hypothalamus, hippocampal formation, striatum, and limbic system (reviewed by Price, 1985; Takagi, 1986). Widespread parts of the olfactory cortex (Fig. 10.1D), including the pyramidal cell layers of the piriform cortex, project to the insular area of neocortex. A portion of the insular cortex also receives an input from the taste pathway, leading to speculation that it is involved in perception of "flavor" (Shipley and Geinisman, 1984). Projection pathways to the thalamus and hypothalamus originate from a relatively small number of predominantly deep cells concentrated in the medial part of the piriform cortex and the olfactory tubercle (Fig. 10.1D). Return projections to the olfactory cortex from the thalamus and hypothalamus are also present, although not from the cell groups that receive inputs (Haberly and Price, 1978a). A major output of special interest with regard to learning and memory (see Chap. 11) is to the hippocampal formation (Fig. 10.1E). Projections to the hippocampal formation originate in the entorhinal cortex and amygdaloid cortex. Return projections from the hippocampal formation to these areas as well as to the olfactory peduncle are also present. The olfactory tubercle, which receives direct and relayed olfactory inputs, has a close association with the corpus striatum (Heimer et al., 1985), perhaps for a direct olfactory input to motor systems. Finally, the amygdala, which receives olfactory input via its cortical portion, provides input to many other parts of the forebrain, especially other limbic areas that are thought to be involved in control of emotions and visceral functions.

PRINCIPAL NEURON

Pyramidal cells in all three types of cerebral cortex have long been considered to be principal neurons by virtue of their long spiny dendrites and projection axons. As detailed below, however, the axonal projections from pyramidal cells in the piriform cortex are largely via tangentially distributed association fibers to other cortical areas rather than deeply directed projection pathways like those that originate in neocortical areas.

As in the neocortex and hippocampus, pyramidal cells in the olfactory cortex have several distinctive features as shown in Figs. 10.2 and 10.3 (Haberly, 1983). At the light microscopic level these include an *apical* dendritic tree directed toward the cortical surface, a *basal* tree radiating from the cell body, a profusion of small dendritic spines, and a myelinated axon that is deep directed.

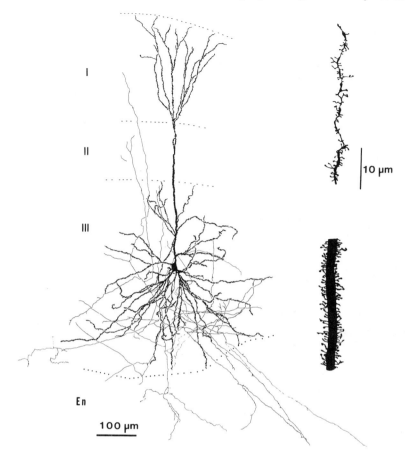

FIG. 10.3. Deep pyramidal cell in layer III of piriform cortex stained
by intracellular injection of Lucifer Yellow with antiserum intensification. Fine processes are axon collaterals. Details of distal and proximal apical dendrites are at right. (From Tseng and Haberly, 1989a.)

Pyramidal cells in all three types of cerebral cortex also display strikingly similar
synaptic relationships and common physiological features.

Two populations of pyramidal cells can be distinguished in the piriform cortex:
superficial pyramidal cells in layer IIb, and *deep pyramidal cells* in layer III
(cells SP and DP in Fig. 10.2; Fig. 10.3). Morphologically these two populations
are virtually indistinguishable except for the lengths of their apical dendritic
trunks, but as described below, they display striking differences in membrane
properties and synaptic responses (Tseng and Haberly, 1989a,b).

The *apical dendrites* of most superficial and deep pyramidal cells aborize into
secondary branches near the border between layers I and II and extend through
the afferent fiber aborization zone in layer Ia. As a consequence, the lengths of
apical trunks are determined by the depths of their somata. Most deep pyramidal

cells have single long apical trunks (Fig. 10.3), whereas superficial pyramidal cells have short apical trunks (Fig. 10.2) or secondary dendrites that extend directly from cell bodies. A comparable difference in morphology of superficial and deep pyramidal cells is also observed in the neocortex (Peters and Kaiserman-Abramof, 1970; Ghosh et al., 1988). *Basal dendrites* of both superficial and deep pyramidal cells are predominantly deep-directed and can extend for several hundred microns.

At the superficial border of layer II there are neurons with apical but no basal dendrites that are commonly termed *semilunar cells* (Calleja, 1893). These cells resemble phylogenetically primitive pyramidal cells (Sanides and Sanides, 1972) and therefore have been termed *pyramidal-type* cells. In the rat, the somata of these cells are segregated in layer IIa (Fig. 10.2) (Haberly and Price, 1978a). Ultrastructurally, somata of semilunar neurons closely resemble pyramidal cells (Haberly and Feig, 1983), and like pyramidal cells they have spiny dendrites and deep directed axons. However, in contrast to the profusion of tiny dendritic spines characteristic of mammalian pyramidal cells, spines on semilunar cells in the opossum are comparatively large; originate in rather widely spaced clusters in layer Ia; and have a distinctive flattened profile in electron micrographs (Haberly, 1983; Haberly and Feig, 1983).

INTRINSIC NEURONS

Several different populations of neurons can be identified in the piriform cortex with morphological characteristics that clearly distinguish them from pyramidal cells. Golgi studies suggest that some of these cells have axons that arborize only locally and thus could be considered to be intrinsic neurons. Indirect evidence from axon tracing studies suggests that others are projection neurons and may therefore qualify as principal neurons, but the lack of direct evidence precludes definite conclusions.

As in all parts of the cerebral cortex, the available Golgi studies (O'Leary, 1937; Cajal, 1955; Valverde, 1965; Haberly, 1983) have revealed a large number of different morphological forms of neurons in the piriform cortex. However, until these morphological data are combined with experimental data on connections, physiological properties, and neurochemistry, it will not be possible to determine how many functionally different neuron populations are present. Given these limitations, the present account will be restricted to a few distinctive types of nonpyramidal cells that clearly subserve different functions.

Multipolar cells. In the deep part of layer III and the subjacent endopiriform nucleus, a large proportion of neurons are of medium to large size and multipolar in form (cells MS and M, Fig. 10.2). A moderate proportion of these neurons have elongated cell bodies and dendritic trees and are commonly termed *fusiform cells.* Dendrites of deep multipolar and fusiform cells can be long, but only rarely extend superficial to layer III (Fig. 10.2). A recent Golgi study in the opossum (Haberly, 1983) has revealed several distinctly different types of dendritic specializations, suggesting a functional diversity. Thus far, however, only a single population of multipolar cells with spiny dendrites has been studied physiologically (see below). Axons of multipolar cells can leave the cell body at any orientation, and can be myelinated and larger in diameter than pyramidal cell axons. Golgi studies (O'Leary, 1937; Valverde, 1965) and intracellular injections (Tseng and Haberly, 1989a) have revealed that axons of these cells give rise to many local collaterals that can extend to the surface of the cortex. Indirect evidence from retrograde transport of horseradish peroxidase (HRP) suggests that many of these cells project over long distances within the piriform cortex and perhaps to other areas. Immunocytochemical studies (see below) indicate that a large proportion (perhaps one third to one half) of these cells use GABA as their neurotransmitter, and that smaller numbers contain a variety of neuropeptides.

Small globular cells. Small stellate cells with globular somata (cell G in Fig. 10.2) can be found in all layers of the piriform cortex, but in layers I and II they are the most abundant form of nonpyramidal cell. Dendritic morphology has been studied only in the opossum (Haberly, 1983); in this species, dendrites of these cells can display an "axoniform" appearance by virtue of fine-caliber varicose shafts that branch at right angles. In spite of this appearance, however, electron microscopic study has failed to reveal any indication that they can serve as presynaptic elements (Haberly and Feig, 1983; Haberly et al., 1987). Axons of small globular cells are fine in caliber, unmyelinated, and often highly branched in the vicinity of the cell body. One variant of this cell type may give rise to the *chandelier axons* that synapse on axon initial segments of pyramidal cells in the piriform cortex (Somogyi et al., 1982), as in the hippocampus and neocortex (Chaps. 11 and 12). A high percentage of small globular soma cells contain GABA and its synthetic enzyme, glutamic acid decarboxylase (GAD) (Haberly et al., 1987), and presumably mediate an inhibitory effect.

Superficial horizontal cells. A distinctive type of horizontal cell is found exclusively in layer Ia of the anterior piriform cortex in the vicinity of the LOT (cell H in Fig. 10.2). In the opossum, where they have been studied with Golgi, immunocytochemical, and electron microscopic methods (Haberly, 1983; Haberly and Feig, 1983; Haberly et al., 1987), they are the largest neurons in the cortex, give rise to the largest diameter axons, have horizontally directed dendrites largely confined to the afferent fiber layer (Ia), and display somatic spines that receive synapses apparently from olfactory bulb afferents. They all contain GABA and GAD. These cells bear a close resemblance to Cajal–Retzius cells that are numerous in embryonic neocortex in layer I, but largely disappear before adulthood (see Chap. 12).

CELL POPULATIONS

The number of pyramidal cells in the piriform cortex has not been rigorously determined, but almost certainly exceeds 10^6 in lower mammals (Haberly, 1985). Because there are only approximately 6×10^4 mitral cells (Chap. 5), it can be concluded that each afferent axon contacts an average of at least 15 principal neurons in the piriform cortex. In fact, given the highly branched nature of mitral cell axons (Ojima et al., 1984) and the distributed projection pattern of the LOT (see below), it is likely that the divergence ratio is very much higher than 1 : 15. This situation contrasts sharply with the olfactory receptor input to the olfactory bulb, where large numbers of receptor neurons converge onto a much smaller number of principal (mitral and tufted) cells (see Chap. 5).

The intrinsic association fiber systems in the piriform cortex bear a likeness to the afferent input in terms of divergence ratio. Although no numbers can yet be given, it is clear from light and electron microscopic studies with extracellularly and intracellularly injected tracers (reviewed in Haberly, 1985) that each pyramidal cell provides synaptic input to a large number of other pyramidal cells within the piriform cortex as well as in other olfactory cortical areas. Conversely, each pyramidal cell receives input from a large number of other pyramidal cells.

SYNAPTIC CONNECTIONS

In general, the synaptic relationships that have been observed in the piriform cortex closely resemble those in both the hippocampal cortex and neocortex. This is especially apparent for pyramidal cells for which the most extensive data are available (Fig. 10.4). As in other parts of the nervous system (see Chap. 1), two major categories of synapses on pyramidal cells can be distinguished: asymmetrical (Gray's type 1) synapses with associated spherical vesicles, and symmetrical (Gray's type 2) synapses with associated pleomorphic vesicles. In some systems asymmetrical synapses have been shown to mediate excitation, and symmetrical synapses to mediate inhibition (Chap. 1). Asymmetrical synapses on pyramidal cells in the piriform cortex are concentrated on dendritic spines and are completely excluded from cell bodies and axon initial segments (Westrum, 1969; Haberly and Behan, 1983; Haberly and Feig, 1983; Haberly and Presto, 1986). Symmetrical synapses are found at high density on axon initial segments and at a lower density on cell bodies and shafts of both apical and basal dendrites out to their distal ends (Westrum, 1970; Haberly and Feig, 1983; Haberly and Presto, 1986.) In absolute terms, the bulk are on the dendritic tree, and the number per unit area of membrane is higher on distal than proximal apical dendrites (Haberly, Presto, and Feig, unpublished). This latter observation is consistent with the finding of a higher concentration of GABAergic synaptic terminals in layer Ia than Ib (see below). Dendritic spines can display considerable variability, but most have small terminal knobs, thin necks (Figs. 10.3 and 10.4), and contain a variable number of small vesicles and flattened sacks termed a *spine apparatus* (Haberly, 1983; Haberly and Feig, 1983; Haberly and Presto, 1986).

Studies with degeneration (Westrum, 1969; Caviness et al., 1977) and EM autoradiographic (Haberly and Behan, 1983) methods have revealed two main types of asymmetrical synapses (Fig. 10.4). Synapses from mitral cell axons onto distal apical dendrites of pyramidal and semilunar cells in layer Ia have a low, regular density of vesicles and are often wrapped by glial processes. By contrast, synapses from pyramidal cell axons onto other pyramidal cells in layers Ib and III have a high density of vesicles near synaptic contacts and lack glial wrapping. It

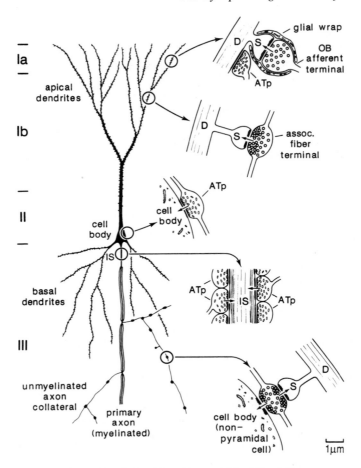

FIG. 10.4. Synaptic relationships of pyramidal cell in piriform cortex.
ATp, axon terminals with pleomorphic vesicles; D, dendritic shaft; IS,
axon initial segment; S, dendritic spines.

has been postulated that the difference in vesicle packing densities is responsible,
in part, for the difference in facilitation properties of these two synapse populations that has been observed in vitro (Bower and Haberly, 1986).

   In contrast to pyramidal cells, most small globular and deep multipolar cells
have asymmetrical synapses on their cell bodies (Haberly and Feig, 1983;
Haberly et al., 1987.) The density of both asymmetrical and symmetrical synapses on somata and dendritic shafts is extremely variable, but can be much
higher than on pyramidal cells. Electron microscopic observations of intracellularly stained cells indicate that some of the asymmetrical synapses on deep
multipolar cells take origin from pyramidal cell axon collaterals (Haberly and
Presto, 1986).

BASIC CIRCUIT

The synaptic organization of the piriform cortex is summarized in the circuit diagrams in Figs. 10.5 and 10.6. Figure 10.5 illustrates the excitatory inputs to pyramidal cells for which there is now a substantial body of supporting anatomical and physiological evidence. Figure 10.6 summarizes inhibitory interactions, for which the evidence concerning the anatomical substrate is more limited and indirect. As will be seen in Chaps. 11 and 12, there is a similarity to the basic circuits for the other main types of cerebral cortex, hippocampus and neocortex.

EXCITATORY SYSTEMS

Laminar organization. Afferent axons arriving in the LOT synapse exclusively in layer Ia on distal apical dendritic segments of pyramidal and semilunar cells (Figs. 10.2 and 10.5) (Heimer, 1968; Haberly and Shepherd, 1973; Price, 1973; Haberly and Behan, 1983). In the entorhinal cortex, LOT inputs to small GABA-containing stellate cells in layer I have been experimentally demonstrated (Wouterlood et al., 1985). In the piriform cortex, indirect anatomical (Haberly et al., 1987) and physiological evidence for feedforward loops (see below) suggests that similar superficial GABA-containing cells also receive direct afferent input (Fig. 10.6).

Axons of superficial pyramidal cells give rise to large numbers of locally arborizing unmyelinated collaterals. These make asymmetrical synapses in layer III on dendritic spines of pyramidal cells (Figs. 10.4 and 10.5) and on dendritic shafts of nonpyramidal cells in the deep part of this layer and in the endopiriform nucleus (Fig. 10.6) (Haberly and Presto, 1986). Deep pyramidal cell axons also give rise to large numbers of local collaterals (Fig. 10.3), but their synaptic targets have not yet been identified.

Most collaterals of superficial and deep pyramidal cell axons remain in layer

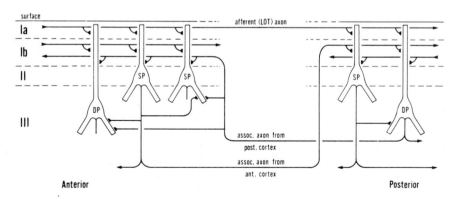

FIG. 10.5. Basic circuit for piriform cortex: excitatory connections of pyramidal cells. SP, superficial pyramidal cells; DP, deep pyramidal cells.

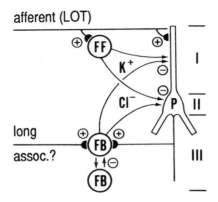

FIG. 10.6. Basic circuit for piriform cortex: inhibitory connections. FB, feedback interneuron; FF, feedforward interneuron; P, pyramidal cell; +, −, excitatory and inhibitory synapses, respectively.

III for 1 mm or more (Haberly, 1985), then ascend to layer I where they make synapses resembling those in layer III on spines of pyramidal cell apical dendrites (Fig. 10.4) (Haberly and Behan, 1983). These synapses are in layer Ib on proximal and middle parts of apical dendrites with minimal overlap into layer Ia where afferent fibers terminate (Fig. 10.5) (Haberly and Shepherd, 1973; Price, 1973; Schwob and Price, 1984; Haberly and Behan, 1983.) Physiological studies have demonstrated that these long collaterals, like afferent fibers, mediate excitatory postsynaptic potentials (EPSPs) in both superficial and deep pyramidal cells (Haberly and Bower, 1984; Tseng and Haberly, 1989a).

Experimental anatomical studies (Haberly and Price, 1978a; Luskin and Price, 1983b) have revealed that pyramidal cells in different parts and layers of the piriform cortex, as well as other olfactory cortical areas, project to different sublaminae within layer Ib. Many details of these associational projections remain to be determined, but Fig. 10.5 illustrates some of the major features demonstrated thus far. Synapses of long collaterals from pyramidal cells in the anterior piriform cortex are concentrated in a superficial sublamina of layer Ib in the posterior piriform cortex as well as in other parts of anterior piriform cortex. By contrast, long collaterals from the posterior piriform cortex project predominantly to the mid to deep part of layer Ib in anterior and posterior parts of the piriform cortex (Fig. 10.5). The projection from the posterior piriform cortex is comparatively heavy locally, but falls off faster with distance than the projection from the anterior piriform cortex.

Study with the collision technique has revealed that single pyramidal cells give rise to both rostrally and caudally directed branches (Haberly, 1978). Axon tracing studies indicate, in addition, that both ipsilateral association fibers and commissural fibers originate from the same pyramidal cells (Haberly and Price, 1978a). A pyramidal neuron thus has a wealth of synaptic targets.

Semilunar cells (Fig. 10.2) also give rise to long associational projections. Available evidence (Haberly and Price, 1978a) indicates that their heaviest projections are to areas outside the piriform cortex. Golgi observations have revealed local axonal arbors (Stevens, 1969; Valverde, 1965), but synaptic targets are unknown.

Horizontal organization. A striking feature of the piriform cortex is that, in contrast to the precise complementary lamination of fiber systems in the vertical (laminar) dimension, in the horizontal dimension (parallel to the cortical surface), projection patterns are highly distributed. Unlike the sensory receiving areas of the neocortex, no point-to-point topographical ordering in either the afferent input or the intrinsic association fiber systems has been detected (Haberly and Price, 1977; Scott et al., 1980; Luskin and Price, 1982, 1983a; Ojima et al., 1984). Nevertheless, these systems are not homogeneously distributed to all parts of the cortex. Broad overlapping patterns have been described in afferent, association, and commissural fiber systems. Although cells are labeled in all parts of the olfactory bulb from injections of retrograde tracers into any area of the olfactory cortex, the density of labeling is not uniform. For commissural projections, the olfactory peduncle (pars lateralis of the anterior olfactory nucleus) projects to the opposite anterior piriform cortex, and the anterior piriform cortex projects to the opposite posterior piriform cortex (Haberly and Price, 1978a,b; Luskin and Price, 1983a). For associational projections, no patterns have been detected in the posterior piriform cortex, but broad overlapping patterns are apparent in inputs to the anterior piriform cortex and neighboring olfactory cortical areas. These associational inputs appear to be organized in terms of distance from the lateral olfactory tract, perhaps for consistent temporal ordering of times of arrival of monosynaptic and disynaptic inputs (Haberly and Price, 1977) (see Dendritic Properties, below).

INHIBITORY SYSTEMS

Recent physiological studies (see Synaptic Actions, below) have shown that synaptically mediated inhibitory processes in the piriform cortex, as in the other types of the cerebral cortex, are diverse and complex. These studies have revealed inhibitory interneurons in feedforward and feedback loops, fast and slow inhibitory processes mediated by Cl^- and K^+, inhibitory inputs to dendrites as well as to cell bodies, and disinhibition (inhibition of inhibitory interneurons). As described earlier (and under Neurotransmitters, below), morphological studies have revealed candidate interneurons and synapses that could mediate the observed effects. Although we are still at an early stage in the synthesis of anatomical and physiological observations that will be required for understanding the functional roles of inhibitory processes in the piriform cortex, we can indicate some possibilities.

GABA-containing cells with dendrites reaching layer Ia are candidates for *feedforward interneurons*—that is, inhibitory cells that are directly excited by afferent fibers in parallel with the cells that they inhibit (Fig. 10.6). Both large superficial horizontal cells and small globular-soma stellate cells (cells H and S in Fig. 10.2) meet these criteria. The large horizontal cells can be postulated to mediate inhibition at very short latency because of their many specializations that would decrease synaptic relay time: afferent input to large somatic spines, thick dendrites, large diameter myelinated axons, and a position subjacent to myelinated afferent fibers in the LOT. Studies in progress (Haberly and Feig) have revealed that their axons give rise to many bouton-studded collaterals in layer I, suggesting that they mediate inhibition onto dendrites. Small stellate cells have no apparent specializations for increasing relay speed; nevertheless, as a consequence of their rather short dendrites and apparent locally arborizing axons, relay times of 3 msec or less are possible (Haberly et al., 1987).

Because pyramidal cell axons are distributed to all layers with the exception of Ia, all but the most superficially placed GABA-containing cells are candidates for *feedback interneurons*—that is, cells that inhibit the cells that excite them (Fig. 10.6). Biedenbach and Stevens (1969) first postulated that deep multipolar cells subserve this function. This hypothesis has been supported by additional indirect physiological evidence (Haberly, 1973a; Satou et al., 1983; Tseng and Haberly, 1988), the presence of GABA in many of these cells (Fig. 10.2) (Haberly et al., 1987), and the demonstration of synaptic input from pyramidal cells (Haberly and Presto, 1986). Based on anatomical considerations, it is likely that there are also interneurons with dendrites that extend into both afferent and intrinsic association fiber layers and therefore participate in both feedforward and feedback pathways (Haberly et al., 1987).

An important question is the extent to which feedback inhibition can be mediated over long distances either by long axons from inhibitory neurons, or indirectly by excitation of inhibitory neurons by long excitatory association fibers (Fig. 10.6). Indirect physiological (Haberly and Bower, 1984) and morphological (Haberly and Behan, 1983) evidence suggests that long projections from inhibitory cells within the piriform cortex are weak, if present, but the demonstration of a projection from a few GABAergic cells in the piriform cortex to the olfactory bulb (Zaborszky et al., 1986) indicates that this possibility should not be ruled out. Electrical stimulation of association axons evokes a strong disynaptic inhibition over long distances (Haberly and Bower, 1984), but it is not known if this effect occurs with normal orthodromic activation because shocks evoke both orthodromic and antidromic propagation of action potentials in association axons.

Physiological evidence indicates that feedback inhibitory interneurons inhibit each other besides inhibiting pyramidal cells (Fig. 10.6) (Satou et al., 1982). The symmetrical synapses that have been observed on somata and dendrites of GABAergic cells (Haberly et al., 1987) could mediate this effect, although the origins of these synapses have not been established.

SYNAPTIC ACTIONS

The lateral olfactory tract is positioned at the surface of the cortex; as a consequence, afferent fibers to the olfactory cortex can be readily activated with shock stimuli. Long association fibers are concentrated in layers Ib and III, so that it is also possible to stimulate this system with appropriately positioned electrodes and thereby examine selectively the actions of pyramidal cell synapses (Haberly and Bower, 1984; Bower and Haberly, 1986; Tseng and Haberly, 1988). As described in previous chapters, the use of shock stimuli provides a powerful tool for analysis of neuronal circuitry because of the synchronous, temporally ordered sequence of synaptic events evoked. Although such impulse stimuli are clearly artificial in nature, recent results described below suggest that findings obtained with this approach in the piriform cortex are directly applicable to understanding the responses to natural stimuli,

EFFECTS OF VOLTAGE-SENSITIVE CHANNELS

Many different voltage-sensitive channels are present in both pyramidal and deep multipolar cells in the piriform cortex. Because these channels can shape responses to synaptic inputs and thereby play a central role in integrative processes, available data will be briefly described before considering synaptic processes.

Plots relating membrane current and voltage (Fig. 10.7B) display nonlinearities in both depolarizing and hyperpolarizing directions for the three types of cells that have been examined: superficial and deep pyramidal cells and deep multipolar cells (Scholfield, 1978a; Constanti and Galvan, 1983a; Tseng and Haberly, 1989b). Although the analysis is far from complete, direct and indirect evidence indicates that voltage-sensitive Ca^{2+}, Na^+, and K^+ channels (Chap. 2) shape these current–voltage curves. In the hyperpolarizing direction, at least two K^+ currents, I_M (Constanti and Galvan, 1983b; Tseng and Haberly, 1989b) and $I_{f.i.r.}$ (Constanti and Galvan, 1983a), contribute to the observed nonlinearities. $I_{f.i.r.}$ may be of limited significance within the normal physiological range, but I_M, which is tonically active at resting membrane potential, induces a "sag" in hyperpolarizing responses, and its blockage by cholinergic agonists may mediate, in part, the synaptic response to acetylcholine (Constanti and

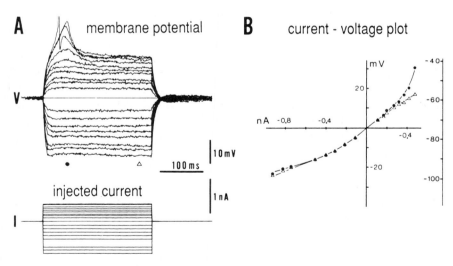

FIG. 10.7. Intracellularly recorded responses of deep pyramidal cell in piriform cortex slice to current injection. **A.** Voltage (V) responses to depolarizing (up) and hyperpolarizing (down) square pulses of current (I). **B.** Plot of voltage versus current at the times indicated by solid circle (●) and open triangle (△) in A. The ordinate is the potential relative to resting membrane potential; the scale at the right is the absolute membrane potential. Note the nonlinear relationship that results from the opening and closing of voltage-sensitive channels (see text). (Modified from Tseng and Haberly, 1989b.)

Galvan, 1983b; Tseng and Haberly, 1989b). In the depolarizing direction, transient and high-threshold Ca^{2+} currents (Halliwell and Scholfield, 1984; Constanti et al., 1985; Tseng and Haberly, 1989b), and one or more unidentified Na^+ currents (Tseng and Haberly, 1989b) appear to contribute. As described below, the long-lasting Ca^{2+} current may exert a strong modulating effect on EPSPs in deep pyramidal and multipolar cells.

There are also channels that are activated by action potentials. These include fast and slow, depolarizing and hyperpolarizing spike afterpotentials (Constanti and Sim, 1987a; Tseng and Haberly, 1988, 1989b). The most prominent of these is a K^+ current activated by Ca^{2+} influx (I_{AHP}). This current is prominent in most deep pyramidal and multipolar cells and generates hyperpolarizing potentials that last for several seconds (Tseng and Haberly, 1989b). Like I_M, it is blocked by cholinergic agonists in the piriform cortex (Constanti and Sim, 1987a; Tseng and Haberly, 1989b). In superficial pyramidal cells, both I_{AHP} and I_M are weak or absent at resting membrane potential in slices (Tseng and Haberly, 1988), but substantial at more depolarized potentials (Constanti and Sim, 1987a).

POSTSYNAPTIC POTENTIALS

Responses of superficial pyramidal cells. Stimulation of either afferent fibers from the olfactory bulb or association fibers from pyramidal cells typically evokes a depolarizing–hyperpolarizing EPSP–IPSP sequence in superficial pyramidal cells, as illustrated in Fig. 10.8 (Biedenbach and Stevens, 1969; Scholfield, 1978b; Satou et al., 1982; Haberly and Bower, 1984). Injection of depolarizing current inverts the late portion of the depolarizing component, revealing two successive hyperpolarizing peaks. Studies in slices with ion substitution and pharmacological agents have revealed that this response consists of at least three components: an initial monosynaptic EPSP that evokes action potentials when of sufficient amplitude, a fast Cl^--dependent IPSP, and a slow K^+-dependent IPSP (Tseng and Haberly, 1988). The first IPSP is depolarizing in polarity at resting membrane potential in slices and therefore exerts its inhibitory effect exclusively via current shunting (Scholfield, 1978b). That is, current from activation of excitatory synapses is diverted by channels opened by inhibitory synapses so that less reaches the site of action potential initiation (see Shunting Inhibition; Chaps. 1, 2, and Appendix). Although the fast IPSP has been reported to be hyperpolarizing in vivo, unpublished observations from our laboratory suggest that when impalements are sufficiently stable to allow recovery from initial damage, membrane potentials are comparable to those observed in vitro and the Cl^--mediated IPSP is depolarizing in sign.

Indirect physiological evidence has been presented that both Cl^-- and K^+-mediated inhibitory processes in the piriform cortex are mediated by both feedforward and feedback inhibitory interneurons (Haberly and Bower, 1984; Tseng and Haberly, 1988). Indirect evidence for feedforward inhibition evoked by LOT axons has also been presented for the entorhinal cortex (Finch et al., 1988). Studies in which synapses on different parts of pyramidal cells were blocked in

Response to stimulation of association fibers

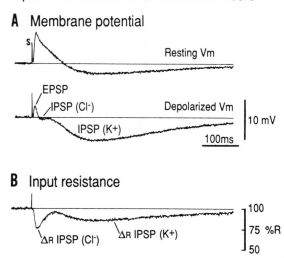

A Membrane potential

B Input resistance

FIG. 10.8. Intracellularly recorded responses of superficial pyramidal cell in piriform cortex slice to shock stimulation of association fibers. Responses to afferent fiber stimulation are similar. **A.** Response at resting membrane potential (top) and at a depolarized membrane potential (bottom) induced by current injection. S, shock artifact. **B.** Percent change in input resistance (R) accompanying responses. Response consists of an initial monosynaptic EPSP and successive Cl^-- and K^+-mediated IPSPs with associated decreases in membrane resistance. Current shunting from the Cl^--mediated IPSP decreases the duration of the EPSP. (Modified from Tseng and Haberly, 1988.)

transversely cut slices (Tseng and Haberly, 1988) have revealed that the K^+-mediated inhibition is concentrated in the dendritic tree whereas a major Cl^--mediated component is generated at or near the cell body (Fig. 10.9).

The basic circuit of Fig. 10.5 leads one to predict that an LOT shock would elicit both monosynaptic and disynaptic EPSPs in superficial pyramidal cells, due to the sequential activation of LOT fibers followed by associational fibers. However, intracellularly recorded responses in vitro (Fig. 10.8) and in vivo (Biedenbach and Stevens, 1969; Haberly and Bower, 1984) display only a single brief EPSP. Studies in which the Cl^--mediated IPSP was delayed or blocked (Haberly and Bower, 1984; Tseng and Haberly, 1988) have revealed that the shunting action of the inhibition evoked by LOT shocks is sufficiently strong to attenuate severely all but the initial part of the monosynaptic EPSP at somatic recording sites. As described below, visualization of dendritic currents by current source–density analysis has confirmed that disynaptic excitatory inputs do occur.

Responses of deep cells. In response to afferent fiber shocks of moderate to high strength, deep pyramidal and multipolar cells display the same three response components observed in superficial pyramidal cells (Fig. 10.8) (Tseng and Haberly, 1989a). In contrast to findings for superficial pyramidal cells, however, the Cl^--mediated IPSP is hyperpolarizing in most deep cells.

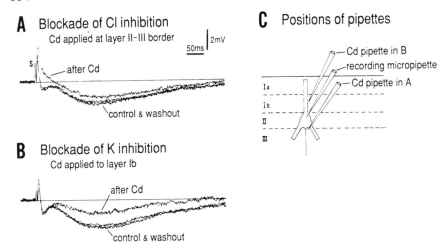

FIG. 10.9. Evidence for spatial segregation of Cl^-- and K^+-mediated IPSPs on super-ficial pyramidal cells in piriform cortex. **A.** Local blockade of synaptic transmission near the cell body by injection of Cd^{2+} from a micropipette suppresses the early Cl^--mediated IPSP (cf. Fig. 10.8). S, shock artifact. **B.** Local blockade of synaptic transmission at the level of the proximal apical dendrite suppresses the slow K^+-mediated IPSP. **C.** Positions of intracellular recording and extracellular Cd^{2+} injection pipettes in the transversely cut piriform cortex slice. (Modified from Tseng and Haberly, 1988.)

The initial EPSP in most deep pyramidal cells is monosynaptically evoked by afferent fiber stimulation as in superficial pyramidal cells. However, in deep multipolar cells whose dendrites do not reach layer Ia, this EPSP is di- or multisynaptically mediated at comparatively long latency. As in superficial pyra-midal cells, direct stimulation of association fibers in layer Ib or deeper layers monosynaptically evokes EPSPs in deep pyramidal and deep multipolar cells (Haberly and Bower, 1984; Tseng and Haberly, 1989a).

A striking feature of both deep pyramidal and multipolar cells that is only rarely observed in superficial pyramidal cells is the presence of EPSPs evoked at long and variable latency by afferent fiber (LOT) stimulation (Fig. 10.10A). These late EPSPs have been recorded intracellularly only in slices, but field potential studies (Ferreyra-Moyana et al., 1988) suggest that they also occur in vivo. The occurrence of these late EPSPs can be greatly facilitated by application of paired shocks (Fig. 10.10B). The mechanism of generation of late EPSPs has not been determined, but an intriguing possibility is that slow regenerative poten-tials in deep pyramidal and multipolar cells contribute to the long delays (Tseng and Haberly, 1989a,b). These slow regenerative potentials, which appear to be due to a long-lasting Ca^{2+} current, can evoke action potentials at long latency which could, in turn, evoke EPSPs in other deep cells. Thus, long response delays could be due to a slow voltage-dependent current in individual cells rather than to a large number of synaptic relays. Such delays could contribute to the analysis of slow temporal patterns in the olfactory code (see Tseng and Haberly, 1989a, and Functional Operations, below).

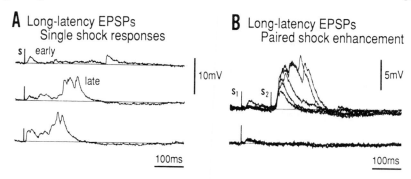

A Long-latency EPSPs
Single shock responses

B Long-latency EPSPs
Paired shock enhancement

FIG. 10.10. Long-latency EPSPs evoked in deep cells in piriform cortex slices by shock stimulation of afferent LOT axons. Intracellularly recorded responses of deep multipolar cells are illustrated; responses of deep pyramidal cells are similar. **A.** Responses to single shocks (s) of identical strength. Note great variability in latency and amplitude of EPSP response components. **B.** Late EPSPs are greatly potentiated in response to paired shocks (S_1, S_2). Top traces are superimposed responses to paired shocks; bottom traces are superimposed responses to first shock alone. Late EPSPs were evoked in this cell only by paired shocks. (Modified from Tseng and Haberly, 1989a.)

SEQUENTIAL ACTIVATION PATTERNS

As a consequence of the horizontal distribution of both afferent and association fiber systems, activation of the piriform cortex and other olfactory cortical areas is sequential rather than synchronous. This has been demonstrated by field potential recordings of responses to shock stimulation. Stimulation of the olfactory bulb output fibers in the LOT evokes a volley that propagates at 6–7 m/sec along the LOT, then spreads out across the remaining surface of the cortex in axon collaterals at 0.8–1.6 m/sec (conduction velocities for cat and opossum; see Kerr and Dennis, 1972; Haberly, 1973b). Shock stimulation of association fibers in layer Ib evokes a high-amplitude field potential associated with the monosynaptic EPSP (Haberly and Bower, 1984; Rodriguez and Haberly, 1989) that propagates across the surface of the piriform cortex at approximately 0.5 m/sec in the opossum with little decrement over distance (Fig. 10.11).

SYNAPTIC PLASTICITY

Another important question is whether memory-related plasticity in synaptic action occurs in the piriform cortex. Recently, it has been reported that when animals are trained to respond to brief trains of LOT shocks delivered at 200-msec intervals (theta frequency; see Chap. 11), long-lasting increases in the amplitudes of monosynaptic EPSPs evoked by single test shocks are induced (Roman et al., 1987). It has also been demonstrated that more sustained shock trains repeated at longer intervals can substantially alter unidentified, long-latency response components (Stripling et al., 1988).

These observations suggest that synaptic plasticity in the piriform cortex can occur, although much work remains to determine which synaptic efficacies can be altered, the mechanism of alteration, and possible roles in memory function.

A Propagated field potential from
association fiber stimulation

B Conduction velocity

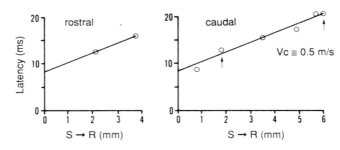

FIG. 10.11. **A.** Stimulation of association fibers in layer Ib of
piriform cortex in vivo evokes a high-amplitude field potential that
propagates over the surface of the cortex. **A.** Extracellularly recorded potentials at cortical surface. Initial negative potential (arrows) is associated with monosynaptic EPSP generated in apical
dendrites. Two responses at different distances from the stimulating electrode are superimposed. S, shock artifact. Afferent fibers
in layer Ia were inactivated by removal of the olfactory bulb one
week earlier. **B.** Latency of the negative peak as a function of
distance in rostral (left) and caudal (right) directions relative to the
stimulation site. Conduction velocity is approximately 0.5 m/sec.
Arrows in caudal plot denote locations of recording sites for potentials in A.

NEUROTRANSMITTERS

EXCITATORY AMINO ACIDS

As for hippocampus and neocortex (Chaps. 11 and 12), there is a rather extensive
literature that is consistent with a role for aspartate and/or glutamate as neurotransmitters in the olfactory cortex. Supporting evidence has been presented from
studies of localization, release, blockade of release by deafferentation, responses
to agonists, and effects of antagonists (reviewed in Collins, 1985).

For synapses of mitral cell afferents in layer Ia there is evidence that suggests a
role for aspartate (Collins and Probett, 1981; Collins, 1985), although a high-
affinity uptake system for aspartate that has been associated with glutamate/
aspartate transmission in other systems does not appear to be present (Walker and

Fonnum, 1983; Fuller and Price, 1988). However, it has been argued that, as for certain GABAergic synapses, this absence may not preclude a neurotransmitter role (Cuénod et al., 1982). A neurotransmitter role for aspartate and/or glutamate in LOT axons has also been questioned on the basis of discrepancies in desensitization (Braitman, 1986) and pharmacological antagonism of responses to LOT stimulation and exogenously applied glutamate and aspartate (Hori et al., 1981, 1982). However, one of the discrepancies in antagonist action can be explained by the recent demonstration of presynaptic receptors (Anson and Collins, 1987), and the concentrations of other antagonists used in some of these experiments may have been too low (Mayer and Westbrook, 1987). The possibility has also been raised that the endogenous dipeptide, *N*-acetylaspartylglutamate (NAAG), is the neurotransmitter for afferent fibers (ffrench-Mullen et al., 1985), but there is, at present, little supporting evidence for this hypothesis (see Collins and Howlett, 1988).

For intrinsic excitatory fiber systems in piriform cortex there is less evidence concerning the role of excitatory amino acids, although studies of Ca^{2+}-dependent release (Collins et al., 1981), effects of antagonists (Collins, 1982; Collins and Howlett, 1988; Tseng and Haberly, 1988), and demonstration of uptake and retrograde transport of D-aspartate by probable pyramidal cell axon collaterals (Watanabe and Kawana, 1984) are consistent with a role for glutamate and/or aspartate.

There have also been a considerable number of studies of the distribution of excitatory amino acid receptor subtypes in the piriform cortex. For afferent fibers from the olfactory bulb, studies with antagonists indicate a role for quisqualate and/or kainate receptors (Hearn et al., 1986; Collins and Howlett, 1988) with some evidence suggesting a preponderance of quisqualate receptors (Collins, 1982; Collins, 1985). This conclusion is supported by the autoradiographic demonstration in layer Ia of a high density of quisqualate receptors (Monaghan et al., 1984; Rainbow et al., 1984) and an apparent low density of kainate receptors (inferred from examination of figures in Monaghan and Cotman, 1982). Studies employing selective antagonists have failed to reveal an *N*-methyl-D-aspartate (NMDA) component in shock-evoked responses to afferent volleys in the LOT (Collins, 1982; Collins and Howlett, 1988), but an anatomical localization study (Monaghan and Cotman, 1985) has revealed a high density of NMDA receptors in layer Ia. Because temporally patterned stimulation is often required to activate NMDA receptors (Mayer and Westbrook, 1987), the lack of response to single shocks does not rule out an NMDA role in synaptic transmission from afferent LOT fibers.

For intrinsic fiber systems in the piriform cortex, there is strong evidence indicating a role for NMDA receptors in synaptic transmission. This includes the demonstration of a late NMDA-mediated component in intracellularly recorded responses to association fiber stimulation (Fig. 10.12) (Hoffman and Haberly, 1989), and evidence from the effects of antagonists on field potentials (Collins, 1982; Collins and Howlett, 1988). Further study will be required to determine if the NMDA-mediated component is directly evoked by synapses from pyramidal cell axons or if a multisynaptic pathway is involved. The initial large EPSP

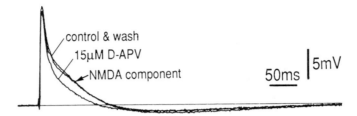

FIG. 10.12. *N*-methyl-D-aspartate (NMDA) receptor-mediated component in response to shock stimulation of association fibers in layer Ib of piriform cortex slice. 2-Amino-5-phospho-D-valerate (D-APV; selective NMDA receptor antagonist) has little effect on the initial part of the response but blocks a late component. (Modified from Hoffmann and Haberly, 1989.)

component that is insensitive to NMDA antagonists (Fig. 10.12) is partially blocked by kynurenic acid (Tseng and Haberly, 1988), suggesting involvement of other subtypes of excitatory amino acid receptors.

GABA

A considerable body of evidence, including studies of postsynaptic action (Scholfield, 1978d; Brown and Scholfield, 1979), Ca^{2+}-dependent release (Collins, 1979; Collins et al., 1981), specific uptake (Brown et al., 1980; Haberly et al., 1987), and effects of antagonists (Scholfield, 1980; Tseng and Haberly, 1988) and other pharmacological agents (Scholfield, 1978c,d; Scholfield, 1980; Brown et al., 1980; Tseng and Haberly, 1988), indicates that GABA acting on $GABA_A$ receptors mediates the Cl^--dependent inhibition observed in neurons in the piriform cortex. An in vitro study (Tseng and Haberly, 1988) has also demonstrated that the slow, K^+-dependent inhibitory process first observed in vivo (Satou et al., 1982) is synaptically mediated, and has provided indirect evidence that this mediation is via $GABA_B$ receptors. Work in our laboratory has subsequently revealed that the slow inhibitory process is blocked by the $GABA_B$ antagonist, phaclofen (Hoffman and Haberly, unpublished).

A study on the opossum piriform cortex using immunocytochemical methods has demonstrated several populations of nonpyramidal cells that contain GABA and its synthetic enzyme glutamate decarboxylase (GAD), as already described (Fig. 10.2, right side). All but the large superficial horizontal cells (cell H in Fig. 10.2) also display high-affinity [^{3}H]GABA uptake (Haberly et al., 1987)—a characteristic of many populations of neurons that are believed to use GABA as their neurotransmitter. Interestingly, pyramidal cells in the olfactory peduncle (pars lateralis of the anterior olfactory nucleus) display high-affinity [^{3}H]GABA uptake and retrograde transport, but do not contain GAD for synthesis of GABA (Haberly et al., 1987).

In a study of layers I and II of the developing and adult rat piriform cortex, Westenbroek and co-workers (Westenbroek et al., 1987, 1988) have described synaptic terminals that are labeled by an antiserum to GAD. These terminals make symmetrical contacts and contain pleomorphic or flattened vesicles. In layer I they contact dendritic shafts and to a lesser extent, necks of dendritic spines; in layer II they contact primarily cell bodies. Such terminals were found at especially high concentrations in layer Ia at the depth of distal dendrites of pyramidal cells. At the light microscopic level, probable GAD-positive terminals have also been demonstrated in layer III of piriform cortex and in the endopiriform nucleus of the rat and opossum (Mugnaini and Oertel, 1985; Haberly et al., 1987). In agreement with Westenbroek and co-workers, these two groups also found many GAD-positive puncta in layer Ia, but they also found a concentration in layer II where Westenbroek and co-workers found few puncta in adult rats using a different GAD antiserum. The finding of many puncta in layer II has been corroborated with an antiserum to GABA, by high-affinity uptake of [^{3}H]GABA (Figs. 3C and 5E in Haberly et al., 1987), and is consistent with physiological findings (Tseng and Haberly, 1988). This suggests that, as in the hippocampus, there is a strong GABAergic inhibitory input to cell bodies or neighboring processes of pyramidal cells as well as to dendrites.

PEPTIDES

Immunocytochemical evidence for the presence of many neuropeptides in neurons intrinsic to olfactory cortical areas has been presented, as for the neocortex and hippocampus. These include somatostatin (Roberts et al., 1982), neuropeptide Y (De Quidt and Emson, 1986), cholecystokinin (CCK) (Greenwood et al., 1981), vasoactive intestinal peptide (VIP) (Loren et al., 1979), neurotensin (NT) (Hara et al., 1982), corticotropin releasing factor (Sakanaka et al., 1987), substance P (Cuello and Kanazawa, 1978; Ljungdahl et al., 1978), and several opioids (e.g., Palkovits and Brownstein, 1985; Fallon and Leslie, 1986; Harlan et al., 1987). Reactivity to an antiserum to neuropeptide K (Valentino et al., 1986) has been demonstrated in axons of extrinsic origin, particularly in the olfactory tubercle. Receptors for CCK (Zarbin et al., 1983), VIP (Magistretti et al., 1988), and somatostatin (Reubi and Maurer, 1985) have also been demonstrated in the olfactory cortex. Indirect evidence has been presented by Inagaki and co-workers (Inagaki et al., 1983a,b) that NT-containing cells in the endopiriform nucleus and adjoining deep part of the piriform cortex project to the medial dorsal nucleus of the thalamus, olfactory peduncle, and nucleus of the diagonal band.

Westenbroek and co-workers have studied neurons and synaptic terminals with CCK-like immunoreactivity in some detail in the piriform cortex of adult and developing rats (Westenbroek et al., 1987, 1988). They have shown that somata of these cells are concentrated in layer II whereas synaptic terminals are concentrated in layers Ib and II. Most CCK-positive synaptic terminals contain pleomorphic vesicles and make symmetrical synaptic contacts onto dendritic shafts and somata, although a few contain spherical vesicles and make asymmetrical contacts onto dendritic spines as also reported for the hippocampus and neocortex (Chaps. 11 and 12).

ACETYLCHOLINE

A series of papers using axon tracing and histochemical methods (reviewed by Luskin and Price, 1982) has established that cholinergic cells in the basal forebrain project to the olfactory bulb and olfactory cortex, as is true for all parts of the cerebral cortex. Recent studies described above have demonstrated that cholinergic agonists acting on muscarinic receptors block I_M and I_{AHP} in superficial and deep pyramidal cells and deep multipolar cells in the piriform cortex, as in the hippocampus (Chap. 11). Blockade of these currents may be responsible, at least in part, for induction of sustained firing and long depolarizing afterpotentials in deep cells (Tseng and Haberly, 1989b), and a variety of excitatory effects described in an early study with extracellular recording methods (Legge et al., 1966).

MONOAMINES

The olfactory cortex, like most parts of the cerebral cortex, receives projections from norepinephrine-, serotonin-, and dopamine-containing neurons in the brainstem (Fig. 10.1F). Evidence includes the presence of axons containing these monoamines (Björklund and Lindvall, 1984; Moore and Card, 1984; Steinbusch, 1984) and retrogradely labeled cells in the appropriate cell groups from HRP injections in the piriform cortex and other olfactory cortical areas (Haberly and Price, 1978a).

A comparatively recent development is the finding of a widely distributed projection from histamine-containing cells in the hypothalamus (Ericson et al., 1987; Watanabe et al., 1984). Support for the presence of this system in the olfactory cortex includes the presence of histaminergic fibers (Airaksinen and Panula, 1988; Inagaki et al., 1988) and a projection to all parts of the olfactory cortex from the tuberomamillary nucleus and other parts of the hypothalamus where histaminergic cells are concentrated (Haberly and Price, 1978a).

DENDRITIC PROPERTIES

SEQUENCE OF SYNAPTIC EVENTS IN PYRAMIDAL CELLS

On the basis of the laminar segregation of excitatory fiber systems in the piriform cortex (see Figs. 10.2 and 10.5), it would be predicted that a series of EPSPs would be generated in different segments of pyramidal cell dendrites following shock activation of the LOT. Application of the current source–density (CSD) technique for computation of membrane currents from depth series of field potentials has confirmed that such a sequence occurs in the opossum (Haberly and Shepherd, 1973; Rodriguez and Haberly, 1989) and rat (Ketchum and Haberly, 1988). The sequence begins with a large-amplitude inward membrane current (sink) in layer Ia that generates the monosynaptic EPSP. This is followed by three sinks at the depths of termination of the three intrinsic association fiber systems from pyramidal cells within the piriform cortex (compare Figs. 10.5 and 10.13). These three sinks are believed to generate disynaptic EPSPs in different dendritic segments of pyramidal cells. A low-amplitude sink in layer III that

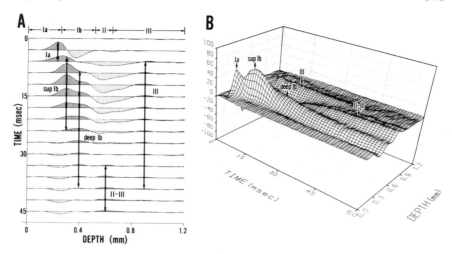

FIG. 10.13. Synaptic events evoked by afferent LOT fiber stimulation in opossum pi-riform cortex in vivo as revealed by current source–density (CSD) analysis. **A.** Plots of net inward (vertical hatching) and net outward (stippling) membrane current as a function of depth in the cortex at a series of times following the LOT shock. Cortical lamination is indicated at the top. **B.** Response in A plotted as a three-dimensional surface. Net inward membrane currents (sinks) in A and B are identified by the layers in which they are maximal (see text). (Modified from Rodriguez and Haberly, 1989.)

starts first is believed to be initiated by local axon collaterals and maintained by long association fibers. A high-amplitude sink that immediately follows in the superficial-most part of layer Ib is generated by the heavy association fiber system that originates in the anterior part of the piriform cortex. Finally, a low-amplitude sink in the mid to deep part of layer Ib is believed to be mediated in part by pyramidal cells in the posterior piriform cortex. These systems can be identified in the basic circuit diagram of Fig. 10.5 and correlated both with the CSD plots of Fig. 10.13 and with the summary diagram of the patterns of extracellular current flows in Fig. 10.14. The three sinks are greatly reduced in amplitude by a preceding conditioning shock timed to block disynaptic re-sponses, which is consistent with mediation by intrinsic association fibers. Fol-lowing the mono- and disynaptically evoked EPSPs, currents associated with the Cl^--mediated IPSP dominate (II–III in Fig. 10.13 and chloride-mediated inhibi-tion in Fig. 10.14).

An intriguing feature is the movement of successive EPSPs in apical dendrites from distal to proximal segments. Assuming passive dendritic properties and generation of action potentials at the axon initial segment, this ordering would, of course, result in an increasing effectiveness of successive EPSPs for action potential generation. It can also be speculated that the occurrence of successive EPSPs in adjacent dendritic segments has special significance in terms of trigger-ing high-threshold, local voltage-dependent conductances—for example, high-threshold Ca^{2+} currents that have been postulated to underlie plastic changes.

CSD analysis has also revealed that both the initial monosynaptic EPSP and

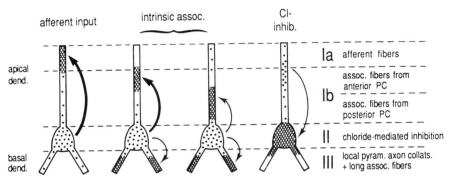

FIG. 10.14. Summary diagram of membrane currents evoked by afferent fiber stimulation in superficial pyramidal cells in piriform cortex as revealed by CSD analysis. Crosshatching denotes sites of net inward membrane current (sinks); stippling denotes sites of net outward current (sources). Layers and dominant fiber systems are indicated at right. (Modified from Rodriguez and Haberly, 1989.)

high-amplitude disynaptic EPSP in the superficial-most part of layer Ib occur at longer latency in the posterior than anterior piriform cortex (Rodriguez and Haberly, 1989), consistent with the sequential activation predicted from the morphology of fiber systems. Based on the reciprocal connections illustrated in Fig. 10.5, one might also expect a return volley from posterior to anterior and perhaps reverberation between anterior and posterior parts of the cortex. However such activity has not yet been detected, and it remains to be determined if the posterior-to-anterior projection is of sufficient strength to allow such reverberating interactions over long distances.

DENDRITIC EVENTS EVOKED BY ODORS

An important question is whether the stereotyped sequence of synaptic events triggered in dendrites by relatively high-strength shocks to afferent fibers is retained when this system is excited by natural stimuli. Given the slow onset and long duration of odor stimuli, it might be predicted that synaptic events would occur in an asynchronous fashion. However, there is a 40- to 60-Hz oscillation induced in the olfactory cortex by odor stimulation (Freeman, 1975), and this provides a possible mechanism for synchronization of neuronal events. It has been postulated that this oscillation reflects a recurring orderly sequence of synaptic potentials similar to that triggered by strong shock stimuli (Haberly, 1985). As a first step in testing this hypothesis, CSD analysis has been performed on responses to very low strength shocks that, as described by Freeman (1975), consist of a long-lasting damped oscillation at approximately the same frequency as odor induced oscillations. Results of this analysis (Fig. 10.15) (Ketchum and Haberly, 1988) suggest that each cycle of the damped oscillation in both anterior and posterior parts of the piriform cortex consists of a monosynaptic EPSP and multiple disynaptic EPSPs, as observed in response to high-strength shocks.

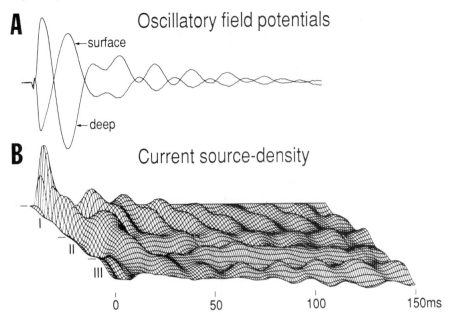

FIG. 10.15. CSD analysis of oscillatory responses to low-strength afferent LOT shocks. **A.** Oscillatory field potentials recorded at surface and deep poles of piriform cortex. **B.** CSD analysis of oscillatory field potentials reveals a stereotyped series of synaptic events within each cycle of the oscillatory response. (Records provided by Dr. Kevin Ketchum.)

Such a periodic temporal convergence of inputs from widespread areas could play a central role in storage and retrieval of olfactory associations (see comments on associative memory, under Functional Operations, below).

INTEGRATION OF SYNAPTIC INPUTS

The studies described above have provided insights into the morphology, membrane properties, and synaptic inputs to pyramidal cells and one population of nonpyramidal cell in the piriform cortex. These insights include considerable detail about the spatial and temporal sequence of excitatory synaptic inputs to dendrites of pyramidal cells. Unfortunately, there are deficiencies in our knowledge that will have to be filled in before it will be possible to fully understand integrative processes in dendrites. One of these is the spatial and temporal distribution of inhibitory inputs. As described above, physiological studies have demonstrated that Cl^-- and K^+-mediated inhibitory processes have different spatial distributions, and morphological studies have revealed a diversity in inhibitory circuits, but essential details are lacking. A second gap in our knowledge of dendritic properties is the spatial distribution of voltage sensitive channels. Studies carried out with micropipettes in or near the soma have revealed the existence of such channels and their strong modulating action on synaptic inputs as also described above, but no data are available on their spatial distributions.

Direct evidence for other cells (e.g., cerebellar Purkinje cells, Chap. 7) and indirect evidence for pyramidal cells including those in piriform cortex (ffrench-Mullen et al., 1983) suggest nonuniform distribution on dendritic spines, dendritic shafts, cell bodies, and axon initial segments. Rigorous modeling of dendritic processes will require determinations of spatial distributions of these channels. See the Appendix for further discussion of these problems.

FUNCTIONAL OPERATIONS

RESPONSES TO ODOR STIMULATION

Responses of the olfactory cortex to odor have been examined with three different experimental approaches: recording of oscillatory field potentials as described above, recording of single "unit" responses (action potentials from cell bodies or axons), and demonstration of spatial activity patterns by autoradiographic visualization of 2-deoxyglucose uptake. In a study in unanesthetized monkeys, more than 50% of single units responded to at least one out of the eight odorants presented and 44% responded to two or more (Fig. 10.16) (Tanabe et al., 1975). As described in Chap. 5, studies with the 2-deoxyglucose method have demonstrated odor-related spatial patterns in the olfactory bulb, but in the

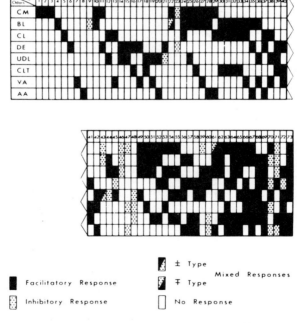

FIG. 10.16. Responses of single units in piriform and amygdaloid cortex of the monkey to eight odorants. Results are arranged from greatest specificity at left (response to one odorant) to lowest specificity at right (response to seven odorants). (From Takagi, 1986.)

cortex, uptake is remarkably uniform in the horizontal dimension (Cattarelli et al., 1988) indicating a lack of spatial segregation of cells with similar response properties. Thus, it appears that at the level of the olfactory cortex each odor is represented by an activity pattern that involves a very large number of spatially distributed cells.

FUNCTIONAL CLUES FROM NEURONAL CIRCUITRY

There are two striking features of the neuronal circuitry in the piriform cortex that provide clues concerning its operation for olfactory discrimination: the spatially distributed, nontopographical afferent input, and the presence of intrinsic associational fiber systems that mediate spatially distributed positive feedback onto pyramidal cells. As described previously in some detail (Haberly, 1985), these circuitry features and the apparent ensemble code for odor quality are reminiscent of "associative" or "content addressable" memory models that have been developed by mathematicians (e.g., Anderson, 1970, 1972; Kohonen, 1977; Hopfield, 1982). These models have shown that a large matrix of "neurons" which receives a spatially distributed input, has an intrinsic distributed positive feedback system, and also "synapses" whose efficacies can be altered by coincident activation (modified Hebb rule; see Hebb, 1949), can store, retrieve, and associate complex spatial patterns. In these networks, memory traces involve the entire network and are addressed by content rather than location. Variations on the original models for spatial pattern recognition can discriminate complex temporal patterns by introduction of delay elements (Tank and Hopfield, 1987). The olfactory cortex would appear to be an ideal system in which to test the applicability of these models for the nervous system because it appears to have the requisite circuitry features, is only one synapse removed from its sensory receptors, and has a structure that is easy to analyze. Initial steps have been taken toward the development of models for the olfactory cortex based on this theoretical framework (Granger et al., 1988; Wilson and Bower, 1988; Chover, 1989), although simplifying assumptions and speculations concerning key features of the circuitry and its operation are still required (Haberly and Bower, 1989). Continued development of these models in parallel with circuitry analysis may elucidate the process of olfactory discrimination as well as perhaps provide clues concerning the operation of higher-order "association areas" of the neocortex that, like the olfactory cortex, have strong spatially distributed connections.

11

HIPPOCAMPUS

THOMAS H. BROWN AND ANTHONY M. ZADOR

The hippocampus, named for its resemblance to the sea horse (*hippo* = horse, *kampos* = sea monster; Greek), is among the best-characterized cortical structures. Its highly regular organization is ideally suited for anatomical and physiological investigations. Much is known about its synaptic organization and the functional characteristics of its neurons. Interest in this structure stems from its involvement in some major neurological disorders as well as its apparent importance in normal cognitive functions.

Neuropsychological studies suggest that the hippocampus plays a key role in certain aspects of learning or memory. Although the exact nature of the mnemonic functions carried out by the circuitry of the hippocampus is still unclear, this region of the temporal lobes seems to be essential for the *declarative memory system* in humans (Squire, 1987). The most commonly cited evidence in support of this claim is the tragic case of patient H.M. In 1953, H.M. underwent a bilateral medial temporal lobectomy for treatment of intractable epilepsy. His IQ remains above average, but since the operation he has had severe anterograde amnesia. He cannot recall episodes from even five minutes into the past, nor can he recognize the medical staff that attend him year after year, but he does remember events from several years before his operation. H.M. is still able to learn some relatively complicated tasks, including solutions to puzzles such as the Tower of Hanoi (Cohen, 1984; Corkin, 1984). The hippocampus does not appear to be essential for the *procedural memory system* or "habit system," as it is sometimes called (Mishkin and Petri, 1984).

Neurophysiological experiments are just beginning to provide glimpses into the nature of the information processing that occurs in the hippocampus and at some of the adaptive cellular mechanisms that might be operating in this brain region. Extracellular recordings from hippocampal neurons in awake animals suggest a role in the processing of complex spatial and temporal patterns. In humans, some hippocampal cells respond specifically to individual words (Heit et al., 1988). Investigations of nonhuman primates have revealed cells that respond selectively to particular faces (Rolls, 1988). In rats, some hippocampal neurons seem tuned to spatial coordinates of their environment (O'Keefe and Nadel, 1978; O'Keefe, 1989).

Intracellular recordings of hippocampal neurons in brain slices have generated a relatively detailed understanding of their cellular neurophysiology. Some of the synapses of the hippocampus display a remarkable form of plasticity—called *long-term potentiation* (LTP)—that could be relevant to the mnemonic functions of this circuitry (Brown et al., 1988b). One form of LTP closely resembles a type of synaptic modification that many learning theorists since Donald Hebb (1949) have postulated serves as a substrate for cognitive learning and the organization of perception (Brown et al., 1990).

Interest in the hippocampus also derives from the fact that pathophysiology in this structure can have such serious clinical consequences. For example, the hippocampus is a target of degenerative disorders such as Alzheimer's disease (Robbins and Kumar, 1987). In addition, it is relatively seizure-prone and is commonly involved in temporal lobe epilepsy. For this reason, there has been much research on the cellular mechanisms underlying normal and abnormal forms of synchronization of hippocampal neuronal activity (Wong et al., 1984; Johnston and Brown, 1984a; Traub et al., 1989). The neuropharmacology of the hippocampus is thus relevant to the search for better anticonvulsants and other therapeutic drugs.

NEURONAL ELEMENTS

The hippocampus is a cylindrical structure whose longitudinal axis forms a semicircle around the thalamus. The position of the hippocampus in relation to other brain structures, and to some degree its internal organization, shows significant interspecies variation. The present discussion will focus primarily on the rodent, because this has been the most common species for experimental analysis. Figure 11.1 shows some of the neuronal elements present in a thin slice of hippocampus taken transversely to its longitudinal axis. Much of the intrinsic circuitry of the hippocampus is preserved in slices taken in this orientation, although there are also significant longitudinal pathways, perpendicular to the plane of the slice (Amaral, 1987).

The hippocampus proper can be divided into four regions (Fig. 11.1), which have been traditionally designated CA1–CA4 (from the Latin *cornu Ammon*, or Ammon's horn, because of its resemblance to a ram's horn). Another nomenclature, introduced by Cajal, divides the hippocampus into *regio superior* and *regio inferior*. The *dentate gyrus*, the *subiculum*, and the *entorhinal cortex* are included in the more general term *hippocampal formation* or *hippocampal region*. The CA1 and CA3 regions constitute most of the hippocampus proper. The CA2 region is so small and indistinct in some species that it is often ignored. The area between the dentate gyrus and the stratum granulosum of the CA3 region is called the *polymorphic* or *hilar region,* or simply the *hilus.*

Both the hippocampus and the dentate gyrus are three-layered cortices. The three fundamental layers of the hippocampus are the polymorphic layer (the *stratum oriens*), the pyramidal layer (the *stratum pyramidale*), and the molecular layer (the *stratum radiatum* and *stratum lacunosum-moleculare*). The dentate gyrus consists of a polymorphic layer (the *hilus*), a granular layer (*stratum*

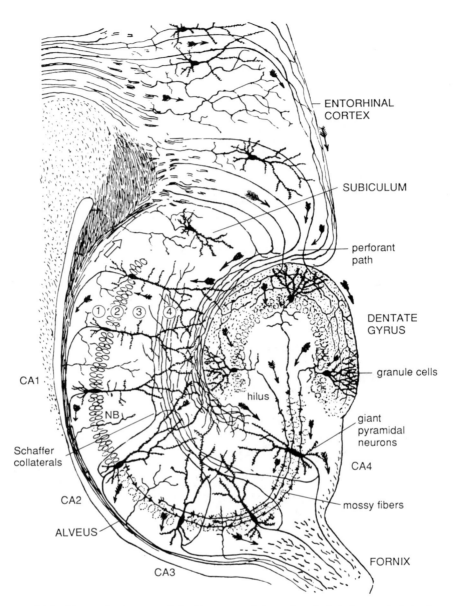

FIG. 11.1. Neuronal elements of the hippocampal formation. Labeled areas include the subiculum, part of the entorhinal cortex, the dentate gyrus, and regions CA1 to CA4. The hippocampus proper is divided into stratum oriens (1), stratum pyramidale (2; cell body layer, drawn as ovals), stratum radiatum (3), and stratum lacunosum-moleculare (4). The giant pyramidal cells in region CA3 project to CA1 pyramidal cells via the Schaffer collaterals, to other CA3 pyramidal cells, and through the commissure via fibers in the alveus to the contralateral hippocampus. Granule cells in the dentate gyrus send mossy fibers to CA3 pyramidal cells. The perforant path input is shown to all regions of the hippocampus and dentate gyrus. Not illustrated are the extent of commissural input to the hippocampus, dentate gyrus and hilus, or the variety of interneurons, including basket cells. (Modified from Cajal, 1911.)

granulosum), and a molecular layer (*stratum moleculare*). The molecular layer of the dentate is continuous with that of the hippocampus.

INPUTS

The main inputs to the hippocampus and dentate gyrus arise from the entorhinal cortex, the septal region, and the contralateral hippocampus. There are also important but less numerous projections from several other regions including the brainstem, hypothalamus, thalamus, and amygdala. The hippocampus falls within the general class of *supramodal association cortex* because its major cortical inputs arise either from polymodal association cortex or else from other supramodal association areas. The hippocampus is sometimes regarded as the most complex of the supramodal associated areas (Swanson, 1983), partly because it receives and integrates information from all the sensory modalities.

Entorhinal cortex. The entorhinal cortex provides a major sensory input to the hippocampus, dentate gyrus, and subiculum via the fibers of the alvear and perforant pathways. The *perforant pathway* originates in layers II and III of the entorhinal cortex and "perforates" the hippocampal fissure before terminating in all parts of the hippocampus. The entorhinal cortex itself receives input from many other regions of the brain. Important inputs to the entorhinal cortex include fibers from the association cortices, the olfactory cortex, several thalamic nuclei, the claustrum, and the amygdala.

The *medial* and *lateral entorhinal cortices* give rise to separate medial and lateral perforant paths. The lateral entorhinal cortex receives input from the olfactory bulb, prepyriform cortex, amygdala, and several other areas (Powell et al., 1965), whereas the medial area receives projections primarily from the presubiculum. The input organization of the entorhinal cortex is preserved in its outputs to the hippocampus, so that the lateral perforant path is largely olfactory, whereas the medial pathway is primarily nonolfactory.

Septum and diagonal band. A second significant input originates in the medial septal nucleus and the nucleus of the diagonal band. It enters the hippocampus by four routes: the fimbria, the dorsal fornix, the supracallosal striae, and the amygdaloid complex. The terminations are distributed throughout the hippocampus, but preferentially in region CA3 and the dentate gyrus. This input includes both cholinergic and GABAergic fibers and may be important in regulating the excitability of the hippocampus (see Neurotransmitters, below).

Commissural connections. A third sizable input consists of commissural fibers from the contralateral hippocampus. These fibers enter the fimbria and then collect in the fornix. The two fornices are connected dorsally by the *hippocampal commissure* or *psalterium,* which crosses the midline underneath the corpus callosum. The extent and organization of these connections are species specific (Amaral et al., 1984; Demeter et al., 1985; Amaral, 1987). In the rat, both regions CA3 and CA1 receive significant, topographically organized projections from the contralateral CA3 region in addition to the ipsilateral input. Similarly, the inner third of the molecular layer of the dentate gyrus receives a powerful, topographically organized input from the contralateral hilus in addition to the projection from the ipsilateral hilus (Amaral, 1987). The CA3 region does not project to either the ipsilateral or the contralateral dentate gyrus (Laurberg and Sorensen, 1981).

Brainstem inputs. Several brainstem nuclei provide inputs to the hippocampus, including an important noradrenergic projection from the locus coeruleus and a serotonergic input from the raphe nuclei. There is also evidence for a dopaminergic input, possibly originating in the ventral tegmental area (see below).

Other inputs. A major *hypothalamic* projection arises from a group of cells near the mamillary bodies. It passes partly through the fimbria and partly through a ventral route to terminate most heavily in the molecular layer of the dentate gyrus and in regions CA2 and CA3. The *anterior thalamic nucleus* projects to the nearby subiculum.

PRINCIPAL NEURONS

Here we consider the output or *principal neurons* of the hippocampus proper and of the dentate gyrus, emphasizing characteristics of the rat and guinea pig.

Hippocampus. The principal neurons of the hippocampus are the *pyramidal* neurons. They are the most numerous class of neuron in the hippocampus. Hippocampal pyramidal neurons are similar in form to their counterparts in other cortices, although they are somewhat smaller.

The pyramidal cell bodies are arranged in a layer called the *stratum pyramidale*. The stratum pyramidale forms a curved sheet that is only two or three cells thick. The pyramidal cell bodies are conical (pyramidal), 20–40 μm at the base, and 40–60 μm in height (Figs. 11.2 and 11.3). Each has a thick apical dendrite, 5–10 μm in diameter, that passes through the *stratum radiatum* to the *stratum lacunosum-moleculare*. There are also several basal dendrites, 3–6 μm in diameter, that arborize over 200–300 μm and form the *stratum oriens*. The apical and basal dendrites are covered with spines.

The morphology of pyramidal cells varies gradually around the hippocampus. Proceeding from region CA3 to region CA1, the cell bodies become smaller and the apical dendrites longer, more slender, and more regular in their branching pattern. The cells in region CA3 are sometimes called *giant pyramidal cells*. An example of a pyramidal neuron from the CA3 region—as viewed with a confocal scanning laser microscope—is illustrated in Fig. 11.2. The low-magnification images (A) are stereo pairs representing three-dimensional optical reconstructions. The so-called *thorny excrescences* are clearly evident in the higher-magnification images (B and C). These "thorns" are unusually large dendritic spines located on the proximal parts of the dendritic tree of CA3 pyramidal neurons. This type of dendritic spine is seen only in regio inferior. In the CA3 region, the

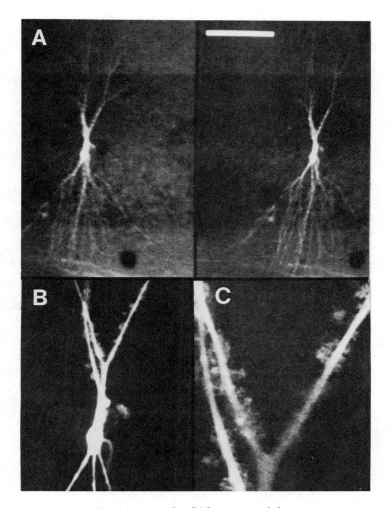

FIG. 11.2. Confocal images of a CA3 neuron and thorny excrescences following injection of Lucifer Yellow. **A.** Stereoimage of cell showing three-dimensional (3-D) structure of the dendrites in stratum lucidum (top of figure) and stratum oriens (bottom). The 3-D image was reconstructed from six optical sections. **B.** Higher magnification of the dendrites near the soma illustrates the distribution, shape, and size of the thorny excrescences with which the mossy fiber expansions from the dentate granule cells form synapses. **C.** Grapelike clusters of small spines with varied shapes show in great detail the complex structure of individual excrescences. Calibration bar: A, 100 μm; B, 45 μm; C, 16.5 μm.

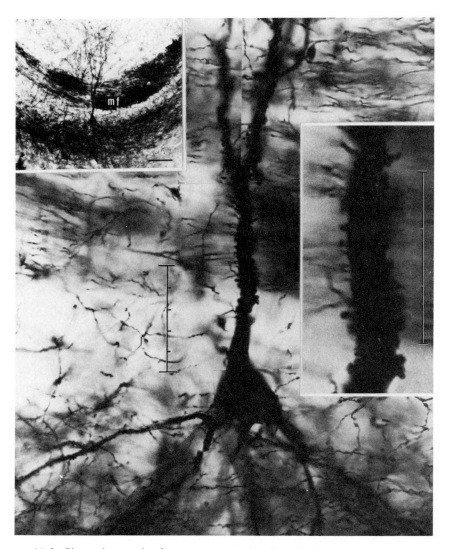

FIG. 11.3. Photomicrograph of a transverse section through the hippocampus that was stained by use of the rapid Golgi method. *Upper left:* Low-power view of a pyramidal neuron from the CA3 region. The dark band passing through the apical region (stratum lacunosum) represents numerous Golgi-stained mossy-fiber (mf) synaptic expansions (scale, 100 μm). *Center:* A higher-power view of the same CA3 pyramidal neuron. Several large mossy-fiber synaptic expansions can be seen within the plane of focus. The dendritic thorny excrescences are also apparent (scale, 50 μm). *Center right:* higher magnification of a portion of the apical dendrite of the same neuron, illustrating the thorny excrescences more clearly (scale, 30 μm). (From Johnston and Brown, 1983.)

thorns constitute the postsynaptic specializations associated with the giant *mossy-fiber synaptic expansions* (see Fig. 11.3).

The hippocampus has several output pathways and targets. One major output proceeds via the *fimbria,* which is a sheet composed of the axons of pyramidal cells and cells in the subiculum (which in turn receive input from the hippocampus proper). These axons then gather to form the *fornix,* which crosses the midline of the brain. *Precommissural* fibers of the fornix, which cross rostrally to the anterior commissure, innervate the lateral nucleus of the septal complex. The *postcommissural* fiber system arises mainly from the subicular complex, rather than from the hippocampus proper, as was previously thought (Swanson and Cowan, 1975). The postcommissural fiber system distributes to the mamillary bodies, the ventromedial nucleus of the hypothalamus, and the anterior thalamus. The other major outputs are to the subiculum and to the deep layers of the entorhinal cortex. Thus the hippocampus projects to many of the same regions that provide its input.

Dentate gyrus. The principal neurons of the dentate gyrus are the *granule cells.* The term "granule" is descriptive and does not imply any functional similarity to the granule cells of other regions, such as the cerebellum and olfactory bulb. The granule cell bodies are considerably smaller (10 μm in diameter) than those of the pyramidal neurons of the hippocampus proper. Each granule cell gives rise to one or several dendrites, 1–2 μm in diameter, that branch extensively and are covered with spines. The outer two-thirds of the dendritic field receives input from layer II of the entorhinal cortex via the perforant path. This projection accounts for the majority of synapses in that region (Matthews et al., 1976). A projection from the cells of the hilus to the inner third of the granule cell dendritic field accounts for the majority of proximal synapses (Amaral, 1987).

The main output of the dentate gyrus is to region CA3 via the mossy fiber axons. As they course toward the CA3 region, the axons of the granule cells give rise to an average of seven primary *collaterals* (Claiborne et al., 1986), which divide into secondary and tertiary branches. These axon collaterals are confined to the hilar region. Synaptic varicosities of various sizes are distributed along the length of the collaterals. The distribution of the *collateral synaptic expansions* in the hilus can be seen by use of the Timm stain, which is selective for axons and terminals with a high zinc content (Fig. 11.4). The Timm stain also shows the distribution of the *mossy-fiber synaptic expansions* in the *stratum lucidum* of the CA3 region. The collateral expansions actually look very much like smaller versions of the mossy-fiber synaptic expansions.

INTRINSIC NEURONS

There are several types of intrinsic neuron in the hippocampus: *Basket cell* bodies range in diameter from 15–30 μm and give off several short, 3- to 6-μm thick dendrites with relatively few spines. Several nonpyramidal, nonbasket cell types have also been identified (Schwartzkroin and Mathers, 1978; Lacaille et al., 1987; Lacaille and Schwartzkroin, 1988), including the *O/A interneuron,* found at the border of *stratum oriens* and the *alveus* in the CA1 region, and the L-M interneuron in the *stratum lacunosum-moleculare* of the CA1 region. The hilus has a complex local circuitry, with at least 21 distinct interneuronal types (Amaral, 1978), including at least five morphologically distinguishable basket cell types (Ribak and Seress, 1983). The most prominent cells of the hilus are the *mossy cells* (Amaral, 1978; Ribak et al., 1985; Scharfman and Schwartzkroin, 1988). Their proximal dendrites are covered with thorny excrescences that are the sites of synaptic input from the granule cell axon collaterals. The distal dendrites have more conventional spines. The dendritic tree of the mossy cell is so large that it can span much of the hilus. Spines can be found on the initial segment of the axon.

QUANTITATIVE ESTIMATES

In the rat there are an estimated 1,000,000 dentate granule cells, 160,000 CA3 pyramidal cells, and 250,000 CA1 pyramidal cells (Squire et al., 1989; cf. Boss et al., 1985, 1987). The number of intrinsic neurons is not known, although they are probably much less numerous than the principal cells. Each CA3 pyramidal neuron averages roughly 80 mossy-fiber synaptic inputs, located mainly on the proximal apical dendrites in the *stratum lucidum* (Squire et al., 1989). A typical CA3 cell may receive as many as 4500 synaptic inputs from the perforant path onto the distal apical dendrites and 7700 recurrents from other CA3 pyramidal neurons (Amaral, personal communication). Input from region CA3 comprises the vast majority of synapses onto CA1 pyramidal neurons, but there are also inputs from the entorhinal cortex and from local circuit neurons. Although accurate estimates are not available, the number of commissural inputs to regions CA3 and CA1 may be of the same order of magnitude as the number of associational inputs.

SYNAPTIC CONNECTIONS

The synaptic connections in the hippocampus are generally axodendritic and axosomatic, although axoaxonic connections have been observed. There does not appear to be the wide variety of synaptic types that characterizes some regions, such as the olfactory bulb. Most of the synapses of the hippocampus appear to be structurally conventional. The notable exceptions are the mossy-fiber synaptic input from the granule cells to the pyramidal neurons of the CA3 region and the collateral synaptic input from the granule cells to the mossy cells of the hilus. What follows focuses mainly on regions CA1 and CA3 of the hippocampus proper and to a lesser extent on the hilus and dentate gyrus.

SCHAFFER COLLATERAL SYNAPSE

The Schaffer collateral axons of the CA3 pyramidal neurons are partially myelinated, with nodes of Ranvier occurring at 60- to 70-μm intervals. One to three unmyelinated axons emerge from each node and then bend to course parallel to the parent fiber (Andersen, 1975). Each of these unmyelinated branches has numerous swellings, about 7 μm apart, each of which makes contact with one or two dendritic spines.

Consistent with their excitatory action, the Schaffer collateral synapses are of the type 1 variety, which means that they have asymmetrical membrane thickenings and round vesicles. The Schaffer collateral synapses are generally made on slender dendritic spines, which are typically unbranched. However, branched spines are also sometimes observed in the CA1 region. In such cases more than one spine emerges from a single protrusion of dendrite. A characteristic feature of Schaffer collateral synapses is the "spinule" that protrudes into the presynaptic terminal.

The Schaffer collateral synapses are among the most extensively studied in the mammalian brain, partly because they are so convenient for in vitro electrophysiological experiments. Extracellular recordings of synaptic field potentials are possible because the geometry and physiology are favorable for distinguishing synaptic current sinks and sources. Such recordings are technically simple and relatively stable over long periods of time (Barrionuevo and Brown, 1983; Kelso and Brown, 1986). This synaptic system has thus been very popular for doing routine pharmacological studies.

The Schaffer collateral synapses are also convenient for intracellular studies of synaptic plasticity (Barrionuevo and Brown, 1983; Kelso et al., 1986). Intracellular recordings are easy to maintain in the brain slice, and the absence of a system of recurrent excitation in the CA1 region is an important consideration for many experiments. The presence of an extensive system of recurrent excitation in the CA3 region is one of the reasons why disinhibition tends to cause epileptiform activity in this area (Johnston and Brown, 1984a, 1986; Traub et al., 1989). Pharmacological disinhibition is often accomplished by adding GABA antagonists such as picrotoxin or bicuculline to the saline in which the slice is bathed. The fact that the CA1 region is less seizure-prone than the CA3 region is useful in experiments that require pharmacological disinhibition. In disinhibited slices, polysynaptic excitatory responses are commonly elicited in the CA3 region by afferent stimulation, but this is less common in the CA1 region.

MOSSY-FIBER SYNAPSE

The mossy-fiber axons arise from the granule cells of the dentate gyrus. These axons are unmyelinated, have an average diameter of 0.2 μm, and tend to course in fascicles that average 1 μm in diameter (Blackstad and Kjaerheim, 1961; Hamlyn, 1962; Amaral and Dent, 1981; Claiborne et al., 1986). At intervals of about 140 μm, the mossy-fiber axons give rise to large "synaptic expansions." In Golgi-stained material, the mossy-fiber synaptic expansions can look like irregular "pieces of moss."

The mossy-fiber synaptic expansions are among the largest presynaptic structures in the brain. They can be more than 5 μm in diameter, which makes them readily identifiable with light microscopy (Fig. 11.3). Each expansion has a complicated structure containing many filamentous protrusions and invaginations. These "giant boutons" are replete with vesicles and mitochondria (Fig. 11.5). The individual synaptic "active zones" appear to be of the usual type 1 variety. The postsynaptic specializations or "thorny excrescences" are also huge. For this reason and because of their characteristic shape, they can be unequivocally recognized by use of light microscopy (Figs. 11.2 and 11.4). Like the

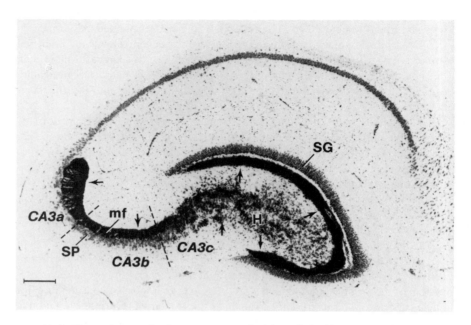

FIG. 11.4. Photomicrograph of a transverse section through the hippocampus in which the mossy-fiber synapses have been stained by use of the Timm method (darker stain), and the cell-body layers have been counterstained with cresyl violet (lighter stain). Stratum granulosum (SG), the mossy-fiber (mf) synaptic band, and regions CA3a, b, and c are labeled. (From Johnston and Brown, 1983.)

synaptic expansions, the thorny excrescences have a complicated three-dimensional structure.

The mossy-fiber synapses (Fig. 11.5) have long fascinated neurobiologists because of their great size and baroque structure and because their strategic location at the entrance to the hippocampus seemed to suggest an important role. They have been called "detonator synapses," based on the assumption that their function matches their form. Although they do indeed look like "throughput" synapses, there is nothing about their known physiology that demonstrates such a function (Brown and Johnston, 1983; Johnston and Brown, 1984b).

The mossy-fiber synaptic system is especially interesting and challenging from an experimental point of view. This system offers a number of advantages, which are reviewed elsewhere (Johnston and Brown, 1984b; Brown et al., 1989), for studies of the microphysiology and biophysics of synaptic transmission. One important advantage is that these synapses are located anatomically and electrotonically very near the cell body—averaging about 3% of a space constant (Johnston and Brown, 1983). This enables more accurate measurement and control of the voltage at the subsynaptic membrane than has been possible for other excitatory synapses of the telencephalon.

Another important experimental advantage is that these synapses can be resolved and identified light microscopically (Fig. 11.3). An effort is now being

FIG. 11.5. Electron micrograph from stratum lucidum in the CA3 region of a BALB/cJ mouse. Spines (S) on an apical dendrite (AD) are seen making synaptic contacts with a mossy-fiber bouton. (From Barber et al., 1974.)

made to develop methods that will allow real-time visualization of the mossy-fiber synapses in the living state. In this way, it may be possible to determine whether use-dependent changes in synaptic function are accompanied by structural changes in the pre- or postsynaptic elements. Like other excitatory synapses of the hippocampal formation, the mossy-fiber inputs display long-term potentiation.

The mossy-fiber synaptic system also has some important experimental limitations. The main reason why there have been so few studies of their physiology is that experiments on the mossy-fiber synapses are generally much more difficult than experiments on other synapses of the hippocampal formation. There are three reasons for this difficulty. First, extracellular field potential recordings, which are so convenient for studying Schaffer collateral synapses with the brain slice technique (Barrionuevo and Brown, 1983; Kelso and Brown, 1986), often do not provide useful information about the mossy-fiber synapses (unpublished observations). Second, it is generally more difficult to obtain good intracellular recordings from pyramidal neurons of region CA3 than from those in region CA1. Finally, the recurrent system makes this region of the hippocampus extremely seizure-prone, especially when the tissue is disinhibited. The presence of epileptiform activity has ruined many otherwise good experiments.

For these reasons the mossy-fiber synapses are much less attractive for certain types of routine studies or for experiments in which a great premium is placed on getting quick results. Thus pharmaceutical researchers might find the mossy-fiber synapses less appealing for some drug screening studies that can be done much more quickly by using extracellular recordings of field potentials in the Schaffer collateral system.

SYNAPSES ONTO GRANULE CELLS

The excitatory synaptic connections onto the granule cells of the dentate fascia are structurally similar to those found on pyramidal neurons of region CA1. Most of the synapses are of the type 1 variety and most are located on dendritic spines. There is often a small spinule protruding into the presynaptic terminal. The perforant path synapses tend to be on the distal portion of the dendritic tree, in contrast to the more proximal locations of synaptic connections from other areas.

SYNAPSES FROM BASKET CELLS

Basket cells and other inhibitory interneurons are thought to form typical type 2 synapses, which means that they have symmetrical junctions and contain smaller, flattened vesicles. Type 2 synapses are found throughout the hippocampus, often on the soma but also sometimes on the dendrites. Axosomatic contacts are estimated to occupy less than 5% of the surface area of pyramidal cells (Gottlieb and Cowan, 1973). This type of synapse is often associated with positive staining with antibodies to glutamic acid decarboxylase (GAD), the final synthetic enzyme for gamma-aminobutyric acid (GABA), which appears to be the most common inhibitory neurotransmitter in the brain (Roberts, 1986).

SYNAPSES FROM HILAR CELLS

Immunohistochemical evidence suggests that about 30–40% of the neurons in the hilus are GABAergic (Squire et al., 1989). Some of the synapses made by hilar interneurons onto dentate granule cells appear to be GAD-positive, but they terminate on spines and otherwise seem similar to excitatory connections in the hippocampus and elsewhere in the cortex. The synapses formed by hilar cells may contain two or even three neuropeptides (Gall et al., 1981).

ELECTRICAL INTERACTIONS

There have been reports of *electrical connections* between pyramidal neurons, although the prevalence and functional significance of gap junctions in the hippocampus remain uncertain (MacVicar and Dudek, 1981; Knowles et al., 1982). *Ephaptic* coupling has also been suggested to be important in the hippocampus (Taylor and Dudek, 1984a,b). The anatomical arrangement of neurons into regular sheets provides an ideal substrate for the action of electric fields on membrane potentials. Such an effect has been suggested to reinforce the synchronization of activity and thus be important in epilepsy (Dudek et al., 1988).

BASIC CIRCUIT

The functional organization of the hippocampus has traditionally been described in terms of the *trisynaptic circuit* (Andersen et al., 1966a,b). This older view is still pedagogically useful for introducing the hippocampus. The current conception of the synaptic organization is considerably more complicated, and vastly more interesting from a computational perspective. In what follows, we start with the classical trisynaptic circuit and then develop a more contemporary way of thinking about the flow of information through the hippocampal region. The contemporary view is built upon five organizational principles: (1) closed-loop control of hippocampal synaptic activity; (2) a staggered projection system that gives rise to short and long loops around the hippocampal region; (3) extensive local circuit interactions within each region of the loop; (4) significant longitudinal circuitry; and (5) neuromodulatory control of activity by extrinsic inputs.

TRISYNAPTIC CIRCUIT

The idea that hippocampal computations can be understood in terms of a simple trisynaptic circuit has had an appealing simplicity. The hippocampus was imagined to be a stack of planar circuits, each containing the same sequence of just three types of excitatory synapses (Fig. 11.6). The trisynaptic sequence is as follows: Fibers of the *perforant path* synapse onto granule cells of the dentate gyrus; these granule cells send their *mossy-fiber axons* to the CA3 region where they synapse on the pyramidal neurons; and these pyramidal neurons send their *Schaffer collateral* axons to the CA1 region, where they form synapses on pyramidal neurons. These three synapses define the trisynaptic circuit. Note that this is a purely feedforward network.

The fibers of the perforant pathway arise from stellate and pyramidal cells in layers II and III of the medial (nonolfactory) and lateral (olfactory) entorhinal cortex (Steward, 1976). They pass through the subicular complex and terminate in the outer two-thirds of the molecular layer of the dentate gyrus. Here they form excitatory synapses onto the dendritic spines of the granule cells. The granule cells of the dentate gyrus send their mossy fiber axons to the CA3 region, where they run primarily in the *stratum lucidum*. The giant mossy-fiber synaptic expansions terminate on the thorny excrescences of the proximal dendrites of the CA3 pyramidal cells. Each mossy-fiber axon makes an average of only 14 expansions at a mean interval of 135 μm (Claiborne et al., 1986). The pyramidal

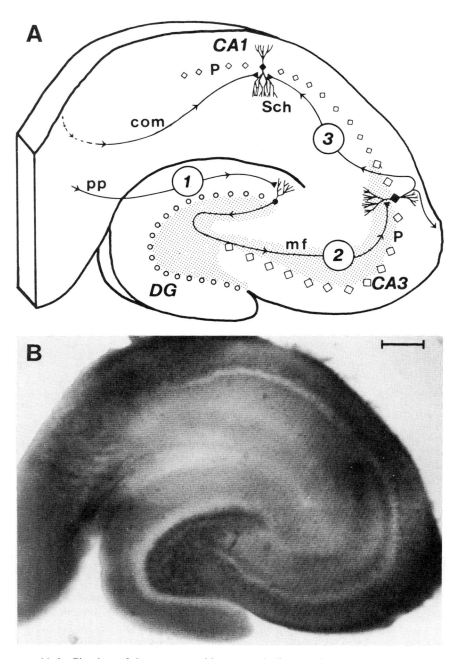

FIG. 11.6. Circuitry of the transverse hippocampal slice. **A.** Schematic diagram of tri-synaptic circuit showing (1) perforant path (pp) input to a granule cell in the dentate gyrus (DG); (2) mossy fiber (mf) input to a pyramidal cell (P) in the CA3 region; and (3) Schaffer collateral (Sch) input to a pyramidal cell in the CA1 region. Also shown is a commissural (com) input. **B.** Video microscopy of the living hippocampal slice from which A was traced. Calibration bar: 55 μm.

neurons of the CA3 region send their axon out of the hippocampus in the fornix. A branch of this axon, the Schaffer collateral, innervates the CA1 region. In strata radiatum and oriens (see Fig. 11.1) the Schaffer collateral forms en passant synapses onto the apical dendrites of the pyramidal neurons.

The organization of the hippocampus is sometimes termed "lamellar," which is meant to imply that the trisynaptic circuitry is organized approximately in a plane that is perpendicular to the long axis of the hippocampus. This is why the transverse hippocampal brain slice has been so useful in evaluating synaptic and circuit properties. However, it is important to remember that the actual synaptic organization is not exclusively lamellar. There are, in fact, significant longitudinal connections that are not preserved within thin transverse brain slices (Amaral, 1987).

CLOSING THE LOOP

The circuitry of the hippocampus cannot be fully appreciated when viewed in isolation from its intimate connections to other structures of the hippocampal region. When these neighboring structures are taken into account, the possibility of closed-loop control of hippocampal activity immediately emerges. The hippocampal region contains additional projection systems that connect one of the major outputs of the hippocampus back to its main sensory input. Axons from CA1 pyramidal neurons project heavily to neurons in the subicular complex and lightly to the entorhinal cortex. There is a major projection from the subiculum to the deep layers (V and VI) of the entorhinal cortex. The deep layers of the entorhinal cortex in turn project to its superficial layers (Kohler, 1986), and the latter give rise to the perforant pathway fibers. Thus the perforant path input to the granule cells originates in the entorhinal cortex, and the output from CA1 is largely to this same area, both directly and via the subiculum.

The circuitry of the hippocampal region thus seems to be organized into a closed loop that enables unidirectional flow of information. Note that this idea of closed-loop control does not imply the existence of a precise topographical mapping of activity in each region onto the next. There is, in fact, little published evidence in regard to this matter. What *is* implied is that the output of each region is a partial determinant of its future input and that some of the postulated feedback occurs within the confines of the hippocampal region itself. Recall that the hippocampal region includes the entorhinal cortices, hippocampus, dentate gyrus, and subiculum.

STAGGERED PROJECTIONS

The circuit described thus far includes a simple sequence of excitatory connections from one region to the next, forming a closed loop. The anatomy is actually considerably more complicated. It turns out that there are several independent paths for the excitation of a "downstream" region by an "upstream" region (Deadwyler, 1988; Fig. 11.7). Thus each region projects not only to the next region in the sequence but also to one or two after it. The only exception is the dentate gyrus, which does not send projections beyond the CA3 region.

The staggered system of projections is illustrated nicely by the perforant path

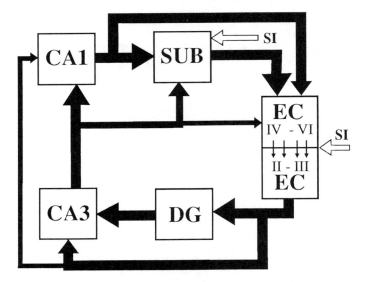

FIG. 11.7. Basic circuit of the hippocampus showing staggered projections. Note how each area, with the exception of the dentate gyrus, projects to at least two other areas. SUB, subiculum; EC, entorhinal cortex; DG, dentate gyrus; SI, sensory inputs. (Modified from Deadwyler et al., 1988.)

input from the entorhinal cortex. In addition to its heavy projection to the dentate gyrus, it also connects directly to regions CA3 and CA1. In fact, the perforant-path projection to region CA3 is quite extensive. There may be as many as 4500 perforant path synapses on each region CA3 pyramidal cell in contrast to only 80 mossy inputs (see above). Similarly, the CA3 region projects not only to the CA1 region but also to the subiculum (Swanson et al., 1981). In the same fashion, the CA1 region sends fibers both to the subiculum and to the entorhinal complex. The result is a series of staggered projections forming overlapping loops, not just a single loop.

Several observations are worth noting regarding the scheme depicted in Fig. 11.7. A minimum of three synapses is required to complete a loop [EC(II) → CA1 → EC(VI) → EC(II)]. The density of the connections is arranged in such a way that the greater part of the activation is to the next region around the loop. The various "short-cuts" provide for the possibility of subtle control of output through timing and gating. There is in fact evidence for gating activation from one region to the next, depending on the behavioral state of the animal (Winson and Abzug, 1978).

LOCAL INTERACTIONS

So far only synaptic connections *among* regions of the hippocampal formation have been emphasized. However, there are also relatively complex synaptic interactions *within* each region. These *local circuits* consist minimally of a principal neuron and an associated inhibitory basket cell, but can also include recurrent excitatory synapses.

Region CA3. The system of recurrent excitation is extremely powerful in the CA3 region, which in part explains why this region of the brain is so seizure-prone (Johnston and Brown, 1981; Johnston and Brown, 1984a; Miles and Wong, 1986; Traub et al., 1989). The recurrent system is formed by axon collaterals of the pyramidal neurons that project back to their neighbors. The probability that a given pyramidal cell contacts another has been estimated to be about .05, assuming a uniform connection probability (Squire et al., 1989). This value is surprisingly close to estimates based on recording simultaneously from pairs of hippocampal neurons in the CA3 region of a brain slice (Miles and Wong, 1986). Such paired recordings have suggested that 1–2% of the recorded pyramidal neurons have a monosynaptic connection between them. Polysynaptic recurrent pathways, unmasked during disinhibition by picrotoxin, are present between about 20% of the sampled cells.

The pyramidal neurons of the CA3 region also receive powerful synaptic inhibition (Brown and Johnston, 1983; Griffith et al., 1986; Miles and Wong, 1984, 1987). Estimates based on paired recordings in the transverse brain slice suggest that each pyramidal cell is innervated by up to 15 inhibitory interneurons (Miles and Wong, 1987). This could be an underestimate for the intact hippocampus because some of the inhibitory connections are organized longitudinally. There is both feedback and feedforward synaptic inhibition of the pyramidal neurons in the CA3 region.

Region CA1. There are at least two different morphological types of inhibitory interneurons in the CA1 region (Knowles and Schwartzkroin, 1981; Schwartzkroin and Kunkel, 1985; Lacaille et al., 1987). The best studied interneuron is the *basket cell,* which is located at the stratum pyramidale/stratum oriens border of CA1. Basket cells form inhibitory contacts onto CA1 pyramidal cells and receive excitatory input from these same CA1 pyramidal cells. Basket cells also receive a direct excitatory input from afferents coursing through strata radiatum and oriens. These interneurons thus appear to mediate both feedforward and feedback inhibition of pyramidal neurons.

Two other interneuronal types have been reported in the CA1 region. The *O/A interneuron,* located at the border of the stratum oriens/alveus, receives excitation from pyramidal cell collaterals and extrinsic afferents. These neurons are thought to mediate both feedforward and feedback inhibition of the pyramidal neurons, and they may also make inhibitory connections with the basket cells. The *lacunosum-moleculare* interneuron mediates only feedforward inhibition. Interestingly, some of the axonal processes of the L-M interneurons cross the hippocampal fissure and enter the dentate gyrus.

Hilar region. There are several types of neuron in the hilus (Amaral, 1978; Ribak and Seress, 1983; Ribak et al., 1985; Claiborne et al., 1986), but their physiology is just beginning to be explored (Scharfman and Schwartzkroin, 1988).

Some hilar cells are interneurons, with axons that arborize locally. Other hilar neurons, such as the *mossy cells,* project to both the ipsilateral and contralateral dentate gyrus. Collaterals of the mossy-fiber axons of the granule cells make en passant synapses with the mossy cells, which in turn project back to synapse on the inner third of the granule cell dendrites, where they are thought to make excitatory connections. In this dentate–hilus–dentate circuit, there must be significant *convergence* to the hilus and *divergence* from the hilus because the number of granule cells vastly exceeds the number of mossy cells.

LONGITUDINAL CIRCUITRY

Because the longitudinal circuits of the hippocampus are less well characterized, discussion has generally focused on the "lamellar" organization presented above. Nevertheless, there is ample anatomical evidence that even distant levels of the hippocampus are connected by important pathways (Swanson et al., 1978; Amaral, 1987; Amaral, personal communication). Interestingly, some of these longitudinal connections appear to be inhibitory. Inhibitory connections arising from the basket plexus of the dentate gyrus often extend some distance longitudinally (Struble et al., 1978).

EXTRINSIC NEUROMODULATION

In addition to the synaptic circuitry described thus far, there are also neuromodulatory inputs from extrinsic sources. Examples include serotonergic fibers from the median raphe nucleus, noradrenergic projections from the locus coeruleus, and cholinergic innervation from the septal nuclei. These inputs generally produce very slow postsynaptic responses and can therefore carry only low-frequency information or slowly changing control signals. Modulatory inputs such as these are usually thought to gate or regulate the overall flow of activity or to control various types of neuronal plasticity. Based on the anatomy, such neuromodulatory influences appear to operate in a spatially diffuse manner. The actions of acetylcholine and norepinephrine, two of the most important neuromodulators, are considered later.

SYNAPTIC ACTIONS

The synaptic organization of the hippocampus is well suited to the brain slice technique (Yamamoto and McIlwain, 1966; Skrede and Westgaard, 1971). Partly for this reason, more is known about the cellular and synaptic neurophysiology of the hippocampus synapses than about these properties in any other part of the forebrain. The transverse hippocampal brain slice was employed in the first demonstration of single quantal events in the mammalian brain (Brown et al., 1979, 1988a,b; Johnston and Brown, 1984b), and in the first voltage-clamp studies of mammalian brain synapses (Johnston and Brown, 1981, 1984b; Brown and Johnston, 1983; Barrionuevo et al., 1986). Hippocampal slices are also proving to be convenient for high-resolution visualization techniques that promise further increases in our understanding of central neurons and their synapses (Keenan et al., 1988, 1989).

ENDOGENOUSLY GENERATED SYNAPTIC RESPONSES

There have been no direct observations of naturally occurring synaptic responses in the hippocampus of conscious, freely behaving animals. The reason of course is that it is technically difficult to obtain stable intracellular recordings under these conditions. Observations of endogenously generated synaptic responses have been facilitated by the use of simpler neurophysiological preparations. With the brain slice preparation it is routinely possible to observe and analyze in detail the biophysical properties of endogenously generated synaptic responses. The relationship of these responses to events occurring in vivo is naturally less certain than one would like. On the other hand, there is no reason to doubt that some biophysical properties of the synapses in a hippocampal brain slice are very similar to those in the hippocampus of a conscious animal. The single quantal conductance is an example of a synaptic property that may be similar in both conditions, but that is difficult or impossible to measure reliably in vivo.

Synaptic potentials. Intracellular recordings from well-impaled neurons of the CA3 region typically reveal an incessant barrage of synaptic activity. All of the spontaneous activity is depolarizing at the normal resting potential (Brown et al., 1979). These excitatory postsynaptic potentials (EPSPs) can be easily resolved above the background noise because the larger ones are several millivolts in amplitude. Hyperpolarizing events can be seen when the cell membrane is de-polarized (Johnston and Brown, 1984b). These inhibitory postsynaptic potentials (IPSPs), which are mediated by a conductance increase to chloride ions, are blocked by picrotoxin or bicuculline, which are antagonists of the $GABA_A$ receptor.

Some of this spontaneous synaptic activity represents evoked (nerve impulse-dependent) release of neurotransmitter. However, the synaptic activity does not cease when the slices are bathed in solutions that contain tetrodotoxin (TTX) and $MnCl_2$, which are known to block evoked release (Brown et al., 1979; Johnston and Brown, 1984b; Brown et al., 1989). This residual synaptic activity represents the spontaneous (nerve impulse-independent) release of single quantal packets of neurotransmitter—so-called miniature postsynaptic potentials. Miniature potentials or "minis" can also be seen in other neurons of the hippocampal formation, but the miniature excitatory postsynaptic potentials (MEPSPs) are largest in pyramidal neurons of region CA3.

At the other end of the size spectrum are the "giant synaptic potentials" associated with interictal epileptiform activity (Fig. 11.8). The *interictal spike* is a field potential that results from the synchronous bursting of a large population of neurons. Its intracellular correlate is a burst discharge called the *paroxysmal depolarizing shift* (PDS) (Ayala et al., 1973). In the CA3 region, the burst appears as a train of action potentials superimposed upon a sudden 20- to 50-mV depolarization that lasts about 90–150 msec. The prolonged depolarization results in part from a giant compound EPSP that seems to be produced by the synchronous activation of the excitatory recurrent system. The PDS is thus a *network-driven* burst discharge (Johnston and Brown, 1981, 1984a, 1986). In the

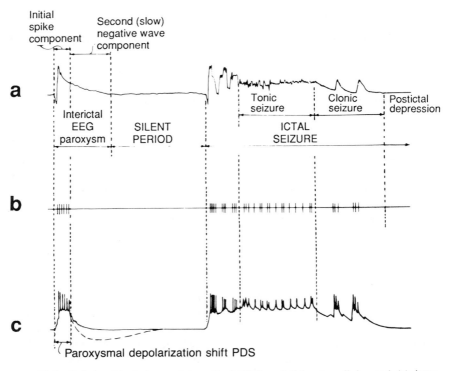

FIG. 11.8. Relationship between (a) cortical EEG and (b) extracellular and (c) intracellular discharges from a feline, penicillin-induced, spontaneous epileptiform discharge. (From Ayala et al., 1970.)

disinhibited brain slice, PDSs can be endogenously generated at intervals of about 10 sec. They can also be elicited by electrical stimulation of an afferent input. Computer simulations of hippocampal networks are contributing to our understanding of network synchronization and epilepsy (Traub et al., 1987, 1989).

Synaptic currents. Using the single-electrode clamp (SEC) device, it has been possible to resolve the currents underlying spontaneous miniature potentials (Johnston and Brown, 1984b; Brown et al., 1988a). Examples of miniature excitatory postsynaptic currents (MEPSCs) are illustrated in Fig. 11.9. In this experiment, the holding potential was −90 mV and the slice was bathed in saline containing TTX to block evoked release and picrotoxin to block the miniature inhibitory postsynaptic currents (MIPSCs). Without this disinhibition, the IPSCs would be reversed at this negative holding potential. Inward currents in the range of 50–200 pA are clearly evident in these records. These MEPSCs correspond to a conductance of the order of 1 nS. Due to the cable properties of the dendrites and the noise levels of the SEC device, the majority of the MEPSCs were probably too small to resolve clearly in these recordings. In the absence of TTX,

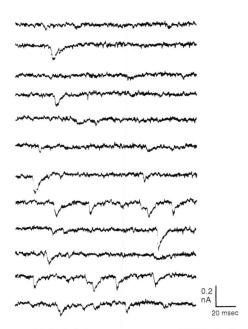

FIG. 11.9. Miniature spontaneous EPSCs in the CA3 region. Typical events are in the range of 1 nS, and larger events are 2–3 nS. The holding potential was -120 mV, the electrode resistance was 22 MΩ, and the slice was bathed in modified saline (1 mM Ca^{2+}, 2 Mn^{2+}, 1 μM TTX) to prevent evoked release. Spontaneous miniature inhibitory currents (MIPSCs) were blocked by 10 μM picrotoxin. In order to reduce the current required to clamp the membrane at very negative holding potentials, 2 mM Cs^{2+} was added to the bath. (From Brown et al., 1988a.)

the frequency of EPSCs seems somewhat higher and the average EPSC amplitude appears to be slightly larger. However, the difference between evoked release and spontaneous quantal release is not as striking as might have been expected, unless epileptiform activity develops. The synaptic currents associated with the PDS are dramatically larger than the normal endogenous EPSCs. The EPSCs underlying the PDS are sometimes so large that they exceed the current-passing capability of the SEC system, which is quite limited when using conventional microelectrodes that have resistances in the range of 30–60 MΩ. The PDS-associated synaptic conductance can be in the range of 50–150 nS.

EXPERIMENTALLY ELICITED SYNAPTIC RESPONSES

Most of what is known about the synaptic physiology is based on experimentally elicited synaptic responses recorded in vitro from the transverse hippocampal brain slice. Usually these synaptic responses are elicited by extracellular electrical stimulation of a population of presynaptic cell bodies or axons using a suitably positioned bipolar electrode. The size of the synaptic responses produced by such extracellular electrical stimulation is a function of the intensity of the electric shock delivered to the tissue. Synchronous activation of a large number of afferent inputs to a postsynaptic neuron will produce a large postsynaptic response. It may also recruit interneurons in a manner that might not commonly occur under normal endogenous conditions.

Synaptic potentials. Extracellular electrical stimulation of afferent inputs to the hippocampus usually elicits a complex response in the postsynaptic neurons that is not fully appreciated until the membrane potential is changed from its resting level. When a cell is depolarized, the response elicited usually becomes clearly biphasic, consisting of a depolarizing phase or EPSP, followed by a hyperpolarizing phase or IPSP.

Results in the CA3 region are typical in this respect. Extracellular electrical stimulation of the granule cells of the dentate gyrus usually elicits an EPSP/IPSP sequence in the pyramidal neurons (Brown and Johnston, 1983). The EPSP becomes smaller with depolarization, whereas the IPSP becomes much larger. With further depolarization, the IPSP dominates the response. When the cell is depolarized by 20–30 mV, the response can appear to be entirely hyperpolarizing, producing the incorrect impression that the reversal potential for the EPSP is as negative as -30 mV.

The hyperpolarizing phase of the synaptic response sometimes consists of two components. The early component of the IPSP often lasts 50–100 msec. It represents a conductance increase to chloride that appears to be controlled by the $GABA_A$ receptor. When the early IPSP is blocked by picrotoxin or bicuculline, the reversal potential of the mossy-fiber EPSP can be seen to be near 0 mV (Brown and Johnston, 1983). The late component of the IPSP is not blocked by $GABA_A$ antagonists and lasts many hundreds of milliseconds (Thalman, 1988; Janigro and Schwartzkroin, 1988). It represents a conductance increase to K^+, probably mediated by the $GABA_B$ receptor. The late component is blocked by compounds, such as pertussis toxin, that inhibit the actions of guanosine triphosphate (GTP)-binding proteins (G-proteins). The late IPSP produced by electrical stimulation of the Schaffer collateral afferent inputs to the CA1 region can be blocked by a high concentration of phaclofen (Dutar and Nicoll, 1988). Phaclofen has also been used in the CA3 region as a selective $GABA_B$ antagonist (Malouf et al., 1988).

Synaptic currents. The most accurate synaptic current measurements have been made on the mossy-fiber synapses because these synapses are located electrotonically near the cell soma, at an average distance, as noted above, of about 3% of a space constant (Johnston and Brown, 1983). The mossy-fiber synaptic input can be stimulated by electrical or chemical excitation of the granule cells of the dentate gyrus. As would be expected from the current-clamp results described above, electrical stimulation of the granule cells does not elicit a pure mossy-fiber EPSC in the CA3 pyramidal neurons, but instead produces a complex EPSC/IPSC sequence (Brown and Johnston, 1983; Griffith et al., 1986). The IPSC is generated by feedforward or recurrent inhibition.

There is an early and a late IPSC, corresponding to the early and late IPSPs described above. The late IPSC is not always present, but the early IPSC is rarely missing. The early IPSC begins only slightly after the EPSC so that the two current waveforms overlap substantially. The conductance associated with the early IPSC is usually larger than that associated with the EPSC (Brown and Johnston, 1983; Griffith et al., 1986). As would be anticipated from the current-clamp measurements described above, the early IPSC can be pharmacologically blocked by adding picrotoxin or bicuculline to the bath (Fig. 11.10). In the presence of these $GABA_A$ antagonists, the mossy-fiber EPSC can be seen uncontaminated by the early IPSC (Brown and Johnston, 1983). Examples of mossy-fiber EPSCs recorded from a disinhibited slice are illustrated in Fig. 11.10. Note that the EPSCs are monophasic and there is a clear reversal potential. When the early IPSC is not completely blocked, biphasic currents can be seen at depolarizing holding potentials and there does not exist a single holding potential at which no current flows for the duration of the mossy-fiber synaptic response.

The illustrated EPSCs are typical in that they get smaller as the holding potential is made less negative, and they reverse direction at potentials more positive than -2 mV. The positive slope of the current–voltage relationship and the clear reversal of the EPSC at a relatively positive potential demonstrate that the mossy-fiber synaptic response is produced by a classical conductance increase mechanism. The mossy-fiber EPSC waveform can be approximated by an "alpha function" of the form $I(t) = Kte^{-\alpha t}$, where t is time, and K and α determine the peak amplitude and rise time (Brown and Johnston, 1983). Such alpha functions are often used in neural simulations to represent synaptic conductances (Jack et al., 1975; Johnston and Brown, 1983; Brown et al., 1988a).

The slope of the current–voltage relationship is about 21 nS in the illustrated examples of mossy-fiber EPSCs (Fig. 11.10). This conductance increase reflects the activity of *many* mossy-fiber synaptic inputs that have been excited by the extracellular stimulating electrode, which was located in the stratum granulosum. The amplitude of the conductance increase produced by a single mossy-fiber synapse appears to be about a decade smaller. This is not the size normally associated with "detonator" or "throughput" synapses, such as the vertebrate motor endplate, where the conductance increase is about a thousand times larger. A curious feature of the mossy-fiber synapses is that they appear structurally to be designed for releasing vast numbers of quanta per nerve impulse and yet they generate such small EPSCs.

Note that the mossy-fiber EPSC does *not* become prolonged as the holding potential is made more positive (Fig. 11.10). This is in contrast to results in some other hippocampal synapses, such as the Schaffer collateral input to the CA1 region or the perforant input to the dentate. In synapses such as these, the EPSC decays with two characteristic rates. The magnitude of the slow component of the decay increases as the membrane potential is made less negative (Forsythe and Westbrook, 1988). This voltage dependence of the slow component is

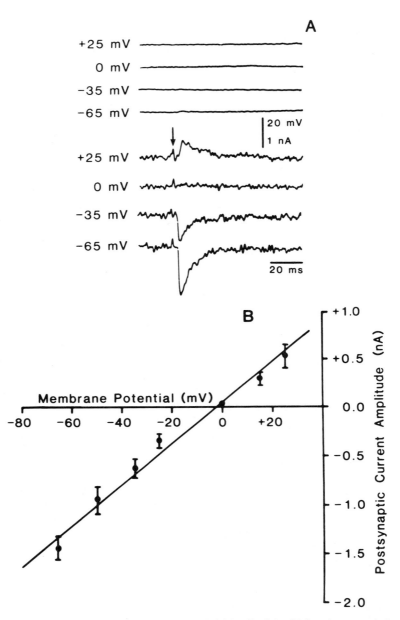

FIG. 11.10. Mossy-fiber EPSCs in a pyramidal cell of the CA3 region recorded in a slice bathed in 10 μM picrotoxin. **A.** Voltage-clamp records at four different holding potentials, indicated at the left of traces. Upper traces show the membrane potential during synaptic input. Lower traces are the corresponding clamp currents. Currents are clearly reversed at a holding potential of +25 mV. **B.** Plot of the mean ± *SE* peak synaptic current as a function of the membrane potential. The regression line gave a reversal potential of −1.9 mV and a conductance increase of 20.6 nS. (From Brown and Johnston, 1983.)

thought to result from an unusual ionic channel, the *NMDA receptor-gated channel,* located on the subsynaptic membrane.

NMDA receptors are not uniformly distributed throughout the hippocampal formation (Monaghan and Cotman, 1985; Cotman and Iversen, 1987; Cotman et al., 1988). The density of NMDA binding sites in the dendritic zones containing the Schaffer collateral synapses (the stratum pyramidale of region CA1) and the perforant pathway synapses (the molecular layer of the dentate gyrus) seems to be the highest in the entire brain. The density of NMDA binding sites is much lower in the dendritic zone containing the mossy-fiber synapses (the stratum lucidum of region CA3). The differential distribution of the NMDA receptor-gated channel is thought to account for some of the physiological and pharmacological differences between the mossy-fiber synapses and other hippocampal synapses.

LONG-TERM POTENTIATION

Long-term synaptic potentiation (LTP) is a persistent increase in synaptic efficacy that can be rapidly induced (Bliss and Lomo, 1973; Bliss and Gardner-Medwin, 1973). Seconds or less of the appropriate activity cause a synaptic strengthening that can last hours, days, or longer. For a variety of reasons, reviewed elsewhere (Brown et al., 1988a,b; 1989, 1990), LTP is a leading candidate for a synaptic mechanism of rapid learning in mammals. This fascinating form of synaptic plasticity has been most thoroughly studied in the mossy-fiber synaptic input to region CA3 and the Schaffer collateral synaptic input to region CA1. There appear to be differences in the LTP mechanisms in these two systems.

Mossy-fiber synapses. An example of LTP in the mossy-fiber synapses is shown in Fig. 11.11. The slice was pharmacologically disinhibited to eliminate any possibility that a *decrease* in the concomitant IPSC might be contributing to the synaptic enhancement. The traces on the left were obtained under current-clamp conditions, whereas those on the right illustrate the corresponding postsynaptic currents recorded under voltage-clamp conditions. During the control period, the EPSP was subthreshold for generating an action potential in the postsynaptic neuron. Following a brief tetanic stimulation (100 Hz for 1 sec), there was a large and persistent increase in synaptic efficacy that lasted for the duration of the experiment. After the tetanic stimulation, the same synaptic input generated a suprathreshold postsynaptic response and the EPSCs were clearly larger.

This experiment illustrates the defining features of LTP. It is a rapid and persistent, use-dependent form of synaptic enhancement. "Rapid" is usually taken to mean that seconds or less of the appropriate stimulation are adequate to induce the synaptic modification. "Persistent" typically means that the modification clearly outlasts another form of use-dependent synaptic enhancement, called *posttetanic potentiation* (PTP). Before LTP was discovered (Bliss and Gardner-Medwin, 1973; Bliss and Lomo, 1973), PTP was the most enduring synaptic modification that was known to be rapidly induced in mammals. The distinction between PTP and LTP is clearly illustrated in Fig. 11.12, which shows post-

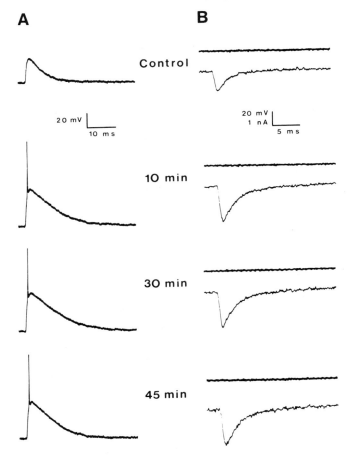

FIG. 11.11. LTP in the mossy-fiber synaptic input to a CA3 pyramidal neuron of the rat hippocampal slice. The membrane potential was maintained at −90 mV in both current-clamp (A) and voltage-clamp (B) modes. **A.** Current-clamp records of EPSPs obtained during the control period and at indicated times after tetanic stimulation of the mossy-fiber synaptic inputs. Illustrated action potential amplitudes are attenuated, owing to filtering and the switching rate of the SEC. **B.** Voltage-clamp recordings of EPSCs obtained from the same neuron at the given times. Top traces show the voltage control; bottom traces are individually recorded EPSCs. (From Barrionuevo et al., 1986.)

tetanic changes in the peak EPSC. These results are from the same cell illustrated in Fig. 11.11. Note that the EPSC amplitude rapidly decays during the first few minutes (PTP) before stabilizing to a new level (LTP). In this and many other cases, the dynamics of the changes can be seen most clearly in voltage-clamp recordings. The use of the voltage clamp improves analysis of the time course by removing secondary nonlinear (voltage-dependent) components of the postsynaptic response.

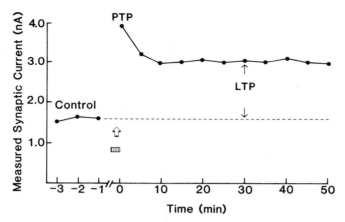

FIG. 11.12. Mossy-fiber EPSC amplitudes plotted over time, before and after the induction of LTP in the same cell as Fig. 11.11. Brief tetanic stimulation was applied at the time indicated (striped bar and arrow). Note the change in the time scale at the time of the stimulation. Each data point is the average of five EPSCs obtained from a holding potential of −90 mV. The tetanic stimulation induced posttetanic potentiation (PTP) and long-term potentiation (LTP). (From Barrionuevo et al., 1986.)

The *expression* of LTP in the mossy-fiber synapses was found (Barrionuevo et al., 1986) not to be accompanied by significant changes in input resistance, spike threshold, or synaptic equilibrium potential. The only detected changes were an increase in the EPSC amplitude and a corresponding increase in the measured synaptic conductance. This increase in the conductance could, in principle, result from pre- and/or postsynaptic modifications. The *induction* of LTP at the mossy-fiber synapses is pharmacologically different from that at the Schaffer collateral synapses. Antagonists of the NMDA receptor, such as DL-2-amino-5-phosphono-pentanoate (AP5), prevent LTP induction in the Schaffer collateral synapses but not in the mossy-fiber synapses (Harris and Cotman, 1986; Cotman et al., 1988). There are other pharmacological differences as well. Antagonists of the beta-noradrenergic receptor interfere with LTP in the mossy-fiber system but seem to have little effect on LTP in the Schaffer collateral system (Hopkins and Johnston, 1988; Johnston et al., 1988, 1989).

Schaffer collateral synapses. LTP has been most extensively studied in the Schaffer collateral synapses, partly because the experiments are easier to perform and often simpler to interpret. Interest in LTP in these synapses has been fueled by the discovery (Kelso et al., 1986; Wigstrom et al., 1986; Malinow and Miller, 1986) that the activity–modification relationship is of a type that many connectionist theories of learning have long postulated to exist. What has come to be known (Stent, 1973) as *Hebb's postulate* for learning is a *synaptic modification rule* that was first clearly stated (Hebb, 1949) in the neuropsychological treatise *The Organization of Behavior.* In this work, the Canadian psychologist Donald Hebb suggested that a synaptic modification based on correlation or conjunction might underlie learning: "When an axon of cell A is near enough to excite cell B or repeatedly or consistently takes part in firing it, some growth process or metabolic changes takes place in one or both cells such that A's efficiency, as one of the cells firing B, is increased" (Hebb 1949, p. 62). Hebb's learning rule can be interpreted and formalized in many ways (Brown et al., 1990). In its simplest form, a *Hebbian synapse* is one that is strengthened when there is a co-occurrence of pre- and postsynaptic activity. Simulations have shown that when synapses of this general type are embedded in the appropriate type of network, interesting learning and self-organizational properties emerge.

The Hebbian nature of LTP induction in the Schaffer collateral synapses was clearly demonstrated in the experiment illustrated in Fig. 11.13. In this experiment, a combination of current- and voltage-clamp techniques was used to force or prevent postsynaptic spiking during tetanic stimulation of an afferent input. Application of a strong depolarizing current step alone, which forced the postsynaptic neuron to fire action potentials, failed to induce LTP. Tetanic stimulation of the afferent input, delivered while applying a voltage clamp that maintained the postsynaptic cell at a negative potential, also failed to induce LTP. However, the same tetanic stimulation delivered during the depolarizing current step did induce LTP in the stimulated input.

Thus some consequence of postsynaptic depolarization enables LTP induction at just those synapses that are concurrently active, owing to presynaptic release of neurotransmitter. By injecting the postsynaptic cells with a local anesthetic, it has been possible to show that the postsynaptic contribution to the interactive mechanism does not require the elicitation of sodium action potentials. The induction step appears instead to be controlled by the NMDA subtype of glutamate receptor. The first evidence came from reports (Collingridge et al., 1983; Harris et al., 1984; Harris and Cotman, 1986) that antagonists of the NMDA receptor such as AP5 prevent LTP induction without blocking synaptic transmission.

The effects of AP5 on LTP and synaptic transmission are clearly illustrated in the experiment shown in Fig. 11.14. In this experiment, two Schaffer collateral synaptic inputs to a neuron, designated S1 and S2, were separately and independently manipulated. LTP was induced in the S2 input (Fig. 11.14B, lower panel) by tetanically stimulating it at 100 Hz for 200 msec. Then the slice was perfused with saline containing 100 μM AP5. Note that the AP5 did not block normal synaptic transmission in the S1 input (Fig. 11.14B,a), nor did it block LTP that had already been induced in the S2 input (Fig. 11.14B,b). But LTP could not be induced in the S1 input when the slice was bathed in AP5. As illustrated, this pharmacological effect was reversible. After the AP5 was washed from the bath, it was possible to induce LTP in the S1 input by using the same stimulation pattern that had previously failed. Thus, blocking the NMDA receptor prevents the induction of LTP but not the expression of LTP that has already been induced. The reason that AP5 does not block normal synaptic transmission or the expression of LTP appears to be that these depend upon a different subtype of glutamate receptor called the kainate–quisqualate (K-Q) receptor.

The voltage- and neurotransmitter-dependent gating of Ca^{2+} influx through the NMDA receptor-gated channels (Mayer et al., 1984; Nowak et al., 1984; Jahr

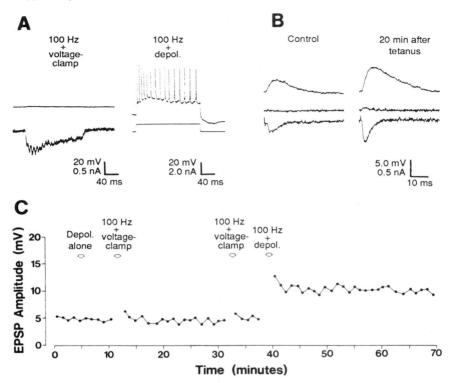

FIG. 11.13. Demonstration of a Hebbian mechanism at Sch/com synapses in the CA1 region. **A.** *Left:* Voltage-clamp record of inward synaptic currents (lower trace) and membrane potential (upper trace) during the synaptic stimulation train. *Right:* Current-clamp recording of postsynaptic action potentials (upper trace) produced by an outward current step (lower trace) that is paired with the synaptic stimulation train. **B.** Current-clamp (top traces) and voltage-clamp (bottom traces) records before and 20 min after pairing synaptic stimulation with the outward current step. The middle trace is the membrane potential during the voltage clamp. **C.** EPSP amplitudes as a function of the time of occurrence (arrows) of three manipulations: an outward current step alone (Depol. alone) or synaptic stimulation trains delivered while applying either voltage clamp (100 Hz + voltage clamp) or an outward current step (100 Hz + depol). Each point is the average of five consecutive EPSP amplitudes. (From Kelso et al., 1986.)

and Stevens, 1987; Ascher and Nowak, 1988) suggests an attractive explanation for the Hebbian interactive mechanism. The NMDA receptor–ionophore complex must receive two signals simultaneously to become highly permeable to Ca^{2+}: An agonist (such as glutamate or NMDA) must be bound to the NMDA receptor, and the membrane must be sufficiently depolarized to relieve the Mg^{2+} block of the channel that occurs at voltages close to the normal resting potential. The common working hypothesis (Brown et al., 1988a) is that Ca^{2+} influx through the NMDA receptor-gated channel and the resultant increase in the intracellular $[Ca^{2+}]$ are partly responsible for triggering the induction of LTP

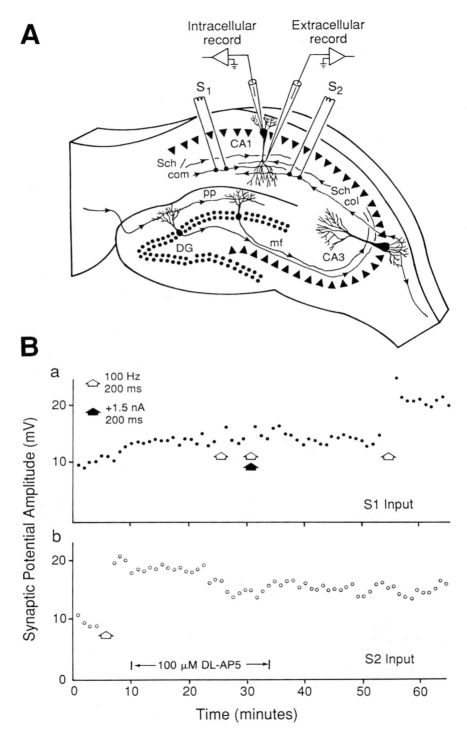

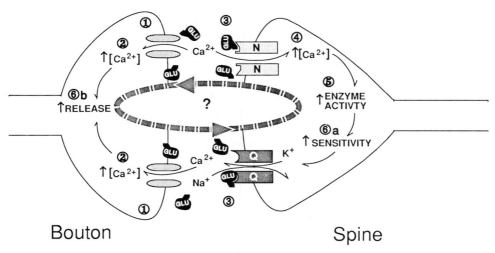

Bouton   **Spine**

FIG. 11.15. Possible mechanisms of LTP at the Sch/com synapses onto CA1 pyramidal cells. Depolarization (1) of the presynaptic bouton causes an influx of calcium through voltage-dependent calcium channels in the presynaptic terminal. The subsequent increase of presynaptic calcium (2) leads to release of glutamate (GLU), which binds (3) to two subtypes of postsynaptic receptor, the NMDA (N) and the kainate-quisqualate (Q) receptors. Postsynaptic depolarization coupled with binding of glutamate relieves the Mg^{2+}-block of the NMDA receptor-associated channel and permits entry of calcium (and monovalent cations, not shown), leading to an increase in the concentration of postsynaptic calcium (4) that is the critical step in the induction of LTP. This increase is thought to trigger a series of biochemical changes (5; see Brown et al., 1988) that ultimately lead to the expression of LTP. The expression of LTP is thought to be mediated by a postsynaptic increase in sensitivity (6a), a presynaptic increase in neurotransmitter release (6b), or both.

(Dunwiddie and Lynch, 1979; Lynch et al., 1983; Malenka et al., 1988). These steps are illustrated in Fig. 11.15.

Research on LTP has suggested a more general lesson regarding synaptic diversity. It is now clear that different classes of hippocampal synapses may be quite different in terms of their physiological and pharmacological properties, even when they use the same neurotransmitter. To understand the computations that occur within the circuitry of the hippocampus, it will be necessary to deter-

---

FIG. 11.14. Demonstration that the AP5 prevents induction of LTP in the Sch/com pathway but does not block synaptic transmission or LTP expression. **A.** Placement of electrodes. A cell in the CA1 region was impaled with an intracellular electrode, while field potentials were monitored with an extracellular electrode. Two sets of synapses could be stimulated independently with two separate electrodes (S1 and S2). **B.** *a:* Bath application 100 μM DL-AP5 did not attenuate the S1-produced EPSPs, but did prevent the *induction* of LTP following either tetanic stimulation (three trains 12 sec apart, 100 Hz for 200 msec each) alone (open arrow) or when paired (open and solid arrows) with simultaneous outward current steps (1.5 nA, 200 msec). *b:* Addition of DL-AP5 did not block the *expression* of LTP in the S2 pathway that was *induced* in the medium prior to the addition of DL-AP5. (From Brown et al., 1989.)

mine the functional properties of each class of synapses. The hippocampal brain slice preparation seems well suited to this challenge (Keenan et al., 1988; Brown et al., 1989).

The theta rhythm is a 5- to 12-Hz pattern of electrical activity observed in the hippocampal region that it is correlated with certain behavioral states in the freely moving animal. It is most pronounced when the animal is alert and interested in its surroundings, and it fades as stimuli are repeated (Vanderwolf et al., 1975). Theta has been proposed to reflect the "gating" of information through the hippocampal circuit (Winson and Abzug, 1978) or a process that serves to "quantize" (Vinogradova, 1975; O'Keefe and Nadel, 1978) or "discretize" time with respect to the passage of information around the circuit.

Theta activity has been studied as a model of rhythmic activity. Its cellular basis is only partially understood, and there may be significant differences among species (Bland, 1986). It has a maximum amplitude in the stratum oriens of the CA1 region and the stratum moleculare of the dentate gyrus (Winson, 1976; Bland, 1986). Theta activity is not unique to the hippocampus, but is also prominent in the entorhinal cortex (Mitchell and Ranck, 1980) and the cingulate cortex (Holsheimer, 1982). The synaptic mechanisms underlying theta activity appear to include cholinergic influences from the septohippocampal pathway (reviewed in Vertes, 1982; see Krnjević et al., 1988).

NEUROTRANSMITTERS

A large number of putative neurotransmitters and neuromodulators have been studied in the hippocampus (Table 11.1). GABA is probably the main inhibitory transmitter, and norepinephrine and acetylcholine are widely distributed neuromodulators; glutamate is the most important excitatory neurotransmitter in this region. Several other substances, including opiates, serotonin, and VIP appear to be present in significant concentrations but are not included in the present discussion.

GABA is thought to be the primary transmitter released by inhibitory interneurons onto both the soma and dendrites of pyramidal and granule neurons. The action of GABA depends on the class of postsynaptic receptor to which it binds (see Chap. 2). As indicated earlier, two main GABA receptor subtypes are distinguished in the hippocampus. The $GABA_A$ subtype is coupled to a fast chloride conductance that can be blocked by picrotoxin and bicuculline. The $GABA_B$ subtype is coupled to a slower potassium conductance. Phaclofen and baclofen are sometimes used, respectively, as antagonists and agonists of the $GABA_B$ receptor, although their degree of specificity is uncertain (Dutar and Nicoll, 1988; Thalman, 1988). These two receptor subtypes may have different spatial distributions, with $GABA_A$ preferentially at the soma and $GABA_B$ preferentially on the dendrites of hippocampal neurons (Janigro and Schwartzkroin, 1988).

Table 11.1. Neurotransmitter Candidates in the Hippocampus

Glutamate	Enkephalin	Dynorphin
Aspartate	GABA	Angiotensin
Acetylcholine	Norepinephrine	Serotonin
Adenosine	Histamine	Dopamine
Somatostatin	Cholecystokinin	Vasoactive intestinal polypeptide
Galanin	Neuropeptide Y	Corticotropin-releasing factor
Substance P		

Source: Nicoll (1988).

GABA can interact with other neurotransmitter or neuromodulatory substances in the hippocampus. Glutamate enhances the $GABA_A$ response (Stelzer and Wong, 1989), and serotonin and $GABA_B$ receptors may interact via potassium channels that are coupled by the same G-protein (Andrade et al., 1986). IPSPs evoked by both $GABA_A$ and $GABA_B$ can be reversibly blocked by an enkephalin analogue (Newberry and Nicoll, 1984), although this effect may be mediated by inhibition of inhibitory interneurons (Masukawa and Prince, 1982). The physiological significance of these interactions remains to be elucidated.

ACETYLCHOLINE

Acetylcholine (ACh) is widely distributed in the hippocampus, where it exerts a number of neuromodulatory effects. The majority of cholinergic inputs to pyramidal cells seem to be in the infrapyramidal layer and to a lesser extent in the suprapyramidal layer, somewhat distal to the soma (Storm-Mathisen, 1977). The cholinergic input to the hippocampus originates primarily in the medial septum and diagonal band, although only about 50% of the fibers from these regions are cholinergic (Amaral, 1987).

The action of ACh on hippocampal pyramidal neurons appears to be mediated via the muscarinic subtype of cholinergic receptor. At least two direct effects of ionophoretic application of ACh to the CA1 region have been reported (Cole and Nicoll, 1984). The first is a depression of the slow afterhyperpolarization that follows a series of spikes evoked by intracellular current injection. This depression arises from reduction of one of the two calcium-activated potassium currents, I_{AHP}. Muscarinic activity has no effect on I_C, the other calcium-activated potassium current. The second effect is a reduction of the accommodation of action potentials that is normally seen in the spike train produced by an outward current step. This action may occur through reduction of a second potassium current, I_M (Halliwell and Adams, 1982). It has also been suggested that ACh may have a presynaptic inhibitory effect on excitatory transmission (Rovira et al., 1983).

NOREPINEPHRINE

An understanding of the importance of the noradrenergic projection to the hippocampus has grown in the last several years. It has long been known that there is an important projection from the locus coeruleus to the hippocampus (Storm-

Mathisen, 1977). These inputs are most dense in the hilus of the dentate gyrus and the stratum lucidum of region CA3, but much less dense to the stratum radiatum of region CA1 (Moore and Bloom, 1979; Loy et al., 1980). An interesting feature of the NE system is that it has distinct effects in different hippocampal regions.

Norepinephrine applied to the CA1 region has at least two separate effects (Madison and Nicoll, 1986, 1988). One pharmacological effect is decrease of inhibition to the pyramidal cells, possibly through an inhibitory effect on the inhibitory interneurons. This action is probably mediated by an α_1 receptor. The other effect is directly on CA1 pyramidal neurons where NE acts on the β_1 receptors to decrease I_{AHP}. The functional consequence of this second effect may be to enhance the response to excitation and reduce accommodation.

Although NE is important in the control of excitability in region CA1, there is no evidence that it has any direct effect on LTP at the Schaffer collateral synapse in this region (Dunwiddie et al., 1982). On the other hand, NE does seem to be required to obtain "nondecremental" LTP in the mossy-fiber synapses in the CA3 region. This β_1 effect seems to be mediated through an increase in intracellular cAMP, which in turn increases the fractional open time of a voltage-dependent calcium channel (Johnston et al., 1989). NE has also been implicated in the modulation of granule cells. The response of the granule cells to perforant path stimulation seems to depend upon the behavioral state of the animal. In the alert state, animals show greater granule cell activity than during slow-wave sleep, but depletion of hippocampal NE eliminates this difference (Dahl et al., 1983).

EXCITATORY AMINO ACIDS

Glutamate is a major excitatory neurotransmitter in the hippocampus. It is known to activate at least two distinct conductances in the hippocampus. The first of these is mediated by the kainate–quisqualate (K-Q) receptor and can be blocked by CNQX (Honore et al., 1988). This fast conductance increase to Na^+ and K^+ can be considered "classical" in the sense that it shows little voltage dependence. The second is mediated by the N-methyl-D-aspartate (NMDA) receptor and, as discussed above, can be blocked by AP5 and MK801 (Watkins and Olverman, 1987). Although the kainate receptor may be distinct from the quisqualate, they are often grouped together as non-NMDA.

The molecular properties of the NMDA receptor-gated channel are especially interesting in regard to activity-dependent neuronal modifications. As indicated earlier, this channel is gated by both membrane voltage and ligand binding. Application of glutamate or NMDA causes current to flow through this channel only if the postsynaptic membrane is sufficiently depolarized. The mechanism for this dual requirement involves block by magnesium of the activated channel at membrane potentials near rest. Depolarization relieves this block and permits current flow. This dual requirement allows the NMDA receptor–ionophore complex to sense the concomitant pre- and postsynaptic activity that is necessary for the induction of LTP in the CA1 region.

In contrast to the channels that are associated with the K-Q receptors, the

NMDA receptor-gated channels have a high permeability to calcium (Ascher and Nowak, 1988). It is the calcium current through these channels that is thought to trigger LTP induction at certain Hebbian synapses. Another difference between the K-Q and the NMDA receptor-mediated currents is their time course (see Cull-Candy and Usowicz, 1987; Jahr and Stevens, 1987). The macroscopic currents generated by NMDA receptor-gated channels can last for 100 msec or longer (Forsythe and Westbrook, 1988). The long time course could allow interaction between two synaptic events separated by a significant interval (Brown et al., 1989).

DENDRITIC PROPERTIES

The passive and active membrane properties of the pyramidal cells of regions CA1 and CA3 and the granule cells of the dentate gyrus have been studied more extensively than those of any other cortical neurons. Most of the work described below was carried out using the hippocampal brain slice preparation.

PASSIVE PROPERTIES

Electrotonic structure of dendrites. There have been several studies of the electrotonic structure of the three types of principal neurons of the hippocampal formation (Traub and Llinás, 1979; Brown et al., 1981; Johnston, 1981; Turner and Schwartzkroin, 1983). Granule cells of the dentate gyrus and pyramidal neurons of regio superior and regio inferior share several features. All three are electronically compact. The electrotonic distance from the soma to the ends of the dendrites appears to average less than one space constant. Part of the explanation for this short electrotonic length is that the specific membrane resistance is relatively high. All three of these principal neurons also have large input resistances (20–100 MΩ) and long time constants (15–30 msec). The cable properties of the principal neurons thus seem well-suited for extensive spatiotemporal processing of synaptic input.

Role of dendritic spines. Dendritic spines are present on most neurons in the hippocampus and are the major postsynaptic targets of excitatory synaptic inputs. Speculation on the functional role of spines in synaptic transmission and plasticity dates back several decades (Chang, 1952), but there is still no firm evidence. Direct physiological studies have not been possible because of their small size. For this reason, computer simulations have been the only available tool for exploring the likely functions of spines (Brown et al., 1988a). With the advent of new visualization methods (Fig. 11.2) and dyes that are sensitive to physiological changes in membrane voltage and cytosolic free calcium, it may become possible to begin experimental tests of ideas that have emerged from the simulations.

Spines could theoretically be involved in the *expression* of LTP (Brown et al., 1989). One way that this might occur is through shape changes that decrease signal attenuation. In particular, an increase in the diameter of the spine shaft or a decrease in its length will reduce the axial impedance and therefore reduce the amount of synaptic current attenuation. LTP has in fact been reported to be accompanied by spine shape changes (Fifkova, 1985; Greenough and Chang, 1985), but computer simulations suggest that the range of reported shape changes is insufficient to account for the expression of LTP (Brown et al., 1988a). These simulations involved *extreme* shape changes—from long and thin to short and fat—that decreased the spine axial resistance fivefold. Even these huge shape changes only increased the synaptic currents by 10–20%, whereas during LTP the synaptic currents commonly increase by as much as 50–100% (Fig. 11.12).

Spines are perhaps more likely to play a role in the *induction* of LTP (Brown et al., 1988a). By physically *restricting* changes in Ca^{2+} or other second messengers, the spine can *amplify* the concentration changes within a limited compartment of the cell. The significance of this local amplification can be appreciated in the case of LTP induction triggered by Ca^{2+} influx through NMDA receptor-gated channels located on the spine head. Computer simulations of this hypothesized mechanism suggest that large increases in $[Ca^{2+}]$ should be limited mainly to the spine region because diffusion through the spine neck is relatively slow (Gamble and Koch, 1987; Zador et al., 1990). The small volume of the spine head amplifies the local increase in $[Ca^{2+}]$ relative to the case of a synapse on the dendritic shaft. By spatially restricting calcium transients, each spine may function to isolate chemical changes from adjacent spines. This could be relevant to the *spatial* or *input specificity* of LTP induction—the fact that only stimulated synapses appear to be potentiated.

ACTIVE PROPERTIES

There are at least a dozen active membrane conductances in hippocampal pyramidal neurons. As in many other systems (Hille, 1984; Kaczmarek and Levitan, 1986), the potassium channels show the greatest diversity. There are also ionic channels that are selective for sodium, chloride, and calcium ions. These various types of ionic channel are controlled by membrane voltage, by changes in the concentration of internal ions, and by neuromodulators. Different types of channels operate on time scales that range over more than four orders of magnitude (see also Chap. 2).

At least six distinct potassium currents are thought to be present in hippocampal neuronal membranes (Halliwell and Adams, 1982; Zbicz and Weight, 1985; Numann et al., 1987; reviewed in Llinás, 1988). These potassium currents can be broadly divided into two classes, according to whether they are voltage or calcium dependent. Four of the potassium currents are purely voltage dependent. The *delayed rectifier* (I_K) inactivates slowly during depolarization and is similar to the current described by Hodgkin and Huxley in the squid axon. As in molluscan neurons (Connor and Stevens, 1971), the *A current* activates and inactivates rapidly. The *anomalous (inward) rectifier* (I_Q) differs from other potassium currents in that it is activated by *hyperpolarization*. The *M current* is a voltage-dependent current that is partially suppressed by muscarine.

There are also two calcium-dependent currents, I_C and I_{AHP}. When the intracellular $[Ca^{2+}]$ is elevated, the *C current* is rapidly activated by depolarization. This voltage dependence appears to result from the Ca^{2+} binding step (Moczydlowski and Latorre, 1983). The *AHP current* is not voltage-dependent and can be partially suppressed by muscarine and norepinephrine. Following a series of action potentials, it gives rise to a characteristic hyperpolarization that can last several seconds.

Several other membrane currents have been described in hippocampal neurons. There is a voltage-dependent sodium current that is rapidly activated and inactivated by depolarization (Kandel et al., 1961; Schwartzkroin, 1975; Wong et al., 1979; Benardo and Masukawa, 1982; Brown and Griffith, 1983b). This TTX-sensitive sodium current is responsible for the rising phase of the action potential. Three types of calcium currents have been observed in cultured hippocampal neurons (Doerner et al., 1988), where it is possible to apply patch-clamp techniques. The *L*- and *N-type* currents have a high threshold for voltage-dependent activation, whereas the *T type* has a lower threshold (Nowycky et al., 1985). Both N- and T-type currents show rapid inactivation. Calcium currents have also been studied in acutely dissociated hippocampal neurons (Kay and Wong, 1987; Doerner et al., 1988). A voltage-dependent chloride conductance has also been reported (Madison et al., 1986). This conductance is said to be active at rest and inactivated by depolarization and protein kinase C.

These various active membrane conductances can give rise to interesting spatiotemporal interactions among synaptic inputs. Responses to synaptic inputs can be highly nonlinear, as in the case of burst discharges (Traub and Llinás, 1979; Traub et al., 1989). If the integrated EPSCs across the dendrites cause sufficient membrane depolarization, a voltage-dependent calcium current will be activated. This slow inward calcium current, possibly in conjunction with a faster sodium current, may give rise to regenerative dendritic spikes. Various types of potassium currents will be activated during this depolarization. Eventually the inward calcium current will cause a sufficient increase in the intracellular $[Ca^{2+}]$ to activate the calcium-dependent potassium currents that will terminate the burst.

There is no reason to assume that the various types of voltage- and/or Ca^{2+}-dependent ionic channels are distributed uniformly across the dendritic membrane. Nonuniform distribution of the various channel types with respect to the electrotonic structure of the neuron will introduce further complexities. The computational implications of such nonuniformity need to be further explored through computer simulations.

HIPPOCAMPAL COMPUTATIONS

In closing this chapter, it is appropriate to ask, What does the hippocampus do? What is its behavioral function or role in the cognitive memory system? Seemingly simple questions such as these are not easy to answer, partly because many traditional psychological functions, such as *working memory,* are probably not neatly localized in individual brain structures.

With the growth of connectionism and computational neuroscience, a different kind of question has been emerging. How does the hippocampus do whatever it does, as one component of a larger system? What kinds of parallel-distributed computations are suggested by the basic circuit and the properties of the neuronal elements and their synapses? How is incoming information transformed by each part of the hippocampal circuitry? It is probably fair to say that none of the answers to the above questions is yet satisfactory. However, the questions are becoming more interesting and the technology for answering them is improving at an impressive rate. What follows are examples of some approaches to the question of what the hippocampus does.

NEUROPSYCHOLOGICAL APPROACHES

Neuropsychological studies have led to a number of intriguing ideas about hippo-campal function and its role in the cognitive memory system. Most of these ideas were derived from lesion studies and in vivo electrophysiological experiments performed on behaving animals. Based on such research, the hippocampus and related structures have variously been proposed to participate in "working memory" (Olton, 1983); "spatial memory" or "contextual memory" or "contextual retrieval" (Hirsch, 1980; Nadel et al., 1985; O'Keefe, 1989); "episodic memory" or the acquisition of "declarative knowledge" (Mishkin and Petri, 1984; Squire, 1987); the detection of "novel" stimuli (Nadel et al., 1985); the "classical conditioning of aversive behavior" (Berger, 1984); and olfactory "tracking" behavior (Lynch and Baudry, 1988).

These ideas are not necessarily mutually exclusive. The hippocampal circuitry is probably engaged during a variety of adaptive behaviors, and these neuropsy-chological functions may all be related to the kinds of computations that the hippocampus and associated structures perform. This level of analysis is a useful starting point for connectionist approaches that attempt to figure out *how* the basic circuit carries out its computations.

CONNECTIONIST APPROACHES

The connectionist approach seeks to infer computational properties from neu-roanatomical and neurophysiological data. Most of the computational ideas de-rive from research on *adaptive neural networks* (reviewed in Hinton and Andersen, 1981; Rumelhart and McClelland, 1986; Brown and Kairiss, 1990). These are systems of mathematically defined *processing elements* that communi-cate with each other through simple but modifiable linkages called *connection weights*. A set of such connection weights is conventionally described in terms of a *connection matrix*. For brevity, what follows will focus on selected computa-tions that can be performed by networks that share certain formal similarities with hippocampal region CA3 (see also Marr, 1971).

Autoassociation. The basic circuit of the CA3 region of a disinhibited brain slice is shown in a highly schematic form in Fig. 11.16. The similarity of this circuit to an *autoassociative neural network* is striking (Kohonen, 1984; Brown and Kairiss, 1990). Autoassociative networks have several properties that would seem to be biologically useful (see Hinton and Anderson, 1981; Hopfield, 1984; Anderson, 1985). If an input pattern that has previously been learned is in-complete or faulty, the autoassociator can complete or restore it (Fig. 11.17a and b). This type of network is thus *noise resistant* and *fault tolerant* with respect to

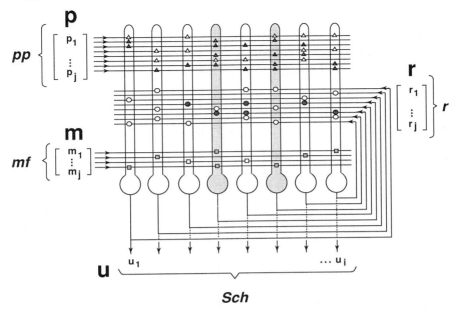

FIG. 11.16. A schematic view of the CA3 hippocampal circuitry in a disinhibited slice. Axons of perforant pathway (pp) synapse directly onto dendrites of pyramidal cells u_i. Axons m_i of the mossy-fiber (mf) projection also form synapses on cells u. Output of the CA3 neurons u_i consists of the Schaffer collateral/commissural fiber system (Sch) to CA1 as well as recurrent projection r_i back onto u_i cells. Inhibitory inputs are not shown, and the synaptic matrix is actually much sparser in reality.

corrupted signals. It affords a *content-addressible* memory that can complete a learned pattern when keyed with a subset of that pattern (Fig. 11.17B–D). Furthermore, if some fraction of the processing elements is damaged or some of the weights are disrupted, the network will only slowly lose its properties of recall. Thus it shows *graceful degradation* with respect to damage to the circuitry or its components.

An autoassociator requires minimally a set of inputs, a set of output elements, and a modifiable feedback system that connects the output elements to each other. The set of inputs in Fig. 11.16 are provided by the mossy fiber (mf) and perforant pathway (pp) projections. Activity in these afferents is denoted by input vectors **m** and **p**. The pyramidal neurons of region CA3 constitute the output elements. Activity in the Schaffer collateral axons (Sch) constitutes the output vector **u** and the recurrents (r) form the feedback vector **r**. In an auto-associator, patterns are stored in the connection matrix formed by the feedback elements. To function as an autoassociative memory, the feedback connections must change according to a Hebb rule (Brown et al., 1989). In its simplest form, the connection weights in the feedback matrix are a function of the outer product of **u** and **r**.

The recurrent system in the CA3 region does not connect every pyramidal neuron to every other pyramidal neuron. The contact probability in this region

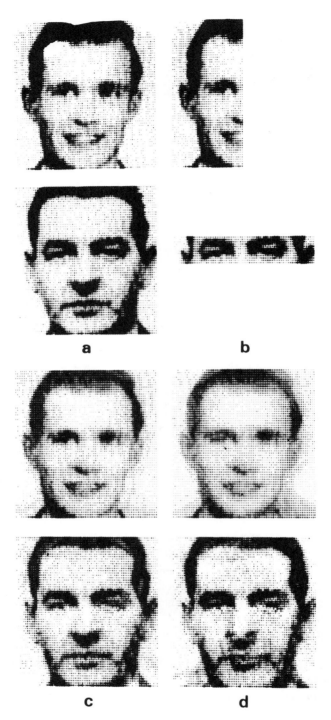

FIG. 11.17. Demonstration of associative recall in a linear autoassociative network. (a) Samples of the original digitized images that were stored in the network. (b) Key patterns

may be about .05. By comparison to other recurrent systems, this represents an extremely high degree of connectivity. Neural networks with recurrent connection probabilities in this range can indeed support reconstructive memories (Brown and Kairiss, 1990). It is not yet known whether the recurrent synapses in the CA3 region are plastic and, if so, whether they obey a Hebbian modification rule.

Temporal sequences of patterns can be learned by an autoassociator if the output is fed back, after a time delay, into the input (Kohonen, 1984). Increasing the number of different time delays can improve the temporal learning. It is clear that the hippocampal region does allow for feedback with several different delays (Fig. 11.7). The spatiotemporal pattern completion that such a network can perform is reminiscent of what some classical learning theorists used to call *redintegration* (Hull, 1929). Redintegration was once considered a fundamental mental process (Hollingworth, 1928), but until recently it was not clear how this might be implemented neurophysiologically.

Novelty filtering. The type of network illustrated in Fig. 11.16 can operate as a *novelty filter* if the feedback connections weaken instead of strengthen (Kohonen, 1984). A novelty filter reconstructs and accentuates those parts of a learned pattern that have changed. Novelty filtering may be relevant to behavioral phenomena such as habituation/dishabituation. The computational properties of this part of the CA3 region thus depend critically on the nature of the recurrent synapses. If they are modifiable, the CA3 system could operate as either an autoassociative memory or a novelty filter, depending on the form of the synaptic modification rule. It is clear that we need to know more about the basic synaptic physiology of this system.

Orthogonalization. The mossy fiber input to CA3 pyramidal cells is very sparse, as compared to the recurrent system. The probability that a given CA3 cell is contacted by a given granule cell is about .00008. This sparseness has been suggested to *orthogonalize* the representation of activity patterns from the dentate region (Rolls, 1989). Stated in the most intuitive terms, the idea implies, for example, that two different spatiotemporal activity patterns that share several neurons in the dentate region would share fewer neurons in the CA3 region. Prior pattern separation or orthogonalization is known to be important for the operation of an autoassociative memory. For a given number of stored patterns, the fidelity of such a memory increases with the degree of orthogonalization of the patterns (Kohonen, 1984). Further theoretical and neurophysiological research will be

---

with partial information that were used to select patterns from the memory. **(c)** Reconstructed patterns following presentation of the key patterns illustrated in b, for a memory with 160 stored images. **(d)** Reconstructed patterns following presentation of the key pattern illustrated in b, for a memory with 500 stored images. Note that the recollected images in d show some degradation compared with those in c, due to partial saturation of the memory. (From Kohonen, 1984.)

needed to evaluate whether the mossy-fiber projection system is in fact best understood as serving an orthogonalizing function.

CONCLUSIONS

Computational approaches may help link behavioral and cellular levels of analysis (Churchland and Sejnowski, 1988). Neural network simulations are useful partly because they offer a third level of analysis that is intermediate between these other two. A dynamic form of connectionism may prove valuable for probing the aggregate properties of very large populations of hippocampal neurons connected by modifiable synapses. The union of behavioral, neurophysiological, and computational approaches promises to furnish a deep understanding of the hippocampus. This will be no meager accomplishment in view of the absolutely essential role of this part of the brain in human cognition.

12

NEOCORTEX

RODNEY J. DOUGLAS and KEVAN A. C. MARTIN

The olfactory bulb, hippocampus, and cerebellum are all examples of cortical ("barklike") structures, which are a major feature of brain organization. The cerebral cortex in mammals is by far the largest of these structures, and as its alternative name "neocortex" implies, it is also the most recent arrival on the evolutionary scene. The fossil record indicates that over the past 3 million years the size of the hominid brain has doubled, largely owing to the increase in the size of the neocortex. A similar progression is seen in modern vertebrates (Fig. 12.1). In modern man, the neocortex and its connections occupy 80% of the brain volume (Passingham, 1982). That the neocortex has expanded so rapidly suggests that it performs its functions extremely well, that its wiring is easy to replicate, and that further expansion requires few additional genetic instructions. The basic task of the cortex is the processing of sense data and the formulation of appropriate motor responses, but to say this disguises the complexity of these tasks and neglects the many other more abstract roles the neocortex plays. The design features of cortex have also made it so adaptable that it can solve a far wider range of problems than those that fueled its evolution. Driving an automobile, for example, presents perceptual and motor tasks that are somewhat removed from the problems facing our hunter-gatherer ancestors.

Psychologists, neurobiologists, and scientists working on machine intelligence have focused on the cerebral cortex because of its role in many high-level functions, such as speech production and comprehension which, like all other cortical functions, appear to involve specific regions of cortex. The concept of cortical localization of function derives from studies that have correlated the site of the cortical lesion in brain-damaged patients with specific deficits in their behavior; for example, damage to a localized area on the left frontal lobe leads to "Broca's" aphasia, where the patient's speech becomes labored and telegraphic, but leaves speech comprehension largely intact. The neurobiologist's interest is obviously in understanding the structure and function of the cortical mirocircuitry in all these different areas, because these microcircuits lie at the heart of the formidable processing power of the cerebral cortex.

The neocortex consists of a laminated sheet of cells, about 2 mm thick on average and covering about 1.5 sq. ft. in man. To fit this sheet into the skull, it is folded into a series of convoluted mountains and valleys that impart its charac-

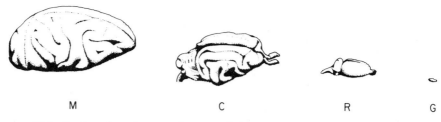

M   C   R   G

FIG. 12.1. Brains of modern vertebrates: goldfish (G), rat (R), cat (C), and Old World monkey (M). Scaled to body weight, the neocortex and its connections form an increasingly greater proportion of the brain volume. The neocortex in monkey completely envelops all the other brain structures. (This and the following figures were prepared from data kindly supplied by members of the Community Workshop and others in the MRC ANU.)

teristic wrinkled appearance. In cross section, the same basic six-layered "isocortex" can be discerned in all mammals. Anatomists have used many different techniques in an attempt to parcellate the cortex on the basis of qualitative differences in the "cytoarchitecture" from region to region. Although clear differences can be seen in the cytoarchitecture of some regions of the cortex (Fig. 12.2), the total number of anatomically distinct areas is a matter of some debate. Estimates for the human cerebral cortex vary between 50 and 100 distinct areas.

The scheme most commonly used for the cytoarchitectural divisions in primates is that devised by Brodmann in 1908. In many cases, these anatomically defined areas correspond to functionally distinct areas, as in the case of area 17, the striate or primary "visual" cortex, or area 4, the "motor" cortex. However, this correspondence is not always found. Some cytoarchitecturally homogeneous areas, such as area 18 in the monkey, are now known to consist of a number of functionally distinct areas. Such observations raise a question as to the functional significance of the cytoarchitectonic differences. One view is that these differences relate more to variations in the size of the efferent and afferent paths to the cortical areas, than to some fundamental differences in the intrinsic circuitry of the cortex (Creutzfeldt, 1977; Powell, 1981). This view has some support from studies of the composition and circuitry of the cortical areas in many species (Shepherd, 1988a,b). An alternative view, that the functional differences between cortical areas are related directly to the cytoarchitectonic differences, has no direct support as yet. If the former view is correct, it then leads to the fascinating question of how the functional differences between the areas arise if they have the same basic microcircuitry.

Most of our knowledge of the neuronal composition and intrinsic circuitry of the cortex comes from the 100-year-old Golgi technique, which is used to impregnate dendritic trees and unmyelinated axons with silver. This technique, so successful in deciphering the basic circuit in the cerebellar cortex and retina, has not provided the same breakthroughs for the neocortex. Since the early 1980s, however, a revolution in the technology has led to significant new understanding of the nature of the intrinsic cortical circuitry. The new methods use tracers that

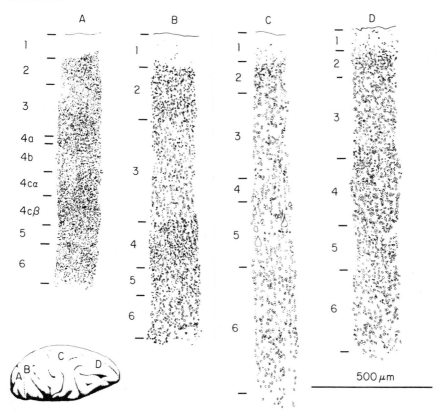

FIG. 12.2. The laminar organization of neurons in different cortical areas of the macaque monkey cortex (inset). **A.** Area 17 (striate visual cortex). **B.** Area 18 (extrastriate cortex). **C.** Area 4 (motor cortex). **D.** Area 9 (frontal cortex). A basic six-layered structure can be identified in all areas. The pia covers layer 1; the white matter is below layer 6. Note the marked difference in cell size and density among the different areas. The giant neurons in layer 5 of area 4 are the Betz cells. (Celloidin-embedded brain, cut in 40-μm-thick parasagittal sections, stained for Nissl substance, uncorrected for shrinkage.)

are transported antero- or retrogradely from the site of deposition. These have enabled the projections to and from the cortex to be traced with relative ease. At the single-neuron level, intracellular injection of tracers like the enzyme horseradish peroxidase (HRP) have revealed the morphology and axonal projections of single neurons in a detail and completeness that was not possible even with the Golgi method. As significant has been the fact that the technique allows the electrical activity of the neuron to be recorded through the enzyme-filled micropipette. Thus for the first time it has been possible to identify morphologically neurons whose functional properties are also known.

The development of immunological methods for identifying different putative neurotransmitters has also provided a technique that has already made a major contribution to our understanding of the organization of immunochemically identified groups of neurons in the cortex. Sophisticated use of these new methods in combination with more traditional methods like electron microscopy has meant that the analysis of cortical circuitry has at last become a soluble problem. In this chapter, we show how these techniques have been applied to the analysis of the neocortex. We emphasize the work on the primary visual cortex (area 17), because this region has for some time been the model for the study of the neocortex and the source of many of our concepts of the functional and structural organization of the cerebral cortex.

NEURONAL ELEMENTS

The most prominent feature of cortex is its lamination. This arises through differences in the packing density of the cells, in combination with their soma size and shape and fiber projections. These laminae are revealed by stains that emphasize the cell bodies (e.g., Nissl stains; see Fig. 12.2) or the fiber composition (e.g., myelin stains), as well as by various stains for enzymes (e.g., cytochrome oxidase). However, the number of neurons in a unit column extending from the cortical surface to the white matter is practically the same for all species and all cortical areas, with the exception of area 17 in primates, which has about double the number (Rockel et al., 1981; Peters, 1987). This might seem surprising, given that the cortical thickness can vary from 0.8 mm in mouse visual cortex to 3 mm thick in motor area 4 in humans, but the compensation occurs through the larger cell-to-cell distances in the thicker cortices (Fig. 12.2). In primates, the packing density of neurons in area 17, the primary visual cortex, is 2.5 times that of other cortical areas. This relative increase in density occurs in all layers, but it is unclear why this rather drastic modification from the basic pattern has occurred only in primate area 17.

On the basis of the laminar organization, three different types of neocortex have been identified. The primary visual cortex is an example of koniocortex, or granular cortex, which is typical of primary sensory areas. Its name arises from the small cell bodies that are densely packed in the middle layer (layer 4). Motor cortex, by contrast, lacks this distinct small cell layer, and is an example of agranular cortex. The third type of cortex contains varying populations of granule cells and is called *eulaminate* or *homotypical* cortex; it includes much of the *association* cortex, which is often a convenient description for cortex whose function has yet to be discovered.

All three divisions of the neocortex contain the same two basic types of neuron: those with spiny dendrites (the *stellate* and *pyramidal* cells, Fig. 12.3), and those whose dendrites are smooth (*smooth* cells, Fig. 12.4). About 90% of cortical neurons fall into these two types. Morphological differences in dendritic structure are only one of many differences, both structural and functional, between these types. For example, there is good evidence that spiny neurons are excitatory in function, whereas the smooth neurons are thought to be the inhibitory cells, although the evidence for this is as yet indirect. The spiny and the smooth neurons have been further subdivided on the basis of their dendritic or axonal morphology (see below).

The proportions of these different morphological types vary among the different cortical layers. Within a single layer, however, the relative proportion of the various types appears remarkably uniform. Layer 1 is relatively cell-free and contains no pyramidal or stellate cells. Layer 4 of the granular cortex is the only layer containing the stellate neurons, which form about 75% of the neurons in layer 4. In all, the stellate cells form only about 10% of the total number of neurons in the granular cortex. In the remaining layers, pyramidal cells constitute about 70–80% of the population (Sloper et al., 1979; Powell, 1981). The smooth neurons form an approximately constant proportion, about 20% in layers 2–6 (Gabbott and Somogyi, 1986).

SPINY NEURONS

The major morphological type is the pyramidal cell (Fig. 12.3C,D), which constitute about two thirds of the neurons in the neocortex. Pyramidal neurons may be found in all cortical layers except layer 1. Their dendritic structure is similar to that of hippocampal pyramidal cells in having spine-covered dendrites, a distinct apical dendrite usually ending in a tuft in layer 1, and usually a pyramidal-shaped soma. All pyramidal cells have an axon collateral system that forms part of the intrinsic cortical circuitry. In addition, many (but not all) pyramidal cells project to other regions of the brain or spinal cord and are the major source of output from the cortex. In the cat visual cortex, pyramidal neurons can have simple or complex receptive fields (Martin, 1984).

A second group of spiny neurons, the stellate cells (Fig. 12.3B), is found exclusively in layer 4 of the granular cortex (Ramón y Cajal, 1911). These also have spiny dendrites but do not have the apical dendrite that is characteristic of the pyramidal cells. Instead, dendrites of approximately equal length radiate out from the soma to give these neurons a starlike appearance, hence their name *stellate* neurons. These neurons occasionally do project to other cortical areas, but the vast majority do not, and their axons form part of the local intrinsic circuitry of the cortex. Stellate cells were previously thought to be the only recipients of thalamocortical inputs in the sensory cortices, but it now appears that many pyramidal cells also receive these direct inputs (Hersch and White, 1981; Hornung and Garey, 1981; Freund, et al., 1985b). In the cat visual cortex, all stellate neurons have simple receptive fields (Martin, 1984).

SMOOTH NEURONS

The class of neurons with spine-free dendrites has been frequently described as *smooth stellates,* but since their dendrites rarely have a stellate appearance, we refer to them here as *smooth neurons.* As shown in Fig. 12.4, these form a very heterogeneous group. Their various dendritic morphologies have been referred to as multipolar, bipolar, bitufted, and stellate, but these classifications have generally not been useful discriminators of the various types. With the advent of staining methods that reveal more of the axonal aborizations, such as intracellular injections of HRP, it has become clear that the morphology of the axons of the smooth neurons provide much better criteria for classification. On the basis of the axonal arborization, at least ten types of smooth neurons have been described (Szentágothai, 1978; Peters and Regidor, 1981). These include the

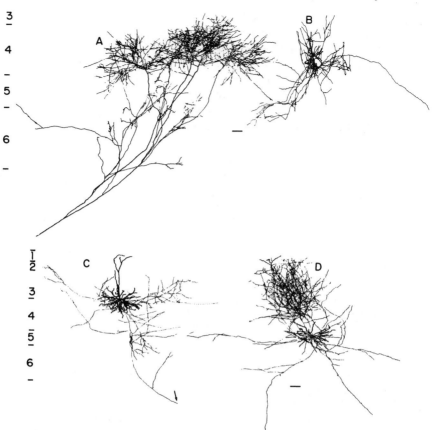

FIG. 12.3. A thalamic afferent and several spiny neurons from cat visual cortex. **A.** Y-type thalamic afferent. Note extensive but patchy arbor in layer 4. This axon formed over 8000 synaptic boutons. **B.** Stellate neuron. **C.** Pyramidal neuron of layer 3. Note characteristic apical dendrite extending to layer 1. Many collateral branches arise from the main axon before it leaves the cortex (arrow). **D.** Pyramidal neuron of layer 5. Note the very rich axon collateral arbor in superficial layers. This neuron did not project out of area 17. The thalamic afferent and neurons were filled intracellularly in vivo with horseradish peroxidase. Cortical layers are as indicated. Bar = 100 μm.

cortical basket cell (Fig. 12.4B), which has a family resemblance to basket cells of the cerebellum and hippocampus, the chandelier or axoaxonic cell (Fig. 12.4A), and the "double-bouquet" cell (Fig. 12.4D). These types have been found in all the cortical areas that have been studied. In cat visual cortex, these neurons have conventional simple or complex receptive fields (Martin, 1984).

Two types of smooth neurons are also found in layer 1, which is relatively cell-free compared to other cortical layers. One type is the Retzius–Cajal neuron, which has a horizontally elongated dendritic tree lying within layer 1, and the small neuron of layer 1, which has a small highly localized dendritic and axonal arbor within layer 1. Layer 1 is linked to layer 6 during development by

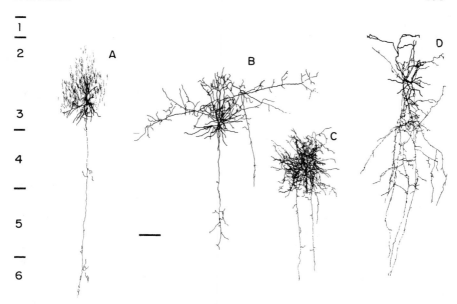

1
2
3
4
5
6

FIG. 12.4. Smooth neurons from cat visual cortex. **A.** Chandelier or axoaxonic cell. **B.** Large basket cell of layer 3. Note lateral axon collaterals. **C.** Clutch cell or small basket cell of layer 4. The major portion of the axonal arbor is confined to layer 4. **D.** Double-bouquet cell. The axon collaterals run vertically. Cortical layers are as indicated. Bar = 100 μm.

the Martinotti cell, whose soma is located in layer 6, but whose processes span the entire depth of the cortex. These processes are thought to form part of the scaffolding for the cell migration that occurs during the genesis of the cortex.

AFFERENTS

The other main neuronal elements are of course the axons of cells that provide inputs to the cortex. These arise from a great many sources.

Thalamus. The thalamus projects to all cortical areas, even the motor cortex. The richness of the thalamic input varies from area to area. Layers 3 and 4 are the main termination zones for the thalamocortical projection, although all other layers may receive a more sparse input. A feature of the thalamic input is that it is highly ordered. In the visual cortex, this means that the topographical arrangement of the cells in the retina is replicated in the lateral geniculate nucleus and in the primary visual cortex. In this way, the visual field is mapped onto the cortical surface. Equivalent topographical cortical maps are found for all the senses. These topographical maps are highly distorted because the regions of high receptor density have a larger representation in cortex. In the retina, the density of receptors and ganglion cells varies considerably from center to periphery. The region of highest density in primates is the fovea, which maps onto a far larger area of cortex than an equivalent-sized patch at the periphery of the retina where the receptor density is considerably lower. This transformation is described by the *magnification factor,* which in the visual cortex is expressed as the number of millimeters of cortex per degree of visual field (Daniel and Whitteridge, 1961). For the foveal representation in the primate, this value is about 30 mm/deg and falls off to about 0.01 mm/deg in the far periphery. In the somatosensory and motor systems of humans, distorted dwarflike creatures called homunculi are used to illustrate the relative cortical surface area representing the skin receptors or movements of particular parts of the body.

In the visual system, the thalamic afferents can be divided into at least two physiologically and morphologically distinct groups. In the primate, these groups arise from different layers in the lateral geniculate nucleus (the parvo- and magnocellular layers) and arborize in separate sublaminae in cortical layer 4, where they are thought to form the source of several distinct streams of processing through the cortex (Martin, 1984, 1988b). In the cat, the thalamic input arises from at least three physiologically and anatomically distinct groups—the W, X, and Y cells—whose laminar zones of termination in the cortex overlap considerably more than in the primate (Freund et al., 1985a; Humphrey et al., 1985). However, even in layer 4, the thalamic axons provide less than 20% of the synapses (LeVay and Gilbert, 1976), the remainder coming mainly from intracortical sources.

The terminal arborizations of single thalamic afferents cover an area several millimeters in diameter, and the collateral branches of a single arbor can bear between 1,000 and 10,000 boutons (Fig. 12.3A). Each afferent makes only a few synapses with any cortical neuron, and each bouton makes on average more than one synapse, so that a single geniculate neuron may connect to several thousand cortical neurons (Freund et al., 1985a). Because the size of the arbors, many overlap at any single point in cortex, and thus a single cortical neuron can potentially receive input from many geniculate neurons.

Other subcortical regions. The thalamus is not the only subcortical source of input to cortical areas. Other subcortical structures include the diagonal band, claustrum, locus coeruleus, the basal forebrain, and the dorsal and median raphe. As has been pointed out in earlier chapters (see olfactory bulb), these pathways have distinct neurochemical signatures, which has made the analysis of their cortical terminations more tractable. The contributions of these different pathways vary considerably from one cortical area to the next, and among species for the same cortical area. In general, in any given cortical area the projections of these nuclei are restricted to particular cortical laminae. As yet it is not possible to provide a simplifying view of the pattern of connections of these "nonspecific" afferents, even for the visual cortex. In addition, their role in the function of the cerebral cortex remains undefined.

Corticocortical connections. The major input to any single cortical area arises from other cortical areas. Indeed, it is this richness of interconnection that has made the analysis of the cortical circuitry so difficult. The interconnections of the different visual areas have been studied in great detail. There are considerable variations among species in the patterns of connections. However, with the exception of some stellate cells, the corticocortical connections are all made by pyramidal cells, which project selectively to particular laminae in their target areas. Slabs of pyramidal cells in the superficial layers of one area project predominantly to layer 4 of another area, whereas pyramidal cells in the deep layers project predominantly outside layer 4 of other areas (Van Essen, 1985; Zeki and Shipp, 1988).

The density of the various corticocortical projections can vary greatly. In the primary visual cortex, intracortical inputs arrive from many different cortical visual areas, and these together project to all six layers of the cortex. Their targets in these layers are mainly other pyramidal cells, although about 15% are smooth neurons (Sloper and Powell, 1979c; Martin, 1988a). The different patterns of interconnections have been used to develop schemes in which the various visual cortical areas are placed in a hierarchical order (Van Essen and Maunsell, 1983; Livingstone and Hubel, 1987). In these schemes, area 17 forms the starting point of an inverted pyramid of hierarchical interrelationships among the various visual areas. However, it is now clear that these hierarchical schemes are highly idealized, because there are extensive parallel connections within and among cortical areas.

SYNAPTIC CONNECTIONS

Two basic types of synapse have been identified in the neocortex (Fig. 12.5; see also Fig. 1.7). The synapses made by thalamic afferents and by the spiny neurons are of one type, whereas those made by smooth neurons are of the second type.

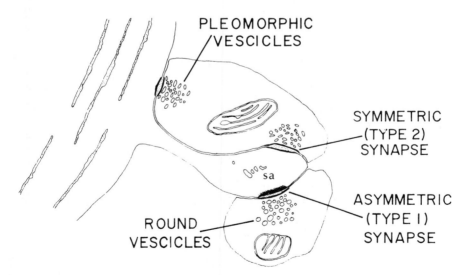

FIG. 12.5. The two synaptic types. The two boutons make synapses with the head of a dendritic spine. One bouton makes an additional synapse on the dendritic shaft. The defining features for type 1 synapses are the asymmetry of the pre- and postsynaptic densities and round vesicles, whereas for type 2 synapses they are pleomorphic vesicles and symmetrical pre- and postsynaptic densities.

Several criteria, based on the appearance of the synapse at the electron microscopic level, have been used to classify these two types. Gray (1959) was the first
to attempt this, and, as noted in Chap. 1, his scheme continues to be the most
widely used. Gray's type 1 synapse shows pronounced electron-dense staining
under the postsynaptic membrane, but not under the presynaptic membrane;
hence they are also called *asymmetrical* synapses. These synapses are associated
with boutons containing round vesicles (if suitable fixatives are used) and are
made by thalamic afferents and by spiny neurons. The type 2 synapse, made by
the smooth neurons, have more *symmetrical* pre- and postsynaptic densities.
They are associated with boutons containing pleomorphic vesicles (Colonnier,
1968). Both types of synapses can be found on all cortical neurons, but their
relative distribution varies. The type 1 synapse is found mainly on dendritic
spines and shafts. The type 2 synapses are concentrated around the soma and
proximal dendrites and form a small percentage of the synapses on spines and
distal dendrites (Gray, 1959; Szentágothai, 1978; White and Rock, 1980; Beaulieu and Colonnier, 1985).

The thalamic afferents and many of the spiny neurons have been shown to be
excitatory. These neurons all make type 1 synapses. By exclusion, the smooth
neurons, such as the cortical basket cells, which make the type 2 synapses, are
thought to be inhibitory. It should be emphasized that the data do not yet allow all
type 1 synapses to be called excitatory, and all type 2 synapses, inhibitory.
However, the association of these two types with neurons of different morphology, connectivity, and neurotransmitters strongly suggests a basic functional
distinction between these types. This conclusion is in line with the general
discussion of these two types in Chap. 1.

SPINY CELLS

The principal postsynaptic targets of both the thalamic afferents and the spiny
cells (forming type 1 synapses) are the dendritic spines (Sloper and Powell,
1979c; see Martin, 1988a). Spiny cells receive very few type 1 synapses on their
dendritic shafts, and none at all on their somata. Each spine head is occupied by
one excitatory synapse, so spine counts provide a useful estimate of the number
of type 1 synapses that contact a spiny neuron. For a typical stellate neuron, the
total number of spines is about 2000, whereas the number for a pyramidal cell in
visual cortex is about 6000 (our estimates). Smooth neurons, by contrast, do
receive type 1 synapses on their somata (e.g., from the thalamic afferents; Freund
et al., 1985b), but their main type 1 input is to the more extensive target provided
by their dendrites. For both spiny and smooth neurons, type 2 synapses are found
on their dendritic shafts (particularly on proximal dendrites), somata, and on the
initial segment of the axon in the case of pyramidal cells. In addition to the type 1
input, some spines (approx. 7% in cat visual cortex; Beaulieu and Colonnier,
1985) also receive a type 2 input. Although the proportion of spines receiving
such a dual input is small, the number of spines on the largest cortical neurons
(e.g., Betz cells in area 4) may be as high as 10,000, so the absolute number of
spines receiving the dual input can be large.

An important functional issue is how many presynaptic neurons contribute the several thousand synapses on a single cortical neuron. In the cerebellum, each Purkinje cell receives specific inputs from only one climbing fiber (see Chap. 6). Similarly, each lateral geniculate neuron receives its retinal drive from a relatively small number of ganglion cells (see Chap. 8). By contrast, the connectivity in the cortex appears to be very different. In the only instance in which it has been examined directly, it has been found that single thalamic afferent fibers provide only 1–10 synapses to any single postsynaptic neuron. In view of the estimates cited above, this is a tiny percentage (0.01–0.10%) of the total input to the neuron (Freund et al., 1985b). The anatomical connections between one cortical cell and another also appear to be weak, and this is supported by physiological results.

Cross-correlation techniques establish the probability that action potentials in one neuron are contingent on action potentials recorded in another neuron. When applied to the cortex, the results of cross-correlation experiments indicate that only a small percentage of the spikes can be accounted for by the activity of the other neuron (Toyama et al., 1981; T'so et al., 1986), indicating a weak connection. The more sensitive and direct technique of spike-triggered averaging, in which action potentials recorded in one neuron are correlated with monosynaptic EPSPs recorded in another, gives similar results (Komatsu et al., 1988; Thomson et al., 1988). The modal amplitude of the correlated EPSPs was about 0.1 mV, which is the order of magnitude anticipated from theoretical studies. The small amplitudes indicate that in order to reach threshold, a cortical cell must receive convergent excitatory input from many other neurons. Cortical cells receive their major excitatory input from other spiny neurons, because even in layer 4 the thalamic input contributes no more than 20% of the type 1 synapses (Le Vay and Gilbert, 1976).

SMOOTH NEURONS

Chandelier or axoaxonic cells. Smooth neurons also provide a convergent input to their targets, but a special feature of these neurons is the selectivity of their innervation. The most selective is the axoaxonic (also called *chandelier*) cell (Fig. 12.4A). It makes its synapses only on the initial segment of the axon of pyramidal neurons, where the action potential is believed to be initiated. Because of this specificity, the axonal morphology of these axoaxonic neurons is very characteristic, consisting of vertical rows of boutons aligned along the initial segment of the pyramidal axon. This radial alignment, seen in Golgi-stained material, initially suggested that the contacts were on the apical dendrites, but electron microscopic examination of identified axoaxonic terminals showed that they were in fact on the initial segment (Somogyi, 1977).

Each pyramidal neuron receives input from three to five axoaxonic cells, and each axoaxonic cell provides approximately 5–10 of the total of about 40 synapses on the initial segment (Somogyi et al., 1982; Peters, 1984). An axoaxonic cell contacts about 300 pyramidal neurons over a surface area of about 0.2–0.4 mm. Superficial layer axoaxonic cells predominantly innervate the pyramidal neurons of the superficial layers, but they also have a small collateral projection to the deep layers. Axoaxonic cells are also found in the deep layers, but pyramidal cells in the superficial layers appear to receive a richer input than those in the deeper layers (Sloper et al., 1979; Peters, 1984).

Basket cells. The cortical basket cells (Fig. 12.4B,C) resemble the basket cells of the cerebellum and hippocampus in that they form contacts around the somata of their targets. It was this aspect that was emphasized in the Golgi studies of Ramón y Cajal (1911). However, electron microscopic studies of a number of different subtypes of basket cell in the visual cortex produced the unexpected finding that a major portion of the basket cell output is onto the dendrites and particularly onto the dendritic spines (Somogyi et al., 1983). A single collateral of a basket cell can provide an input both to the soma and the dendrites of the same target cell. The sphere of influence of the basket cell varies with the layer in which it is found. In the superficial and deep layers of the cortex, the lateral spread of the arbors is largest, having a diameter of 1–1.5 mm (Fig. 12.4B). In the granular layer (layer 4), the basket cells (also called *clutch* cells) have a much more compact arbor, extending about 0.3–0.5 mm (Mates and Lund, 1983; Kisvárday et al., 1985). The basket cells project predominantly to the layer in which their somata are located. However, most basket cells have a small projection to other layers as well. Indirect calculations suggest that in all layers, each basket cell contacts about 300 target neurons. Each neuron receives input from about 10–30 basket cells.

Double-bouquet cells. Another smooth neuron that has a very characteristic axonal arborization is the double-bouquet cell (Fig. 12.4D). Unlike the basket cell, the predominant orientation of the axon of the double-bouquet cell is vertical. Regardless of where the soma is located, the axon of the double-bouquet cell spans all layers (Ramón y Cajal, 1911; Somogyi and Cowey, 1981). It was initially thought that the vertical arrangement of the axonal arbor meant that the axon was making multiple contacts along the apical dendrites of pyramidal cells. However, electron microscopic studies indicate that this is not the case. Instead, the double-bouquet axons form synapses onto the small- or medium-sized dendrites of smooth neurons (Somogyi and Cowey, 1981). The synapses, in common with those made by other smooth cells, are type 2. The extent of the axons of these cell types has not been mapped because all the material has been drawn from Golgi-stained material in which the axons are incompletely stained with silver. A spiny version of the double-bouquet cell has also been found in layer 4 of monkey area 17 (Somogyi and Cowey, 1981), but little is known of the full extent of its axonal arbor. As with all spiny neurons, this spiny double-bouquet cell also makes type 1 synapses.

Although other smooth cell types have been described at the light microscopic level with the use of Golgi-stained material (e.g., Lund, 1973; Peters and Regidor, 1981), the synaptic connections of these types await examination under the electron microscope.

BASIC CIRCUIT

Intrinsic cortical circuits have been examined from a number of different angles. The first attempts to decipher the connections were made by Ramón y Cajal (1911) using the Golgi technique. This approach relies on matching Golgi-stained pre- and postsynaptic elements on the basis of their complementary shapes, much as one would assemble a jigsaw puzzle. This method can provide a useful starting point if the region contains a small number of different neuronal types, and if these elements are relatively segregated and have a very characteristic morphology, as in the retina, hippocampus, or cerebellum. However, as we have seen for the examples of the chandelier cell and the double-bouquet cells, attempts to construct neocortical circuits on the basis of jigsaw-style matching have been unsuccessful. The reason is simply that the large number of different elements are intermingled, so that any axon has a large number of potential postsynaptic targets.

Golgi-stained cells can be examined under the electron microscope, and this combination has provided important information as to the connectivity of the different types of cortical neurons. However, neurons whose axons are myelinated are inevitably incompletely stained, so much information as to the extent and laminar specificity of the connections is missing. A further drawback of this purely anatomical method is that functional interpretations are necessarily speculative. Since the late 1970s, the introduction of new techniques has overcome the former limitations and revolutionized the analysis of the cortical circuitry. In particular, the method of recording intracellularly with dye-filled micropipettes has enabled the physiology, morphology, and connections of single neurons to be studied in a detail and with an accuracy that was not previously possible.

COLUMNAR ORGANIZATION

Functional methods of examining cortical circuits began with single-unit recordings of the receptive field properties. Recordings from somatosensory cortex showed that cells with similar properties were arranged in vertical columns extending radially from the pial surface to the white matter (Powell and Mountcastle, 1959). Similar findings were obtained in the motor and visual cortices, suggesting that cortical circuits are arranged in vertical columns with relatively little interconnection between the columns. As can be seen in Fig. 12.7, the dominant systems in the visual cortex are the ocular dominance and orientation columns (Hubel and Wiesel, 1977). Neurons lying within a single orientation column all respond selectively to bar stimuli of the same orientation. Similarly, cells lying at the center of an ocular dominance column are driven from only one eye. Binocular cells are found at the borders of the left- and right-eye ocular dominance columns. The columns dominated by the left and right eye alternate like the black and white stripes of a zebra. The constraints this columnar structure imposes can be seen even at the level of individual thalamic arborizations. Single arbors are generally larger than the width of a single ocular dominance column. In order to produce the structure of the ocular dominance columns, a single arbor therefore has to skip the inappropriate columns. The result is that a single arbor is not homogeneous, but forms patches with intervals between patches corresponding to the spacing between the columns (see Fig. 12.3A).

An ocular dominance "hypercolumn" is formed by one left-eye and one adjacent right-eye column. The orientation columns are also arranged in a regular manner, wherein each column has a slightly different orientation preference from its neighbor. As can be seen in Fig. 12.6, an orientation hypercolumn contains one complete set of all orientations. Viewed from the surface, each hypercolumn

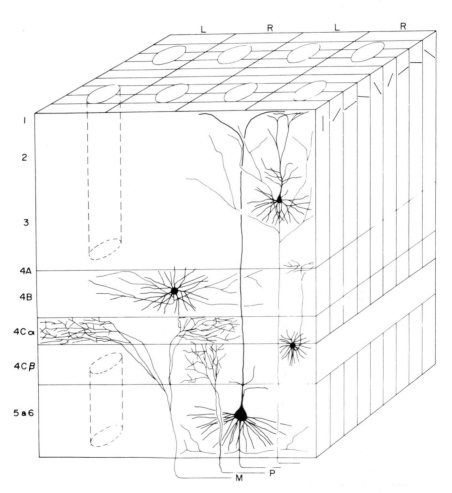

FIG. 12.6. "Ice-cube" model of visual cortex in the macaque monkey. L and R indicate ocular dominance "columns" or "slabs." The narrower orientation columns run orthogonally. The cytochrome oxidase rich "blobs" appear as cylinders in the center of ocular dominance columns. M, P: Thalamic afferents originating from the magno- and parvocellular layers of the lateral geniculate nucleus, respectively, and terminating in separate subdivisions of layer 4, within appropriate ocular dominance slabs.

system appears as a series of parallel slabs, stacked together like sliced bread. The two columnar systems have a complex interrelationship but for convenience are depicted as intersecting orthogonally. In the primate area 17, another prominent columnar system has recently been discovered. This system was revealed with stains for the enzyme cytochrome oxidase. The cytochrome oxidase-rich zones (Fig. 12.6) appear as a series of blobs (Hendrickson et al., 1981; Horton and Hubel, 1981; Wong-Riley and Carroll, 1984). The blobs lie in the middle of the ocular dominance columns and so are spaced about 0.5 mm apart. Neurons in the cytochrome oxidase columns lack orientation selectivity but are often selective for the wavelength of the stimulus.

SERIAL AND PARALLEL CIRCUITS

Further functional clues to the circuitry have been obtained by examining in more detail the receptive fields of neurons within a single column. Most neurons in layer 4 of the cat visual cortex are selective for the orientation of the stimulus, despite the fact that they receive a major input from thalamic neurons whose receptive fields are not orientation-selective. Nevertheless, the receptive field substructure of these neurons is closely related to the receptive fields of the thalamic afferents. These receptive fields are called "simple." Outside layer 4, the receptive fields appeared to be more "complex," and it was suggested that these arose through the convergence of input from many simple cells. These observations were used to develop the hypothesis of serial or hierarchical processing, in which the thalamic input arriving in the middle layers was further processed in other cortical layers before being relayed out again (Hubel and Wiesel, 1962).

Support for the notion of parallel processing in the cortex arose from studies in which the thalamic afferents were stimulated electrically and either the latency of response of single cortical neurons (e.g., Hoffmann and Stone, 1971; Bullier and Henry, 1979), or the field potential, was measured in different layers (Mitzdorf and Singer, 1978). Among the neurons with the shortest latencies were those lying in middle layers. However, neurons were found in both deep and superficial layers that also responded with short latencies, indicating that they were receiving direct input from the thalamic afferents (Martin, 1984). From the morphology of the neurons concerned, this may not be surprising: cortical neurons, pyramidal cells in particular, have dendrites that span many laminae. The thalamic input is also not homogeneous (Fig. 12.6; see Chap. 8), but is subdivided into several functionally distinct groups whose terminals tend to be segregated in different sublaminae of the middle cortical layers. Thus, in addition to serial processing, there is parallel processing of a number of separate incoming streams from the thalamus. In the cat, these are the X, Y, and W streams. In the primate, the basic parvo- and magnocellular streams may be further subdivided into functionally separate streams: those involving the cytochrome blob system and those that lie outside the blobs (Fig. 12.6; Livingstone and Hubel, 1987). The afferents of the magnocellular layers of the lateral geniculate nucleus form arbors that are much larger than those of the parvocellular layers, thus compensating to some extent for the fewer numbers of magnocellular neurons.

The Synaptic Organization of the Brain

CORTICAL OUTPUT

We have seen that the "output" neurons from the cortex are generally pyramidal cells. These same cells however, may also be "input" neurons, in that they may receive direct input from the thalamus. The pyramidal cells that project out of their cortical area are arranged in laminae, according to the position of their targets. Because all projection neurons have extensive intracortical collaterals (see Fig. 12.3C,D), there are no layers that have exclusively output functions. A simplified view of the laminar organization is provided in the basic circuit diagram of Fig. 12.7. As can be seen, the *corticocortical* connections ("association" fibers) arise from the neurons of layers 2 and 3. In all cortical areas, deep layer neurons (usually layer 5) provide *subcortical* projections to regions that are associated with motor activity—for example, basal ganglia, midbrain, brainstem, and spinal nuclei. Thus every cortical area has an output to a subcortical structure that is concerned with motor functions. The *corticothalamic* projections arise from the layer 6 neurons. It should be noted that there are exceptions to these general patterns. For example, in area 4, the subcortical projections (e.g., corticospinal) include neurons in layer 3, and in many areas corticocortical connections arise from neurons in the deep layers. However, the simplifications made here are generally useful, because they place constraints on the connections that can be made within the basic circuits. For example, if a circuit in cat visual cortex requires an output to the eye-movement maps of the superior colliculus, then necessarily it will have to connect to the pyramidal cells of layer 5.

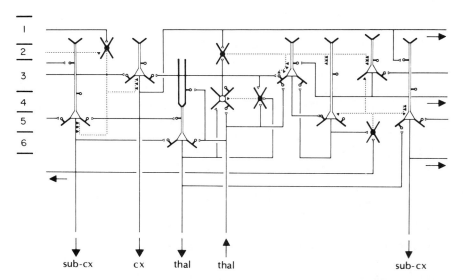

FIG. 12.7. Basic circuit for visual cortex. GABAergic neurons are indicated by black (filled) somata; their axons are indicated by dotted lines, their synaptic boutons by solid symbols. Spiny neurons are indicated by open somata; their axons are indicated by solid lines, their synaptic boutons by open symbols. thal: thalamus, cx: cortex, sub-cx: subcortical. Cortical layers are as indicated.

Neurons that project to other cortical areas also appear to be organized into columnar systems. For example, the neurons that provide the output from area 17 to area 18 are arranged in a series of slabs reminiscent of the ocular dominance hypercolumns (Gilbert and Kelly, 1975; Gilbert and Wiesel, 1981). Thus, as with the laminar constraints, circuits that require particular thalamic inputs or particular projection neurons will be constrained to form particular patterns of patchy connections. This may be the reason why the axons of cortical neurons do not fill a homogeneous region of cortex, but form patchy connections, as if they are skipping inappropriate columns in a manner analogous to the thalamic afferent innervation of the ocular dominance columns. The reasons for the patchy axonal arborizations of the cortical neurons remain speculative, but in most of the cases that have been studied the connections are made between columns that share stimulus preferences, for example, for the orientation or wavelength of the stimulus (T'so et al., 1986; T'so and Gilbert, 1988).

In elaborating the basic circuit (Fig. 12.7), both the vertical and the horizontal organization of the connections have to be taken into account. Historically, these two dimensions have been treated separately. The temporal aspects of processing have emerged from studies using electrical stimulation of the afferents. These, like the extracellular receptive field studies, have placed emphasis on the vertical connections and serial or hierarchical processing. The horizontal connections have come to prominence only recently through the use of the intracellular injections of horseradish peroxidase. Such intracellular staining has revealed how extensive the local collateral projections of thalamic afferents and cortical neurons really are. The spread of the thalamic and cortical arbors ensures that many neurons receive simultaneous activation from a single source, thus distributing their output for processing by many neurons in parallel. The horizontal connections necessarily involve spatial aspects of the receptive fields because of the topographical mapping of the thalamic and corticocortical inputs. This provides a general introduction to basic circuits in the cortex. Specific types of microcircuits will be discussed below, as well as a canonical circuit that may be present in all cortical areas.

SYNAPTIC ACTIONS

In considering the action of synapses, there are two key issues. One is the effect of synapses on the neuron at the site of the synapses; the other, the response of the whole neuron to the local synaptic actions. The latter issue includes the attributes of the neuron such as its membrane properties, the ionic currents involved, and the shape and cable properties of the neuron. In this section we follow the same outline as in Chap. 2, by discussing first the intrinsic membrane properties of cortical neurons, and then proceed to consider the effects of specific synapses and their receptors. The influence of shape and cable properties of the neuron will be covered in the section on Dendritic Properties.

INTRINSIC MEMBRANE PROPERTIES

Present methods of investigating whole-neuron function depend strongly on the measurement of electrical properties associated with membrane excitability. The most useful of these have been the measurement of membrane potentials and currents, and their response to various electrical and chemical manipulations of the neuronal membrane. The membrane potential depends on the relative conductance of the membrane for the various ion species that are present in the extra- and intracellular fluids, and also on the transmembrane concentration gradients of those ions. Some of the ionic conductances are modified during the events associated with synaptic transmission and action potential generation, and the effects of these modifications are reflected as changes in membrane ionic currents and potential. Changes in membrane conductances entail changes in membrane currents that in turn determine changes in membrane potential. In this sense, the characteristic patterns of membrane potential shifts observed by intracellular recordings in cortical neurons can be explained by the behavior of the underlying conductances and the currents they generate. These basic properties were explained in Chap. 2, and the reader should refer to that chapter as a foundation for the following discussion of specific properties that are important for the functions of cortical circuits.

The classical action potential in isolated squid axon depends on changes in just two ionic conductances: one for sodium, the other for potassium. Whole neurons are considerably more complex (Fig. 12.8). They contain a variety of additional conductance channels that vary in respect of ion permeability, voltage dependence, receptor coupling, and so on. As discussed in Chap. 2, each class of neuron has a characteristic set of these conductance channels, and each channel type has a characteristic spatial density distribution over the neuronal surface. The resulting dynamics of the conductance will determine the neuron's threshold

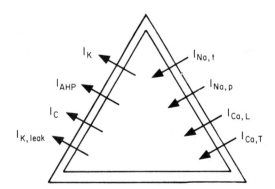

FIG. 12.8. Schematic showing ionic currents identified in cortical pyramidal neurons. Inward currents: $I_{Na,t}$, sodium action potential current; $I_{Na,p}$, persistent sodium current; $I_{Ca,L}$, high-threshold calcium current; $I_{Ca,T}$, low-threshold calcium current. Outward currents: I_K; delayed-rectifier potassium current; I_{AHP}, slow calcium-dependent potassium current; I_C, transient calcium-dependent potassium current; $I_{K,leak}$, leak current.

for action potential generation, efficacy of synaptic inputs, signal transmission in dendrites, and repetitive discharge properties.

Sodium currents. Generation of the action potential entails regenerative depolarization followed by a restorative repolarization. In cortical neurons, as in most other neurons, these two phases are mediated by a fast voltage-dependent inactivating sodium current (Connors et al., 1982), and a delayed voltage-dependent noninactivating potassium current (Prince and Huguenard, 1988), respectively. However, cortical neurons also exhibit a fast, voltage-dependent sodium current that does not inactivate (Stafstrom et al., 1982, 1985). The current is similar to the noninactivating sodium current observed in cerebellar Purkinje cells (Llinás and Sugimori, 1980a) and hippocampal pyramidal cells (Hotson et al., 1979), and analogous to the slow inward calcium current (I_i) seen in spinal motoneurons (Schwindt and Crill, 1980). In cortical neurons this "persistent" sodium current ($I_{Na,p}$) is activated about 10–20 mV positive to the resting potential and attains steady-state conductance within about 4 msec. The fast response, magnitude, and the range of voltage sensitivity suggest that $I_{Na,p}$ acts as an inward current amplifier for perithreshold depolarizing inputs such as EPSPs. Indeed, $I_{Na,p}$ can provide regenerative depolarization that is able to drive the membrane into the range in which the larger spike-generating sodium current is activated (Stafstrom et al., 1982).

Potassium currents. The inward currents that induce spike depolarization are opposed by an increase in outward currents that ultimately restore the neuronal membrane to its resting level. The action potential mechanism in the squid axon provides restorative outward current by just one delayed voltage-dependent potassium conductance. By contrast, the restorative outward current of cortical neurons is enhanced by several additional potassium currents. These currents influence the dynamics of membrane during the subthreshold response to depolarizing inputs, and also during postspike recovery. Consequently, they affect the repetitive discharge behavior of the neuron. These mechanisms are of particular interest because neurons respond repetitively to natural stimuli.

In the simplest case, a sustained inward current that depolarizes the neuron beyond threshold will evoke a train of action potentials. Each action potential ends with a repolarization that drives the membrane potential below threshold. The duration of the subsequent interspike interval will depend on the rate of postspike depolarization, because this rate will determine the latency to the next threshold crossing. If the time constants of the membrane currents are all short (i.e., of the order of an action potential duration), then the interspike intervals will be of equal duration, and the neuron will exhibit a rapid and regular discharge. In practice, some of the potassium conductances have much longer time constants and so their outward currents span successive interspike intervals. Because these outward potassium currents oppose the depolarizing inward currents, they retard threshold crossing and so increase the interspike interval. These interactions are the basis of *adaptation*—that is, the progressive lengthening of interspike interval that occurs during a sustained depolarizing input to some cortical neurons.

The outward potassium currents underlie the afterhyperpolarizations (AHP) seen in cortical neurons both in vivo and in vitro. Three separate AHPs have been identified in layer 5 neurons of sensorimotor cortex: a fast, medium, and slow AHP (Schwindt et al., 1988a). The fast AHP (fAHP) has a duration of milliseconds, and follows spike repolarization. It is often followed by a transient delayed afterdepolarization (ADP). The medium AHP (mAHP) follows a brief train of spikes. It has a duration of tens of milliseconds, and its amplitude and duration are increased by the frequency and number of the spikes in the train. The slow AHP (sAHP) is evoked by sustained discharge, and has a duration of seconds. All three hyperpolarizations are sensitive to extracellular potassium concentration, but they have different sensitivities to divalent ion substitutions and pharmacological manipulations (Schwindt et al., 1988a,b). This suggests that they are mediated by at least three distinct potassium conductances. However, the individual potassium conductances have not been identified completely. This is due to several problems, including the difficulty in comparing the characteristics of the many different potassium conductances found in various excitable cells, the many different methods of investigation, and the nonuniformities of nomenclature.

In summary, the following potassium conductances have been identified in neocortical cells: (1) a delayed rectifier; (2) a fast transient voltage-dependent (A-like) current; (3) a slow calcium-mediated (AHP-like) current; and (4) a slow receptor-modulated voltage-dependent (M-like) current (Connors et al., 1982; Schwindt et al., 1988a,b). Although the potassium currents of neocortical neurons are qualitatively similar to those reported in hippocampal neurons of archicortex, the cortical currents have rather different pharmacological sensitivities (Madison and Nicoll, 1984; Schwindt et al., 1988a,b). These, and other conductance differences, may point to important functional constraints on the discharge of neocortical neurons that are different from the discharge requirements of hippocampal neurons.

Calcium currents. Calcium currents also contribute to the dynamics of cortical neurons. Their effects may be direct, by contributing to the electrical behavior of the membrane; or indirect, by changing the internal calcium concentration, which in turn affects potassium conductance. Voltage-dependent calcium currents are inward currents and so are able to contribute to spike generation. If the calcium conductance is much larger than that for potassium, then it is possible to initiate a calcium-dependent action potential. Such calcium spikes have been evoked in hippocampal pyramidal cells, after blockade of the sodium currents alone (Schwartzkroin and Slawsky, 1977; Wong and Prince, 1978; Wong et al., 1979). However, the calcium currents of cortical neurons appear to be much smaller, and are revealed only after the sodium currents are blocked (with tetrodotoxin, TTX) and the potassium currents are depressed (with tetraethylammonium, TEA). Under these abnormal conditions, a calcium spike can be elicited from some cortical neurons (Stafstrom et al., 1985; Connors et al., 1982). The threshold for this spike is about 30–40 mV positive to the resting potential, and therefore well above the activation thresholds for the sodium currents, $I_{Na,p}$ and $I_{Na,t}$.

The need to depress the potassium conductance in order to reveal the calcium conductance implies either that it is larger in cortical neurons than in other calcium-spiking cells, or that the calcium conductance is smaller. An alternative explanation is that the site of the calcium conductance is located in dendrites that are electronically distant from the soma. In this case, any polarization large enough to activate the calcium conductance would also strongly activate the voltage-dependent potassium conductances that are located more proximally in the dendrites or soma. The resulting increase in potassium conductance would shunt depolarizing current injected into the soma, and so prevent the more distal dendritic membrane from reaching the threshold for calcium current activation.

Stafstrom et al. (1985) provide evidence that there are two calcium conductances in cortical neurons, and that these are differentially distributed over the neuronal surface. The somatic calcium conductance is slow and small, and has a high threshold. The dendritic conductance is larger than its somatic counterpart. Its threshold is also high, but this is partly due to electrotonic distance from the soma. Both currents contribute to the calcium spike. Somatic depolarization activates the somatic calcium current, and that in turn activates the more distal dendritic calcium conductance that powers the calcium spike. Both currents are probably persistent, and so they require activation of an outward current to effect the recovery phase of the spike. This outward current would be provided by the slow potassium (AHP) current that is activated by the influx of calcium in cortical and hippocampal neurons (Hotson and Prince, 1979; Madison and Nicoll, 1984; Lancaster and Adams, 1986).

Intracellular calcium. Both sodium and calcium can provide the inward currents necessary for membrane excitability. However, unlike calcium, increases in intracellular sodium have no striking effects on cellular metabolism. This suggests that neurons have separated the purely electrical requirements of signaling by action potentials from the metabolic requirements of neuronal communication. Voltage-dependent calcium channels probably arose in the earliest eukaryotes as a mechanism for coupling active cellular responses to events impinging on the external surface of the cell membrane (Hille, 1987). They are present in most neurons, where they contribute to the control of cellular functions by modification of the intracellular free calcium.

The normal intracellular concentration of ionized calcium is about 0.1 µM, which is about 10,000 times smaller than the extracellular calcium concentration. This differential is maintained by a number of buffering mechanisms that include uptake by the endoplasmic reticulum, mitochondria, and possibly the spine apparatus; binding to proteins such as calbindin and parvalbumin; and transport across the cell membrane (Burgoyne et al., 1983; Fifkova et al., 1983; McBurney and Neering, 1987). However, the intracellular free calcium can be increased by the influx that occurs through the voltage-dependent calcium channels of the membrane, and also by release from intracellular sequestration sites such as the endoplasmic reticulum.

Depolarization increases calcium influx through voltage-dependent calcium channels. Most of this influx is buffered, and less than 1% accumulates in the cytoplasm as free calcium (McBurney and Neering, 1985). However, even this small accumulation can result in a significant transient rise in the free calcium during a train of action potentials. When vertebrate neurons discharge at 30/sec for 5 sec, the ionized calcium rises to micromolar levels (Smith et al., 1983). The rise in internal calcium concentration activates the calcium-mediated potassium conductances discussed above, but it can also activate neuronal metabolic mechanisms.

On the most trivial level, the calcium influx accompanying membrane depolarization acts as a simple messenger that links membrane electrical events to cell responses—for example, the release of synaptic vesicles at axonal end terminals, or the activation of outward potassium currents to restore the polarization of the membrane. Here, the effects of calcium are short term, but unlike sodium and potassium, calcium entry also has a more lasting influence on the behavior of the neuron. The simplest example of this would be the changes in the intensity of discharge, and adaptation, discussed above. A more intriguing possibility is that calcium entry is coupled to learning/memory. Support for the latter notion has been obtained from studies of long-term potentiation.

Long-term potentiation. Brief tetanic stimulation of a set of input fibers potentiates synaptic activity of hippocampal excitatory synapses for days to weeks (Bliss and Lømo, 1973). This long-term potentiation (LTP) can be evoked by just one short burst of afferent activity, and so it offers a possible mechanism for memory storage. Hippocampal LTP depends on both calcium (Lynch et al., 1983) and depolarization (Wigstrom et al., 1986; Gustafsson et al., 1987), and it is blocked by N-methyl-D-aspartate (NMDA) receptor antagonists (Collingridge et al., 1983). Because LTP depends on both a presynaptic neurotransmitter release (implying presynaptic excitation) and a postsynaptic depolarization (implying postsynaptic excitation), it appears to be a realization of the Hebb rule for synaptic modification (Hebb, 1949). This rule asserts that the coincidence of pre- and postsynaptic activity induces a modification of synaptic efficacy. The coincidence of neurotransmitter release and postsynaptic depolarization is rather common in the nervous system, but LTP is not. So the NMDA channel, or the NMDA channel's association with calcium influx, may be the crucial factor in LTP.

Not all calcium influx can contribute to LTP, because not all voltage-dependent calcium conductances are sensitive to NMDA antagonists. On the other hand, if LTP is calcium dependent, then calcium influx via the normal voltage-dependent channels (e.g., in the soma and dendrites, Fig. 12.8) must be prevented from constantly activating the LTP mechanisms. This could be achieved by restricting the LTP mechanisms to a compartment where the calcium ingress is exclusively via NMDA channels. The dendritic spines are the obvious candidates for such a compartment. The narrow and extended spine neck could provide a diffusion barrier for calcium, and so restrict the flux of calcium between the trunk dendrite and the spine head. Calcium entering via the somatic or the dendritic voltage-dependent channels would not then affect LTP mechanisms located in the spine heads. If the only calcium channels on the spine head were the NMDA-mediated channels, then the spine head would fulfill the LTP requirement for calcium entry, depolarization, and sensitivity to NMDA antagonists. Moreover, spines contain a unique organelle, the spine apparatus, that has the cytochemical characteristics of a calcium store (Burgoyne et al., 1983; Fifkova et al., 1983) and accumulates calcium after synaptic activity (Andrews et al., 1988). This sequestration mechanism would prevent calcium from diffusing from one spine to the next and influencing the LTP mechanisms in adjacent spines. In this way, LTP could be restricted to particular synaptic inputs.

In this scheme, the metabolic process of LTP is activated within individual spine heads. However, spatiotemporal associative effects are possible between spine heads. Synaptic activation of a given spine head results in current injection into the trunk dendrite. The resultant depolarization of the trunk dendrite will depolarize also other spines on that trunk that are within short electrotonic distances. The passive depolarization of these heads predisposes them to LTP. The associative LTP is realized by synchronous synaptic excitation of the depolarized heads, which then activates NMDA receptors and so admits the necessary calcium. Thus the electrotonic depolarization could act synergistically with an excitatory input that was too weak to trigger LTP alone (Wigstrom et al., 1986).

The neocortex is assumed to be crucial to learning and memory, and so we might anticipate that LTP would be a prominent feature of cortical neurons. In fact, this is not the case (Komatsu et al., 1981; Bindman et al., 1988; Artola and Singer, 1987). LTP can be induced easily in cortical neurons that exhibit bursting (Artola and Singer, 1987), but other neocortical neurons must be transformed into hyperexcitable cells, or bursting cells, in order to induce LTP. For example, LTP can only by elicited in regular discharging neocortical neurons if the GABA antagonist bicuculline is present (Artola and Singer, 1987). Similar facilitation of LTP by blockade of inhibition occurs in hippocampus (Wigstrom and Gustafsson, 1983), and as in hippocampus, the LTP of neocortex is blocked by low concentrations of the NMDA receptor antagonist 2-amino-5-phosphonovalerate (AP5) (Artola and Singer, 1987). It is possible that LTP of the form seen in hippocampus is confined to a small subclass of neocortical neurons.

Repetitive discharges. Three types of cortical neurons have been differentiated on the basis of their action potential discharge in response to injection of sustained current. The types are fast spiking, regular spiking, and bursting cells (McCormick et al., 1985). When identified morphologically, the fast-spiking cells are the smooth neurons, whereas the regular-spiking and bursting cells are pyramidal neurons. Regular firing is the predominant behavior of pyramidal neurons; bursting responses are found only occasionally, usually in layer 5 pyramidal neurons.

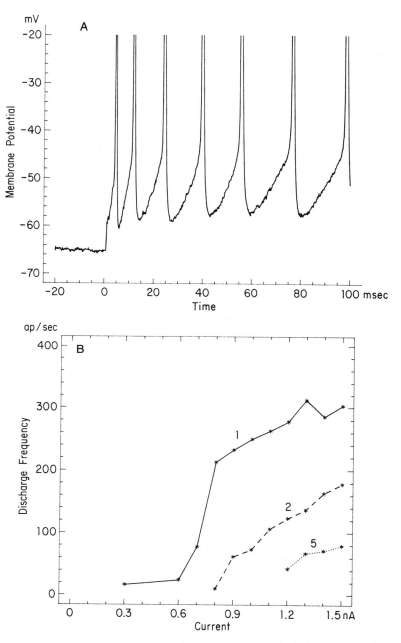

FIG. 12.9. Response of pyramidal neuron to constant depolarizing current. The neuron was recorded from layer 5 of rat visual cortex, in vitro. **A.** Record for single sweep, depolarizing constant current (0.6 nA) injected at time = 0. Note the slowing of the spike discharge as the neuron adapts. The tops of spikes are truncated. **B.** Current–discharge relationship for a pyramidal neuron, measuring the instantaneous spike frequency (ap/sec) for the first, second, and fifth interspike intervals. The relationship becomes flatter and more linear for later intervals.

Similar distinctions between pyramidal and smooth cells have been reported in turtle cortex (Connors and Kriegstein, 1986). The electrophysiology of the stellate neurons has not been reported.

The action potentials of fast-spiking cells have short durations in comparison with those of regular-firing pyramidal neurons. The repolarization phase of the action potential is rapid and followed by a significant undershoot. This indicates the presence of a large and fast repolarizing potassium current. Fast-spiking cells have a higher density of "delayed-rectifier" potassium currents than do pyramidal neurons (Prince and Huguenard, 1988). The spike repolarization is followed by a transient afterhyperpolarization, but outward currents with longer time-constants are not prominent, and so the membrane depolarizes quickly to threshold again. Consequently, a cardinal attribute of these cells is that they are capable of high discharge rates with little or no adaptation. Because adaptation places limits on the duration that a neuron can fire at its maximum discharge rate, the correlation of the degree of adaptation with a cell's morphological type may have important functional consequences. The fast-spiking and minimal adaptation of the smooth (presumptive inhibitory) neurons may reflect their need to sustain maximum inhibition of their target cells. By contrast, the excitatory (pyramidal) neurons respond less strongly for a given current input, and have lower maximal discharge rates.

A feature of the pyramidal cells is that they adapt (Fig. 12.9)—that is, they respond initially with a rapid discharge—but then relax to more moderate rates. The instantaneous frequency curves (Fig. 12.9B) show that the magnitude of the synaptic current required to bring the neuron to firing threshold is small compared to that required to maintain a reasonable rate of discharge. The adaptation and current–discharge relations of these cells are sensitive to neurotransmitters that are able to influence the outward current conductances. For example, activation of muscarinic receptors of pyramidal neurons reduces their outward potassium M current, and so enhances the effect of depolarizing currents (McCormick and Prince, 1986). Similarly, cholinergic stimulation decreases the slow calcium-activated potassium (AHP) current of cortical pyramidal neurons (McCormick and Prince, 1986). These modulations of slow outward currents provide a means whereby the cholinergic afferents can control both the current discharge relation and the adaptation of cortical cells. Similar modulations have been observed in hippocampal neurons (Bernardo and Prince, 1982; Cole and Nicoll, 1984; Madison and Nicoll, 1984).

Bursting neurons (Connors et al., 1982; McCormick et al., 1985) respond to depolarization by generating a short burst of about three spikes, followed by a constant, low-frequency discharge for as long as the depolarizing current is maintained. In simulation experiments, the regular discharge of a model neuron can be transformed into a bursting discharge simply by reducing the fast outward current (I_C, Fig. 12.8) in the presence of a dominant AHP current (Berman et al., 1989). Reducing the fast outward current encourages a short interspike interval and consequently a rapid discharge. The discharge is ultimately terminated by the growing AHP current. Thus, variations in the parameters of the same outward current conductances can determine whether a pyramidal neuron discharges in regular or burst mode. This model is supported by the report of Schwindt et al. (1988a) that a reduction of the transient fast potassium conductance converts normal firing into burst firing. The bursting mode of cortical pyramidal neurons may be required to achieve sustained depolarization of presynaptic terminals involved in NMDA transmission, as discussed above.

EXCITATORY SYNAPSES

For the purposes of this discussion we consider the action of two synaptic types: those that tend to produce action potentials and those that tend to inhibit action potential discharge. Action potentials are initiated when the depolarizing currents induced by excitatory synapses reach a threshold. The inward (depolarizing) currents that produce the excitatory postsynaptic potentials (EPSPs) are contributed by synaptic excitatory currents, and nonsynaptic sodium and calcium currents. The outward (hyperpolarizing) currents that produce inhibitory postsynaptic potentials (IPSPs) are contributed by synaptic inhibitory currents, and nonsynaptic potassium currents. The triggering of an action potential depends on the balance of inward and outward currents, as well as the form and time-dependence of the current–voltage relation (Jack et al., 1975; see Chap. 2).

Mechanisms for excitation. For simplification we will divide the receptors involved in excitation into those that have NMDA as their specific agonist, and those that do not. Activation of the non-NMDA receptors evokes a short duration EPSP (milliseconds) that is driven by an increase in conductance to sodium and, to a lesser extent, potassium (Hablitz and Langmoen, 1982; Crunelli et al., 1984). The non-NMDA channel has a low permeability to calcium, and is not voltage dependent. By contrast, activation of the NMDA receptor evokes a long-duration EPSP (tens to hundreds of milliseconds) with a reversal potential of approximately 0 mV. The EPSP is driven by a voltage-dependent conductance that provides a generalized increase in conductance to sodium, potassium, and calcium (Ascher and Nowak, 1987; Jahr and Stevens, 1987; MacDermott and Dale, 1987). Although a significant flux of calcium can occur through NMDA channels, they are much less conductive for calcium than the voltage-dependent calcium channels. Consequently, sodium and potassium carry a large fraction of the total current delivered through NMDA channels.

At the resting potential, the NMDA channels are blocked by magnesium ions. As the membrane is depolarized, the magnesium ions are displaced from the NMDA channels, allowing the influx of other ions like calcium (see Chap. 2). Thus, the contribution of the NMDA channels to the amplitude of unitary synaptic potentials will depend on the degree to which synaptic currents are able to depolarize the postsynaptic membrane and thereby reduce the magnesium blockade. Increasing depolarization will cause the voltage-dependent NMDA channels to become more conductive, and so contribute to further depolarization, to the extent of inducing a negative slope conductance as the NMDA channels are released from the magnesium block. The necessary synaptic depolarization could be generated by the non-NMDA component of the afferent excitation, or by non–NMDA-mediated excitation from other afferents. Conversely, inhibitory processes that hyperpolarize the membrane would disable the NMDA mechanism.

Locations of excitatory inputs. The major fraction (65–85%) of excitatory input to cortical spiny cells is via their spines, the remainder being onto the shafts of dendrites. Smooth cells receive excitatory contacts directly on their somata, but the major input is to their dendritic shafts. The amplitudes in the soma of single-spine EPSPs are not known. However, the small amplitudes of the EPSPs recorded from the soma in cortical slices after activation of single afferents (Komatsu et al., 1988; Thomson et al., 1988), suggest that in the order of tens to hundreds of synapses must be activated simultaneously to reach and cross the threshold for action potentials. The problem of threshold crossing is complicated by the temporal form of the PSP in the soma. The time course depends on the electrotonic properties of the dendritic tree, the distance of the synapse from the soma, and the duration and form of the synaptic current (Jack et al., 1975; see Appendix).

The somatic PSP that is obtained by the same brief pulse of synaptic current applied at various sites on a pyramidal neuron is shown in Fig. 12.10. The peak amplitude decreases, and the half-width and time-to-peak increase with distance from the soma. Thus, proximal synapses would provide more transient inputs than distal synapses, which have a longer time response and so provide relatively sustained inputs. The responses shown in Fig. 12.10 give theoretical values for the Rall-type cable model (see Appendix). Experimentally, the picture may vary for different neurons. For example, the EPSP amplitudes in the somata of cat spinal motoneurons and hippocampal pyramidal neurons have been found to be independent of the distance of the excitatory synapse from the soma (see Chaps. 4 and 11; Andersen et al., 1980b; Jack et al., 1981a; Redman and Walmsley, 1983a). This suggests that the current injected by synapses on distal dendrites must be much greater than that at more proximal sites (Jack et al., 1981a; Redman and Walmsley, 1983a). One hypothesis is that the increase in synaptic current in distal dendrites is due to an increase in the number of activated receptors per synapse (Jack et al., 1981a,b; Redman and Walmsley, 1983a,b; see Chaps. 4 and 11 for further discussion of these issues).

INHIBITORY SYNAPSES

The existence of inhibition in the cortex has been demonstrated repeatedly by means of intracellular recording (Phillips, 1959; Creutzfeldt et al., 1966; Toyama et al., 1974). Immunocytochemical techniques indicate that 15–20% of cortical neurons contain GABA (Gabbott and Somogyi, 1986), and there is ample anatomical evidence that the smooth (GABAergic) neurons of cortex make type 2 synapses on other cortical neurons (Le Vay, 1973; Ribak, 1978; White and Rock, 1980; Freund et al., 1983; Peters, 1987). There is also good agreement between the membrane effects of exogenous GABA and those of electrically evoked

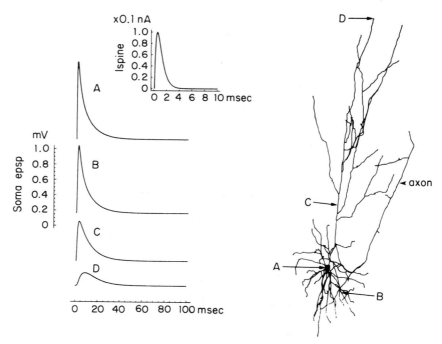

FIG. 12.10. Simulation of excitatory synaptic potentials evoked in the soma of a pyramidal neuron (recorded and filled in vivo with horseradish peroxidase in layer 6 of cat visual cortex). An idealized synaptic current (inset) was applied to a model spine placed at locations A–D (arrows) on the dendritic tree, and the resultant EPSP amplitudes detected by a recording electrode at the soma are indicated on the left. Compare the time-to-peak, duration, and amplitude of the potential with those of the synaptic current. For this example, a large value was assumed for the synaptic current; if the actual value was lower, the EPSP amplitudes were scaled down proportionally.

IPSPs in cortex (Krnjević and Schwartz, 1967), whereas the application of GABA antagonists disrupts normal cortical processing (Sillito, 1975; Tsumoto et al., 1979; Vidyasagar and Heide, 1986). Despite the impressive case for the presence of GABAergic inhibition in cortex, however, the topic of inhibition remains of intense interest, because the mechanisms of inhibition are considerably more complex than originally supposed, and because the precise role of inhibition in generating stimulus-specific responses in cortical neurons remains unresolved.

Mechanisms: linear and nonlinear inhibition. Inhibition implies that the action of a presynaptic neuron causes its postsynaptic target to become less sensitive to excitatory input. The orthodox view of inhibitory mechanisms is that they operate by one of two simple electrical principles (Bloomfield, 1974). In the first case, the membrane potential is polarized further away from threshold; in the second case, the membrane conductance is increased, and more of the excitatory current leaks out of the neuron. The first strategy is variously called additive, subtractive, hyperpolarizing, or linear inhibition; the second is variously called multiplicative, divisive, shunting, silent, or nonlinear inhibition.

For its effect, linear inhibition relies entirely on increasing the difference between the resting membrane potential and the threshold for action potential generation. The maximum difference that could be achieved would be between the membrane potential and the IPSP reversal potential. In a typical cortical neuron this amounts to adding about 10 mV to the 15–20 mV normally required to reach threshold. Thus the synaptic current required to drive the membrane to threshold is increased by about 50%. But the current–discharge relations of pyramidal cells (Fig. 12.9) show that the dynamic range of the neuron is about 3–4 times the threshold current. Consequently, linear inhibition could suppress only a small excitatory load. To achieve inhibition over the full range of excitatory load the neuron must either reduce the excitatory input current at its source, or shunt the input current before it reaches the axon initial segment. In this case, the inhibitory mechanism effectively multiplies the excitatory input by a fraction (equivalent to division). The IPSP reversal potential may be identical with or close to the resting membrane potential, and the conductance change is much larger than is the case for maximal linear inhibition—that is, hyperpolarizing the membrane to the reversal potential.

Nonlinear postsynaptic inhibition of this kind was reported first by Fatt and Katz (1953) in a crustacean muscle. More recently, conductance increases associated with inhibition have been demonstrated in a variety of neurons. These shunting conductances appear most marked in brain-slice preparations of neocortex and hippocampus, but the functional role of the shunting inhibition is far from clear. Large conductance changes occur only transiently, shortly after the onset of the IPSPs. Consequently, the nonlinear interaction is confined to the initial 15–25 msec of the IPSP (i.e., the period of maximum conductance increase) and is relatively weak by comparison with the hyperpolarizing inhibition, which far outlasts the shunting phase (Dingledine and Langmoen, 1980; Ogawa et al., 1981). Moreover, in vivo recordings from the soma and proximal dendrites of neurons in the visual cortex have shown that the inhibition of action potential discharge that occurs during visual stimulation is not associated with large and sustained conductance increases, despite the obvious potency of the inhibition (Douglas et al., 1988). One interpretation of this surprising finding is that the inhibitory synapses are located in positions that make them undetectable by recordings in the soma.

Locations of inhibitory inputs. Anatomical and immunocytochemical studies have shown that there are a number of morphological classes of inhibitory neurons in cortex, and that these classes have characteristic patterns of projection. The major concentration of their symmetrical (Gray's type 2) GABAergic synapses are located on the proximal dendrites and soma (Le Vay, 1973; Ribak, 1978; White and Rock, 1980; Freund et al., 1983; Peters, 1987), which is consistent with the electrophysiological findings. It is not known whether inhibitory neurons selectively activate $GABA_A$ or $GABA_B$ receptors.

The presynaptic bouton is the earliest point in the excitation sequence where inhibitory control could be applied (see Fig. 1.2, Chap. 1). Here, the release of a inhibitory neurotransmitter could disable the mechanism of excitatory neurotransmitter release and so prevent or reduce excitation of the postsynaptic target. Presynaptic inhibition is interesting from a computational point of view because it offers a means of applying inhibition selectively to specific excitatory afferents, while leaving the remainder unaffected. However, there is no anatomical evidence for presynaptic inhibitory circuits in cortex such as those found in the spinal cord (see Chap. 4).

Some degree of selective inhibition could also be obtained by placing inhibitory and excitatory inputs on the same dendritic spine (Fig. 12.5). This arrangement could permit inhibitory control of the excitatory synapse either by hyperpolarizing the spine head to a level that inactivates the excitatory process (NMDA channels, for example), or by increasing the local membrane conductance enough to shunt some of the excitatory current and so reduce the excitatory current that is injected into the trunk dendrite. Although the anatomical preconditions for selective inhibitory control of spines exist in cortex, it is not clear whether this is a characteristic of cortical processing.

Only about 7% of all spines in the visual cortex receive a dual input from asymmetrical and symmetrical synapses (Beaulieu and Colonnier, 1985) suggesting that only a small fraction of cortical processing could depend on spine-based selective inhibition. Of course, it may be the case that this mode of inhibition is applied only to a specific subset of excitatory afferents—for example, the geniculocortical afferents. However, the available anatomical evidence does not support this proposal (C. Dehay, R. Douglas, K. Martin, C. Nelson, and D. Whitteridge, unpublished observations). Single axon collaterals of the GABAergic basket cells also tend to make multiple contacts with their postsynaptic targets; for example, a single basket cell collateral can form synapses on the soma, the proximal dendritic shafts, and the spines of the same pyramidal cell (Somogyi et al., 1983). This appears to make selective inhibition of spines less likely. So, if inhibition is to act selectively rather than globally, that selectivity must be at the level of whole dendrites, or portions of them.

Strategies of inhibitory actions. An analysis of the inhibitory potential that follows electrical stimulation of cortical afferents shows that the IPSP is composed of at least two components that have different temporal, biophysical, and pharmacological characteristics (Avoli, 1986). A fast component is prominent in the early phase of the IPSP, whereas a slower component dominates the late phase. The fast component has a reversal potential that is close to the resting membrane potential and is accompanied by a sharp increase in neuronal input conductance that decays away with a time constant of about 20–30 msec. The magnitude of this early conductance change is more dramatic in vitro than in vivo. Neurons in slices show an increase of about ten fold (Ogawa et al., 1981), whereas neurons in vivo show a twofold increase, or less (Dreifuss et al., 1969; R. Douglas and K. Martin, in preparation). The slow component involves a hyperpolarization of long duration (100–300 msec) and large amplitude (5–15 mV; Fig. 12.11). The

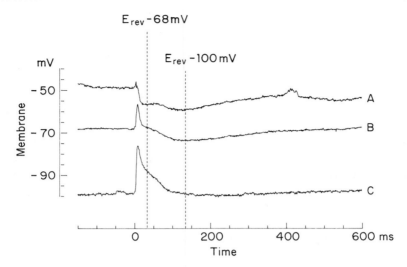

FIG. 12.11. Postsynaptic potentials evoked by electrical stimulation (at time = 0) of the white matter. The sequence of EPSP followed by IPSP is typical of all cortical neurons, here recorded intracellularly in rat visual cortex, in vitro. The IPSP has two components, as shown by their different reversal potentials (E_{rev}) and time courses. **A.** Membrane depolarized by constant current. **B.** Resting potential. **C.** Membrane hyperpolarized.

amplitude of the late component can be reduced by hyperpolarization, but usually does not reverse completely. The apparent reversal potential is about 15 mV more negative than resting potential, and so more negative than that of the fast component. This suggests either that the two components depend on different ion conductances, or that the later component arises at a site that is electrotonically distant from that of the early component. The incomplete reversal may indicate that the conductance that mediates the slow component is voltage dependent. The conductance changes associated with the slow component are very much smaller (about 20%) than those of the fast component.

In hippocampal slices the fast and slow components of the IPSP show differential sensitivity to $GABA_A$ and $GABA_B$ antagonists (Dutar and Nicoll, 1988); the early component is selectively blocked by bicuculline, whereas the later component is selectively blocked by phaclofen. Neocortical neurons exhibit similar sensitivities in vitro and in vivo. These findings suggest the following arrangements. GABAergic afferents acting on the soma operate via $GABA_A$ receptors, producing a relatively large but short-lived increase in chloride conductance that polarizes the soma toward the equilibrium potential for chloride. The $GABA_A$ phase has a short latency, and a duration of a few tens of milliseconds. GABAergic afferents acting on the dendrites operate via $GABA_B$ receptors. These receptors effect a relatively small increase in potassium conductance that hyperpolarizes the dendrites toward the equilibrium potential for that ion.

The GABA$_B$ phase may have a short latency, but it evolves slowly and persists for approximately 100–300 msec (Connors et al., 1988; N. Berman, R Douglas, and K. Martin, unpublished observations).

It is not known whether the fast and slow components are always present in concert during natural stimulation. Their simultaneous activation may be an artifact of nonselective electrical stimulation of input fibers. Nor is it known whether the two components are separately controlled by different classes of GABAergic smooth neurons. If we assume that the two components are separable, then several different roles in cortical processing seem possible. The proximal location, larger conductance changes, and metabolic independence of the GABA$_A$-mediated inhibition suggest a straightforward biophysical role in inhibiting any excitation arriving on that neuron. Of course, the GABA$_A$-mediated inhibition may be activated in a very precise way by the cortical circuitry. In contrast, the dendritic location, more negative reversal potential, and calcium sensitivity of the GABA$_B$-mediated inhibition suggest a role in the control of calcium spiking, or perhaps NMDA transmission. The GABA$_A$ and GABA$_B$ receptors may have opposite dependencies on calcium. The GABA$_A$ receptor sensitivity is reduced in the presence of the raised intracellular calcium (bullfrog sensory ganglion; Inoue et al., 1986), whereas reduction in intracellular calcium blocks the GABA$_B$ response (hippocampal pyramidal neurons; Blaxter et al, 1986).

NEUROTRANSMITTERS

EXCITATORY AMINO ACIDS

In the cortex there are four receptor types for excitatory amino-acids: *N*-methyl-D-aspartate (NMDA), quisqualate, kainate, and L-2-amino-4-phosphonobutyrate (L-AP4). Because of the lack of specific receptor antagonists they can, for all practical purposes, be grouped into just two functional classes: the NMDA receptor, and the non-NMDA receptors. The NMDA receptor agonists of note are NMDA itself, aspartate, and L-glutamate; a specific antagonist is D-2-amino-5-phosphonovaleric acid (D-AP5). Important agonists of the non-NMDA receptors are kainate, quisqualate, and L-glutamate: The receptors are antagonized by the nonspecific excitatory amino acid antagonists such as kynurenate.

Both glutamate and aspartate are putatative excitatory neurotransmitters in neocortex (Krnjević and Phillis, 1963b; Baughman and Gilbert, 1981; Ottersen et al., 1983; Hicks, 1987). Both glutamate and aspartate are released when the cortex is stimulated (Hicks et al., 1985). Stimulation of the fibers of the corpus callosum, which link the two hemispheres, evokes an EPSP with NMDA characteristics in layer II/III neurons of neocortical slices (Thomson, 1986). The EPSPs increase in duration and amplitude with depolarization and decrease with hyperpolarization around the resting potential. In the absence of magnesium, they become larger and display a normal dependence on membrane potential. The presence of both NMDA and non-NMDA receptors has been demonstrated pharmacologically and in binding studies (Thomson, 1986; Hagihara et al., 1988). D-2-Amino-5-phosphonovalerate (D-AP5) selectively blocks the NMDA receptors of cat visual cortex, and kynurenate appears to be a nonspecific blocker of all cortical excitatory amino-acid receptors (Perkins and Stone, 1985; Tsumoto et al., 1986; Huettner and Baughman, 1986). So far, however, the role of the NMDA receptors in the adult visual cortex appears rather minor.

Glutamate activates both non-NMDA and NMDA receptors, and so one might expect to see a postsynaptic EPSP composed of two components: a fast transient component with normal EPSP characteristics (due to non-NMDA channels), and a more protracted D-AP5 sensitive component with the voltage-dependent characteristics, but in practice it is difficult to elicit the NMDA component in cortical neurons. However, EPSPs are not readily antagonized by D-AP5 except in the presence of GABA blockade and low extracellular magnesium (Herron et al., 1985; Jones and Baughman, 1988).

GABA

When GABA is applied to either hippocampal or neocortical neurons, it produces several separable responses. The reason is that GABA activates more than one mechanism, and that these mechanisms are located in different regions of the neuronal membrane. Application of GABA to the somatic region of hippocampal or cortical neurons evokes a hyperpolarization that is chloride dependent, has a reversal potential of about -70 mV, and is accompanied by a large decrease in neuronal input resistance (Alger and Nicoll, 1979, 1982; Andersen et al., 1980a; Thalman et al., 1981; Scharfman and Sarvey, 1987; Connors et al., 1988). The somatic chloride channel is activated by the $GABA_A$ receptor. This receptor is selectively blocked by bicuculline, whereas the chloride channel itself is blocked by picrotoxin.

The dendritic response to GABA is more complicated than that of the soma. It can be hyperpolarizing, depolarizing, or both. Because the recording site is usually intrasomatic, *dendritic hyperpolarization* or *depolarization* refers to the response in the soma when GABA is applied to the dendrites. Consequently, the electrical observations of these potentials may be distorted by electrotonic effects, and results of pharmacological manipulations that depend on intracellular injection may be distorted by intracellular diffusion delays.

Dendritic hyperpolarization is pharmacologically and biophysically distinct from somatic hyperpolarization. It is not blocked by the $GABA_A$ antagonists bicuculline and picrotoxin (Alger and Nicoll, 1982; Blaxter et al., 1986; Connors et al., 1988). Instead, the response depends on activation of the $GABA_B$ receptors, which are weakly antagonized by phaclofen (Kerr et al., 1987; Dutar and Nicoll, 1988; Karlsson et al., 1988). The dendritic hyperpolarization is due to an increase in potassium conductance and has a reversal potential of about -80 mV (Blaxter and Cottrell, 1985; Blaxter et al., 1986). The $GABA_B$-mediated conductance is probably located on the dendrites and not on the soma because the hyperpolarizing response is optimized by GABA application to the dendritic field, and patch-clamp studies confirm that the somatic GABA-sensitive conductance is dominated by chloride permeability (Ashwood et al., 1987). Although both the $GABA_B$ response and the (somatic) AHP are sensitive to extracellular potassium, the slope dependence of the $GABA_B$ response is somewhat less than that anticipated from the Nernst relationship (Blaxter et al., 1986). This reduction in slope may be due to the electrotonic distance of the site of the $GABA_B$-mediated conductance from the somatic recording site. Finally, the blockade of $GABA_B$ response by intracellular EGTA* (injected at the soma) takes about twice as long as block of the AHP (Blaxter et al., 1986).

The dendritic hyperpolarization is sometimes replaced, or preceded by dendritic depolarization that has a reversal potential of about -50 mV (Alger and Nicoll, 1979; Andersen et al., 1980a; Thalman et al., 1981; Blaxter et al., 1986; Scharfman and Sarvey, 1987). The response is accompanied by a moderate decrease in input resistance. It is chloride dependent and is blocked by the $GABA_A$ antagonists bicuculline and picrotoxin (Alger and Nicoll, 1982). Thus, the properties of this mechanism are similar to those of the somatic $GABA_A$ conductance. The fact that depolarization occurs in the dendrites but not in the soma is explained by a spatial nonuniformity in the equilibrium potential for chloride. The dendritic equilibrium potential for chloride is less negative than the resting potential (Blaxter and Cottrell, 1985), whereas the somatic equilibrium potential for chloride is the same, or more hyperpolarized, than the resting potential, as discussed above.

The depolarizing response may have no significant role in normal GABAergic synaptic transmission, because the receptors appear to be extrasynaptic (Alger and Nicoll, 1982; Brown, 1979; Brown et al., 1981). Moreover, the depolarizing mechanism has no direct interaction with the $GABA_B$-activated hyperpolarization. The dendritic hyperpolarizing response can occur in the absence of the dendritic depolarizing response (Blaxter et al., 1986), and reduction in extracellular chloride increases the size of the depolarizing GABA response without affecting the hyperpolarizing response (Blaxter et al., 1986).

ACETYLCHOLINE

Cholinergic input to cortex is derived from a small population of brainstem reticulocortical projections (Vincent et al., 1983) and a more profuse projection from the basal forebrain cholinergic neurons (Mesulam et al., 1983) that are themselves innervated by cholinergic afferents from the brainstem (Woolf and Butcher, 1986). The onset of muscarinic excitation is slow, and the response is sustained for many seconds. Ionophoretic application of ACh modifies the response to visual stimulation of most neurons in the cat visual cortex (Sillito and Kemp, 1983). The effect is usually facilitatory and seems to enhance the signal-to-noise ratio of the neuron, rather than being generally excitatory. The ACh control is mediated by a muscarinic receptor.

The peptide somatostatin selectively enhances ACh-induced excitation, although somatostatin itself inhibits spontaneous firing (Mancillas et al., 1986). ACh induces a depolarization of deep layer cortical neurons. The depolarization is accompanied by an increase in resistance. The reversal potential of the response is about that for potassium, and this suggests that the depolarization is due

---

*Ethyleneglycolbis(β-aminoethyl ether)-*N,N'*-tetraacetic acid.

to a decrease in the conductance for potassium (Krnjević et al., 1971) by modulation of the outward potassium M current (McCormick and Prince, 1986).

The slow depolarization of cortical neurons by ACh is preceded by a short-latency hyperpolarization and a decrease in resistance. This is probably due to rapid muscarinic excitation of the inhibitory smooth cells (McCormick and Prince, 1985, 1986). Some of the effects of acetylcholine are probably mediated by cGMP (Stone and Taylor, 1977). Neurotransmitters may also modulate currents that interact with the slow hyperpolarizing potassium currents. Schwindt et al. (1988b) have shown that low concentrations of muscarine abolish the slow AHP (sAHP), but at higher concentrations the sAHP is replaced by a slow afterdepolarization (sADP) that is not sensitive to potassium, nor to the sodium channel blocker TTX. The mechanism of the sADP is unknown.

In addition to the above effects, acetylcholine also evokes a transient early inhibition of pyramidal neurons. However, two findings indicate that this inhibition is probably an indirect effect of the excitation of inhibitory interneurons. Firstly the inhibition is mediated by a chloride conductance similar to that activated by GABA. Second, ACh has a rapid excitatory effect on the fast-spiking (presumably GABAergic) cortical neurons (McCormick and Prince, 1986), and smooth cells are known to have cholinergic afferents (Houser et al., 1985).

BIOGENIC AMINES

Norepinephrine depresses the spontaneous extracellular activity of most cortical neurons (Reader et al., 1979; Armstrong-James and Fox, 1983), but some deep cortical cells are excited by low concentrations and inhibited by higher concentrations (Armstrong-James and Fox, 1983).

The intracellular reasons for these effects is not known. In hippocampal neurons, norepinephrine induces a hyperpolarization that is accompanied by an increase in input resistance, but occasionally depolarization occurs instead (Madison and Nicoll, 1986a,b). Serotonin also has a predominantly inhibitory effect on cortical cells (Reader, 1978; Olpe, 1981).

NEUROPEPTIDES

Synaptic boutons are thought to release only one type of nonpeptide neurotransmitter. In addition to this, they may also release one or more neuropeptides that affect the performance of the postsynaptic neuron. The neuropeptides may have neurotransmitter-like properties, but more commonly they potentiate or block the effects of other transmitters. They may also produce changes in the metabolism of the postsynaptic target. Their response usually has a slow onset and a sustained time course, and may act at some distance from the release site. In neocortex, subsets of the GABAergic neurons may contain one or more of the neuropeptides somatostatin (SSt), cholecystokinin (CCK), neuropeptide Y (NPY), and substance P (Hendry et al., 1984; Schmechel et al., 1984; Somogyi et al., 1984; Demeulemeester et al., 1988). Vasoactive intestinal polypeptide (VIP) and substance P are usually associated with cholinergic axons (Vincent et al., 1983; Ekenstein and Baughman, 1984). The physiological role of neuropeptides on cortical neurons is largely unknown. Salt and Sillito (1984) showed that SSt sometimes inhibited and sometimes excited cortical neurons, but they were unable to demonstrate a modulatory effect of SSt on either GABAergic or cholinergic neurotransmission in cat cortex. Mancillas et al. (1986) found that somatostatin inhibited rat cortical neurons. The variety of actions of SSt could be explained by a condition-dependent interaction. Equally unremarkable effects have been obtained with CCK (Grieve et al., 1985b) and VIP (Grieve et al., 1985a,b). Both of these produce mild excitation in some neurons.

DENDRITIC PROPERTIES

The morphology of the dendritic trees may be important in so far as they affect the passive spread of synaptic currents to the initial segment. However, with the possible exception of the apical dendrite, there is no reason to believe that the detailed morphology of the dendritic tree is correlated with the computational role of the neuron. Most cortical neurons are electrotonically compact, whether determined electrophysiologically (Stafstrom et al., 1985) or anatomically (Fig. 12.10). Sustained (DC) inputs to the periphery of cortical dendritic trees (excluding the apical dendrites) are only moderately attenuated on route to the soma. Because the majority of synapses are located more centrally on the dendritic tree, they are affected even less. It appears that the soma dendrite does not cause a significant loss of a sustained signal. However, individual synaptic inputs are not sustained, but transient, and so contain a spectrum of frequency components. Electrotonic length increases with frequency, and thus the dendrites will attenuate more strongly the higher-frequency components of transient synaptic events (Fig. 12.12).

Linear cable models that are purely passive (Rall, 1977; see Appendix) assume that the effects on the soma of current injections at various locations on the soma dendrite summate linearly. This implies that the current sources do not interact with one another and that the properties of the cable are not modified by synaptic action. Neither of these two requirements is strictly upheld in real neurons. The synaptic current depends on driving potential, and thus on local membrane potential. To the extent that an active synapse is able to modify the local membrane potential of neighboring synapses, it will affect the driving potential at those synapses, and so may modify the synaptic current that can be delivered there. This raises the possibility of nonlinear interactions between the active synapse and its neighbors (Rall, 1964, 1967, 1977; see Appendix). These interactions will be particularly dramatic if the neighboring synaptic conductance is voltage dependent. In the case of NMDA channels, for example, the sensitivity of the NMDA synapse to its normal excitatory input could be sharply reduced by an adjacent active synapse that hyperpolarizes the membrane.

The conductance changes associated with the activation of synapse can modify the local electrotonic properties of the cable, and so change the contribution to the soma of neighboring synaptic inputs, as well as those located more distally on the same dendritic subtree. Of course, the magnitude of these interactions will depend on the timing of the various inputs. Shunting inhibition is a particular

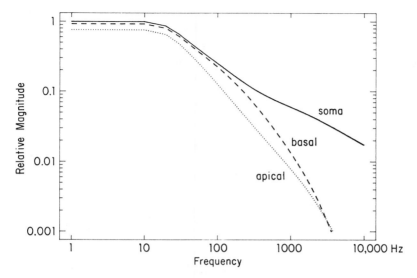

FIG. 12.12. Frequency–response relationship for a simulation of a current pulse input to the pyramidal neuron shown in Fig. 12.10. The pulse was applied at positions A, B, and D in Fig. 12.10, and the response was recorded in the soma. Note the marked attenuation in the response to frequency components above 30 Hz.

case of nonlinear interaction, in which a large conductance change (without a change in membrane potential) leads to a large local electrotonic decay.

The inhibitory inputs can have various degrees of selectivity. When they are applied proximally (perisomatically), they will be equally effective against all excitatory inputs. Those that are located more distally will be most effective against the excitatory inputs that occur distally on the same dendrite. If the excitatory and inhibitory inputs are widely separated (e.g., on different dendritic branches), then the interaction will be minimized (Rall et al., 1967; Koch et al., 1982). The distribution of inhibitory synapses on cortical neurons can be interpreted in terms of the above principles. The clustering of $GABA_A$ receptor effects around the soma suggests that $GABA_A$ mechanisms provide a global inhibition of cortical neurons. By contrast, $GABA_B$ receptor effects appear to be on the dendrites, and so are better placed for selective inhibition of individual dendritic branches. Some of the principles governing these interactions are discussed in Chaps. 1 and 2, and in the Appendix.

APICAL DENDRITES

A special feature of the pyramidal cell is its apical dendrite. Electrotonically, the apical dendrite presents a problem. If we make the usual assumption that the neuronal membrane has a uniform resistance, then the electrotonic length (L) of the apical dendrites is about 2–3, about three times greater than that of the basal dendrites. This implies that signals injected at the distal end of the apical dendrite will be greatly attenuated before reaching the soma (Fig. 12.10). This apparent ineffectiveness of the apical dendrite in contributing to the somatic potentials is counterintuitive. There has been no phylogenetic trend to dispense with apical dendrites (with the possible exception of layer 4 stellate cells). Therefore, it seems reasonable to suppose that the apical dendrite does have a significant function. One possibility is that the apical dendrite makes use of active conductances to enhance its signal transmission.

SPINES

One of the most prominent features of cortical neurons is their dendritic spines. Because they are the major recipients of excitatory input in the cortex, a consideration of their function is of some importance. One extreme view is encapsulated in the "spines-only-connect" hypothesis (Swindale, 1981), in which the spine is thought to be a device whereby en passant axonal boutons can more easily connect to dendrites (Peters and Kaiserman-Abramof, 1970). In the second view, spines are thought to be designed for important functional tasks, including the selective modification of transmission (Rall, 1974a,b). In theory, spines may respond to excitation in two possible ways. In the first condition the spine is electrically passive, and is simply a specialized site for synaptic input. In the second condition, the spine is the site of action potential generation, and these action potentials are the foundation of signal conduction and computation by the dendrites.

EXCITATORY INPUT

The membrane area of the spine neck is very small, and consequently little synaptic current flows through the neck membrane. Therefore, most of the synaptic current injected into the spine head reaches the trunk dendrite via the spine neck. But the resistance to passage of current through the neck is still rather high, perhaps of the order of 100 MΩ or more (Segev and Rall, 1988). The input resistance of a typical spiny dendrite at $L = 0.5$ is also of the order of 100 MΩ. So the spine neck will attenuate by about half the voltage applied at the spine head. Because synaptic currents must pass through the spine neck, it offers a mechanism for controlling the efficacy of the synapse simply by changing the impedance (Rall, 1962). Thus, changes in spine neck diameter and length (Rall, 1974a,b), or partial occlusion of the neck by the spine apparatus (Rall and Segev, 1987), could provide the basis of synaptic plasticity (Fifkova and Anderson, 1981).

The "twitching spine" hypothesis of Crick (1982) proposes that a change in spine length could be brought about by calcium activation of myosin and actin localized in the spine neck. Although the notion that spines "twitch" seems farfetched, it turns out that actin is present in spines and is orientated parallel to the long axis of the neck (Fifkova and Delay, 1982; Markham and Fifkova, 1986). In theory, burst discharges of excitatory afferents could raise the free calcium concentration to the micromolar levels required to activate the actin in the spine

(Gamble and Koch, 1987). Although cortical neurons are certainly capable of generating burst discharges, this mode of behavior occurs rather rarely, however (see above).

SATURATION

The flow of synaptic current through the spine input resistance will shift the spine head potential toward the PSP reversal potential and reduce the driving potential for the synaptic current. For small conductances the synaptic current increases approximately linearly with conductance, but at high conductances the synaptic current saturates. This interdependence gives rise to a sigmoid relationship between synaptic conductance and current. We do not know where on this relationship the operating range of the synapse lies. The observation that spines generally receive only one excitatory synapse can be interpreted in two opposing ways. It may reflect the need to avoid saturating the synapse. On the other hand, it may indicate that a single synapse is sufficient to saturate the synapse so that additional inputs would be redundant. If the synapse is driven into saturation (Fig. 12.13), then the spine potential will be relatively insensitive to the exact amount of neurotransmitter delivered to the synapse. The spine head will simply turn on to a repeatable voltage level. Because axial resistance of the spine neck is

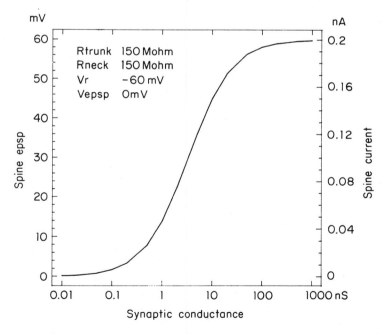

FIG. 12.13. Effect of synaptic conductance on the amplitude of EPSP in the spine head, and the magnitude of the current delivered through the spine neck resistance (R_{neck}) to the trunk dendrite (resistance, R_{trunk}) for a simulated spine and dendrite. V_r, resting potential; V_{epsp}, reversal potential of EPSP. Simulation of steady-state activation of synapse on the spine head.

at least as large as that of the dendritic trunk, the spine resembles a constant current source attached to the dendrite. The synapse on the spine head is less susceptible to changes in the dendritic input resistance than is a synapse located on the dendritic trunk.

The advantage of having spines that saturate is that it provides a means of simplifying the operation of the dendrites in two important ways. First, excited spines will approximate load-independent current sources, because the saturated driving potential is followed by the large-series resistance of the spine neck (Fig. 12.14). Second, the dendrites will approximate the behavior of a linear cable, because they are protected from large excitatory conductance changes by the spine neck resistances. Thus, whereas individual spines operate nonlinearly, their currents will sum approximately linearly in the dendrites. Moreover, because the spines operate in saturation, a wide range of input conductance amplitudes will be compressed into a narrow range of output. This will restrict the influence of any one synapse on the soma, and so prevent any single spine synapse from dominating transmission (Koch and Poggio, 1983).

SPINE ACTION POTENTIALS

A special case of saturation behavior arises if the spine head membrane contains active mechanisms for generating an action potential. Spinous action potentials restricted to the spine head must be distinguished from the dendritic action

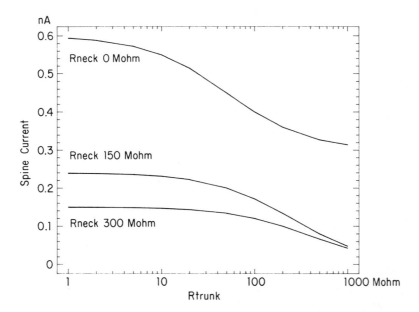

FIG. 12.14. Relationship of spine current to input resistance of trunk dendrite (R_{trunk}) for different neck resistances. Simulation of steady-state activation of synaptic conductance = 10 nS. Note that the current becomes more constant with higher spine neck resistances.

potentials that occur on dendritic branches or at nodes of cerebellar Purkinje neurons (Llinás and Sugimori, 1980b). Simulations have shown that the voltage-gated ion channels in the spine membrane could amplify the response to synaptic input (Jack et al., 1975; Miller et al., 1985), enhancing the amplitude of the dendritic voltage signal by a factor of 6 compared to the passive case (Rall and Segev, 1987). However, this effect is strongly dependent on the spine neck resistance. If the spine neck resistance is less than 400 MΩ, the differences in response between the passive and active cases are negligible (Segev and Rall, 1988).

Spine action potentials provide a mechanism for saltatory conduction of impulses from spine to spine, which is able to enhance the effectiveness of distal spines in controlling the output of a neuron (Shepherd et al., 1985; Rall and Segev, 1987). Spine action potentials also present interesting possibilities for dendritic processing. Rall and Segev (1987) and Shepherd and Brayton (1987) have shown how clusters of excitable spines on the same dendritic tree could interact cooperatively. The spatiotemporal ordering of cluster activation offers dramatic differences in somatic effect and exhibits properties analogous to logic operations such as AND, OR, AND-NOT (Shepherd and Brayton, 1987). These cooperative interactions could theoretically provide a logical foundation for dendritic processing (Rall and Segev, 1987).

If action potentials occur on the dendrites of neocortical cells, they should then be visible from recording sites in the soma, because cortical neurons are electrotonically compact. If the spikes occur only on spines, then these signals could be partially masked from the soma by the axial resistance of the spine neck. Nonetheless, even the attenuated spinous spike that occurs in the dendritic trunk could be relatively large in amplitude (Shepherd and Brayton, 1987), in which case it ought to be visible from the soma, particularly if it arose in the most proximal spines. There have been no experimental observations of action potential phenomena in normal cortical dendrites, despite considerable evidence for active dendritic properties in slice recordings (e.g., Masukawa and Prince, 1984). However, it may be difficult to distinguish attenuated dendritic and spinous action potentials from each other, and from EPSPs, in somatic recordings (Shepherd et al., 1989).

NONSATURATING SPINES

If the synapse on the spine is not driven into saturation, but operates instead on the linear segment of Fig. 12.13, then the synapse is particularly susceptible to nonlinear interactions with other spines. The resistance of the spine head membrane is much greater than the axial resistance of the spine neck: A potential change in the trunk dendrite therefore spreads into a spine with little attenuation. Thus, depolarizing of the trunk dendrite will depolarize the spine head, so reducing its driving potential. Because the original depolarization of the trunk dendrite was itself induced by synaptic input at other spines, it follows that spines within the same electrotonic neighborhood are able to modify the performance of their neighbors. These spine-to-spine interactions are possible only if spines operate in their linear range. If the individual spines operate in saturation, then the spinous synapses would be insensitive to voltage-mediated interactions of the kind described.

INHIBITION ON SPINES

A small proportion (about 10%) of neocortical spines receive input from a type 2 (GABAergic) synapse in addition to the type 1 (excitatory) synapse (Jones and Powell, 1969; Peters and Kaiserman-Abramof, 1970; Sloper and Powell, 1979a-c; Somogyi et al., 1983; Beaulieu and Colonnier, 1985; Somogyi and Soltez, 1986). This arrangement raises the possibility that some excitatory inputs receive selective inhibition. Nonlinear inhibition of excitatory inputs on the same spine can be large, and is essentially limited to the affected spine (e.g., Koch and Poggio, 1983). The spine necks interpose a large series resistance between the spine head and the trunk dendrite. This resistance masks changes that occur in the spine head from the trunk dendrite, and so restrict the inhibitory control to the affected spine. The magnitude of hyperpolarization in the spine head will be attenuated in proportion to the ratio of the spine neck resistance to the input resistance of the trunk dendrite (Jack et al., 1975; Rall, 1981). The magnitude of conductance presented by the affected spine to the trunk dendrite will be limited to that of the neck resistance. Inhibitory input to a spine head could allow relatively selective inhibition of an excitatory input to the same spine (Jack et al., 1975). This excitatory/inhibitory interaction on a single spine has attracted much theoretical interest (e.g., Koch and Poggio, 1983; Segev and Rall, 1988) because of its computational possibilities, but only a small percentage of the excitatory input to a single neuron could be gated in this way.

Less selective locations of inhibitory synapses may also allow strong nonlinear effects. For example, inhibitory synapses on the trunk dendrite would reduce interspine communication and could control saltatory conduction between spines (Shepherd and Brayton, 1987). This raises the possibility of switching on/off selected branches of the dendrites. However, logical computations in spines do not necessitate inhibitory inputs. Triggering an action potential could be conditioned by activation of the spine head synapse (e.g., the NMDA receptor is voltage dependent only if gated by neurotransmitter). If saltatory conduction were conditional on excitatory input, then this arrangement would provide an elegant means of signal gating that depends on the coincidence of excitatory inputs, rather than the interaction of excitatory and inhibitory inputs. Unfortunately, experimental tests of these hypotheses are some distance off. The spines are too small to permit the use of conventional electrophysiological recording techniques. However, the advent of voltage- and ion-sensitive fluorescent dyes, and sophisticated imaging techniques, may make such investigations feasible. For the present, we are confined to observations on the dendrites that can be made from the soma or proximal dendrites.

FUNCTIONAL OPERATIONS

SINGLE NEURONS OR NEURAL NETWORKS?

The history of ideas of cortical function makes a fascinating account of the interplay of hypothesis and experiment (Martin, 1988a,b). Since about the

1960s, the experimental results from microelectrode recordings from single cortical units (neurons) have had a deep influence on our ideas of cortical function. Much of the motivation for studying the functional properties of single units in the cortex in such detail arises from the fact that the activity of single cortical cells is thought by many neuroscientists to be related in a very simple way to our subjective experiences. This idea was formalized in the well-known "neuron doctrine" proposed by Barlow (1972). He proposed five dogmas that encapsulate the powerful idea that percepts are the product of the activity of certain individual cortical neurons, rather than by some more complex (and obscure) properties of "combinatorial rules of usage of nerve cells." The force of Barlow's thesis in molding our ideas is evident in most textbooks of psychology and neurobiology, which are well-stocked with illustrations showing how the selectivity of neurons arises from a hierarchical sequence of processing through the cortical circuits.

Recently, however, the pendulum has begun to swing back, and the antithetical proposition—that perceptual processing occurs through the collective properties of parallel cortical networks rather than through the activity of single units—has begun to receive close attention from theoreticians developing "neural network" or "connectionist" models of cortical function. Results obtained from computer simulations of these hypothetical nerve circuits have led to a model of cortical function that is quite different from that proposed in the neuron doctrine. In this section, we consider how the physiological and structural data we have described in the previous sections fits with these two opposing models of cortical function.

Essentially, what has to be decided is what weight has to be given to the activity of a single neuron. For the neuron doctrine, the activity of a single cortical neuron can be highly significant in causing a percept. For neural-network theories the activity of individual neurons in cortex is relatively insignificant. These opposing views of the operation of the cortex have a long and distinguished history. Sherrington (1941), for example, contrasted the notion of "one ultimate pontifical nerve-cell . . . [as] the climax of the whole system of integration" with the concept of mind as "a million-fold democracy, whose each unit is a cell." The dialectic of one versus many neurons is best considered in the context of the visual system, where the physiology and anatomy are known in great detail, and where the behavioral performance is well established.

A NEURONAL BASIS FOR VISUAL PERCEPTION

It is evident that visual perception is a complex task. We need not only to determine the form, movement, and position in space, of the objects we encounter, but also to recognize them as being particular objects. Solving this key problem was central to Barlow's development of the neuron doctrine. He proposed that the primary visual cortex dealt only with the elemental building blocks of perception, the detection of orientated line segments, or the local motion of these segments, for example. In order to build these into neurons that were selective for a cat, chair, or grandmother, he proposed a hierarchical sequence of processing within single cortical areas and through the many visual areas (Fig. 12.15). The neurons at each stage of the hierarchy would become progressively

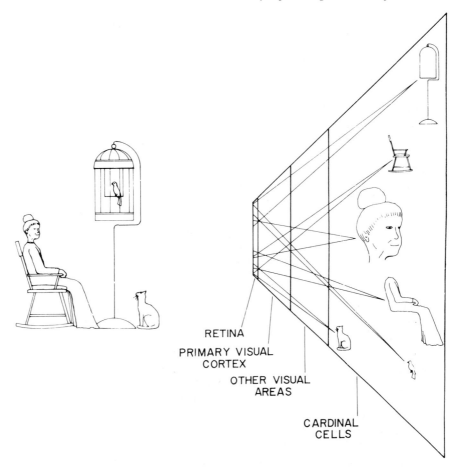

FIG. 12.15. Pictorial representation of Barlow's neuron doctrine. From retina to visual cortex, progressively more neurons are available to represent a visual scene, but the increasing stimulus selectivity of the neurons means that at each stage of processing, fewer and fewer neurons are active. The cardinal cells are at the highest level, and their extreme selectivity (e.g., for grandmother's face) means that only 1000 or so need to be active to represent a given scene. Large numbers of cardinal cells are necessary to represent all possible scenes.

more selective to the attributes of the stimulus, so that while the neurons in the primary visual cortex would respond to many different objects, neurons at the highest level of the hierarchy of processing would respond only to particular objects. Barlow (1972) suggested that the activity of about 1000 of these high-level "cardinal" neurons would be sufficient to represent a single visual scene. Because the number of possible percepts is very large, however, the total number of cardinal neurons would have to be "a substantial fraction of the 10^{10} cells of the human brain" (Barlow, 1972).

Since 1972, many of the visual areas outside area 17 have been explored in some detail. Efforts to discover whether such cardinal cells reside in these visual areas have met with mixed success. In most areas, the stimulus requirements for activating neurons are not very different from those for area 17. If anything, the requirements are less restrictive, in that only one property of the stimulus might be important, such as its direction of motion, or color, or depth in visual space (see Barlow, 1981). Only in the inferotemporal region of cortex have neurons with higher-order properties been found (Gross et al., 1972). These neurons respond to parts of the body, especially faces. However, they do not respond in an all-or-none fashion to particular faces, for example, but respond over a range of different stimuli. Thus, even these relatively complex responses do not have the attributes that might be expected for cardinal cells. It should be noted that the so-called face cells form only about 5% of the population in the region of inferotemporal cortex in which they are found. The other neurons in this area respond to much less specific stimuli (Gross et al., 1972; Bruce et al., 1981; Richmond et al., 1983). The general conclusion from these studies is that individual neurons do not respond selectively to single "trigger features." Instead, each neuron is sensitive to a number of different stimulus characteristics, such as contrast, dimension, depth, and orientation. Single cortical neurons appear unable to signal unambiguously the presence of a particular stimulus or "feature" and therefore cannot act as cardinal cells. An important reason why such cardinal cells are not found may lie in the basic organization of the cortical circuitry.

NEW DESIGNS FOR THE VISUAL CIRCUITS

When Barlow proposed his neuron doctrine in 1972, the modern study of cortical microcircuitry was in its infancy. Anatomical studies had emphasized the vertical, columnar organization of the cortex. This view was reinforced by many electrophysiological studies, which showed vertical functional columns (Fig. 12.2). Technical advances since the late 1970s have resulted in a wealth of new information about the cortical microcircuitry. The technique of intracellularly injecting neurons with horseradish peroxidase (e.g., Figs. 12.3 and 12.4) has revealed an extensive system of horizontal connections within the cortex. Because the horseradish peroxidase fills the entire axonal arborization, the very large number of synapses made by a single neuron is also now evident for the first time.

The horizontal spread of connections means that each point in cortex is covered by the axons of a very large number of neurons. For example, estimates for the number of geniculate X cells that provide input to any point in cat area 17 range from 400 to 800 (Freund et al., 1985b), whereas the figure for the Y axons may be even higher. The geniculate axons form only 10–20% of the synapses in layer 4, so the number of cortical neurons providing the input to a single point must be considerably higher. Because one cortical neuron supplies only a few synapses to any other cortical neuron, each neuron can potentially be activated by hundreds of other neurons. Conversely, each neuron connects to hundreds of other neurons. It is this highly divergent and convergent connectivity that is a feature of the neocortex, and differs considerably from that of the lateral geniculate nucleus (Chap. 8), where there is a much tighter coupling between neurons.

The widespread and rich connections of the thalamic afferents ensure that even the smallest detectable disturbance at the retinal receptor layer—for example, that induced by a dim flash of light—alters the probability of firing of thousands of neurons in the primary visual cortex. The signal is then amplified by the divergent axonal arbors of the cortical neurons, which ensure that many thousands more neurons are activated, both within area 17 and in the other cortical areas to which these neurons project. Thus, although there certainly is the convergence of many inputs that is required to create the cardinal cells, the considerable divergence of the connections ensures the simultaneous activation of many neurons. In such a context, it is difficult to see how the activity of any single neuron can be isolated from that of its companions in order to signal a unique percept. Instead, the combined activity of large numbers of cortical neurons seems more likely to be the basis of our perceptual experience. The problem now is to determine the precise form and function of these circuits.

PARALLEL PROCESSING IN NEURAL NETWORKS

It is evident from the above discussion that normal vision involves the activity of very large numbers of cortical neurons. These large numbers do not simply reflect redundancy, but are a necessary part of the process of perception. This is evident in the example of color vision, where both behavioral and theoretical studies show that the relative stability of the perceived color of objects in the face of changing illumination (e.g., moving from indoors to outdoors) requires the comparison of the reflected wavelengths over a large region of the visual field (Jameson and Hurvitch, 1959; Land, 1959, 1983). This phenomenon of "color constancy" necessarily involves large numbers of cortical neurons. Similar considerations apply in the case of binocular vision, where two slightly different views of the same complex scene have to be fused to produce single vision and stereopsis. Attempts to replicate this performance, using computer models, have shown that this is a difficult task (e.g., Marr and Poggio, 1979). Yet we fuse the image and extract the three-dimensional information effortlessly and far more rapidly than any computer yet can. One problem is that the strict hierarchy of serial processing, such as that used in computers designed with von Neumann architecture, is extremely slow. In the cerebral cortex, by contrast, the higher degree of divergence in the connections make it likely that much of the processing occurs simultaneously through parallel pathways.

The realization that parallel processing offers a great increase in speed has now been exploited in a growing number of models of visual processing. The most common form of the connectionist models is the three-layered "neural network" (Fig. 12.16). The first layer, the "input" layer, connects extensively to units of the second layer, the "intermediate" or "hidden" layer, which in turn sends highly divergent connections to the third layer, the "output" layer. The responses of the units in the hidden and output layers are determined by summing the activities of all the units in the previous layer, which are weighted and can be positive ("excitatory") or negative ("inhibitory"). The activity of the units varies between 0 and 1. A learning algorithm called "back-propagation" is usually

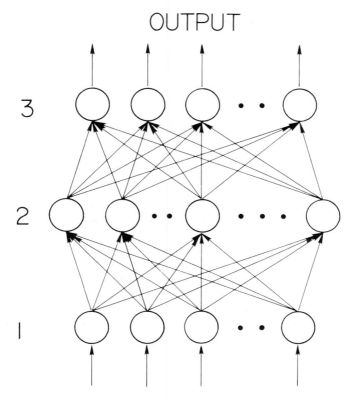

FIG. 12.16. Neural-network model. The three-layered network consists of an input layer (1), an intermediate hidden layer (2), and an output layer (3). The number of units within each layer can be varied according to the particular requirements. Such a network was used by Sejnowski and Lehky (1988) to extract shape from shading (see text).

used to allow the connections between the units to develop into the mature network.

Such networks turn out to be powerful and are capable of solving some of the perceptual problems of depth, form, and motion perception. In one recent demonstration, the network (Lehky and Sejnowski, 1988) was able to extract the shape of an object from the gradations of shading that are found on curved surfaces. Remarkably, the "receptive fields" of the second layer units in this particular network were similar to those of neurons in area 17. This discovery raises two important points with respect to the alternative view offered by the neuron doctrine. One is that it is impossible to predict what the neural network is doing by examining the "receptive fields" of the units involved. In the shape-from-shading network of Lehky and Sejnowski, units were found that resembled "edge" and "bar" detectors, but instead these were involved in extracting information about curvature. Second, these networks do not require the equivalent of a hierarchy of visual areas to perform what is a difficult perceptual task. Instead, a single, structurally simple three-layered network is sufficient. However, these two points are strictly relevant to questions of cortical function only if some equivalence has been established between the circuitry of the neural networks and the actual circuitry of cortex.

ARE NEURAL NETWORKS LIKE CORTICAL CIRCUITS?

The neural network models, besides being functionally successful, have a strong appeal because of their resemblance to the structure of cortex: They are layered and have highly interconnected units. However, it is worth examining this resemblance in more detail in light of our knowledge of the structure of area 17. The basic circuit diagram of Fig. 12.7 illustrates some of the main neuronal components and their connections within area 17. Comparing Fig. 12.7 with the neural network circuit shown in Fig. 12.16 shows a number of similarities. The first layer units in the neural network correspond to the map of the lateral geniculate terminals, the second layer units to the neurons in layer 4 that project to the superficial cortical layers, and the third layer to the pyramidal neurons projecting from cortical layers 2 and 3. In respect of this laminar organization, the pattern of the model corresponds to that of cortex in that very few neurons in layer 4 provide an output to other cortical areas, whereas a large proportion (70%+) of the pyramidal neurons in layers 2 and 3 do project to other areas. However, it is evident at a glance that the organization of cortex shown in Fig. 12.7 is in many important respects different from the neural network circuit of Fig. 12.16.

Unlike the neural network, the primate visual cortex receives at least two physiologically and anatomically distinct inputs from the thalamus, which are laminar specific. The cortex has twice as many or more layers, particularly if the subdivisions of the six basic layers are taken into account (Fig. 12.6). This may in part be a requirement of the cortex to divide its output destinations (addresses) into lamina-specific zones. However, the internal connectivity and physiology differ between layers, indicating that there may be important differences in the processing within layers. In contrast to the units within a layer of the neural network, there are extensive lateral connections within a single lamina or sublamina of cortex. Similarly, the interconnections between cortical laminae are highly specific and do not simply connect adjacent layers as in the units of the neural network. Both the vertical and the lateral connections in cortex are highly clustered, focusing on discrete zones. This columnar structure is a feature of cortical organization, but is the exception (Linsker, 1986) rather than the rule in most neural-network models.

The point is readily made that these and many other differences show that the neural networks are different in important respects from the cerebral cortex. Neural networks are really not neural at all; they are just networks operating according to a specific algorithm, and it would be rash to press their analogy to cortical circuits too far. Nevertheless, the potential usefulness of network models that are biologically based cannot be overestimated. The major problem lies in trying to bridge the gap between the experimental data and the theory. Our knowledge of the complexity of the cortical microcircuitry outstrips our understanding of the function of these circuits. This disparity, together with the sheer complexity of the cortical circuits, is a significant barrier to moving from networks that are simply "neurally inspired" to those that actually incorporate basic features of the biology. To achieve this step, the diagram shown in Fig. 12.7, and its associated physiology, has to be simplified. Given the outline of the preceding sections, one such simplification can now be suggested. This simple circuit was arrived at from an analysis of the structure and function of local circuits in the visual cortex. However, an analysis of the circuits of other cortical structures such as the olfactory cortex (paleocortex) and hippocampus (archicortex) reveals that they, too, bear many striking resemblances to the circuits of the neocortex (Shepherd, 1988a,b). Thus, it is tempting to suppose that there may be some common basic principles that underlie the organization and operation of all cortical circuits.

A CANONICAL CORTICAL CIRCUIT

From the anatomy, several components and connections seem to dominate in most cortical areas. These are summarized in a simplified model of cortex (Fig. 12.17).

Any realistic model must separate inhibitory (GABAergic) and excitatory neurons into distinct populations. The excitatory group (80% of the cortical neurons) can be subdivided into two major pools, one being the group found in the granular and supragranular layers (layers 2–4), the other in the deep layers (layers 5 and 6). Although these groups are extensively interconnected, this division is made because their output targets are quite distinct, and because inhibition appears to be stronger in the deep layers (Douglas et al., 1989). The different types of GABAergic smooth neurons cannot yet be distinguished on functional grounds; they are therefore represented in the diagram of Fig. 12.17 by a single group.

Neurons within each division form connections with other members of that division. The dominant interlaminar connections are between the superficial and the deep layer groups of spiny neurons, whereas the inhibitory neurons connect across the laminae to both groups of spiny neurons. All three groups receive direct activation from the thalamic afferents, but because the thalamic input provides only about 10% of the excitatory input, 90% of the excitation is provided here by the intracortical connections between pyramidal neurons. This tremendous recurrent excitation may be required to provide the synaptic current necessary to maintain reasonable rates of spike discharge (Fig. 12.9B). Inhibition acts by preventing the recurrent excitation, and so is effective even though it is relatively weak.

The excitatory neurotransmitters act on two receptor types, the NMDA and the non-NMDA receptors. The inhibitory neurotransmitter GABA acts via the $GABA_A$ and the $GABA_B$ receptors. These distinctions are made because the

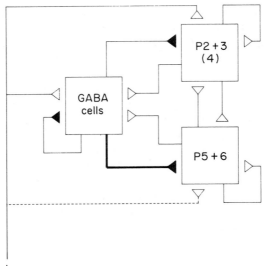

Thalamus

FIG. 12.17. The canonical cortical circuit. It is applicable to all cortical areas so far examined. Three populations of neurons interact with one another: One population is inhibitory (GABAergic cells, solid synapses), and two are excitatory (open synapses) representing superficial (P2+3) and deep (P5+6) pyramidal neurons. The properties of layer 4 stellates (4), which contribute 10% of neurons in granular cortex, less elsewhere, are similar to those of the superficial pyramids. The thickness of the connecting lines indicates the functional strength of the input. Note that the dominant connection is between excitatory neurons, so that a relatively weak thalamic input can be greatly amplified by the recurrent excitation of the spiny neurons.

receptor types have distinctly different kinetics. The biophysical characteristics of the cortical neurons, as outlined in the preceding sections, also need to be incorporated to give the appropriate response characteristics.

This model provides the minimum specifications that seem necessary for the basic cortical circuit. The form of this simplified model is sufficiently general that it can apply equally well to visual cortex as to motor cortex, and as such has the properties of a canonical circuit. This circuit forms only a basic building block. Obviously, each cortical area has individual features that need to be incorporated—for example, the spatial mapping of the thalamic inputs to visual cortex. However, the convergence of theory and biology is imperative if we are to understand how the synaptic organization of the neocortex produces the complexity of cortical function.

13

APPENDIX: DENDRITIC ELECTROTONUS AND SYNAPTIC INTEGRATION

GORDON M. SHEPHERD AND CHRISTOF KOCH

As is abundantly evident throughout this book, neurons have dendrites that greatly increase the surface area available for synaptic connections. The study of synaptic organization is, therefore, a study not only of the generation of synaptic potentials at given sites on the dendritic tree, but also of the properties that determine the spread of the responses through the dendritic tree to link them to the sites of output. This linkage depends on both active and passive properties of the dendritic membrane. Active properties are discussed in Chap. 2, as well as in most of the other chapters, and constitute an area of intensive current research. Passive properties, by contrast, are less well understood by most neuroscientists, yet they are the foundation upon which all the interactions of synaptic and active properties depend. The fact that passive properties can be described in rigorous mathematical terms, although enhancing their power, has hardly made them popular! The aim of this chapter is therefore to provide a brief introduction to the passive spread of electric current in neuronal dendrites, and to show how this provides a basis for interpreting the synaptic and voltage-gated properties involved in the integrative activity of the neuron.

HISTORICAL BACKGROUND

The terms used in this work come from a long tradition in neurophysiology and biophysics, and it will be useful to indicate their origin. The analysis of current spread in neuronal processes began with the very earliest electrophysiological studies of activity in peripheral nerves in the 1840s. Subsequent work in the nineteenth century was devoted to the problem of distinguishing the passive spread of current in peripheral nerves from the "action" currents associated with the action potential. From this work arose the concept of the nerve fiber as a "core conductor," due to its high membrane resistance and relatively low internal resistance. Passive spread through the core conductor was termed *electrotonus,* and the changes in membrane potential caused by the passive spread were termed *electrotonic potentials.* These current flows and potential changes were described mathematically, and it was soon realized that the equations were essentially similar to those for the flow of current in an electrical cable, the flow of heat in a metal rod, and the diffusion of substances in a solute. The equations as applied to nerve cells are often referred to as the *cable equations*.

Application of the cable equations to experiments on the nerve axon awaited the modern era, beginning with the analysis of the action potential mechanism and the development of the Hodgkin–Huxley model. The equations were adapted for use in studying conduction of the action potential in muscle fibers, and they were also important in Katz's analysis of the end-plate potential at the neuromuscular junction.

The application of the equations to these relatively simple cases (in a geometrical sense) was no easy task, and it therefore seemed that neuronal dendrites, with their limitless variety of shapes and branching patterns and their inaccessibility to experimental approaches, were beyond the reach of rigorous analysis. Beginning in 1957, however, the systematic adaptation of the equations to dendrites was carried out by Wilfrid Rall (see 1957; 1959a,b; 1967; 1970; reviewed in 1977 and 1989). Here we describe only the essentials of the methods, in order to provide the student with the basic tools for understanding the spread of synaptic responses and the functional properties of dendrites. For further orientation see Jack et al. (1975) and Jack (1979).

STEADY-STATE ELECTROTONUS

The basic characteristics of electrotonic potentials are illustrated in Fig. 13.1. We consider a length of neuronal process (either axon or dendrite) as a series of segments. A steady depolarization is imposed on the membrane of segment 1. This places a persisting excess of positive charge inside the membrane. In this steady-state condition, the membrane capacitance acts as an open circuit, and we need only consider the current through the resistance pathways of the neuron. Some of the internal current (I_0) at point a will leak out (i_m) through the membrane resistance (r_m) of segment 1, but some current (I) will flow through the small internal resistance (r_i) to point b. At point b, the current has two similar alternative paths: out through the membrane resistance of segment 2 or onward through the small internal resistance to point c. The sequence repeats itself further along the neuronal process until all the current has leaked out across the membrane and returned to segment 1 to complete the circuit.

DERIVING THE CABLE EQUATION

The way the potential due to the steady current decays in spreading from segment to segment is described as follows, according to Rall (1958). We define the electrotonic potential (V) as the difference between the instantaneous value of the membrane potential (V_m) and the resting value of the membrane potential (E_r) set by the leak resistance (this is mostly due to a steady potassium conductance; see Chap. 2):

$$V = V_m - E_r \qquad (13.1)$$

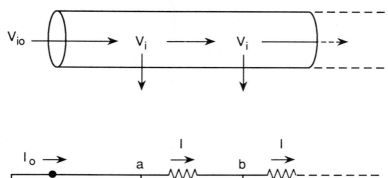

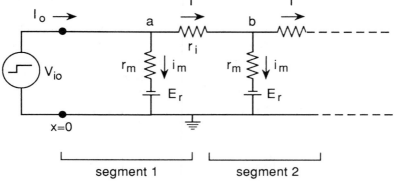

FIG. 13.1. Principles of spread of electrotonic potential under steady-state conditions. *Top:* Schematic representation of a nerve process (axon or dendrite). At one end, a steady-state depolarization (V_{io}) is imposed, which spreads as a change in the internal potential (V_i) along the process. *Bottom:* Representation of the dendrite as a cable. Steady-state current (I_0) generates V_{io} at $x = 0$. The current inside the dendrite spreads from segment to segment; at each site (a,b), some current (I) continues through the internal resistance (r_i) within the dendrite, and some current (i_m) leaks out across the membrane resistance (r_m). The resting membrane potential is represented by E_r. The cable is considered to extend infinitely to the right.

Since E_r is assumed to be constant along the dendrite, it is neglected. V_m is taken to be the same as V_i, the potential just inside the membrane (the potential just outside, V_e, being assumed to be ground). The change of electrotonic potential (V) with distance (x) along the dendrite is determined by substituting V_i for V_m and differentiating Eq. 13.1 with respect to x:

$$\frac{dV}{dx} = \frac{dV_i}{dx} \tag{13.2}$$

This states simply that the gradient of electrotonic potential (V) along the dendrite is equal to the gradient of the internal potential (V_i).

Next we must define the gradient of the electrotonic potential (V) in terms of the current it generates and the resistance through which it spreads. This is given by Ohm's law, one of the basic equations of electric circuit theory:

$$v = ir \tag{13.3}$$

where v is voltage across, and i is current through, the resistance r.

As applied to the gradient of V, this gives:

$$\frac{dV}{dx} = -Ir_i; \quad I = \frac{-1}{r_i} \cdot \frac{dV}{dx} \tag{13.4}$$

where I is the internal current (in amperes) spreading through r_i (the axial resistance associated with segment 1, per unit length (ohms/cm)); the sign is negative to indicate the decreasing gradient of electrotonic potential with distance.

In addition to spreading internally through the dendrite, the current also flows outward through the membrane resistance of segment 1. We can describe this current precisely by means of Kirkhoff's laws, another workhorse of electrical circuit design. This states simply that at any point in a circuit the current flowing to that point has to equal the current flowing away from it. For segment 1, this means that i_m, the outward membrane current per unit length (ampere-cm), is equal to the decrease in axial current per unit length (ampere/cm):

$$i_m = -\frac{dI}{dx} \tag{13.5}$$

Substituting Eq. 13.4 in Eq. 13.5 gives:

$$i_m = \frac{-d}{dx}\left[\frac{-I}{r_i} \cdot \frac{dV}{dx}\right] = \frac{1}{r_i}\left[\frac{d^2V}{dx^2}\right] \tag{13.6}$$

From Ohm's law,

$$i_m = \frac{V}{r_m} \tag{13.7}$$

Equating Eqs. 13.6 and 13.7 gives:

$$V = \frac{r_m}{r_i} \cdot \frac{d^2V}{dx^2} \tag{13.8}$$

This is a fundamental equation of cable theory. As applied to the neuron, it states that, given a steady-state current input at $x = 0$, the spread of electrotonic potential (V) along an axon or dendrite with invariant passive properties is proportional to the second spatial derivative of the potential (V) times the ratio of the membrane resistance (r_m) to the internal resistance (r_i).

SOLVING THE CABLE EQUATION

The cable equation in Eq. 13.8 is usually rearranged as follows:

$$\lambda^2\left(\frac{d^2V}{dx^2}\right) - V = 0; \quad \lambda = \sqrt{\frac{r_m}{r_i}} \tag{13.9}$$

where lambda (λ) is referred to as the *length constant* of the cable. We will return to its significance below.

One solution of this equation gives:

$$V = V_0 e^{-x/\lambda} \tag{13.10}$$

where V is the electrotonic potential, as defined previously, and V_0 is V at $x = 0$. Rearranging, we have

$$\frac{V}{V_0} = e^{-x/\lambda} \tag{13.11}$$

This equation is plotted in Fig. 13.2. In A, a dendrite is injected with steady current (i) at $x = 0$, with spread of current indicated by arrows. The dendrite is represented by a series of segments in B, each segment consisting of an equivalent circuit as in Fig. 13.1. This is a convenient form for computing the spread of electrotonic potential given by Eq. 13.11, the result being plotted in Fig.13.2C. Here the electrotonic potential V is shown relative to its maximum value at V_0, and distance along the dendrite is given in terms of λ units, the significance of which will soon be apparent.

Let us pause to discuss this curve. First, we note that at $x = 0$ the exponent of e will be 0; e raised to the zeroth power is 1, so that V must equal V_0 at $x = 0$, which of course it does. Next, we know by intuition that as we move away from $x = 0$, V will decay toward zero. A very useful way to characterize this decay is to note that when $x = \lambda$, Eq. 13.10 becomes:

$$\frac{V}{V_0} = e^{-1} = \frac{1}{e} \approx 0.37 \tag{13.12}$$

This means that if we know λ, then we know the distance along the axon or dendrite at which the electrotonic potential will have decayed to .37 of its value at the origin (see Fig. 13.2). Like the notion of half-life for radioactive decay, it is a way of characterizing and comparing electrotonic spread in axons or dendrites of different size and shape. Another term for λ is *characteristic length*, reflecting this use.

This plot illustrates the salient features of electrotonic decay with distance. The electrotonic potential falls to 37% of its original amplitude at $\lambda = 1$, and 15% at $\lambda = 2$; by $\lambda = 3$, it is only about 6%. It is usually considered that an electrotonic distance within 1λ is within range for quite effective interactions within a dendrite or dendritic tree, as, for example, the effect of a synaptic potential on the site of impulse generation at the soma. An electrotonic distance of 2λ is usually considered to be near the limit for effective interactions.

THE SIGNIFICANCE OF THE CHARACTERISTIC LENGTH

In view of the definition of λ (see Eq. 13.9 above), it is clear that the ratio r_m/r_i is a fundamental determinant of the effectiveness of spread of electric current and potentials in neurons. In order to make use of this ratio to analyze axons or dendrites of any arbitrary size, we need to be very careful about units. In the form employed for deriving Eqs. 13.1–13.8 above, we have

$$r_m = \frac{\text{specific membrane resistance}}{\text{circumference}}$$

$$r_m = \frac{R_m \, (\Omega \, \text{cm}^2)}{2\pi a(\text{cm})} = [\Omega \, \text{cm}] \tag{13.13}$$

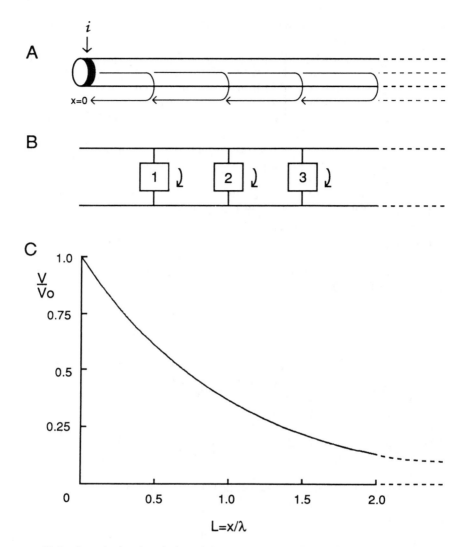

FIG. 13.2. Quantitative description of the steady-state electrotonic potential in a cable of semiinfinite extension. **A.** Schematic representation of cable and current path during steady-state injection of current i at $x = 0$. **B.** Schematic representation of cable as a series of segments, as defined in Fig. 13.1. **C.** Distribution of electrotonic potential according to Eq. 13.11. Ordinate: Amplitude of the electrotonic potential (V) relative to the steady-state potential imposed at V_0. Abscissa: Length (L), defined as real length (x) relative to the characteristic length (λ) (see Eqs. 13.9 and 13.18).

where r_m is the membrane resistance for a unit length of cylinder in Ω centimeters, and a is the radius of the cylinder; and

$$r_i = \frac{\text{specific internal resistance}}{\text{cross-sectional area}}$$

$$r_i = \frac{R_i(\Omega \text{ cm})}{\pi a^2 (\text{cm}^2)} = [\Omega \text{ cm}^{-1}] \qquad (13.14)$$

where r_i is the internal resistance per unit length of cylinder in $\Omega \text{ cm}^{-1}$.

Dividing Eq. 13.13 by Eq. 13.14 gives r_m/r_i as follows:

$$\frac{r_m}{r_i} = \frac{R_m}{R_i} \cdot \frac{a}{2} \qquad (13.15)$$

This means that r_m/r_i is proportional to the radius of the cylinder; the larger the radius, the larger the ratio r_m/r_i, and the farther the spread of electrotonic potential (V). In addition, by rearranging Eqs. 13.13 and 13.14 we have the definitions:

$$R_m = (\text{membrane resistance per unit length}) \cdot (\text{circumference})$$
$$R_m = r_m(\Omega \text{ cm}) \cdot 2\pi a(\text{cm}) = [\Omega \text{ cm}^2] \qquad (13.16)$$

where R_m is the specific membrane resistance across a unit area of membrane in $\Omega \text{ cm}^2$; and

$$R_i = (\text{internal resistance per unit length}) \cdot (\text{cross-sectional area})$$
$$R_i = r_i (\Omega \text{ cm}^{-1}) \cdot \pi a^2(\text{cm}^2) = [\Omega \text{ cm}] \qquad (13.17)$$

where R_i is the specific resistance across a unit cross-sectional area of the internal medium. (The units are indicated in the equations as an aid to avoid confusion.)

These definitions are important, because the specific values are independent of the size of the dendrite. Putting these values into the equation for λ (see Eq. 13.9 above) gives

$$\lambda = \sqrt{\frac{r_m}{r_i}} = \sqrt{\frac{R_m}{R_i} \cdot \frac{d}{4}} \qquad (13.18)$$

where $d = 2a$ is the diameter of the process.

In this form, we have the fundamental relation that the characteristic length, which defines the spread of electrotonic potential, is dependent on the square root of the ratio of R_m to R_i, scaled for the diameter of the axon or dendrite.

EFFECT OF DENDRITIC DIAMETER ON ELECTROTONIC SPREAD

We are now in a position to understand the factors that control electrotonic spread. The three main factors are the diameter of the dendrite, its membrane resistance, and its modes of branching and termination. Let us begin with diameter.

Neuronal dendrites vary in diameter over a considerable range, from the thinnest of approximately 0.1 μm to the thickest of approximately 20 μm. Intuitively, it is obvious that electrotonic spread will depend on diameter. Spread of current will be enhanced by a large diameter, since the effective resistance (r_i) inside a dendrite decreases as the diameter increases, as defined by Eq. 13.14 above. From Eq. 13.18, we know that the characteristic length (λ) varies with the square root of the diameter, given fixed values for R_m and R_i.

The relation between electrotonic spread and diameter is illustrated graphically in Fig. 13.3. In A, a dendrite is shown as a cable which receives a steady current input at $x = 0$. Let us make some starting assumptions, that $R_m = 8,000 \ \Omega\text{cm}^2$ and $R_i = 80 \ \Omega\text{cm}$. Then, from Eq. 13.18,

$$\lambda = \sqrt{\frac{R_m}{R_i} \cdot \frac{d}{4}} = \sqrt{100 \cdot \frac{d}{4}} = 5\sqrt{d} \tag{13.19}$$

For a dendrite of diameter 1 μm, this gives $\lambda = 500 \ \mu$m; the electrotonic spread for this case, under the assumption of infinite extension of the cable, is shown in (a) in Fig. 13.3A. The characteristic length increases by the square root of the diameter, and the electrotonic spread increases proportionately, as shown in (b) and (c) for dendrites that are 4 μm and 16 μm in diameter. The relative sizes of the dendrites are shown in B.

EFFECT OF MEMBRANE RESISTANCE ON ELECTROTONIC SPREAD

We turn next to the electrical properties that affect electrotonic spread. It should be recalled that we made an initial assumption to ignore any effect of *extracellular resistance* along the dendrite, all points being considered at ground. With regard to the specific *intracellular resistance* (R_i), this is commonly estimated at 80–100 Ωcm (see above). It is assumed to be invariant with different diameters, with different neurons, and under different conditions of activity, though of course these assumptions deserve testing.

The remaining electrical property to consider is the specific *membrane resistance* (R_m). This varies over a considerable range in different neurons, and therefore is an important variable in determining electrotonic spread. From Eq. 13.18 we know that λ varies with the square root of R_m, so the same relations hold that applied to the diameter. Thus, we can take the three curves from Fig. 13.3C, and show in Fig. 13.4 that they describe the electrotonic spread for three R_m values that span a corresponding 16-fold range. These curves describe the decay in distance (x) for the case of a 4-μm-diameter dendrite, as indicated below the graph.

It is obvious that variations in both diameter and R_m provide for many possible combinations determining the value of λ. A convenient way to summarize these multiple relations is in the graph of Fig. 13.5, which provides a starting point for assessing the range of effectiveness for electrotonic spread in dendrites for different sizes.

By way of general comment, it may be noted that λ has a relatively high value even in very fine processes. Thus, the finest dendritic branches in the brain are of the order of 0.1 μm in diameter; as can be seen in Fig. 13.5, one might expect a λ of several hundred micrometers for them. Such branches are, in fact, rarely this long, which indicates that current spread may well be quite effective in small

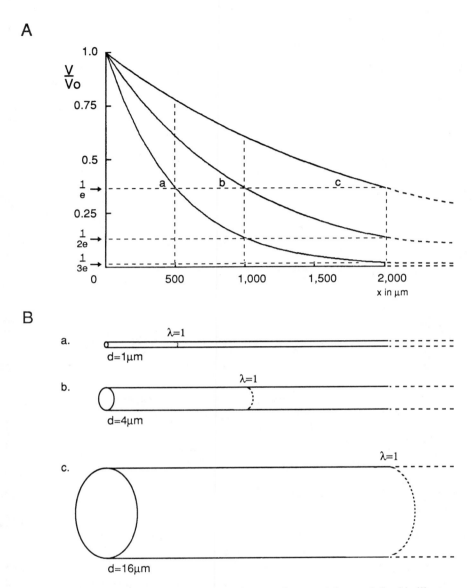

FIG. 13.3. Effect of dendritic diameter on electrotonic potential spread. In this illustration, $R_m = 8000 \ \Omega \ cm^2$ and $R_i = 80 \ \Omega cm$. **A.** Electrotonic spread in semiinfinite cables with diameters of 1 μm (a), 4 μm (b), and 16 μm (c). Dashed lines indicate distances in terms of multiples of characteristic lengths. **B.** Diagrammatic representation, to scale, of the three sizes of dendrites, showing λ lengths (note that these are in register with vertical dashed lines in A).

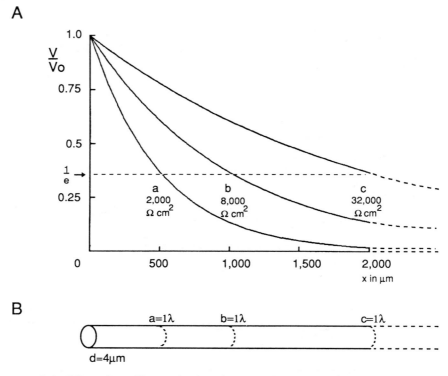

FIG. 13.4. Effect of specific membrane resistance (R_m) on electrotonic potential spread. This example assumes a dendrite of 4-μm diameter, with $R_i = 80\ \Omega$cm, and R_m of 3 different values: (a) 2,000 Ωcm^2; (b) 8,000 Ωcm^2; (c) 32,000 Ωcm^2. The curves for the spread of electrotonic potential (**A**) and distances for λ (**B**) are identical to those in Fig. 13.3. This illustrates the equivalence of changes in diameter and membrane resistance in controlling electrotonic current spread (cf. Eq. 13.18).

dendrites and dendritic trees. On the other hand, large dendritic trunks (as, for example, in motoneurons and cortical pyramidal cells) are 10–15μm in diameter. From Fig. 13.5 we might expect a dendrite of 10-μm diameter to have λ of the order of several millimeters. This is longer than the length of most large dendrites (but see the neocortex, Chap. 12), which indicates that the spread of synaptic current through them may also be relatively effective. Thus, there is a *scaling factor* which appears to adjust the length of a dendrite to its diameter and characteristic length. By contrast, the large axons of principal neurons are many times longer than their characteristic lengths; the longest motoneuron axons to the lower limb, for example, are over a meter (1000 mm) in length. Communication through these axons must therefore be by action potentials, because the attenuation of passive potentials would be too great (Eq. 13.10).

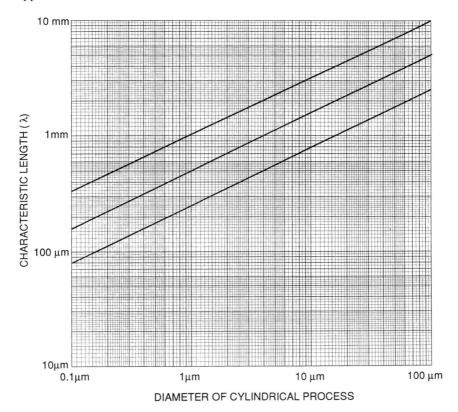

FIG. 13.5. Graph illustrating the dependence of the characteristic length (λ) on the diameter and electrical properties of a nerve process, as expressed in Eq. 13.18. Ordinate: logarithm of the characteristic length; abscissa: logarithm of the diameter of a nerve process. The lines plot three values of the ratio R_m/R_i: 25, 100, 400 (e.g., $R_i = 80$ Ωcm; $R_m = 2,000$, $8,000$, and $32,000$ Ωcm^2). (Modified from Shepherd, 1979.)

EFFECT OF DENDRITIC BRANCHING AND TERMINATION ON ELECTROTONIC SPREAD

The third main factor that affects electrotonic spread is the manner in which a dendrite branches and terminates. This factor is called the *boundary condition* of the cable. The solution to the basic cable equation (Eq. 13.11) describes the decay of electrotonic potential with distance (x) only for a cable with a boundary condition of infinite extension in the x direction. Obviously, this does not apply in the real nervous system. Any departure from this assumption influences the actual amount of spread. Let us first consider the effect of the mode of termination.

Dendritic termination. This influence is best appreciated by considering extreme examples, as illustrated in Fig. 13.6. In A are shown three dendritic cables in

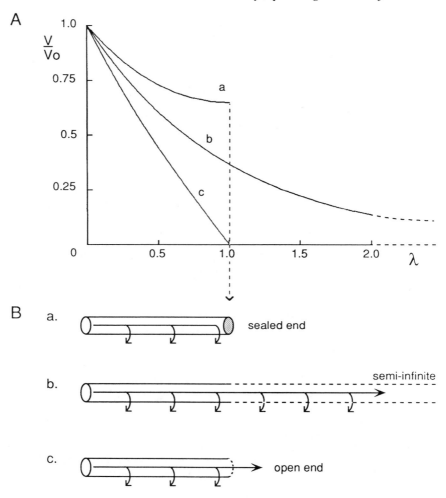

FIG. 13.6. Effect of different modes of termination on the spread of electrotonic potential.
A. Graph of steady-state potential spread for the case of a sealed end at $\lambda = 1$ (a), an
infinite extension of the cable (b), and an open end (short-circuit) at $\lambda = 1$ (c). **B.**
Diagrams illustrating each of the boundary conditions in A. (Modified from Rall, 1958.)

which there is electrotonic current spread (dotted lines and arrows). At point $\lambda =$
1, three different boundary conditions apply. In (b), the cable continues as a
direct extension with the same properties; this then is the standard case of a semiinfinite cable, with the decrement shown again in curve (b) of the graph.

By contrast, consider, as in (a), that the cable terminates at x with an infinitely
high resistance (an open circuit). This boundary condition is called a "sealed-
end." Intuitively, one can see that all the current that otherwise would have gone
on down the cable is now added to current crossing the membrane before reaching x. The effect will be to decrease the gradient of potential along the cable, and
increase the amplitude of the potential. The result is the curve shown in (a).

The other possible mode of termination at x is a very low resistance pathway to the extracellular fluid (a "killed-end" boundary condition). This in effect places a short-circuit to ground; current is drawn from the inside of the dendrite, and one says that a high-conductance load is placed on the cable. Intuitively, one can see that, since the terminal is essentially at ground, the electrotonic potential must be zero at that point; there is therefore a very steep decay of electrotonic potential because much of the current, instead of crossing the membrane, is siphoned off through the open end [see (c)].

Dendritic branching. Dendritic branching can be considered as a special case of a boundary condition. Thus, if we take the infinite cable defined in (b) of Fig. 13.6, what will be the effect of branching at point x? Figure 13.7 provides several examples. If the branches are thick relative to the stem, the effect will be to place a conductance load on the stem, and the decay of potential in the stem will move toward the short-circuit condition [curve (c) in A; see B,c]. If the branches are relatively thin, the effect will move toward the open-circuit condition, as in (a) of Fig. 13.7A (see also B,a). If the branches are electrotonically equivalent to the stem, the decay should resemble that of an infinite cable, as illustrated by curve (b) in A, and the diagram in B,b.

The effect of branching can therefore be seen to involve a match between the electrotonic properties of the stem and the electrotonic properties of the branch; in other words, it is a question of impedance matching. For the steady-state case, this involves the relative input conductances of the stem and branches. Following Rall (1959b), the input conductance of a dendritic segment is given by Ohm's law:

$$G_D = \frac{I_0}{V_0} = \frac{1}{\sqrt{r_m r_i}} \tag{13.20}$$

where I_0 is the current entering the dendrite and V_0 is the electrotonic potential at that point. Substituting Eqs. 13.13 and 13.14 for r_m and r_i, we get

$$G_D = \frac{\pi}{2} \frac{1}{\sqrt{R_m R_i}} d^{3/2} \tag{13.21}$$

This important result states that the input conductance of a dendrite varies with the diameter raised to the 3/2 power. Rall (1959b) went on to show that there is electrotonic continuity at a branch point if the 3/2 power law is satisfied; thus

$$d_0^{3/2} = \sum_i d_i^{3/2} \tag{13.22}$$

where d_0 is the diameter of the parent dendrite and d_i is the diameter of each of the daughter branches.

Let us apply this equation to the examples in Fig. 13.7. Assume that the diameter of the stem dendrites is 4 μm; then $4^{3/2}$ is 8. For the constant decay condition of (b) to obtain, each branch would have to have a diameter of approximately 2.5 μm ($2.5^{3/2} \simeq 4$; 2 branches $\simeq 8$). Under these conditions, the summed conductance of the branches is equivalent to that of an extension of the

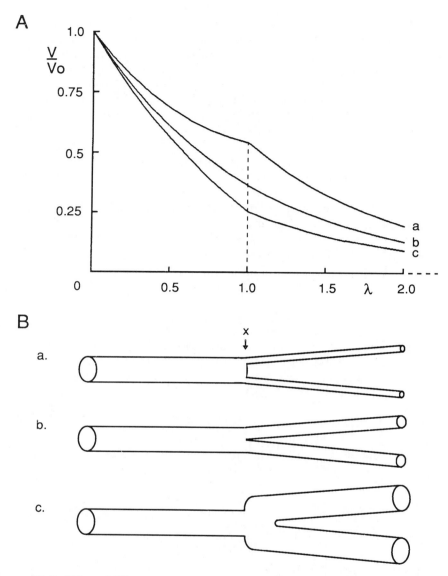

FIG. 13.7. Effect of different modes of dendritic branching on the spread of electrotonic potential. **A.** Graph of steady-state potential spread for the three cases illustrated in B. **B.** Diagrams illustrating three basic modes of branching; in each case the stem diameter is 4 μm : (a) Each daughter branch is 1 μm in diameter; (b) each daughter branch is approximately 2.5 μm, so that sum of $d^{3/2}$ equals $d^{3/2}$ of the stem; (c) each daughter branch has the same diameter as the stem. (Modified from Rall, 1958.)

stem dendrite. The stem plus branches therefore constitute an *equivalent cylinder*, in Rall's terminology. If the branch diameters in this example were smaller than 2.5 μm, then condition (a) would prevail; if larger than 2.5 μm, then condition (c) would prevail, given equal branch lengths.

Just as the characteristic length provides a means for comparing spread of synaptic potentials in single dendrites of different size, so does the concept of equivalent cylinder provide the means for comparing spread in branching dendrites of different size. It may be asked whether this 3/2 power relationship between a dendritic process and its branches has a general validity in the brain. In fact, it turns out to be a reasonable first approximation for a number of dendritic trees. The equations are simple under this constraint (because the cylinder has a uniform radius throughout its length), but any branching relationship or changes in diameter can be incorporated into the equivalent cylinder model with suitable assumptions about amounts of taper (see Rall, 1959a,b). To emphasize the general applications, one may refer to it simply as an electrotonic model for the dendritic system one is interested in.

SIGNIFICANCE OF THE INPUT RESISTANCE OF A NEURON

Before moving on to consider transient potentials in dendrites, it will be useful to mention an electrotonic property that is widely used in electrophysiological studies of neurons. This is the *whole neuron input resistance* (R_N), defined as the resistance measured when current is passed between an intracellular electrode tip at the cell body and external ground. The input resistance is a basic reflection of the electrotonic structure of a neuron, as well as an important tool in analyzing synaptic properties (see Chap. 2). It is determined by multiple factors, which give insight into the integrative organization of the neuron.

The multiple conductances of the neuron. We begin by considering a simple situation, as in Fig. 13.8A,B, in which an electrode is placed inside a spherical cell body that lacks any process (axon or dendrite). Current passed between this electrode and ground generates a voltage, which gives, by Ohm's law, the input resistance, R_N. If R_N is due entirely to the passive membrane properties of this soma, we may represent it by G (conductance of a passive soma), where

$$G = \frac{1}{R} = \frac{4\pi a^2}{R_m} \qquad (13.23)$$

The unit of conductance (G) is ohm^{-1}, or, by convention, siemens (S); for example, a resistance of 10^6 ohms equals 10^{-6} S (1 microsiemen).

This simple case obviously does not accurately portray most neurons. Closest are certain types of sympathetic ganglion cells (Chap. 3) and dorsal root ganglion cells that lack dendrites, although one must still take into account the axon extending from the cell body. Most neurons, however, have extensive dendritic trees. The electrotonic effect of the tree is given by the summed input conductance of all the dendritic trunks arising from the soma. This can be calculated, as Rall (1959b) showed, by summing the input conductances of each trunk weighted by its $d^{3/2}$ and by a factor (B) reflecting its terminal boundary condition:

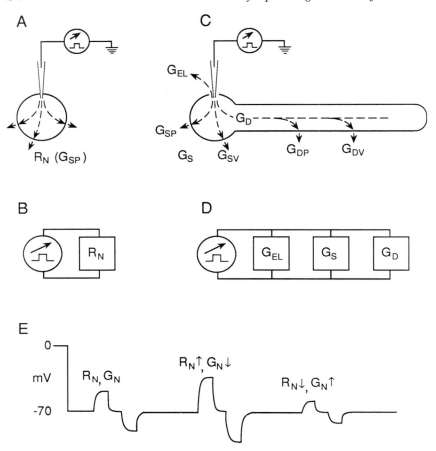

FIG. 13.8. Dependence of input resistance (R_N) on the multiple conductances of the neuron. **A.** Oversimplified representation of a neuron as consisting of only a cell body, with no axon or dendrites. The diagram shows an intracellular or whole-cell patch electrode, for injecting current and for recording the input resistance (R_N), which is the inverse of the input conductance (G). This example assumes only a passive soma membrane (G_{SP}). **B.** Electrical diagram of A. **C.** Complete representation of a neuron. The soma conductance (G_S) consists of a passive component (G_{SP}) and a variable component (G_{SV}), the latter composed of voltage-gated and ligand-gated conductances. The dendritic conductance (G_D) consists of a passive component (G_{DP}) and a variable component (G_{DV}). There is, in addition, a leak around the electrode (G_{EL}), present in conventional intracellular recordings but not in the whole-cell patch mode. **D.** Electrical summary of C. **E.** Illustration of an intracellular recording, during which R_N (G_N) is tested at rest, during decreased synaptic conductance, and during increased synaptic conductance.

$$\sum_i G_D = \sum_i \frac{\pi}{2} \frac{1}{\sqrt{R_m R_i}} d_i^{3/2} \cdot B_i \qquad (13.24)$$

For an infinite cable, B obviously is 1 (cf. Eq. 13.21); for open or sealed boundary conditions, B is greater or smaller than 1, respectively.

By making detailed measurements of the dendritic branches of Golgi-stained neurons, Rall (1959b) showed that the input conductance is commonly 5–10 times greater for the dendrites than the soma. This is an electrical expression of the far greater amount of dendritic than somatic membrane. Rall (1959b) expressed this as a ratio, *rho* (ρ), reflecting dendritic dominance of the whole neuron conductance. This is shown schematically in Fig. 13.8C,D, where it can be seen that with respect to the measuring electrode, G_D is in parallel with G_S to ground. The enlarged view of the dendritic as well as somatic contributions to whole neuron conductance is shown in Fig. 13.8C,D.

The next complication to deal with is the fact that neurons do not have purely passive membrane: Their membranes contain channels that are voltage-gated (excitable) and ligand-gated (synaptic). In both cases, activation is accompanied by a change in membrane conductance, which affects the input resistance. We therefore need to add variable conductances to the soma (G_{SV}) and dendrites (G_{DV}), in parallel with the passive conductances (G_{SP} and G_{DP}), as shown in Fig. 13.8C,D.

Input resistance and synaptic actions. We are now in a position to understand how measurements of input resistance are used in analyzing the synaptic inputs to a neuron. The input resistance is first measured in the resting state by passing small alternating depolarizing and hyperpolarizing current pulses through the electrode while recording the voltage deflections (Fig. 13.8E). Dividing the voltage change by the injected current gives the input resistance (Ohm's Law, Eq. 13.3).

The neuron is then activated by a synaptic input pathway. The alternating pulses are continued during the responses to determine if the voltage deflections show any change. If the deflections get smaller (see Fig. 13.8E), it means that the synaptic potential is associated with an increased conductance—that is, by an opening of ion channels. If the synaptic potential is a depolarizing EPSP, this implicates the opening of Na^+ or Na^+/K^+ channels; if the synaptic potential is a hyperpolarizing IPSP, it implicates the opening of Cl^- or K^+ channels. However, if the deflections get larger (see Fig. 13.8E), signaling a decrease in G_N (increase in R_N), it means that the synaptic potential is associated with a decreased conductance—that is, a closing of ion channels. For a depolarizing EPSP, this could be produced by a closing of K^+ channels; for a repolarizing IPSP, this could be due to a closing of Na^+ channels.

Measurement of input conductance is therefore a basic first step in characterizing the ionic mechanisms underlying synaptic actions. The application of this method is described in Chap. 2 and at various places throughout this book.

Input resistance and membrane resistance. Measurement of input resistance is also an essential step toward obtaining experimental evidence for the value of the specific membrane resistance (R_m). As we have seen, this is one of the basic parameters governing the electrotonic spread of potentials in a neuron. If a neuron consisted only of a sphere, then R_m could be calculated directly from R_S, by a variation of Eq. 13.13, as applied to a sphere of radius a instead of a cylinder; thus:

$$R_S = \frac{R_m}{4\pi a^2}; R_m = R_S \cdot 4\pi a^2 \qquad (13.25)$$

However, this does not take account of the dendrites, which, as we have seen, actually account for the major share of the input conductance and current spread because of their large expanse of membrane. The specific resistance of this membrane cannot be calculated directly, as in the case of the soma, because it is not everywhere equidistant from the current-passing electrode. Its relation to the electrode is governed by the electrotonic properties of the dendrites.

Equations have been derived by Rall that take account of these properties and permit estimates to be made of the value of R_m. These estimates depend on several simplifying assumptions, among them that the conditions of an equivalent cylinder roughly apply, and that R_m is everywhere constant. In recent years increasing attention has been given to factors that complicate these estimates. These include (1) deviations from the equivalent cylinder, based on more accurate measurements of dye-filled neurons; (2) variations in R_m, especially the likelihood that soma R_m is lower than dendritic R_m (see Chap. 4); (3) the contribution of slow voltage-gated conductances to whole-neuron measurements (see G_{SV} and G_{DV} in Fig. 13.8C; and Chap. 4); and (4) the complication of a leak around the intracellular electrode, introducing an additional conductance to be taken into consideration (G_{EL} in Fig. 13.8C). Determinations of R_m will have to take these factors into account. Computational neuronal models are a valuable adjunct in exploring these properties and their functional consequences.

TRANSIENT ELECTROTONUS

Thus far we have described the spread of electrotonic current under steady-state conditions. This is a necessary starting point for characterizing the electrotonic properties of dendrites. Background depolarizations (as in cells of the retina) are essentially steady states, and for many types of slow synaptic potentials the assumption of a steady state is reasonable.

There are, however, many cases in which the essence of synaptic action is rapid transmission between neurons, with immediate triggering of synaptic output or impulse generation. The synaptic potentials for these functions have a rapid onset and decay, and the description of these rapid signals requires an elaboration of the cable equations that is the foundation for the study of transient electrotonus.

SPREAD OF TRANSIENT POTENTIALS

Conceptually, the analysis of transient electrotonus is simple enough; it involves putting the capacitance of the membrane back into the picture. For a passive membrane containing a capacity (c_m) in parallel with the membrane resistance (r_m), we have

$$i_m = c_m \frac{\partial V}{\partial t} + V \qquad (13.26)$$

Substituting into Eq. 13.6, modified for partial derivatives because of variations in both space and time, gives

$$\frac{r_m}{r_i} \cdot \frac{\partial^2 V}{\partial x^2} = r_m c_m \cdot \frac{\partial V}{\partial t} + V \qquad (13.27)$$

where the length constant (lambda) is defined as $\sqrt{\frac{r_m}{r_i}}$ (see Eq. 13.9), and the membrane time constant (tau) is defined as

$$\tau = r_m c_m \qquad (13.28)$$

Thus, we have

$$\lambda^2 \frac{\partial^2 V}{\partial x^2} = \tau \frac{\partial V}{\partial t} + V \qquad (13.29)$$

which is the fundamental equation underlying passive cable theory. In mathematical terms, this is a linear, second-order, partial differential equation of the parabolic type. In this form it governs the spread of electrical potential in a passive cable; a very similar equation governs the diffusion of substances in a cable (see Rall, 1959b).

It should be pointed out that the cable equation of Eq. 13.8 not only governs linear spread of potential, but also provides the necessary foundation for analysis of nonlinear properties as well. Thus, for the case of the Hodgkin-Huxley equations describing the generation of the action potential in the squid giant axon, one has

$$i_m = c_m \frac{\partial V}{\partial t} + \bar{g}_{Na} m^3 h(V - E_{Na}) + \bar{g}_k n^4 (V - E_k) + \bar{g}_r(V - E_r) \qquad (13.30)$$

where $\bar{g}$ is the maximum conductance for Na, K and r (the resting leak current), and the rate constants m, h, and n are complex functions of time and voltage as defined by another set of differential equations (Hodgkin and Huxley, 1952d). This expression for the membrane current is then substituted into Eq. 13.6 in order to arrive at a nonlinear version of Eq. 13.29.

The membrane capacitance is incorporated into the equivalent circuit as shown in Fig. 13.9, which depicts two dendritic segments, A and B. It can be seen that this differs from the equivalent circuit for the steady-state case (Fig. 13.1 above) by the inclusion of the membrane capacitance in parallel with the membrane resistance. In addition, we represent the membrane resistance by two conductance pathways, one for depolarizing the membrane and the other for hyperpolarizing it.

We begin by considering closely the relation between an ionic conductance change and the resultant change in membrane potential. For an example, we take the case of a brief increase in conductance to Na ions. As shown in Fig. 13.9, Na^+ moves inward through its conductance channel at the activity site (A). In the electrical circuit, the current reaches a point on the inside, where it can travel in two directions. Some current passes onto the inner surface of the membrane capacitance, where it adds positive charge that depolarizes the membrane. The

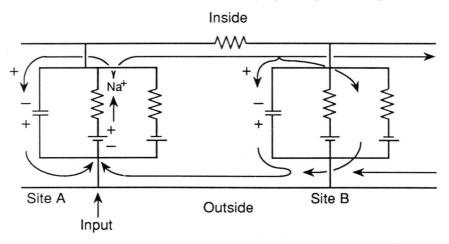

FIG. 13.9. Equivalent circuit of a dendrite, for describing the transient spread of electrotonic potential. The diagram illustrates the pathways for current induced by an input at
site A, which increases net inward current across the membrane at that site. The pathways
illustrate how the membrane capacitance is charged up (depolarized) by *inward* current at
A; electrotonic current flowing longitudinally through the dendrite then produces depolarization by *outward* flow at neighboring site B. Note that the external current path
completes the circuit back to site A. (Modified from Shepherd, 1979.)

rest passes along the inside of the membrane to the next segment (B), where it
can follow three paths: onto the membrane capacitance, through the membrane
resistance, or further along the fiber. Ultimately, all the current must pass out
across the membrane and spread back along the outside of the cell to the negative
pole of the Na battery.

Careful study of the diagram and current flows will help answer two questions
that are often puzzling to the student. The first is, How can inward and outward
currents both depolarize the membrane? As can be seen, the reason is that inward
current at an active site and outward current at a neighboring site both have the
same effect, of putting positive charge on the inside of the membrane capacitance. The same reasoning applies to the relation between oppositely directed
current and hyperpolarization.

The second question is, what are the time relations between the current and the
potential changes? Rapid changes in electric current across the membrane must
first charge or discharge the membrane capacitance. This results in a time-
dependent storage of electrical energy that delays, distorts, and attenuates the
imposed signal. This is seen clearly in experiments in which a step of current is
introduced through an intracellular electrode, and the voltage change is recorded
through the same electrode (Fig. 13.10). Compared with the abrupt rise in
injected current, the voltage rises more slowly; one calls it a *charging transient*.
The amount of slowing of the voltage depends in the first instance on the time
constant (τ_m) of the membrane, as defined above (Eq. 13.28). If the cell were
only a soma, as in Fig. 13.8A,B above, then the value of τ would equal the time

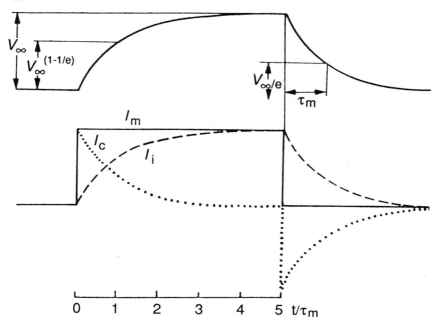

FIG. 13.10. Time course of charging and discharging of a passive membrane capacitance in parallel with a passive membrane resistance, in response to a current step. Abbreviations: I_m, applied membrane current: I_c, current through the capacitance; I_i, current through the ionic leak; V_∞, steady-state voltage, τ_m, membrane time constant. (Modified from Jack et al., 1975.)

to reach $1/e$ of the final steady-state voltage, as shown graphically in Fig. 13.10. However, for a real neuron, the electrotonic properties of the dendrites must again be taken into account. In general, the effect of the dendrites is to siphon current quickly from the soma until the electrotonic potential is uniform throughout the cell. We shall discuss this point further below.

SYNAPTIC POTENTIALS IN DENDRITIC TREES

The membrane time constant together with the electrotonic properties of the dendritic tree determines the time course of a synaptic potential. Mathematically the description of these relationships is quite complicated. However, computational models based on compartmental analysis, introduced by Rall (1964) for this purpose, make it possible to explore quite readily any arbitrary system. The essence of the methods, as they apply to the analysis of synaptic potentials, may be illustrated with respect to the diagram of Fig. 13.11. Consider a cell with an extensively branched dendritic tree, in which we wish to compare the synaptic potentials recorded in the cell body when the synaptic inputs are at three different sites: the cell body itself, midway in the dendritic tree, and at the far end of the tree. What will be the effect of the electrotonic properties of the dendrites on the characteristics of the synaptic potentials as they reach the cell body?

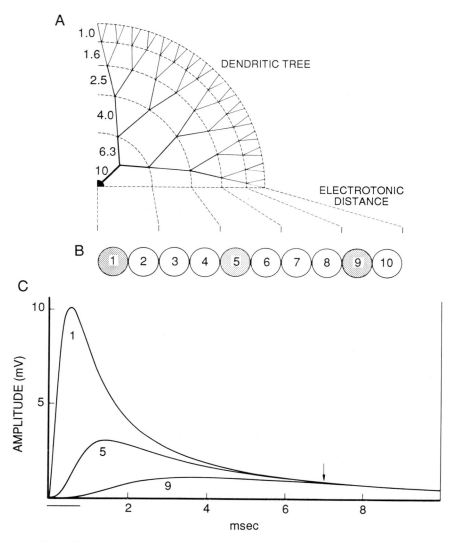

FIG. 13.11. Transient electrotonus in a dendritic tree. **A.** Diagrammatic representation of an extensively branched dendritic tree, in which a 3/2 power constraint applies to all branch points. **B.** Equivalent cylinder for the dendritic tree, modeled as a chain of compartments. Transient electrotonic potentials that would be recorded from the cell body (1) for the cases of brief synaptic conductance change (bar) in compartments (1), (5), or (9). Ordinate: amplitude (millivolts); abscissa: time (milliseconds). (After Rall, 1964; computer simulation by K. L. Marton.)

For simplicity, it is assumed that each dendrite gives rise to two daughter branches that meet the 3/2 power constraint previously described. By suitable assumptions for the branch lengths, an equivalent cylinder is obtained (Fig. 13.11A) in which increments of electrotonic length (ΔL) correspond to successive levels of branching. This modeling of an extensively branched system illustrates the power of the equivalent cylinder concept more persuasively than the previous simple case.

For computational purposes, the equivalent cylinder is put into the form of a chain of compartments, as shown in Fig. 13.11B. Each compartment represents a segment of the simple electrical circuit of Fig. 13.9. Each segment is joined to its neighbor by an ohmic resistance (R_i) that is scaled for the diameter of the segment. An excitatory synapse is modeled as a brief increase in membrane conductance in series with a battery representing the synaptic reversal potential. This places an excess positive charge inside the synaptic membrane, giving rise to a depolarizing EPSP. We assume the three different sites of synaptic input shown in Fig. 13.11 and ask the computer to compute the resulting voltage transients in the cell body due to electrotonic spread from these sites.

The results are shown in Fig. 13.11C. When the synaptic locus is in the soma, the EPSP at that site has an immediate sharp rise and a rapid decay. For the middle dendritic input, the soma transient is reduced in amplitude and has a slower time course. This effect becomes more extreme with the peripheral input site. In these respects, the results agree with our intuitive expectations. The computer, however, allows precise measurements of these different EPSPs to be made and compared. For example, Rall has shown that certain kinds of shape index, such as the time-to-peak and the half-width, are sensitive indicators of the sites of input. This has been a valuable tool in the analysis of synaptic loci in motoneurons.

The model also permits one to be quite precise in describing synaptic current flows. For the soma input (trace 1), we can say that the decay of the EPSP is more rapid than would be the case for a single RC element (i.e., single membrane segment) because, in addition to leakage of the charge locally across the somatic membrane, there is a rapid electrotonic spread of the charge from there to the rest of the cylinder. Rall conceives of this as equalizing spread, for which equalizing time constants can be obtained through standard techniques of peeling exponentials from semilogarithmic plots of the decays (Rall, 1970, 1977). As time goes by, the charge gets evenly distributed throughout the cylinder, as is shown by the fact that the final decays of all the transients, regardless of their initial input locus, become similar. Thus, beginning at the arrow in Fig. 13.11C, the decays everywhere are governed solely by the membrane time constant.

Consider now the soma EPSP generated by the distant dendritic input (trace 9). Not only is this transient slow and attenuated, but it is very delayed (i.e., a long time-to-onset). In fact, during the time of the synaptic conductance change in the periphery, no change in soma membrane potential is observed at all! This, of course, is a property of dendritic electrotonus, in that it takes time for the charge to spread along the capacitance of the equivalent cylinder.

Specific transmission through distal dendrites. The traditional conclusion from this type of analysis is that local inputs provide for large and rapid responses, whereas distant inputs provide for weaker and slower background modulation of activity at the soma. However, there is increasing awareness that distal dendrites nonetheless may mediate EPSPs that are large and rapid at the soma and clearly are involved in transmitting precise information. This is seen, for example, in olfactory cortex (Chap. 10), where pyramidal neurons receive all their specific sensory information through their distal dendrites. How does the neuron compensate for the slowing and decay imposed by the electrotonic properties of the dendrites?

One way is by having a relatively *high specific membrane resistance,* to minimize the decay of electrotonic potential. With improved recording methods, including the use of patch recordings, and taking into account the presence of a significant G_{EL} in conventional intracellular recordings, the estimated value of R_m has increased so that in many neurons it appears to be in the range of 30–60,000 Ωcm^2 (cf. olfactory mitral cells, Chap. 5). With these values, the equivalent cylinder of the dendritic tree has a characteristic length that is usually below 1, and may be below 0.5 (cf. hippocampal pyramidal neurons, Chap. 11). This provides for very effective coupling of even the most distal synaptic responses to the site of impulse generation in the soma.

A second way is to increase the amplitude of the synaptic response. Studies of the motoneuron (Chap. 4) show that unitary EPSPs generated at a distal and a proximal dendritic site differ in time course, as in the examples of (1) and (9) in Fig. 13.11, but have similar amplitudes. An obvious possibility is that the distal synapse is associated with a *larger synaptic conductance;* it has been estimated that this may be up to ten times as great in the distal as the proximal synapses (Redman et al., 1983a,b; see Chap. 4).

A third way to increase the amplitude of distal synaptic responses is by means of *active sites* on the dendritic membrane that boost the responses. As is evident throughout this book, there is increasing evidence for voltage-gated properties of dendrites, and for the role they may play in amplifying and shaping dendritic synaptic potentials.

These mechanisms are not mutually exclusive. As explained in Chap. 2, it is likely that different combinations are present in different neurons to enhance the contributions of all parts of the dendritic tree to the computational complexity of the neuron.

SYNAPTIC POTENTIALS IN DENDRITIC SPINES

The smallest type of dendritic branch is the spine. As already discussed in Chaps. 1 and 2 and in many other chapters of this book, spines are found on many neurons, and provide targets for specific synaptic inputs and, in some cases, outputs as well. The spread of potentials into and out of a spine has been an intriguing question, particularly in view of the very thin spine neck in many cases. The small size of the spine has put this question out of reach of direct electrophysiological analysis.

In the absence of direct experimental data from spines, there has been considerable exploration of their electrotonic properties, both mathematically (Rall and Rinzel, 1973; Jack et al., 1975) and by computational models (Shepherd and Brayton, 1979; Koch and Poggio, 1983; Gamble and Koch, 1986). These studies have given insight into the principles that underlie the input–output operations of a spine. We can illustrate some of these principles by the model of olfactory granule cell spines shown in Fig. 13.12.

We consider here a compartmental model of a length of granule cell dendrite (GR) from which arise two spines. Each spine consists of a head (compartments 1 and 3), commonly 0.5–2.0 μm in diameter, connected to the dendritic branch by a thin neck (compartments 2 and 4), commonly 0.1–0.2 μm in diameter and 0.5–3.0 μm in length. In this example, the branch itself has a diameter of 1 μm, but in different neurons, spines may arise from very thin branches down to 0.2 μm in diameter as well as thick dendritic trunks of 10-μm diameter or more.

The synaptic input to a dendritic spine can come from an axonal bouton or a presynaptic dendrite. In this example, the granule spines are interconnected with mitral cell dendrites by reciprocal synapses. Through these synapses the granule cell spines mediate feedback inhibition of an excited mitral cell (M_1) and lateral inhibition of neighboring mitral cells (M_2) (see B,a). This mechanism is dis-

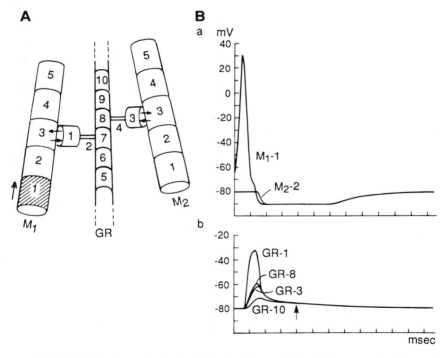

FIG. 13.12. Model of spread of potential in dendrites and dendritic spines. **A.** Compartmental model of two mitral cell dendrites (M_1 and M_2) and granule cell dendrite (GR) with two spines. The impulse is generated in compartment 1 of M_1 (shaded). **B.** Potentials recorded in model: (a) potential in mitral cells; (b) potentials in granule cell. Time scale in milliseconds. (From Shepherd and Brayton, 1979.)

cussed in Chaps. 2 and 5; here we wish to focus on the electrotonic properties of the granule spines (Fig. 13.12B,b), as a model for understanding dendritic spine functions.

Most of the interesting properties of spines are due to their very small size. A first consequence is that the spine has a very high input resistance (R_N), since R_N for a spheroidal spine head will, like a cell soma, vary inversely with the membrane area (i.e., the smaller the membrane area, the higher the total resistance across it; see Eq. 13.25 above). This high resistance means that, for a given synaptic conductance, there will be a correspondingly greater synaptic potential than for a synapse located on a larger dendrite or on the cell soma. This is shown in the model, where the EPSP has an amplitude of over 40 mV (GR-1 in B,b). Associated with the large amplitude is a rapid time course, reflecting the tiny capacitance of the small area of spine head membrane. This is typical of a locally generated potential (see trace 1 in Fig. 13.11), and is made briefer by the large conductance load of the dendrite, which siphons away the decaying current. The computational model thus allows us to look inside even the smallest parts of a dendrite and see that distal events are not weak and slow as they appear when viewed from the soma (see Fig. 13.11, trace 9), but large and fast when viewed at their site of generation (see also Chap. 4, motoneuron distal dendrites).

The high input resistance of the spine and the large conductance load of the branch mean that there is an impedance mismatch between them, so that the EPSP spreading electrotonically through the spine neck (compartment 2) into the branch (compartment 8) suffers a decrement (approximately half in the example of trace GR-8). The amount of decrement is dependent on the resistance of the neck and the impedance mismatch between spine and branch. Rall (1974a,b) has shown that there is an impedance range within which the spread of potential is maximally sensitive to very small changes in stem resistance. He has pointed out that if such changes were activity-dependent, they would provide a mechanism for memory storage. This hypothesis has possible application to several brain regions (see Chaps. 11 and 12).

In contrast to the decrement of an EPSP in spreading *out* of a spine, there is very little decrement in spreading in the other direction, from the branch *into* a spine (compare traces GR-8 and GR-3 in B,b). This reflects the fact that the spine neck, although thin, is so short that it is only a tiny fraction of the characteristic length (λ). In addition, the spine head approximates to a sealed end boundary condition, which elevates the curve for electrotonic decay (cf. Fig. 13.6 above, upper trace). As a consequence, the spine follows rather closely any changes in branch membrane potential. It can therefore transmit specific signals from neighboring spines, as in the example of Fig. 13.12. It also will register the weighted average activity of all spines within effective electrotonic distance. These properties are further illustrated in the models of Chap. 5 (Fig. 5.13).

Spine properties and spine functions. These considerations indicate that spines have complex electrotonic relations with their parent branch and with other spines. These relations are only a beginning for understanding spine functions. The list of spine properties and their possible functions is growing and full of surprises. For example, spines contain several types of filaments, including actin (see Fifkova and Delay, 1982; Landis and Reese, 1983); does this confer on them the ability to "twitch" (Crick, 1982)? The small volume of the spine head means that the effects of transmembrane ion flows on ion concentrations within the head are greatly magnified (cf. Qian and Sejnowski, 1989). This could provide a powerful link between inflow of Ca^{2+} and activation of Ca^{2+}-dependent structural and metabolic processes that may underlie memory (Gamble and Koch, 1987). The thin neck restricts diffusion of substances, thereby creating a microcompartment that is metabolically as well as electrically isolated from the parent dendrite (Shepherd, 1974; Harris and Stevens, 1988).

Finally, the presence of any voltage-dependent channels in the spine membrane adds considerably to the computational complexity of the distal dendrites. For example, such channels would enhance an EPSP (Miller et al., 1985). The effectiveness of passive spread between spines, as shown in Fig. 13.12B,b, would enable an action potential to be conducted in a pseudosaltatory manner from spine to spine (Shepherd et al., 1985), or from spine cluster to spine cluster (Rall and Segev, 1987). Interactions between excitable spines could be the basis for simple logic operations (Shepherd and Brayton, 1987), greatly increasing the computational complexity of distal dendrites. As discussed in Chap. 2 and elsewhere, NMDA channels in spines also have a nonlinear voltage-gated Ca^{2+} conductance (see Ascher and Nowak, 1987). Thus, if spine interactions can mediate computations for on-line information processing (through their excitability thresholds), they may also have the capacity for storage of information (through interactions dependent on their voltage-gated synaptic conductances). Assessment of all these possible functions will require a clear understanding of the underlying electrotonic properties of spine–branch relations.

SYNAPTIC INTEGRATION

Thus far we have focused on the electrotonic properties of current spread from a single synaptic site. We conclude this overview by considering the electrotonic interactions between two synaptic sites that are activated simultaneously. These interactions lie at the heart of the mechanisms of synaptic integration.

Interactions of conductances. Let us begin by considering two nearby excitatory synapses that are active either singly or simultaneously. We represent them with our equivalent circuit diagram of two neighboring segments of dendrite (A and B), as in Fig. 13.13A. When only the synapse at A is activated, there is a net inward current through the increased conductance at that point, producing a depolarizing EPSP (see Fig. 13.13B, trace a). This is precisely the situation previously described (Fig. 13.9 above).

Now we consider the case when both A and B are activated simultaneously. As shown in Fig. 13.13B, the combined EPSP is bigger, but it is not quite twice as big as expected, despite the fact that the conductance increase is twice as big. This is because the increased conductance of each synapse lowers the input resistance, siphoning current that otherwise would go to changing the charge on the membrane capacitance (see Fig. 13.13A). In addition, each active synapse

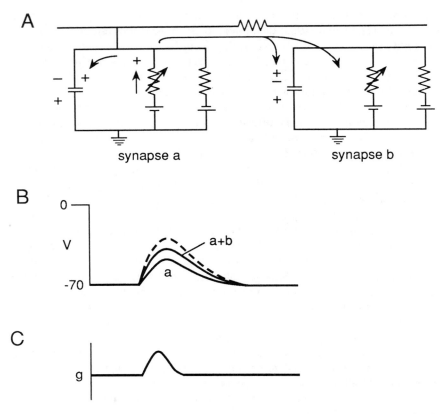

FIG. 13.13. Interactions between synaptic conductances mean that synaptic integration in general is not linear. **A.** The basic two-segment equivalent circuit model of a dendrite (cf. Fig. 13.9). Activation of synapse (a) increases the conductance for net inward current at (a), giving rise to current paths as shown (similar to Fig. 13.9). Simultaneous activation of synapse (b) adds further depolarizing effect, but lessens the effect of synapse (a) by lowering the effective membrane resistance. **B.** Schematic representation of recordings, showing the effect of (a) and (a+b); the dashed line shows the linear addition of (a+b). **C.** Time course of transmembrane conductance change (and ionic currents).

moves the membrane potential further toward the synaptic reversal potential, reducing the driving force associated with the conductance change. Thus, activating two synapses simultaneously will always produce a smaller potential than adding the potentials produced by the two synapses acting individually (cf. Fig. 13.13B). This is what is meant by the "nonlinear," or, more specifically, the "sublinear" interaction between conductance inputs. If, by contrast, the EPSPs were due to current injection (as from an intracellular electrode), the resulting potential would be exactly the sum of the individual potentials (because current injection does not change the input conductance nor is it dependent on the level of the membrane potential). Such a system, in which current is the input variable and voltage the output variable, is linear.

The general rule to take away from this discussion is that a synaptic system is not really linear. When synaptic conductances change, the system is perturbed; mathematically it can be shown that it becomes a different system (Rall, 1964). This means that synaptic interactions are much more dynamic than is usually recognized; they have the capacity to make nonlinear contributions to the functional operations of synaptic circuits, in parallel with the nonlinear activity of voltage gated intrinsic membrane properites, as discussed in Chap. 2 (see also Torre and Poggio, 1981). These nonlinear contributions will be larger the larger are the conductance changes and the closer together are the synapses. We therefore must consider the importance of synaptic location for these interactions.

Significance of dendritic branching. From Fig. 13.13 it is clear that a neuron cannot simply increase its inputs by adding synapses on its soma, because the conductances would increasingly interact to reduce the synaptic efficacy. The neuron wants to be able to minimize this effect. How does it do it?

The main way is through the branching structure of the dendrites. Figure 13.14 illustrates the strategy. We consider a dendritic tree with four terminal branches, each of which can receive an excitatory input (a–d). A single input produces the EPSP shown in B, trace (a), as recorded at the soma. If EPSPs summed linearly, two simultaneous inputs at (a) would produce twice the EPSP amplitude at the soma, as shown by the top trace (dashed line). The closest we can approach this is to place the two synapses as far apart as possible (a+d); as the synaptic sites move closer together (a+b, a+a), their summed EPSP at the soma falls, owing to the increasing interactions between the conductances.

We conclude that an effective way to maintain EPSP amplitudes is to place synapses on separate branches. This could underlie the division of many dendritic trees into subunits or compartments (see Chap. 1 and elsewhere). However, if a cell has many inputs, it is expensive to elaborate separate branches for each of them. What matters is to separate them by an equivalent electrotonic distance, by interposing a high internal resistance between them. This of course is precisely the property of the thin neck of a dendritic spine. Thus, one could crowd four inputs together onto a single dendritic branch, as in Fig. 13.14C, and achieve the same optimal summation of EPSPs as in the case of inputs to separate branches. Seen from this perspective, the dendritic spine appears to be a superb example of miniaturization of circuit components.

Thus far, we have discussed ways that conductance increases are minimized by geometrical means alone. In addition, the neuron has other mechanisms. These include synapses that work by conductance *decreases* rather than conductance increases (see Chap. 3). We have also seen that there is a variety of voltage-gated channels that can provide for background modulation of the membrane conductance at any given site on a dendritic tree (see Chap. 2). The neuron thus possesses a rich repertoire of mechanisms for regulating the electrotonic properties of its dendrites, and thereby the integration of signals within them.

Inhibitory conductances and inhibitory actions. We turn next to consider inhibitory synaptic conductances. To begin with, it should be clear that all of the properties governing electrotonic interactions between excitatory synaptic con-

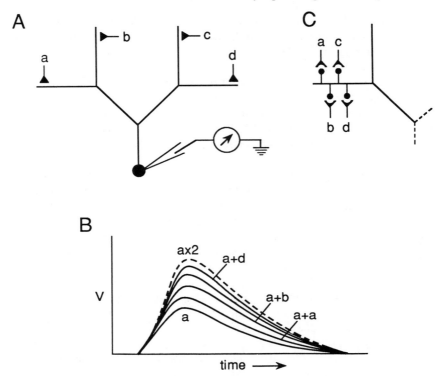

FIG. 13.14. Dendritic branching reduces interactions between synaptic conductances. **A.**
Schematic representation of a neuron and dendritic tree with four possible sites (a–d) of
synaptic input. The problem is to achieve the largest summation of two simultaneous
inputs. **B.** Recordings of response to different combination of inputs: (a) single response at
(a); (a+a) double the conductance at (a); (a+b, a+c, a+d) simultaneous inputs at these
sites; (a×2) linear doubling of (a). Note that the largest summation is achieved with the
widest separation, which minimizes the interaction of conductances. These nonlinear
effects are magnified in the drawing for purposes of illustration. **C.** Placement of the four
possible synaptic sites (a–d) on four spines on a single dendritic branch. With appropriate
assumptions about electrotonic properties, the responses in B apply similarly to these
interactions. See text.

ductances outlined above apply equally to inhibitory conductances. In addition,
inhibitory synapses may have two distinct types of effect on the membrane
potential, and hence on synaptic integration, depending on the relation of the
equilibrium potential for the conductance to the membrane potential level (see
Koch and Poggio, 1987).

The two effects are shown in Fig. 13.15A, where an intracellular electrode
records the response of an inhibitory synapse (s). We assume that the resting
membrane potential is at -70 mV, at the level of the equilibrium potential for
chloride ion (E_{Cl}), but above the level of the equilibrium potential for potassium
ion (E_K). If the synaptic response is Cl-mediated, as at a GABA$_A$ synapse (see
Chap. 2), there would be no change in the membrane potential, even though the

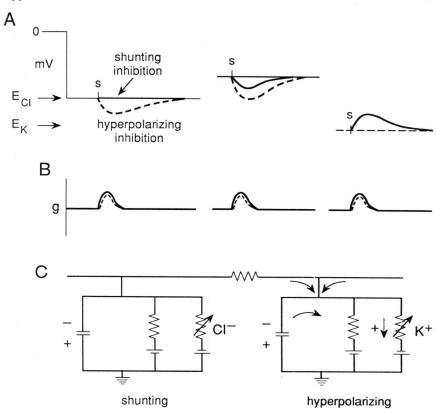

FIG. 13.15. Two types of synaptic inhibition. **A.** Schematic representations of intracellular voltage recordings: s, start of inhibitory conductance change; solid line, response of shunting inhibition (increased g_{Cl}); dashed line, response of hyperpolarizing inhibition (increased g_K). Left trace: E_m at E_{Cl}; middle trace: E_m above E_{Cl}; right trace: E_m at E_K. **B.** Synaptic conductance changes are the same for all trials. **C.** Two-segment dendritic model, showing current paths for the two types of inhibition for the case of $E_m = E_{Cl}$.

conductance increases (see B), because the membrane is already at the equilibrium potential for the ions involved. Because the only effect of the synapse is to increase the conductance across the membrane, this is commonly referred to as *shunting inhibition;* because it is not associated with a change in membrane potential, it is also called *silent inhibition.* If, on the other hand, the response is K^+-mediated, as at a $GABA_B$ synapse, the membrane will be hyperpolarized toward E_K. This is therefore referred to as *hyperpolarizing inhibition.* However, these distinctions must always be made relative to the resting membrane potential level; both types involve a shunt, and either type may be depolarizing or hyperpolarizing. If the resting membrane potential is raised, both responses are hyperpolarizing; if it is lowered to the level of E_K, then the K^+-mediated response is flat and the Cl-mediated response is depolarizing, as shown in the traces to the right in Fig. 13.15A.

The mechanisms of these actions are illustrated in Fig. 13.15C, where we see our two familiar segments of dendrites, in this case representing the sites of the two types of synapse. At the resting membrane potential level, increasing the Cl^- conductance does not induce any current flow, as shown on the left. By contrast, increasing the K^+ conductance increases the charge on the membrane, inducing the current flow shown on the right. The reader may deduce the current flows that would result from the shifts in resting membrane potential shown in A.

Integration of excitation and inhibition. The functional significance of these two types of inhibition is expressed in the way they interact with excitatory synaptic conductances. This is illustrated in Fig. 13.16. We consider first, in A, the case of simultaneous activation of an excitatory and inhibitory synapse when the

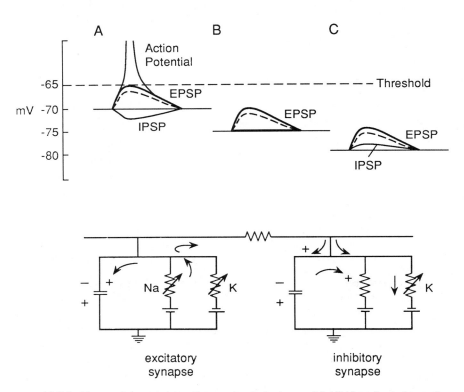

FIG. 13.16. Neuronal integration of synaptic excitation and inhibition depends on the relation of the membrane potential to excitatory and inhibitory equilibrium potentials. **A.** EPSP alone generates an action potential; a hyperpolarizing IPSP reduces the EPSP below threshold (dashed lines). **B.** Lowering E_m to the inhibitory equilibrium potential converts the IPSP to purely shunting inhibition, which still reduces the concurrent EPSP. **C.** Lowering E_m further converts the IPSP to a depolarizing response, but the shunt still reduces the concurrent EPSP. *Below:* Two-segment electrotonic model illustrates the basic interaction between excitatory synapses (involving increase in g_{Na} and g_K) and inhibitory synapse (example of a g_K increase). (A–C modified from Shepherd, 1979.)

membrane potential is above the level of the inhibitory equilibrium potential. This is the case of hyperpolarizing inhibition. As shown in A, the effect of the IPSP is to reduce the EPSP (dashed line) below the threshold for activation of an impulse. The reduced response is not exactly an algebraic summation of the EPSP and IPSP, because of the nonlinear nature of the interaction between the underlying conductances, as discussed above.

This effect is seen very clearly in the case of shunting inhibition, illustrated in B. Here, the membrane potential is at the level of the inhibitory equilibrium potential. The "silent" inhibition produces no potential response by itself, but significantly reduces a concurrent EPSP (dashed line) through its shunting effect. The interaction is even more nonlinear when the membrane potential is more hyperpolarized than the inhibitory equilibrium potential, as in C; in this case, the IPSP is *depolarizing*, yet reduces the concurrent EPSP owing to the shunting effect of the conductance increase. The mechanism of the interaction in B is illustrated in the equivalent circuit diagram below.

These examples should leave no doubt as to the underlying nonlinear character of the effect of synaptic inhibition on concurrent synaptic excitation. The effect is more linear, the more hyperpolarizing is the inhibition. This confers on hyperpolarizing inhibition the ability to be used in operations equivalent to addition and subtraction within a dendritic tree. By contrast, the effect is most nonlinear for purely shunting or silent inhibition. This type of inhibition is therefore believed to have the property of performing the computational equivalent of division and multiplication in its interactions with excitatory inputs (see Koch and Poggio, 1987, for full discussion of these operations).

Significance of synaptic placement. The computational operations that may arise in the course of synaptic integration are not due solely to the interactions themselves, but depend in addition on the geometrical relations between the sites of the synapses and the site where the signals are integrated. This idea has already been introduced in Chap. 1; following Rall (1964) and Koch et al. (1982), we consider here more closely its mechanism.

We assume a chain of compartments (see A in Fig. 13.17), which may represent a single dendrite, or the equivalent cylinder of a branching dendritic tree. A single excitatory synapse (e) is placed in the middle compartment (3), halfway between the soma (1) and the dendritic terminals (5). When this synapse is activated, it produces an EPSP which spreads electrotonically through the equivalent cylinder, producing the response at the soma shown in trace A (solid and dashed line) in the diagram below. We now ask, How will the effect of an inhibitory synapse vary with different possible sites in the dendritic tree?

The first possible site is further out on the dendrites, near their terminals (i into compartment 5 in A). As shown in the graph below (solid line in trace A), there is essentially no effect on the peak amplitude of the EPSP recorded at the soma. By contrast, when the inhibitory synapse is at the same site as the excitatory synapse (B), the EPSP is reduced to almost half. Inhibition sited at the soma (C) is even more effective. In fact, one can prove in a vigorous fashion that the

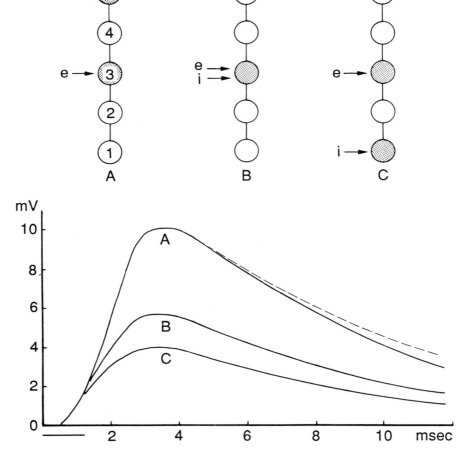

FIG. 13.17. *Top:* Dependence of excitatory and inhibitory synaptic integration on the electrotonic relation between input sites. In all cases an excitatory synapse (e) is located in the middle of an equivalent cylinder (position 3). An inhibitory synapse (i) is located at one of three possible sites: peripherally (A), same site (B), or proximally (C). *Bottom:* Resulting synaptic potentials recorded proximally (at 1) for these three cases. Dashed line: Control response to an excitatory synapse alone. (After Rall, 1964.)

location for inhibition maximally to block or reduce an EPSP is between the site of the excitatory synapse and the soma. These interactions have been fully explored by Koch et al. (1982) in their model of the dendritic trees of retinal ganglion cells.

    The general conclusion from this analysis is that the specific relations of synaptic sites to each other and to sites of output are crucial in determining the type and complexity of the neural computation carried out in the dendritic tree. This is discussed in Chap. 1 with reference to the generation of simple logic operations; the situations depicted in Fig. 13.17 are similar to those in the diagram of Fig. 1.5. It can be seen that case C represents the "on-path" condition, whereas (A) is the "off-path" condition, discussed in relation to Fig. 1.5.

These results should not give the impression that inhibition should always be sited preferentially at the soma. In many neurons inhibition is indeed strong in the soma, where it can mediate global control of all excitatory responses spreading to the cell body. However, it is clear that inhibitory synapses sited in middle and distal dendrites are also extremely effective against excitatory synapses *at those same sites*. This provides a means for specific and local computations to be carried out independently of computations in other parts of the tree. Furthermore, the effectiveness of these interactions is clearly exquisitely dependent on "on-the-path" and "off-the-path" relations. This implies that the specific target of an ingrowing axon in making a synapse during development would be critical in determining the kind of computation in which that synapse would be involved. Similarly, small adjustments in the site or efficacy of a synapse might be critical in processes of learning or memory. There is obviously much fruitful ground to be explored here, both by experimental analysis, and by computational models that accurately reflect the importance of electrotonic properties for these functional operations.

REFERENCES

Adams, P.R., and Galvan, M. 1986. Voltage-dependent currents of vertebrate neurons and their role in membrane excitability. Adv. Neurol. 44: 137–170.

Adams, P.R., Jones, S. W., Pennefather, P., Brown, D.A., Koch, C., and Lancaster, B. 1986. Slow synaptic transmission in frog sympathetic ganglia. J. Exp. Biol. 124: 259–285.

Adrian, E.D. 1950. Sensory discrimination with some recent evidence from the olfactory organ. Br. Med. Bull. 6: 330–333.

Aghajanian, G.K. 1985. Modulation of a transient outward current in serotonergic neurones by alpha-adrenoceptors. Nature 315: 501–503.

Aghajanian, G.K., and Rasmussen, K. 1988. Basic electrophysiology. In: Psychopharmacology: The Third Generation of Progress (Meltzer, H., ed.). New York: Raven Press, pp. 67–74.

Airaksinen, M.S., and Panula, P. 1988. The histaminergic system in the guinea pig central nervous system: An immunocytochemical mapping study using an antiserum against histamine. J. Comp. Neurol. 273: 163–186.

Akaike, A., Ohno, Y., Sasa, M., and Takaori, S. 1987. Excitatory and inhibitory effects of dopamine on neuronal activity of the caudate nucleus neurons in vitro. Brain Res. 418: 262–272.

Akeson, R.A. 1988. Primary olfactory neuron·subclasses. In: Molecular Neurobiology of the Olfactory System (Margolis, F.L., and Getchell, T.V., eds.). New York: Plenum Press, pp. 297–318.

Alger, B.E., and Nicoll, R.A. 1979. GABA mediated biphasic inhibitory response in the hippocampus. Nature 281: 315–317.

Alger, B.E., and Nicoll, R.A. 1982. Feed-forward dendritic inhibition in rat hippocampal pyramidal cells studied in vitro. J. Physiol. (Lond.) 328; 105–123.

Allison, A.C. 1953. The structure of the olfactory bulb and its relationship to the olfactory pathways in the rabbit and the rat. J. Comp. Neurol. 98: 309–353.

Amaral, D.G. 1978. A Golgi study of cell types in the hilar region of the hippocampus in the rat. J. Comp. Neurol. 1824: 851–914.

Amaral, D.G. 1987. Memory: Anatomical organization of candidate brain regions. In: Handbook of Physiology, Section 1: The Nervous System, Vol. V: Higher Functions of the Brain, Part 1 (Plum, F., ed.). Bethesda: American Physiological Society, pp. 211–294.

Amaral, D.G., and Dent, J.A. 1981. Development of the mossy fibers of the dentate gyrus: I. A light- and electron microscopic study of the mossy fibers and their expansions. J. Comp. Neurol. 195: 51–86.

Amthor, R.R., Oyster, C.W., and Takahashi, E.S. 1984. Morphology of ON–OFF direction-selective ganglion cells in the rabbit retina. Brain Res. 298: 187–190.

Andersen, P. 1975. Organization of hippocampal neurons and their interconnections. In: The Hippocampus (Isaacson, R.L., and Pribram, K.H., eds.). New York: Plenum Press, Vol. 1, pp. 155–174.

Andersen, P., Blackstad, T.W., and Lømo, T. 1966a. Location and identification of excitatory synapses on hippocampal pyramidal cells. Exp. Brain Res. 1: 236–248.

Andersen, P., Dingledine, R., Gjerstad, L., Langmoen, I.A., and Laursen, A.M. 1980a.

Two different responses of hippocampal pyramidal cells to application of gamma-aminobutyric acid. J. Physiol. (Lond.) 305: 279–296.

Andersen, P., Eccles, J.C., and Voorhoeve, P.E. 1964. Postsynaptic inhibition of cerebellar Purkinje cells. J. Neurophysiol. 27: 1139–1153.

Andersen, P., Holmqvist, B., and Voorhoeve, P.E. 1966b. Entorhinal activation of dentate granule cells. Acta. Physiol. Scand. 66: 448–460.

Andersen, P., Silvenius, H., Sunberg, S.H., and Sveen, O. 1980b. A comparison of distal and proximal dendritic synapses on CA 1 pyramids in guinea pig hippocampal slices in vitro. J. Physiol. (Lond.) 307: 273–299.

Anderson, J.A. 1970. Two models for memory organization using interacting traces. Math. Biosci. 8: 137–160.

Anderson, J.A. 1972. A simple network generating an interactive memory. Math Biosci. 14: 197–220.

Anderson, J.A. 1985. What Hebb synapses build. In: Synaptic Modification, Neurons Selectivity and Nervous System Organization (Levy, W.B., Anderson, J.A., and Lehmkuhle, S., eds.). Hillsdale, NJ: Lawrence Erlbaum Associates, pp. 153–174.

Andrade, R., Malenka, R.C., and Nicoll, R.A. 1986. A G protein couples serotonin and GABA B receptors to the same channels in hippocampus. Science 234: 1261–1265.

Andrade, R., and Nicoll, R.A. 1987. Pharmacologically distinct actions of serotonin on single pyramidal neurones of the rat hippocampus recorded in vitro. J. Physiol. (Lond.) 394: 99–124.

Andres, K. 1965. Der feinaudes bulbos olfactorius der ratte inter besonderer Berksichtigung der verbindungen. Z. Zellforsch. Mikrosk. Anat. 65: 530–561.

Andrews, S.B., Leapman, R.D., Landis, D.M.D., and Reese, T.S. 1988. Activity-dependent accumulation of calcium in Purkinje cell dendritic spines. Proc. Natl. Acad. Sci. USA 85: 1682–1685.

Anson, J., and Collins, G.G.S. 1987. Possible presynaptic action of 2-amino-4-phosphonobutyrate in rat olfactory cortex. Br. J. Pharmacol. 91: 753–761.

Armstrong, D.M. 1974. Functional significance of connections of the inferior olive. Physiol. Rev. 54: 358–417.

Armstrong-James, M., and Fox, K. 1983. Effects of ionophoresed noradrenaline on the spontaneous activity of neurones in rat primary somatosensory cortex. J. Physiol. (Lond.) 335: 427–447.

Artola, A., and Singer, W. 1987. Long-term potentiation and NMDA receptors in rat visual cortex. Nature 330: 649–652.

Ascher, P., and Nowak, L. 1987. Electrophysiological studies of NMDA receptors. Trends Neurosci. 10: 284–287.

Ascher, P., and Nowak, L. 1988. The role of divalent cations in the N-methyl-D-aspartate responses of mouse central neurones in culture. J. Physiol. (Lond.) 399: 247–266.

Ashmore, J.F., and Copenhagen, D.R. 1980. Different postsynaptic events in two types of retinal bipolar cell. Nature 288: 84–86.

Ashwood, T.J., Collinridge, G.L., Herron, C.E., and Wheal, H.V. 1987. Voltage-clamp analysis of somatic γ-aminobutyric acid responses in adult rat hippocampal CA1 neurones in vitro. J. Physiol. (Lond.) 384: 27–87.

Attwell, D. 1986. Ion channels and signal processing in the outer retina. Q. J. Exp. Physiol. 71: 497–536.

Avoli, M. 1986. Inhibitory potentials in neurons of the deep layers of the in vitro neocortical slice. Brain Res. 370: 165–170.

Ayala, G.F., Dichter, M., Gumnit, R.J., Matsumoto, H., and Spencer, W.A. 1973.

Genesis of epileptic interictal spikes. New knowledge of cortical feedback systems suggests a neurophysiological explanation of brief paroxysms. Brain Res. 52: 1–17.

Ayala, G.F., Matsumoto, H., and Gumnit, R.J. 1970. Excitability changes and inhibitory mechanisms in neocortical neurons during seizures. J. Neurophysiol. 33: 73–85.

Ayoub, G.S., Korenbrot, J.I., and Copenhagen, D.R. 1989. The release of endogenous glutamate from isolated cone photoreceptors of the lizard. Neurosci. Res. [Suppl.] 10: 547–556.

Baker, H. 1988. Neurotransmitter plasticity in the juxtaglomerular cells of the olfactory bulb. In: Molecular Neurobiology of the Olfactory System (Margolis, F.L., and Getchell, T.V., eds.). New York: Plenum Press, pp. 185–216.

Baker, H., Kawano, T., Margolis, F.L., and Joh, T.H. 1983. Transneural regulation of tyrosine hydroxylase expression in olfactory bulb of mouse and rat. J. Neurosci. 3: 69–78.

Baldissera, F., Hultborn, H., and Illert, M. 1981. Integration in spinal neuronal systems. In: Handbook of Physiology, Section 1: The Nervous System, Vol. II: Motor Control, Part 1 (Brooks, V.B., ed.). Bethesda: American Physiological Society, pp. 509–595.

Banks, M.S., Geisler, W.S., and Bennett, P.J. 1987. The physical limits of grating visibility. Vision Res. 27: 1915–1924.

Barber, P.C., and Lindsay, R.M. 1982. Schwann cells of the olfactory nerves contain glial fibrillary acidic protein and resemble astrocytes. Neuroscience 7: 2687–2695.

Bargas, J., Galarraga, E., and Aceves, J. 1988. Electrotonic properties of neostriatal neurons are modulated by extracellular potassium. Exp. Brain Res. 72: 390–398.

Barker, P.R., Vaugn, J.E., Wimer, R.E., and Wimer, C.C. 1974. Genetically associated variations in the distribution of dentate granule cell synapses upon the pyramidal cell dendrites in mouse hippocampus. J. Comp. Neurol. 156: 417–434.

Barlow, H.B. 1953. Summation and inhibition in the frog's retina. J. Physiol. (Lond.) 119: 69–88.

Barlow, H.B. 1972. Single units and sensation: A neuron doctrine for perceptual psychology. Perception 1: 371–394.

Barlow, H.B. 1981. Critical factors limiting the design of the eye and visual cortex. The Ferrier lecture. Proc. R. Soc. Lond. [B] 212: 1–34.

Barlow, H.B. 1982. General principles: The senses considered as physical instruments. In: The Senses (Barlow, H.B., and Mollon, J.D., eds.). Cambridge: Cambridge University Press, pp. 1–33.

Barlow, H.B., Fitzhugh, B.R., and Kuffler, S.W. 1957. Change of organization in the receptive fields of the cat's retina during dark adaptation. J. Physiol. (Lond.) 137: 338–354.

Barlow, H.B., Hill, R.M., and Levick, W.R. 1964. Retinal ganglion cells responding selectively to direction and speed of image motion in the rabbit. J. Physiol. (Lond.) 173: 377–407.

Barlow, H.B., and Levick, W.R. 1969. Changes in the maintained discharge with adaptation level in the cat retina. J. Physiol. (Lond.) 202: 699–718.

Barlow, H.B., Levick, W.R., and Yoon, M. 1971. Responses to single quanta of light in retinal ganglion cells of the cat. Vision Res. 3: 87–101.

Barrett, E.F., and Barrett, J.N. 1976. Separation of two voltage-sensitive potassium currents, and demonstration of a tetrodotoxin-resistant calcium current in frog motoneurones. J. Physiol. (Lond.) 255: 737–774.

Barrett, E.F., Barrett, J.N., and Crill, W.E. 1980. Voltage-sensitive outward currents in cat motoneurones. J. Physiol. (Lond.) 304: 251–276.

Barrett, E.F., and Magleby, K.L. 1976. Physiology of cholinergic transmission. In: Biology of Cholinergic Function (Goldberg, A.M., and Hanin, E., eds.). New York: Raven Press, pp. 29–100.

Barrett, J.N., and Crill, W.E. 1974. Specific membrane properties of cat motoneurons. J. Physiol. (Lond.) 239: 301–324.

Barrionuevo, G., and Brown, T.H. 1983. Associative long-term potentiation in hippocampal slices. Proc. Natl. Acad. Sci. USA 80: 7347–7351.

Barrionuevo, G., Kelso, S., Johnston, D., and Brown, T.H. 1986. Conductance mechanism responsible for long-term potentiation in monosynaptic and isolated excitatory synaptic inputs to hippocampus. J. Neurophysiol. 55: 540–550.

Baughman, R.W., and Gilbert, C.D. 1981. Aspartate and glutamate as possible neurotransmitters in the visual cortex. J. Neurosci. 1: 427–439.

Baylor, D.A., Fuortes, M.G.F., and O'Bryan, P.M. 1971. Receptive fields of cones in the retina of the turtle. J. Physiol. (Lond.) 214: 265–294.

Baylor, D.A., Nunn, B.J., and Schnapf, J.F. 1984. The photocurrent, noise and spectral sensitivity of rods of the monkey *Macaca fasicularis*. J. Physiol. (Lond.) 375: 575–607.

Beaulieu, C., and Colonnier, M. 1983. The number of neurons in the different laminae of the binocular and monocular regions of area 17 in the cat. J. Comp. Neurol. 217: 337–344.

Beaulieu, C., and Colonnier, M. 1985. A laminar analysis of the number of round–asymmetrical and flat–symmetrical synapses on spines, dendritic trunks and cell bodies in area 17 of the cat. J. Comp. Neurol. 231: 180–189.

Bell, C.C., and Grimm, R.J. 1969. Discharge properties of Purkinje cells recorded on single and double microelectrodes. J. Neurophysiol. 32: 1044–1055.

Benardo, L.S., Masukawa, L.M., and Prince, D.A. 1982. Electrophysiology of isolated hippocampal pyramidal dendrites. J. Neurosci. 2: 1614–1622.

Benardo, L.S., and Prince, D.A. 1982. Cholinergic excitation of hippocampal pyramidal cells. Brain Res. 249: 315–333.

Bennett, M.V.L. 1977. Electrical transmission: A functional analysis and comparison to chemical transmission. In: Handbook of Physiology, Section 1: The Nervous System, Vol. I: Cellular Biology of Neurons, Part 1 (Kandel, E.R., ed.). Bethesda: American Physiological Society, pp. 357–416.

Benson, T., Burd, G.D., Greer, C.A., Landis, D., and Shepherd, G.M. 1985. High-resolution 2-deoxyglucose autoradiography in quick-frozen slabs of neonatal rat olfactory bulb. Brain Res. 339: 67–78.

Benson, T., Ryugo, D., and Hinds, J. 1984. Effects of sensory deprivation on developing mouse olfactory system. J. Neurosci. 4: 638–653.

Berger, T.W. 1986. Neural representations of associative learning in the hippocampus. In: Neuropsychology of Memory (Squire, L.R., and Butters, N., eds.). New York: Guilford Press, pp. 443–461.

Berman, N.J., Bush, P.C., and Douglas, R.J. 1989. Adaptation and bursting may be controlled by a single fast potassium current. Q. J. Exp. Physiol. 74: 223–226.

Berman, N.J., Douglas R., and Martin, K.A.C. 1990. The conductances associated with inhibitory postsynaptic potentials are larger in visual cortical neurons in vitro than in similar neurons in intact, anaesthetized rats. J. Physiol. (Lond.)

Biedenbach, M.A., and Stevens, C.F. 1969. Synaptic organization of the cat olfactory cortex as revealed by intracellular recording. J. Neurophysiol. 32: 204–214.

Bindman, L.J., Meyer, T., and Pockett, S.J. 1988. Postsynaptic control of the induction of long-term changes in efficacy of transmission at neocortical synapses in slices of rat brain. J. Neurophysiol. 60: 1053–1065.

Bishop, G.A., Chang, H.T., and Kitai, S.T. 1982. Morphological and physiological properties of neostriatal neurons: An intracellular horseradish peroxidase study in the rat. Neuroscience 7: 179–191.

Bisti, S., Iosif, G., Marchesi, G.F., and Strata, P. 1971. Pharmacological properties of inhibitions in the cerebellar cortex. Exp. Brain Res. 14: 24–37.

Björklund, A., and Lindvall, O. 1984. Dopamine-containing systems in the CNS. In: Handbook of Chemical Neuroanatomy, Vol. 2: Classical Transmitters in the CNS, Part I (Björklund, A., and Hökfelt, T., eds.). Amsterdam: Elsevier, pp. 55–122.

Blackman, J.G., Ginsborg, B.L., and Ray, C. 1963. Synaptic transmission in the sympathetic ganglion of the frog. J. Physiol. (Lond.) 167: 355–373.

Blackstad, T.W., and Kjaerheim, A. 1961. Special axo-dendritic synapses in the hippocampal cortex: Electron and light microscopic studies on the layer of the mossy fibers. J. Comp. Neurol. 177: 133–159.

Bland, B.H. 1986. The physiology and pharmacology of hippocampal formation theta rhythm. Prog. Neurobiol. 26: 1–54.

Blaxter, T.J., Carlen, P.L., Davies, M.F., and Kujtan, P.W. 1986. γ-Aminobutyric acid hyperpolarizes rat hippocampal pyramidal cells through a calcium-dependent potassium conductance. J. Physiol. (Lond.) 373: 181–194.

Blaxter, T.J., and Cottrell, G.A. 1985. Actions of GABA and ethylenediamine on CA1 pyramidal neurones in the rat hippocampus. Q. J. Exp. Physiol. 70: 75–93.

Bliss, T., and Rosenberg, M. 1979. Activity-dependent changes on conduction velocity in the olfactory nerve of the tortoise. Pflugers Archiv 381: 209–216.

Bliss, T.V.P., and Gardner-Medwin, A.R. 1973. Long-lasting potentiation of synaptic transmission in the dentate area of the unanaesthetized rabbit following stimulation of the perforant path. J. Physiol. (Lond.) 232: 357–374.

Bliss, T.V.P., and Lomo, T. 1973. Long-lasting potentiation of synaptic transmission in the dentate area of the unanaesthetized rabbit following stimulation of the perforant path. J. Physiol. (Lond.) 232: 331–356.

Bloedel, J.R., and Courville, J. 1981. Cerebellar afferent systems. In: Handbook of Physiology, Section 1: The Nervous System, Vol. II: Motor Control, Part 2 (Brooks, V.B., ed.). Bethesda: American Physiological Society, pp. 735–829.

Bloom, F.E., Costa, E., and Salmoiraghi, G.C. 1964. Analysis of individual rabbit olfactory bulb neuron responses to the microelectrophoresis of acetylcholine, norepinephrine and serotonin synergists and antagonists. J. Pharmacol. Exp. Ther. 146: 16–23.

Bloom, F.E., Hoffer, B.J., and Siggins, G.R. 1971. Studies on norepinephrine-containing afferents to Purkinje cells of rat cerebellum. I. Localization of the fibers and their synapses. Brain Res. 25: 501–521.

Bloomfield, S. 1974. Arithmetical operations performed by nerve cells. Brain Res. 69: 115–124.

Bloomfield, S.A., Hamos, J.E., and Sherman, S.M. 1987. Passive cable properties and morphological correlates of neurones in the lateral geniculate nucleus of the cat. J. Physiol. (Lond.) 383: 653–692.

Bloomfield, S.A., and Sherman, S.M. 1988. Postsynaptic potentials recorded in neurons of the cat's lateral geniculate nucleus following electrical stimulation of the optic chiasm. J. Neurophysiol. 60: 1924–1945.

Bloomfield, S.A., and Sherman, S.M. 1989. Dendritic current flow in relay cells and interneurons of the cat's lateral geniculate nucleus. Proc. Natl. Acad. Sci. USA. 86: 3911–3914.

Bogan, N., Brecha, N., Gall, C., and Karten, H.J. 1982. Distribution of enkephalin-like immunoreactivity in the rat main olfactory bulb. Neuroscience 7: 895–906.

Bolam, J.P., Clark, D.J., Smith, A.D., and Somogyi, P. 1983. A type of aspiny neuron in the rat neostriatum accumulates [³H]gamma-aminobutyric acid: Combination of Golgi-staining, autoradiography and electron microscopy. J. Comp. Neurol. 213: 121–134.

Bolam, J.P., Ingham, C.A., Izzo, P.N., Levy, A.I., Rye, D.B., Smith, A.D., and Wainer, B.H. 1986. Substance P-containing terminals in synaptic contact with cholinergic neurons in the neostriatum and basal forebrain: A double immunocytochemical study in the rat. Brain Res. 397: 279–289.

Bolam, J.P., Izzo, P.N., and Graybiel, A.M. 1988. Cellular substrate of the histochemically defined striosome/matrix system of the caudate nucleus: A combined Golgi and immunocytochemical study in cat and ferret. Neuroscience 24: 853–875.

Bolam, J.P., Powell, J.F., Totterdell, S., and Smith, A.D. 1981. A second type of striatonigral neuron: A comparison between retrogradely labelled and Golgi-stained neurons at the light and electron microscopic levels. Neuroscience 6: 2141–2157.

Bolam, J.P., Wainer, B.H., and Smith, A.D. 1984. Characterization of cholinergic neurons in the rat neostriatum. A combination of choline acetyltransferase immunocytochemistry, Golgi-impregnation and electron microscopy. Neuroscience 12: 711–718.

Bolz, J., and Thier, P. 1985. Phototropic action of thyrotropin-releasing hormone in the cat retina. Proc. Roy. Soc. London [B] 224: 463–473.

Bolz, J., Thier, P., Voight, T., and Wässle, H. 1985. Action and localization of glycine and taurine in the cat retina. J. Physiol. (Lond.) 362: 395–413.

Boss, B.D., Peterson, G.M., and Cowan, W.M. 1985. On the numbers of neurons in the dentate gyrus of the rat. Brain Res. 338: 144–150.

Boss, B.D., Turlejski, K., Stanfield, B.B., and Cowan, W.M. 1987. On the numbers of neurons in fields CA1 and CA3 of the hippocampus of Sprague-Dawley and Wistar rats. Brain Res. 406: 280–287.

Bower, J.M., and Haberly, L.B. 1986. Facilitating and nonfacilitating pyramidal cell synapses: A correlation between physiology and morphology. Proc. Natl. Acad. Sci. USA 83: 1115–1119.

Bowery, N.G., Hudson, A.L., and Price, G.W. 1987. GABA$_A$ and GABA$_B$ receptor site distribution in the rat central nervous system. Neuroscience 20: 365–383.

Bowling, D.B., and Michael, C.R. 1984. Terminal patterns of single, physiologically characterized optic tract fibers in the cat's lateral geniculate nucleus. J. Neurosci. 4: 198–216.

Boycott, B.B., and Kolb, H. 1973. The connections between the bipolar cells and photoreceptors in the retina of the domestic cat. J. Comp. Neurol. 148: 91–114.

Boycott, B.B., Peichl, L., and Wässle, H. 1978. Morphological types of horizontal cell in the retina of the domestic cat. Proc. R. Soc. Lond. [B] 203: 229–245.

Boycott, B.B., and Wässle, H. 1974. The morphological types of ganglion cells of the domestic cat's retina. J. Physiol. (Lond.) 240: 397–419.

Boyd, I.A., and Davey, M.R. 1968. Composition of Peripheral Nerves. Edinburgh: E. and S. Livingstone.

Braitenberg, V., and Atwood, R.P. 1958. Morphological observations in the cerebellar cortex. J. Comp. Neurol. 109: 1–34.

Braitenberg, V., and Onesto, N. 1958. The cerebellar cortex as a timing organ: Discussion of a hypothesis. Proc. 1st Int. Conf. Med. Cibernetics (Naples), pp. 1–19.

Braitman, D.J. 1986. Desensitization to glutamate and aspartate in rat olfactory (prepyriform cortex slice). Brain Res. 364: 199–203.

Brazier, M.A.B. 1960. The historical development of neurophysiology. In: Handbook of Physiology, Section 1: Neurophysiology, Vol. I (Magoun, H.W., ed.). Bethesda: American Physiological Society, pp. 1–58.

Brock, L.G., Coombs, J.S., and Eccles, J.C. 1952. The recording of potentials from motoneurons with an intracellular electrode. J. Physiol. (Lond.) 117: 431–460.

Brodal, A. 1981. Neurological Anatomy, 3rd ed. New York: Oxford University Press.

Brown, A.G. 1981. Organization in the Spinal Cord: The Anatomy and Physiology of Identified Neurones. New York: Springer-Verlag.

Brown, A.G., and Fyffe, R.E.W. 1981. Direct observations on the contacts made between Ia afferents and α-motoneurones in the cat's lumbosacral spinal cord. J. Physiol. (Lond.) 313: 121–140.

Brown, A.G., Rose, P.K., and Snow, P.J. 1977. The morphology of spinocervical tract neurones revealed by intracellular injection of horseradish peroxidase. J. Physiol. (Lond.) 270: 747–764.

Brown, A.M., and Birnbaumer, L. 1988. Direct G protein gating of ion channels. Am. J. Physiol. 254: H401–H410.

Brown, D.A. 1979. Extrasynaptic GABA systems. Trends Neurosci. 2: 271–273.

Brown, D.A. 1988a. M currents. In: Ion Channels, Vol. 1 (Narahashi, T., ed.). New York: Plenum Press.

Brown, D.A. 1988b. M-currents: An update. Trends Neurosci. 44: 294–299.

Brown, D.A., and Adams, P.R. 1980. Muscarinic suppression of a novel voltage-sensitive K^+ current in a vertebrate neurone. Nature 283: 673–676.

Brown, D.A., Collins, G.G.S., and Galvan, M. 1980. Influence of cellular transport in the interaction of amino acids with γ-aminobutyric acid. GABA-receptors in the isolated olfactory cortex of the guinea pig. Br. J. Pharmacol. 68: 251–262.

Brown, D.A., and Griffith, W.H. 1983a. Calcium-activated outward current in voltage-clamped hippocampal neurones of the guinea pig. J. Physiol. (Lond.) 337: 287–301.

Brown, D.A., and Griffith, W.H. 1983b. Persistent slow inward calcium current in voltage-clamped hippocampal neurones of the guinea pig. J. Physiol. (Lond.) 337: 303–320.

Brown, D.A., Higgins, A.J., Marsh, S., and Smart, T.G. 1981. Actions of GABA on mammalian neurones, axons, and nerve terminals. In: Amino Acid Neurotransmitters (DeFeudis, F.V., and Mandel, P., eds.). New York: Raven Press, pp. 321–326.

Brown, D.A., and Scholfield, C.N. 1979. Depolarization of neurones in the isolated olfactory cortex of the guinea pig by γ-aminobutyric acid. Br. J. Pharmacol. 65: 339–345.

Brown, T.H., Chang, V.C., Ganong, A.H., Keenan, C.L., and Kelso, S.R. 1988a. Biophysical properties of dendrites and spines that may control the induction and expression of long-term synaptic potentiation. In: Long-Term Potentiation: From Biophysics to Behavior (Landfield, P.W., and Deadwyler, S.A., eds.). New York: Alan R. Liss, pp. 201–264.

Brown, T.H., Chapman, P.F., Kairiss, E.W., and Keenan, C.L. 1988b. Long-term synaptic potentiation. Science 242: 724–728.

Brown, T.H., Fricke, R.A., and Perkel, D.H. 1981. Passive electrical constants in three classes of hippocampal neurons. J. Neurophysiol. 46: 812–827.

Brown, T.H., Kairiss, E.W., and Keenan, C.L. 1990. Hebbian synapses—biophysical mechanisms and algorithms. Annu. Rev. Neurosci. 13.

Brown, T.H., Ganong, A.H., Kairiss, E.W., Keenan, C.L., and Kelso, S.R. 1989. Long-term potentiation in two synaptic subsystems of the hippocampal brain slice. In: Neural Models of Plasticity: Experimental and Theoretical Approaches (Byrne, J.H., and Berry, W.O., eds.). San Diego: Academic Press, pp. 266–306.

Brown, T.H., and Johnston, D. 1983. Voltage-clamp analysis of mossy fiber synaptic input to hippocampal neurons. J. Neurophysiol. 50: 487–507.

Brown, T.H., and Kairiss, E.W. 1990. Neural Networks. Cambridge, MA: MIT Press, in preparation.

Brown, T.H., Wong, R.K.S., and Prince, D.A. 1979. Spontaneous miniature synaptic potentials in hippocampal neurons. Brain Res. 177: 194–199.

Bruce, C., Desimone, R., and Gross, C. 1981. Visual properties of neurons in a polysensory area in superior temporal sulcus of the macaque. J. Neurophysiol. 46: 369–384.

Buchsbaum, G., and Gottschalk, A. 1983. Trichromacy, opponent colours coding and optimum colour information transmission in the retina. Proc. R. Soc. Lond. [B] 220: 89–113.

Bullier, J., and Henry, G.H. 1979. Neural path taken by afferent streams in striate cortex of the cat. J. Neurophysiol. 42: 1271–1281.

Burd, G. 1980. Myelinated dendrites and neuronal perikarya in the olfactory bulb of the mouse. Brain Res. 181: 450–454.

Burgoyne, R.D., Gray, E.G., and Barron, J. 1983. Cytochemical localization of calcium in the dendritic spine apparatus of the cerebral cortex and at synaptic sites in the cerebellar cortex. J. Anat. 136: 634–635.

Burke, R., Lundberg, A., and Weight, F. 1971. Spinal border cell origin of the ventral spinocerebellar tract. Exp. Brain Res. 12: 283–294.

Burke, R.E. 1967. The composite nature of the monosynaptic excitatory postsynaptic potential. J. Neurophysiol. 30: 1114–1137.

Burke, R.E. 1968a. Group Ia synaptic input to fast and slow twitch motor units of cat's triceps surae. J. Physiol. (Lond.) 196: 615–631.

Burke, R.E. 1968b. Firing patterns of gastrocnemius motor units in the decerebrate cat. J. Physiol. (Lond.) 196: 631–654.

Burke, R.E. 1981. Motor units: Anatomy, physiology and functional organization. In: Handbook of Physiology, Section 1: The Nervous System, Vol. II: Motor Control, Part 1 (Brooks, V.B., ed.). Bethesda: American Physiological Society, pp. 345–422.

Burke, R.E. 1986. The control of muscle force: Motor unit recruitment and firing patterns. In: Human Muscle Power (Jones, N.L., McCartney, N., and McComas, A.J., eds.). Champaign, IL: Human Kinetics Press, pp. 97–106.

Burke, R.E. 1987. Synaptic efficacy and the control of neuronal input–output relations. Trends Neurosci. 10: 42–45.

Burke, R.E., Dum, R.P., Fleshman, J.W., Glenn, L.L., Lev-Tov, A., O'Donovan, M.J., and Pinter, M.J. 1982. An HRP study of the relation between cell size and motor unit type in cat ankle extensor motoneurons. J. Comp. Neurol. 209: 17–28.

Burke, R.E., Fedina, L., and Lundberg, A. 1971. Spatial synaptic distribution of recurrent and group Ia inhibitory systems in cat spinal motoneurones. J. Physiol. (Lond.) 214: 305–326.

Burke, R.E., Fleshman, J.W., and Segev, I. 1989. The control of synaptic efficacy: Lessons from the Ia synapse. J. Physiol. (Paris) 83: 133–140.

Burke, R.E., Jankowska, E., and ten Bruggencate, G. 1970. A comparison of peripheral and rubrospinal synaptic input to slow and fast twitch motor units of triceps surae. J. Physiol. (Lond.) 207: 709–732.

Burke, R.E., and Nelson, P.G. 1971. Accommodation to current ramps in motoneurons of fast and slow twitch motor units. Int. J. Neurosci. 1: 347–356.

Burke, R.E., and Rudomin, P. 1977. Spinal neurons and synapses. In: Handbook of Physiology, Section 1: The Nervous System, Vol. I: The Cellular Biology of Neurons, Part 2 (Kandel, E.R., ed.). Bethesda: American Physiological Society, pp. 877–944.

Burke, R.E., Rymer, W.Z., and Walsh, J.V. 1973. Functional specialization in the motor unit population of cat medial gastrocnemius muscle. In: Control of Posture and Locomotion (Stein, R.B., Pearson, K.B., Smith, R.S., and Redford, J.B., eds.). New York: Plenum Press, pp. 29–44.

Burke, R.E., Rymer, W.Z., and Walsh, J.V. 1976. Relative strength of synaptic input from short latency pathways to motor units of defined type in cat medial gastrocnemius. J. Neurophysiol. 39: 447–458.

Burke, R.E., Strick, P.L., Kanda, K., Kim, C.C., and Walmsley, B. 1977. Anatomy of medial gastrocnemius and soleus motor nuclei in cat spinal cord. J. Neurophysiol. 40: 667–680.

Burke, R.E., and Tsairis, P. 1977. Histochemical and physiological profile of a skeletofusimotor (beta) unit in cat soleus muscle. Brain Res. 129: 341–345.

Burke, R.E., Walmsley, B., and Hodgson, J.A. 1979. HRP anatomy of group Ia afferent contacts on alpha motoneurons. Brain Res. 160: 347–352.

Bührle, C.P., and Sonnhof, U. 1985. The ionic basis of postsynaptic inhibition of motoneurons of the frog spinal cord. Neuroscience 14: 581–592.

Cajal, S. Ramón y. 1888. Estructura de los centros nerviosos de los aves. Rev. Trimestr. Histol. Normal Patol. 1: 305–315.

Cajal, S. Ramón y. 1904. S. La Textura del Sistema Nervioso del Hombre y los Vertebrados. Madrid: Moya.

Cajal, S. Ramón y. 1893. La retiné des vertébrés. Cellule 9: 217–257.

Cajal, S. Ramón y. 1911. Histologie du Système Nerveux. Paris: A. Maloine.

Cajal, S. Ramón y. 1955. Studies on the Cerebral Cortex (Limbic Structures), trans. L.M. Kraft. London: Lloyd-Luke.

Calabresi, P., Mercuri, N., Stanzione, P., Stefani, A., and Bernardi, G. 1987. Intracellular studies on the dopamine-induced firing inhibition of neostriatal neurons in vitro: Evidence for D1 receptor involvement. Neuroscience 20: 757–771.

Caldwell, J.H., Daw, N.W., and Wyatt, H.J. 1978. Effects of picrotoxin and strychnine on rabbit retinal ganglion cells: Lateral interactions for cells with more complex receptive fields. J. Physiol. (Lond.) 276: 277–298.

Calleja, C. 1893. La Region Olfactoria del Cerebro. Madrid: Moya.

Calvin, W.H., and Schwindt, P.C. 1972. Steps in production of motoneuron spikes during rhythmic firing. J. Neurophysiol. 35: 297–310.

Carbonne, E., and Lux, H.D. 1984. A low voltage-activated calcium conductance in embryonic chick sensory neurons. Biophys. J. 46: 413–418.

Cattarelli, M., Astic, L., and Kauer, J.S. 1988. Metabolic mapping of 2-deoxyglucose uptake in the rat piriform cortex using computerized image processing. Brain Res. 442: 180–184.

Caviness, V.S., Korde, M.G., and Williams, R.S. 1977. Cellular events induced in the molecular layer of the piriform cortex by ablation of the olfactory bulb in the mouse. Brain Res. 134: 13–34.

Celio, M.R. 1986. Parvalbumin in most γ-aminobutyric acid-containing neurons of the rat cerebral cortex. Science 231: 995–997.

Chan-Palay, V. 1977. Cerebellar Dentate Nucleus. New York: Springer-Verlag.

Chan-Palay, V., Palay, S.L., and Wu, J.Y. 1979. Gamma-aminobutyric acid pathways in the cerebellum studied by retrograde and anterograde transport of glutamic acid decarboxylase antibody after in vivo injections. Anat. Embryol. 157: 1–14.

Chang, H-T. 1952. Cortical neurons with particular reference to the apical dendrites. Cold Spring Harbor Symp. Quant. Biol. 17: 189–202.

Chang, H.T., and Kitai, S.T. 1982. Large neostriatal neurons in the rat: An electron microscopic study of gold-toned Golgi-stained cells. Brain Res. Bull. 8: 631–643.

Chang, H.T., Wilson, C.J., and Kitai, S.T. 1981. Single neostriatal efferent axons in the globus pallidus: A light and electron microscopic study. Science 213: 915–918.

Chang, H.T., Wilson, C.J., and Kitai, S.T. 1982. A Golgi study of rat neostriatal neurons: Light microscopic analysis. J. Comp. Neurol. 208: 107–126.

Cherubini, E., Herrling, P.L., Lanfumey, L., and Stanzione, P. 1988. Excitatory amino acids in synaptic excitation of rat striatal neurones in vitro. J. Physiol. (Lond.) 400: 677–690.

Cherubini, E., and Lanfumey, L. 1987. An inward calcium current underlying regenerative calcium potentials in rat striatal neurons in vitro enhanced by K 8644. Neuroscience 21: 997–1005.

Chesselet, M.F., and Graybiel, A.M. 1985. Striatal neurons expressing somatostatin-like immunoreactivity: Evidence for a peptidergic interneuronal system in the cat. Neuroscience 17: 547–571.

Chover, J. 1989. Modeling afferent connectivity, postsynaptic plasticity and signal discrimination. Synapse 3: 101–116.

Chujo, T., Yamada, Y., and Yamamoto, C. 1975. Sensitivity of Purkinje cell dendrites to glutamic acid. Exp. Brain Res. 23: 293–300.

Chun, M.H., and Wässle, H. 1989. GABA-like immunoreactivity in the cat retina: Electron microscopy. J. Comp. Neurol. 279: 55–67.

Churchland, P.S., and Sejnowski, T.J. 1988. Perspectives on cognitive neuroscience. Science 242: 741–745.

Claiborne, B.J., Amaral, D.G., and Cowan, W.M. 1986. A light and electron microscopic analysis of the mossy fibers of the rat dentate gyrus. J. Comp. Neurol. 246: 435–458.

Clark, W.E. le Gros. 1957. Inquiries into the anatomical basis of olfactory discrimination. Proc. R. Soc. Lond. [B] 146: 299–319.

Cleland, B.G., Dubin, M.W., and Levick, W.R. 1971. Sustained and transient neurons in the cat's retina and lateral geniculate nucleus. J. Physiol. (Lond.) 217: 473–496.

Cleland, B.G., Harding, T.H., and Tulunay-Keesey, U. 1979. Visual resolution and receptive field size: Examination of two kinds of cat retinal ganglion cell. Science 205: 1015–1017.

Cohen, E., and Sterling, P. 1986. Accumulation of [3H]glycine by cone bipolar neurons in the cat retina. J. Comp. Neurol. 250: 1–7.

Cohen, E., and Sterling, P. 1990a. The mosaic of five types of cone bipolars innervating sublamina *b* of the inner plexiform layer. In preparation.

Cohen, E., and Sterling, P. 1990b. Microcircuitry for the receptive field center of the on-beta ganglion cell. In preparation.

Cohen, N.H. 1984. Observed learning capacity in amnesia: Evidence for multiple memory systems. In: Neuropsychology of Memory (Squire, L.R., and Butters, N., eds.). New York: Guilford Press, pp. 83–103.

Cole, A.E., and Nicoll, R.A. 1984. Characterization of a slow cholinergic postsynaptic potential recorded in vitro from rat hippocampal pyramidal cells. J. Physiol. (Lond.) 352: 173–188.

Cole, K.S. 1968. Membrane, Ions and Impulses: A Chapter of Classical Biophysics. Berkeley: University of California Press.

Collingridge, G.L., and Bliss, T.V.P. 1987. NMDA receptors—their role in long-term potentiation. Trends Neurosci. 10: 288–293.

Collingridge, G.L., Kehl, S.J., and McLennan, H. 1983. Excitatory amino acids in synaptic transmission in the Schaffer collateral–commissural pathway of the rat hippocampus. J. Physiol. (Lond.) 334: 33–46.

Collins, G.G.S. 1979. Evidence of neurotransmitter role for aspartate and γ-aminobutyric acid in the rat olfactory cortex. J. Physiol. (Lond.) 291: 51–60.

Collins, G.G.S. 1982. Some effects of excitatory amino acid receptor antagonists on synaptic transmission in the rat olfactory cortex slice. Brain Res. 244: 311–318.

Collins, G.G.S. 1985. Excitatory amino acids as transmitters in the olfactory system. In: Excitatory Amino Acids (Roberts, P.J., ed.). New York: Macmillan, pp. 131–142.

Collins, G.G.S., Anson, J., and Probett, G.A. 1981. Patterns of endogenous amino acid release from slices of rat and guinea pig olfactory cortex. Brain Res. 204: 103–120.

Collins, G.G.S., and Howlett, S.J. 1988. The pharmacology of excitatory transmission in the rat olfactory cortex slice. Neuropharmacology 27: 697–705.

Collins, G.G.S., and Probett, G.A. 1981. Aspartate and not glutamate is the likely transmitter of the rat lateral olfactory tract fibers. Brain Res. 209: 231–234.

Colonnier, M. 1968. Synaptic patterns on different cell types in the different laminae of the cat visual cortex. An electron microscope study. Brain Res. 9: 268–287.

Connor, J.A., and Stevens, C.F. 1971. Voltage-clamp studies of a transient outward current in a gastropod neural somata. J. Physiol. (Lond.) 213: 21–30.

Connors, B.W., Gutnick, M.J., and Prince, D.A. 1982. Electrophysiological properties of neocortical neurons in vitro. J. Neurophysiol. 48: 1302–1320.

Connors, B.W., and Kriegstein, A.R. 1986. Cellular physiology of the turtle visual cortex: Distinctive properties of pyramidal and stellate neurons. J. Neurosci. 6: 164–177.

Connors, B.W., Malenka, R.C., and Silva, L.R. 1988. Two inhibitory postsynaptic potentials, and $GABA_A$ and $GABA_B$ receptor-mediated responses in neocortex of rat and cat. J. Physiol. (Lond.) 406: 443–468.

Conradi, S., Cullheim, S., Gollnik, L., and Kellerth, J.-O. 1983. Electron microscopic observations on the synaptic contacts of group Ia and muscle spindle afferents in the cat lumbosacral spinal cord. Brain Res. 265: 31–40.

Conradi, S., Kellerth, J.-O, and Berthold, C.-J. 1979a. Electron microscopic studies of cat spinal α-motoneurons: II. A method for the description of neuronal architecture and synaptology from serial sections through the cell body and proximal dendritic segments. J. Comp. Neurol. 184: 741–754.

Conradi, S., Kellerth, J.-O., Berthold, C.-H., and Hammarberg, C. 1979b. Electron microscopic studies of cat spinal alpha-motoneurons: IV. Motoneurons innervating slow twitch (type S) units of the soleus muscle. J. Comp. Neurol. 184: 769–782.

Constanti, A., and Galvan, M. 1983a. Fast inward-rectifying current accounts for anomalous rectification in olfactory cortex neurones. J. Physiol. (Lond.) 335: 153–178.

Constanti, A., and Galvan, M. 1983b. M-current in voltage-clamped olfactory cortex neurones. Neurosci. Lett. 39: 65–70.

Constanti, A., Galvan, M., Franz, P., and Sim, J.A. 1985. Calcium-dependent inward currents in voltage-clamped guinea pig olfactory cortex neurones. Pflügers Arch. 404: 259–265.

Constanti, A., and Sim, J.A. 1987a. Calcium-dependent potassium conductance in guinea pig olfactory cortex neurones in vitro. J. Physiol. (Lond.) 387: 173–194.

Constanti, A., and Sim, J.A. 1987b. Muscarinic receptors mediating suppression of the M-current in guinea pig olfactory cortex neurones may be of the M_2-subtype. Br. J. Pharmacol. 90: 3–5.

Coombs, J.S., Eccles, J.C., and Fatt, P. 1955. Excitatory synaptic actions in motoneurons. J. Physiol. (Lond.) 130: 374–395.

Cooper, J.R., Bloom, F.E., and Roth, R.H. 1987. The Biochemical Basis of Neuropharmacology, 5th ed. New York: Oxford University Press.

Copenhagen, D.R., Ashmore, J., and Schnapf, J. 1983. Kinetics of synaptic transmission from photoreceptors to horizontal and bipolar cells in turtle retina. Vision Res. 23: 363–369.

Copenhagen, D.R., and Jahr, C.E. 1989. Release of endogenous excitatory amino acids from turtle photoreceptors. Nature 341: 536–539.

Corkin, S. 1984. Lasting consequences of bilateral medial temporal lobectomy: Clinical course and experimental findings in H.M. Semin. Neurol. 4: 249–259.

Cotman, C.W., and Iversen, L.L. 1987. Excitatory amino acids in the brain—Focus on NMDA receptors. Trends Neurosci. 7: 263–265.

Cotman, C.W., Monaghan, D.T., and Ganong, A.H. 1988. Excitatory amino acid neurotransmission: NMDA receptors and Hebb-type synaptic plasticity. Annu. Rev. Neurosci. 11: 61–80.

Cotman, C.W., Monaghan, D.T., Ottersen, O.P., and Storm-Mathisen, J. 1987. Anatomical organization of excitatory amino acid receptors and their pathways. Trends Neurosci. 10: 273–279.

Crepel, F., and Penit-Soria, J. 1986. Inward rectification and low threshold calcium conductance in rat cerebellar Purkinje cells. An in vitro study. J. Physiol. (Lond.) 372: 1–23.

Crepel, F., Dhanjal, S.S., and Sears, T. 1982. Effect of glutamate, aspartate and related derivatives on cerebellar Purkinje cell dendrites in the rat: An *in vitro* study. J. Physiol. (Lond.) 329: 297–317.

Creutzfeldt, O.D. 1977. Generality of the functional structure of the neocortex. Naturwissenschaften 64: 507–517.

Creutzfeldt, O.D., Lux, H.D., and Watanabe, S. 1966. Electrophysiology of cortical nerve cells. In: The Thalamus (Purpura, D.P., and Yahr, M.D., eds.). New York: Columbia University Press, pp. 209–235.

Crick, F. 1982. Do dendritic spines twitch? Trends Neurosci. 5: 44–46.

Crill, W.E., and Schwindt, P.C. 1983. Active currents in mammalian central neurons. Trends Neurosci. 6: 236–240.

Crone, C., Hultborn, H., Kiehn, O., Mazieres, L., and Wigström, H. 1988. Maintained changes in motoneuronal excitability by short-lasting synaptic inputs in the decerebrate cat. J. Physiol. (Lond.) 405: 321–343.

Crunelli, V., Forda, S., and Kelly, J.S. 1984. The reversal potential of excitatory amino acid action on granule cells of the rat dentate gyrus. J. Physiol. (Lond.) 351: 327–342.

Cuello, A.C., and Kanazawa, I. 1978. The distribution of substance P immunoreactive fibers in the rat central nervous system. J. Comp. Neurol. 178: 129–156.

Cuénod, M., Bagnoli, P., Beaudet, A., Rustioni, A., Wiklund, L., and Streit, P. 1982. Transmitter-specific retrograde labeling of neurons. In: Cytochemical Methods in Neuroanatomy (Chan-Palay, V., and Palay, S.L., eds.). New York: Alan R. Liss, pp. 17–44.

Cull-Candy, S.G., and Usowicz, M.M. 1987. Glutamate and aspartate activated channels and inhibitory synaptic currents in large cerebellar neurons grown in culture. Brain Res. 402: 182–187.

Cull-Candy, S.G., and Usowicz, M.M. 1987. Multiple-conductance channels activated by excitatory amino acids in cerebellar neurons. Nature 325: 525–528.

Cullheim, S., Fleshman, J.W., Glenn, L.L., and Burke, R.E. 1987. Membrane area and dendritic structure in type-identified triceps surae alpha-motoneurons. J. Comp. Neurol. 255: 68–81.

Cullheim, S., and Kellerth, J.-O. 1978a. A morphological study of the axons and recurrent axon collaterals of cat sciatic α-motoneurons after intracellular staining with horseradish peroxidase. J. Comp. Neurol. 178: 537–558.

Cullheim, S., and Kellerth, J.-O. 1978b. A morphological study of the axons and recurrent axon collaterals of cat α-motoneurons supplying different functional types of muscle unit. J. Physiol. (Lond.) 281: 301–313.

Cullheim, S., and Kellerth, J.-O. 1981. Two kinds of recurrent inhibition of cat spinal α-motoneurones as differentiated pharmacologically. J. Physiol. (Lond.) 312: 209–224.

Cullheim, S., Kellerth, J., and Conradi, S. 1977. Evidence for direct synaptic interconnections between cat spinal α-motoneurons via the recurrent axon collaterals: A morphological study using intracellular injection of horseradish peroxidase. Brain Res. 132: 1–10.

Curtis, D.R., and Eccles, J.C. 1960. Synaptic action during and after repetitive stimulation. J. Physiol. (Lond.) 150: 374–398.

Curtis, D.R., and Felix, D. 1971. The effect of bicuculline upon synaptic inhibition in the cerebral and cerebellar cortices of the cat. Brain Res. 34: 301–321.

Curtis, D.R., Gynther, B.D., and Malik, R. 1986. A pharmacological study of group I muscle afferent terminals and synaptic excitation in the intermediate nucleus and Clarke's column of the cat spinal cord. Exp. Brain Res. 64: 105–113.

Curtis, D.R., and Johnston, G.A.R. 1974. Amino acid transmitters in the mammalian central nervous system. Ergeb. Physiol. 69: 98–188.

Curtis, D.R., Lodge, D., Bornstein, J.C., and Peet, M.G. 1981. Selective effects of (−)baclofen on spinal synaptic transmission in the cat. Exp. Brain Res. 42: 158–170.

Czarkowska, J., Jankowska, E., and Sybirska, E. 1981. Common interneurones in reflex pathways from group Ia and Ib afferents of knee flexors and extensors in the cat. J. Physiol. (Lond.) 319: 367–380.

Dacey, D. 1988. Dopamine-accumulating retinal neurons revealed by in vitro fluorescence display a unique morphology. Science 240: 1196–1198.

Dahl, D., Bailey, W.H., and Winson, J. 1983. Effect of norepinephrine depletion of hippocampus on neuronal transmission from perforant pathway through dentate gyrus. J. Neurophysiol. 49: 123–133.

Dale, H.H. 1935. Pharmacology and nerve endings. Proc. R. Soc. Med. 28: 319–332.

Daniel, P.M., and Whitteridge, D. 1961. The representation of the visual field on the cerebral cortex in monkeys. J. Physiol. (Lond.) 159: 203–221.

Dann, J.F., Buhl, E.H., and Peichl, L. 1988. Postnatal dendritic maturation of alpha and beta ganglion cells in cat retina. J. Neurosci. 8: 1485–1499.

Darian-Smith, I. 1984. The sense of touch: Performance and peripheral neural processes. In: Handbook of Physiology, section 1: The Nervous System, Vol. III: Sensory Processes, Part 2 (Darian-Smith, I., ed.). Bethesda: American Physiological Society, pp. 739–788.

Datta, A.K., and Stephens, J.A. 1981. The effects of digital nerve stimulation on the firing of motor units in human first dorsal interosseous muscle. J. Physiol. (Lond.) 318: 501–510.

Davies, S.N., Lester, R.A.J., Reymann, K.G., and Collingridge, G.L. 1989. Temporally distinct pre- and post-synaptic mechanisms maintain long-term potentiation. Nature 338: 500–503.

Davis, B.J., Burd, G.D., and Macrides, F. 1982. Localization of methionine-enkephalin, substance P and somatostatin immunoreactivities in the main olfactory bulb of the hamster. J. Comp. Neurol. 204: 377–383.

Daw, N., Brunken, W., and Parkinson, D. 1989. The function of synaptic transmitters in the retina. Annu. Rev. Neurosci. 12: 205–225.

De Monasterio, F.M., Schein, S.J., and McCrane, E.P. 1981. Staining of blue-sensitive cones of the macaque retina by a fluorescent dye. Science 213: 1278–1281.

De Montigny, C., and Lamarre, Y. 1974. Rhythmic activity induced by harmaline in the olivo-cerebellar-bulbar systems of the cat. Brain Res. 53: 81–95.

De Quidt, M.E., and Emson, P.C. 1986. Distribution of neuropeptide Y-like immunoreactivity in the rat central nervous system. II. Immunocytochemical analysis. Neuroscience 18: 545–618.

Deadwyler, S.A., Hampson, R.E., Foster, T.C., and Marlow, G. 1988. The functional significance of long-term potentiation: Relation to sensory processing by hippocampal circuits. In: Long-Term Potentiation: From Biophysics to Behavior (Landfield, P.W., and Deadwyler, S.A., eds.). New York: Alan R. Liss, pp. 499–534.

DeLong, M.R. 1973. Putamen: Activity of single units during slow and rapid arm movements. Science 179: 1240–1242.

Demeter, S., Rosene, D.L., and Van Hoesen, G.W. 1985. Interhemispheric pathways of the hippocampal formation, presubiculum, and entorhinal and posterior parahippocampal cortices in the rhesus monkey: The structure and organization of the hippocampal commissures. J. Comp. Neurol. 233: 30–47.

Demeulemeester, H., Vandesande, F., Orban, G.A., Brandon, C., and Vanderhaegen, J.J. 1988. Heterogeneity of GABAergic cells in cat visual cortex. J. Neurosci. 8: 988–1000.

Deniau, J.M., and Chevalier, G. 1985. Disinhibition as a basic process in the expression of striatal function. II. The striatal-nigral influence on thalamo-cortical cells of the ventromedial thalamic nucleus. Brain Res. 334: 227–233.

Denny-Brown, D. 1929. On the nature of postural reflexes. Proc. R. Soc. Lond. [B] 104: 252–301.

Denny-Brown, D. 1949. Interpretation of the electromyogram. Arch. Neurol. Psychiatr. 61: 99–128.

Derrington, A.M., and Lennie, P. 1982. The influence of temporal frequency and adaptation level on receptive field organization of retinal ganglion cells in the cat. J. Physiol. (Lond.) 333: 343–366.

Desmedt, J.E., and Godaux, E. 1977. Ballistic contractions in man: Characteristic recruitment pattern of single motor units of the tibialis anterior muscle. J. Physiol. (Lond.) 264: 673–694.

Devor, M. 1976. Fiber trajectories of olfactory bulb efferents in the hamster. J. Comp. Neurol. 166: 31–48.

DiFiglia, M., and Aronin, N. 1982. Ultrastructural features of immunoreactive somatostatin neurons in the rat caudate nucleus. J. Neurosci. 2: 1267–1274.

DiFiglia, M., and Carey, J. 1986. Large neurons in the primate neostriatum examined with the combined Golgi-electron microscope. J. Comp. Neurol. 244: 36–52.

DiFiglia, M., Pasik, P., and Pasik, T. 1976. A Golgi study of neuronal types in the neostriatum of monkeys. Brain Res. 114: 245–256.

DiFiglia, M., Pasik, P., and Pasik, T. 1980. Ultrastructure of Golgi-impregnated and gold-toned spiny and aspiny neurons in the monkey neostriatum. J. Neurocytol. 6: 311–337.

DiFiglia, M., and Rafols, J.A. 1988. Synaptic organization of the globus pallidus. J. Electron Microsc. 10: 247–263.

DiFrancesco, D. 1985. The cardiac hyperpolarization-activated current, I_f: Origins and development. Prog. Biophys. Mol. Biol. 46: 163–183.

Dingledine, R., and Langmoen, I.A. 1980. Conductance changes and inhibitory actions of hippocampal recurrent IPSPs. Brain Res. 185: 277–287.

Dodd, J., and Horn, J.P. 1983a. A reclassification of B and C neurones in the 9th and 10th paravertebral sympathetic ganglia of the bullfrog. J. Physiol. (Lond.) 334: 225–269.

Dodd, J., and Horn, J.P. 1983b. Muscarinic inhibition of sympathetic C neurones in the bullfrog. J. Physiol. (Lond.) 334: 271–291.

Dodge, F.A. 1979. The nonuniform excitability of central neurons as exemplified by a model of the spinal motoneuron. In: The Neurosciences: Fourth Study Program (Schmitt, F.O., ed.). Boston: MIT Press, pp. 439–455.

Dodt, H.U., and Misgeld, U. 1986. Muscarinic slow excitation and muscarinic inhibition of synaptic transmission in the neostriatum. J. Physiol (Lond.) 380: 593–608.

Doerner, D., Pitler, T.A., and Alger, B.E. 1988. Protein kinase C activators block specific calcium and potassium current components in isolated hippocampal neurons. J. Neurosci. 11: 4069–4078.

Dolphin, A.C., and Scott, R.H. 1986. Inhibition of calcium currents in cultured rat dorsal root ganglion neurones by (−)-baclofen. Br. J. Pharmacol. 88: 213–220.

Donoghue, J.P., and Herkenham, M. 1986. Neostriatal projections from individual cortical fields conform to histochemically distinct striatal compartments in the rat. Brain Res. 365; 397–403.

Douglas, R.J., Martin, K.A.C., and Whitteridge, D. 1988. Selective responses of visual cortical cells do not depend on shunting inhibition. Nature 332: 642–644.

Douglas, R.J., Martin, K.A.C., and Whitteridge, D. 1989. A canonical microcircuit for neocortex. Neural Comput. 1.

Dowling, J.E. 1986. Dopamine: A retinal neuromodulator? Trends Neurosci. 9: 236–240.

Dowling, J.E. 1987. The Retina. An Approachable Part of the Brain. Cambridge, MA: Harvard University Press.

Dowling, J.E., and Boycott, B.B. 1966. Organization of the primate retina: Electron microscopy. Proc. R. Soc. Lond. [B] 166: 80–111.

Dowling, J.E., and Cowan, W.M. 1966. An electron microscope study of normal and degenerating centrifugal fiber terminals in the pigeon retina. Z. Zellforsch. Mikrosk. Anat. 71: 14–28.

Dreifuss, J.J., Kelly, J.S., and Krnjevic, K. 1969. Cortical inhibition and γ-aminobutyric acid. Exp. Brain Res. 9: 137–154.

Dubé, L., Smith, A.D., and Bolam, J.P. 1988. Identification of synaptic terminals of thalamic or cortical origin in contact with distinct medium-size spiny neurons in the rat neostriatum. J. Comp. Neurol. 267: 455–471.

Dubin, H. 1976. The inner plexiform layer of the vertebrate retina: A quantitative and comparative electron microscopic analysis. J. Comp. Neurol. 140: 479–506.

Dudek, F.E., Gribkoff, V.K., and Christian, E.P. 1988. Mechanisms of potentiation independent of chemical synapses. In: Long-Term Potentiation: From Biophysics to Behavior (Landfield, P.W., and Deadwyler, S.A., eds.). New York: Alan R. Liss, pp. 439–464.

Dunlap, K., and Fischbach, G.D. 1981. Neurotransmitters decrease the calcium conductance activated by depolarization of embryonic chick sensory neurons. J. Physiol. (Lond.) 317: 519–535.

Dunwiddie, T.V., and Lynch, G. 1979. The relationship between extracellular calcium concentrations and the induction of long-term potentiation. Brain Res. 169: 103–110.

Dunwiddie, T.V., Roberson, N.L., and Worth, T. 1982. Modulation of long-term potentiation: Effects of adrenergic and neuroleptic drugs. Pharmacol. Biochem. Behav. 17: 1257–1264.

Dupont, J.L., Crepel, F., and Delhaye-Bouchaud, N. 1979. Influence of bicuculline and picrotoxin on reversal properties of excitatory synaptic potentials in cerebellar Purkinje cells of the rat. Brain Res. 173: 577–580.

Durand, D. 1984. The somatic shunt cable model for neurons. Biophys. J. 46: 645–653.

Dutar, P., and Nicoll, R.A. 1988. A physiological role for GABA$_B$ receptors in the central nervous system. Nature 332: 156–158.

Dykes, R.W. 1983. Parallel processing of somatosensory information: A theory. Brain Res. Rev. 6: 47–115.

Easter, S.S., and Stuermer, C. 1984. An evaluation of the hypothesis of shifting terminals in goldfish optic tectum. J. Neurosci. 4: 1052–1063.

Ebner, F.F. 1969. A comparison of primitive forebrain organization in metatherian and eutherian mammals. Ann. N.Y. Acad. Sci. 167: 240–262.

Eccles, J.C. 1964. The Physiology of Synapses. New York: Academic Press.

Eccles, J.C., Eccles, R.M., and Lundberg, A. 1957. The convergence of monosynaptic excitatory afferents onto many different species of alpha-motoneurones. J. Physiol. (Lond.) 137: 22–50.

Eccles, J.C., Eccles, R.M., Iggo, A., and Lundberg, A. 1960a. Electrophysiological studies on gamma motoneurones. Acta Physiol. Scand. 50: 32–40.

Eccles, J.C., Eccles, R.M., and Lundberg, A. 1960b. Types of neurone in and around the intermediate nucleus of the lumbosacral cord. J. Physiol. (Lond.) 154: 89–114.

Eccles, J.C., Fatt, P., and Koketsu, K. 1954. Cholinergic and inhibitory synapses in a pathway from motor axon collaterals in motoneurones. J. Physiol. (Lond.) 126: 524–562.

Eccles, J.C., Fatt, P., and Landgren, S. 1956. The central pathway for the direct inhibitory action of impulses in the largest afferent fibers to muscle. J. Neurophysiol. 19: 75–98.

Eccles, J.C., Llinás, R., and Sasaki, K. 1966a. The excitatory synaptic action of climbing fibers on the Purkinje cells of the cerebellum. J. Physiol. (Lond.) 182: 268–296.

Eccles, J.C., Llinás, R., and Sasaki, K. 1966b. The inhibitory interneurons within the cerebellar cortex. Exp. Brain Res. 1: 1–16.

Eccles, J.C., Llinás, R., and Sasaki, K. 1966c. Parallel fiber stimulation and the responses induced thereby in the Purkinje cells of the cerebellum. Exp. Brain Res. 1: 17–39.

Eccles, J.C., Llinás, R., and Sasaki, K. 1966d. The mossy fibre–granule cell relay of the cerebellum and its inhibitory control of Golgi cells. Exp. Brain Res. 1: 82–101.

Eccles, R.M., and Lundberg, A. 1958. Integrative pattern of Ia synaptic actions on motoneurones of hip and knee muscles. J. Physiol. (Lond.) 144: 271–298.

Eccles, R.M., and Lundberg, A. 1959. Synaptic action in motoneurones by afferents which may evoke the flexion reflex. Arch. Ital. Biol. 97: 199–221.

Eckenstein, F., and Baughman, R.W. 1984. Two types of cholinergic innervation in the cortex, one co-localised with vasoactive intestinal polypeptide. Nature 309: 153–155.

Edwards, F.R., Redman, S.J., and Walmsley, B. 1976. Statistical fluctuations in charge transfer at Ia synapses on spinal motoneurones. J. Physiol. (Lond.) 259: 665–688.

Emonet-Denand, F., Jami, L., and Laporte, Y. 1975. Skeletofusimotor axons in hindlimb muscles of the cat. J. Physiol. (Lond.) 249: 153–166.

Eng, D.L., and Kocsis, J.D. 1987. Activity dependent changes in extracellular potassium and excitability in turtle olfactory nerve. J. Neurophysiol. 57: 740–754.

Engberg, I., and Marshall, K.C. 1979. Reversal potential for Ia excitatory postsynaptic potentials in spinal motoneurons of cats. Neuroscience 4: 1583–1591.

Enroth-Cugell, C., Hertz, B.G., and Lennie, P. 1977. Convergence of rod and cone signals in the cat's retina. J. Physiol. (Lond.) 269: 297–318.

Enroth-Cugell, C., and Robson, J.G. 1966. The contrast sensitivity of retinal ganglion cells of the cat. J. Physiol (Lond.) 187: 517–552.

Eranko, O., and Harkonen, M. 1965. Monoamine-containing small cells in the superior cervical ganglion of the rat and an organ composed of them. Acta Physiol. Scand. 63: 511–512.

Ericson, H., Watanabe, T., and Köhler, C. 1987. Morphological analysis of the tuberomammillary nucleus in the rat brain: Delineation of subgroups with antibody against L-histidine decarboxylase as a marker. J. Comp. Neurol. 263: 1–24.

Fahrenbach, W.H. 1985. Anatomical circuitry of lateral inhibition in the eye of the horseshoe crab, *Limulus polyphemus*. Proc. R. Soc. Lond. [B] 225: 219–249.

Fallon, J.H., and Leslie, F.M. 1986. Distribution of dynorphin and enkephalin peptides in the rat brain. J. Comp. Neurol. 249: 293–336.

Famiglietti, E.V. 1983. 'Starburst' amacrine cells and cholinergic neurons: mirror symmetric ON and OFF amacrine cells of rabbit retina. Brain Res. 261: 138–144.

Famiglietti, E.V., and Kolb, H. 1975. A bistratified amacrine cell and synaptic circuitry in the inner plexiform layer of the retina. Brain Res. 84: 293–300.

Famiglietti, E.V., and Kolb, H. 1976. Structural basis for "ON" and "OFF" -center responses in retinal ganglion cells. Science 194: 193–195.

Famiglietti, E.V., and Peters, A. 1972. The synaptic glomerulus and the intrinsic neuron in the dorsal lateral geniculate nucleus of the cat. J. Comp. Neurol. 144: 285–334.

Farbman, A.I. 1986. Prenatal development of mammalian olfactory receptor cells. Chem. Senses 11: 3–18.

Fatt, P., and Katz, B. 1953. The effect of inhibitory nerve impulses on a crustacean muscle fibre. J. Physiol. (Lond.) 121: 374–389.

Feldman, A.G., and Orlovsky, G.N. 1975. Activity of interneurons mediating reciprocal Ia inhibition during locomotion. Brain Res. 84: 181–194.

Feldman, M.L. 1984. Morphology of the neocortical pyramidal neuron. In: Cerebral

Cortex, Vol. 1: Cellular Components of the Cerebral Cortex (Peters, A., and Jones, E.G., eds.). New York: Plenum Press, pp. 123–200.

Ferreyra-Moyana, H., Cinelli, A.R., Molina, J.C., and Barragán, E. 1988. Current generators and properties of late components evoked in rat olfactory cortex. Brain Res. Bull. 20: 433–446.

Ferster, D., and Koch, C. 1987. Neuronal connections underlying orientation selectivity in cat visual cortex. Trends Neurosci. 10: 487–492.

Fetcho, J.R. 1987. A review of the organization and evolution of motoneurons innervating the axial musculature of vertebrates. Brain Res. Rev. 12: 243–280.

ffrench-Mullen, J.M.H., Hori, N., Nakanishi, H., Slater, N.T., and Carpenter, D.O. 1983. Asymmetric distribution of acetylcholine receptors and M channels on prepyriform neurons. Cell. Mol. Neurobiol. 3: 163–181.

ffrench-Mullen, J.M.H., Koller, K., Zaczek, R., Coyle, J.T., Hori, N., and Carpenter, D.O. 1985. N-Acetylaspartylglutamate: Possible role as the neurotransmitter of the lateral olfactory tract. Proc. Natl. Acad. Sci. USA 82: 3897–3900.

Fifkova, E. 1985. Synaptic hypertrophy in the dentate fascia of the hippocampus. In: Recent Achievements in Restorative Neurology (Dimitrijevic, M., and Eccles, J.C., eds.). Basel: Karger, pp. 263–271.

Fifkova, E., and Anderson, C.L. 1981. Stimulation induced changes in the dimensions of stalks of dendritic spines in the dentate molecular layer. Exp. Neurol. 74: 621–627.

Fifkova, E.F., and Delay, R. 1982. Cytoplasmic actin in dendritic spines as possible mediator of synaptic plasticity. J. Cell Biol. 95: 345–350.

Fifkova, E., Markham, J.A., and Delay, R.J. 1983. Calcium in the spine apparatus of dendritic spines in the dentate molecular layer. Brain Res. 266: 163–168.

Finch, D.M., Tan, A.M., and Isokawa-Akesson, M. 1988. Feedforward inhibition of the rat entorhinal cortex and subicular complex. J. Neurosci. 8: 2213–2226.

Finkel, A.S., and Redman, S.J. 1983. The synaptic current evoked in cat spinal motoneurones by impulses in single group Ia axons. J. Physiol. (Lond.) 342: 615–632.

Fisher, S., and Boycott, B. 1974. Synaptic connections made by horizontal cells within the outer plexiform layer of the retina of the cat and the rabbit. Proc. R. Soc. Lond. [B] 186: 317–331.

Fitzpatrick, D., Conley, M., Luppino, G., and Diamond, I.T. 1988. Cholinergic projections from the midbrain reticular formation and the parabigeminal nucleus to the lateral geniculate nucleus in the tree shrew. J. Comp. Neurol. 272: 43–67.

Fitzpatrick, D., Lund, J.S., Schmechel, D.E., and Towles, A.C. 1987. Distribution of GABAergic neurons and axon terminals in the macaque striate cortex. J. Comp. Neurol. 264: 73–91.

Fitzpatrick, D., Penny, G.R., and Schmechel, D.E. 1984. Glutamic acid decarboxylase-immunoreactive neurons and terminals in the lateral geniculate nucleus of the cat. J. Neurosci. 4: 1809–1829.

Fitzpatrick, D., Raczkowski, D., and Diamond, I.T. 1989. Cholinergic and monoaminergic innervation of the cat's thalamus: Comparison of the lateral geniculate nucleus with other principle sensory nuclei. J. Comp. Neurol.

Fleshman, J.W., Munson, J.B., Sypert, G.W., and Friedman, G.A. 1981a. Rheobase, input resistance and motor-unit type in medial gastrocnemius motoneurons in the cat. J. Neurophysiol. 46: 1326–1338.

Fleshman, J.W., Munson, J.B., and Sypert, G.W. 1981b. Homonymous projection of individual group Ia-fibers to physiologically characterized medial gastrocnemius motoneurons in the cat. J. Neurophysiol. 46: 1339–1348.

Fleshman, J.W., Rudomin, P., and Burke, R.E. 1987. Supraspinal control of a short-latency cutaneous pathway to hindlimb motoneurons. Exp. Brain Res. 69: 449–459.

Fleshman, J.W., Segev, I., and Burke, R.E. 1988. Electrotonic architecture of type-identified alpha-motoneurons in the cat spinal cord. J. Neurophysiol. 60: 60–85.

Fonnum, F., Storm-Mathisen, J., and Walberg, F. 1970. Glutamate decarboxylase in inhibitory neurons. A study of the enzyme in Purkinje cell axons and boutons in the cat. Brain Res. 20: 259–275.

Fonnum, F., and Walberg, F. 1973. An estimation of the concentration of gamma-aminobutyric acid and glutamate decarboxylase in the inhibitory Purkinje axon terminals in the cat. Brain Res. 54: 115–127.

Foote, S.L., Bloom, F.E., and Aston-Jones, G. 1983. Nucleus locus ceruleus: New evidence of anatomical and physiological specificity. Physiol. Rev. 63: 844–914.

Forehand, C.J. 1985. Density of somatic innervation on mammalian autonomic ganglion cells is inversely related to dendritic complexity and preganglionic convergence. J. Neurosci. 5: 3403–3408.

Forsythe, I.D., and Westbrook, G.L. 1988. Slow excitatory postsynaptic currents mediated by N-methyl-D-aspartate receptor on cultured mouse central neurones. J. Physiol. (Lond.) 396: 515–533.

Fox, C.A., Andrade, A.N., Hillman, D.E., and Schwyn, R.C. 1971. The spiny neurons in the primate striatum: A Golgi and electron microscopic study. J. Hirnforsch. 13: 181–201.

Fox, C.A., Hillman, D.E., Seigesmund, K.A., and Dutta, C.R. 1967. The primate cerebellar cortex: A Golgi study and electron microscopy study. In: Progress in Brain Research, Vol. 25 (Fox, C.A., and Sneider, R., eds.). Amsterdam: Elsevier, pp. 174–225.

Fox, S., Krnjević, K., Morris, M.E., Puil, E., and Werman, R. 1978. Action of baclofen on mammalian synaptic transmission. Neuroscience 3: 495–515.

Fox, C.A., and Rafols, J.A. 1976. The striatal efferents in the globus pallidus and in the substantia nigra. In: The Basal Ganglia (Yahr, M.D., ed.). New York: Raven Press, pp. 37–55.

Frank, B., and Hollyfield, J. 1987. Retinal ganglion cell morphology in the frog, *Rana pipiens*. J. Comp. Neurol. 266: 413–434.

Frank, K., and Fuortes, M.G.F. 1957. Presynaptic and postsynaptic inhibition of monosynaptic reflexes. Fed. Proc. 16: 39–40.

Frederickson, R.C.A., Neuss, M., Morzorati, S.L., and McBride, W.J. 1978. A comparison of the inhibitory effects of taurine and GABA on identified Purkinje cells and other neurons in the cerebellar cortex of the rat. Brain Res. 145: 117–126.

Freed, M.A., Nakamura, Y., and Sterling, P. 1983. Four types of amacrine in the cat retina that accumulate GABA. J. Comp. Neurol. 219: 295–304.

Freed, M.A., Smith, R.G., and Sterling, P. 1987. The rod bipolar array in cat retina: Pattern of input from rods and GABA-accumulating amacrine cells. J. Comp. Neurol. 266: 445–455.

Freed, M.A., and Sterling, P. 1988. The ON-alpha ganglion cell of the cat retina and its presynaptic cell types. J. Neurosci. 8: 2303–2320.

Freeman, W.J. 1974. Topographic organization of primary olfactory nerve in cat and rabbit as shown by evoked potentials. Electroencephalogr. Clin. Neurophysiol. 36: 33–45.

Freeman, W.J. 1975. Mass Action in the Nervous System. New York: Academic.

Freeman, W.J. 1983. Dynamics of image formation by nerve cell assemblies. In: Synergetics of the Brain (Basar, E., Flohr, H., and Mandell, A.J., eds.). New York: Springer-Verlag.

Freund, T.F., Martin, K.A.C., Smith, A.D., and Somogyi, P. 1983. Glutamate decarboxylase-immunoreactive terminals of Golgi-impregnated axo-axonic cells and of presumed basket cells in synaptic contact with pyramidal cells of the cat's visual cortex. J. Comp. Neurol. 221: 263–278.

Freund, T.F., Martin, K.A.C., Somogyi, P., and Whitteridge, D. 1985b. Innervation of cat visual areas 17 and 18 by physiologically identified X- and Y-type afferents. II. Identification of postsynaptic targets by GABA immunocytochemistry and Golgi impregnation. J. Comp. Neurol. 242: 275–291.

Freund, T.F., Martin, K.A.C., and Whitteridge, D. 1985a. Innervation of cat visual areas 17 and 18 by physiologically identified X- and Y-type thalamic afferents. I. Arborization patterns and quantitative distribution of postsynaptic elements. J. Comp. Neurol. 242: 263–274.

Freund, T.F., Powell, J.F., and Smith, A.D. 1984. Tyrosine hydroxylase-like immunoreactive boutons in synaptic contact with identified striatonigral neurons with special reference to dendritic spines. Neuroscience 13: 1189–1215.

Friedman, W.A., Sypert, G.W., Munson, J.B., and Fleshman, J.W. 1981. Recurrent inhibition in type-identified motoneurons. J. Neurophysiol. 46: 1349–1359.

Fukuda, Y., Hsiao, C., Watanabe, M., and Ito, H. 1984. Morphological correlates of physiologically identified Y-, X-, and W-cells in the cat retina. J. Neurophysiol. 52: 999–1013.

Fuller, T.A., and Price, J.L. 1988. Putative glutamatergic and/or aspartatergic cells in the main and accessory olfactory bulbs of the rat. J. Comp. Neurol. 276: 209–218.

Fyffe, R.E.W. 1984. Afferent fibers. In: Handbook of the Spinal Cord, Vol. I: Physiology (Davidoff, R.E., ed.). New York: Marcel Dekker, pp. 79–136.

Fyffe, R.E.W., and Light, A.R. 1984. The ultrastructure of group Ia afferent fiber synapses in the lumbosacral spinal cord of the cat. Brain Res. 300: 201–209.

Gabbott, P.L.A., and Somogyi, P. 1986. Quantitative distribution of GABA-immunoreactive neurons in the visual cortex (area 17) of the cat. Exp. Brain Res. 61: 323–331.

Gall, C., Brecha, T., Chang, T., and Karten, H. 1981. Localization of enkephalins in rat hippocampus. J. Comp. Neurol. 198: 335–350.

Gamble, E., and Koch, C. 1987. The dynamics of free calcium in dendritic spines in response to repetitive synaptic input. Science 236: 1311–1315.

Garnett, R., and Stephens, J.A. 1980. The reflex responses of single motor units in human first dorsal interosseous muscle following cutaneous afferent stimulation. J. Physiol. (Lond.) 303: 351–364.

Garnett, R., and Stephens, J.A. 1981. Changes in the recruitment threshold of motor units produced by cutaneous stimulation in man. J. Physiol. (Lond.) 311: 463–473.

Gerfen, C.R. 1984. The neostriatal mosaic: Compartmentalization of corticostriatal input and striatonigral output systems. Nature 311: 461–464.

Gerfen, C.R. 1985. The neostriatal mosaic. I. Compartmental organization of projections of the striatonigral system in the rat. J. Comp. Neurol. 236: 454–476.

Gerfen, C.R., Baimbridge, K.G., and Miller, J.J. 1985. The neostriatal mosaic: Compartmental distribution of calcium binding protein and parvalbumin in the rat and monkey. Proc. Natl. Acad. Sci. USA 82: 8780–8784.

Gerschenfeld, H.M., Piccolino, M., and Neyton, J. 1980. Feed-back modulation of cone synapses by L-horizontal cells of turtle retina. J. Exp. Biol. 89: 177–192.

Gershon, M.D. 1987. The enteric nervous system. In: Encyclopedia of Neuroscience, Vol. 1 (Adelman, G., ed.). Boston: Birkhauser, pp. 398–399.

Gesteland, R.C. 1986. Speculation on receptor cells as analyzers and filters. Experientia 42: 287–291.

Getchell, T.V., and Shepherd, G.M. 1975a. Synaptic actions on mitral and tufted cells elicited by olfactory nerve volleys on the rabbit. J. Physiol. (Lond.) 251: 497–522.

Getchell, T.V., and Shepherd, G.M. 1975b. Short-axon cells in the olfactory bulb: Dendrodendritic synaptic interactions. J. Physiol. (Lond.) 251: 523–548.

Getting, P. 1983. Mechanisms of pattern generation underlying swimming in *Tritonia*. III. Intrinsic and synaptic mechanisms for delayed excitation. J. Neurophysiol. 49: 1036–1050.

Ghosh, S., Fyffe, R.E.W., and Porter, R. 1988. Morphology of neurons in area 4γ of the cat's cortex studied with intracellular injection of HRP. J. Comp. Neurol. 269: 290–312.

Gilbert, C.D., and Kelly, J.P. 1975. The projection of cells in different layers of the cat's visual cortex. J. Comp. Neurol. 163: 81–106.

Gilbert, C.D., and Wiesel, T.N. 1981. Laminar specialization and intracortical connections in cat primary visual cortex. In: The Organization of the Cerebral Cortex (Schmitt, F.O., Worden, F.G., Adelman, G., and Dennis, S.G., eds.). Cambridge: MIT Press, pp. 163–191.

Giuffrida, R., and Rustioni, A. 1988. Glutamate and aspartate immunoreactivity in corticothalamic neurons of rats. In: Cellular Thalamic Mechanisms (Macchi, G., Bentivoglio, M., and Spreafico, R., eds.). Amsterdam: Elsevier, pp. 311–320.

Gobel, S., Falls, W.M., Bennett, G.J., Abhelmoumene, M., Hayashi, H., and Humphrey, E. 1980. An EM analysis of the synaptic connections of horseradish peroxidase-filled stalked cells and islet cells in the substantia gelatinosa of adult cat spinal cord. J. Comp. Neurol. 194: 781–807.

Gogan, P., Gueritaud, J.P., Horchelle-Bossavit, G., and Tyc-Dumont, S. 1977. Direct excitatory interactions between spinal motoneurones of the cat. J. Physiol. (Lond.) 272: 755–767.

Goldman, D.E. 1943. Potential, impedance and rectification in membranes. J. Gen. Physiol. 27: 37–60.

Goldman, P.S., and Nauta, W.J. 1977. An intrincately patterned prefrontocaudate projection in the rhesus monkey. J. Comp. Neurol. 72: 369–386.

Goldman-Rakic, P.S. 1982. Cytoarchitectonic heterogeneity of the primate neostriatum: Subdivision into island and matrix cellular compartments. J. Comp. Neurol. 205: 398–413.

Golgi, C. 1885. Opera Omnia.

Gottlieb, D.I., and Cowan, W.M. 1973. Autoradiographic studies of the commissural and ipsilateral association connections of the hippocampus and dentate gyrus of the rat. J. Comp. Neurol. 149: 393–420.

Granger, R., Ambros-Ingerson, J., and Lynch, G. 1988. Derivation of encoding characteristics of layer II cerebral cortex. J. Cognit. Neurosci. 1: 61–87.

Graveland, G.A., and DiFiglia, M. 1985a. The frequency and distribution of medium-sized neurons with indented nuclei in the primate and rodent neostriatum. Brain Res. 327: 308–311.

Graveland, G.A., and DiFiglia, M. 1985b. A Golgi study of the human neostriatum: Neurons and afferent fibers. J. Comp. Neurol. 234: 317–333.

Gray, E.G. 1959. Axo-somatic and axo-dendritic synapses of the cerebral cortex: An electron-microscope study. J. Anat. 93: 420–433.

Graybiel, A.M., Baughman, R.W., and Eckenstein, F. 1986. Cholinergic neuropil of the striatum observes striosomal boundaries. Nature 323: 625–627.

Graybiel, A.M., and Ragsdale, C.W., Jr. 1983. Biochemical anatomy of the striatum. In: Chemical Neuroanatomy (Emson, P.C., ed.). New York: Raven Press, pp. 427–503.

Graybiel, A.M., Ragsdale, C.W., Jr., and Mood Edley, S. 1979. Compartments in the striatum of the cat observed by retrograde cell labeling. Exp. Brain Res. 34: 189–195.

Graybiel, A.M., Ragsdale, C.W., Jr., Yoneoka, E.S., and Elde, R.P. 1981. An immunohistochemical study of enkephalin and other neuropeptides in the striatum of the cat with evidence that opiate peptides are arranged to form mosaic patterns in register with striosomal compartments visible by acetylcholinesterase staining. Neuroscience 6: 377–397.

Graziadei, P.P.C., and Monti-Graziadei, G.A. 1979. Neurogenesis and neuron regeneration in the olfactory system of mammals. I. Morphological aspects of differentiation and structural organization of the olfactory sensory neurons. J. Neurocytol. 8: 1–18.

Greenough, W.T., and Chang, F-LF. 1985. Synaptic structural correlates of information storage in the mammalian nervous system. In: Synaptic Plasticity (Cotman, C.W., ed.). New York: Guilford Press, pp. 335–372.

Greenwood, R.S., Godar, S.E., Reaves, T.A., Jr., and Hayward, J.N. 1981. Cholecystokinin in hippocampal pathways. J. Comp. Neurol. 203: 335–350.

Greer, C.A. 1984. A Golgi analysis of granule cell development in the neonatal rat olfactory bulb. Soc. Neurosci. Abstr. 10: 531.

Greer, C.A. 1987. Golgi analyses of dendritic organization among denervated olfactory bulb granule cells. J. Comp. Neurol. 257: 442–452.

Greer, C.A. 1988. High voltage electromicroscopic analyses of olfactory bulb granule cell spine geometry. Chem. Senses 13: 693.

Greer, C.A., and Halasz, N. 1987. Plasticity of dendrodendritic microcircuits following mitral cell loss in the olfactory bulb of the murine mutant PCD. J. Comp. Neurol. 256: 284–298.

Greer, C.A., Kaliszewski, C.K., and Cameron, H.A. 1989. Ultrastructural analyses of local circuits in the olfactory system. Proc. EMSA 47: 790–791.

Greer, C.A., and Shepherd, G.M. 1982. Mitral cell degeneration and sensory function in the neurological mutant mouse Purkinje cell degeneration (PCD). Brain Res. 235: 156–161.

Grieve, K.L., Murphy, P.C., and Sillito, A.M. 1985a. The actions of VIP and ACh on the visual responses of neurones in the striate cortex. Br. J. Pharmacol. [Suppl.] 85: 253.

Grieve, K.L., Murphy, P.C., and Sillito, A.M. 1985b. An evaluation of the role of CCK and VIP in the cat visual cortex. J. Physiol. (Lond.) 365: 42P.

Griffith, W.H., Brown, T.H., and Johnston, D. 1986. Voltage-clamp analysis of synaptic inhibition during long-term potentiation in hippocampus. J. Neurophysiol. 55: 767–775.

Grillner, S. 1981. Control of locomotion in bipeds, tetrapods and fish. In: Handbook of Physiology, Section 1: The Nervous System, Vol. II: Motor Control, Part 2 (Brooks, V.B., ed.). Bethesda: American Physiological Society, pp. 1179–1236.

Gross, C., Rocha-Miranda, C., and Bender, D. 1972. Visual properties of neurons in inferotemporal cortex of the macaque. J. Neurophysiol. 35: 96–111.

Groves, P.M. 1980. Synaptic endings and their postsynaptic targets in neostriatum: Synaptic specializations revealed from analysis of serial sections. Proc. Natl. Acad. Sci. USA 77: 6926–2629,

Guillery, R.W. 1966. A study of Golgi preparations from the dorsal lateral geniculate nucleus of the adult cat. J. Comp. Neurol. 128: 21–50.

Guillery, R.W. 1971. Patterns of synaptic interconnections in the dorsal lateral geniculate nucleus of cat and monkey: A brief review. Vision Res. [Suppl.] 3: 211–227.

Gustafsson, B., Wigstrom, H., Abraham, W.C., and Huang, Y.Y. 1987. Long-term potentiation in the hippocampus using depolarizing current pulses as the conditioning stimulus to single volley synaptic potentials. J. Neurosci. 7: 774–780.

Haas, H.L., and Konnerth, A. 1983. Histamine and noradrenaline decrease calcium-activated potassium conductance in hippocampal pyramidal cells. Nature 302: 432–434.

Haberly, L.B. 1973a. Unitary analysis of the opossum prepyriform cortex. J. Neurophysiol. 36: 762–774.

Haberly, L.B. 1973b. Summed potentials evoked in opossum prepyriform cortex. J. Neurophysiol. 36: 775–778.

Haberly, L.B. 1978. Application of collision testing to investigate properties of multiple association axons originating from single cells in the piriform cortex of the rat. Soc. Neurosci. Abstr. 4: 75.

Haberly, L.B. 1983. Structure of the piriform cortex of the opossum. I. Description of neuron types with Golgi methods. J. Comp. Neurol. 213: 163–187.

Haberly, L.B. 1985. Neuronal circuitry in olfactory cortex: Anatomy and functional implications. Chem. Senses 10: 219–238.

Haberly, L.B. 1990. Comparative aspects of olfactory cortex. In: Cerebral Cortex, Vol. 8 (Jones, E.G., and Peters, A., eds.). New York: Plenum Press.

Haberly, L.B., and Behan, M. 1983. Structure of the piriform cortex of the opossum. III. Ultrastructural characterization of synaptic terminals of association and olfactory bulb afferent fibers. J. Comp. Neurol. 219: 448–460.

Haberly, L.B., and Bower, J.M. 1984. Analysis of association fiber system in piriform cortex with intracellular recording and staining methods. J. Neurophysiol. 51: 90–112.

Haberly, L.B., and Bower, J.M. 1989. Olfactory cortex: Model circuit for study of associative memory? Trends Neurosci. 12: 258–264.

Haberly, L.B., and Feig, S. 1983. Structure of the piriform cortex of the opossum. II. Fine structure of cell bodies and neuropil. J. Comp. Neurol. 216: 69–88.

Haberly, L.B., Hansen, D.J., Feig, S.L., and Presto, S. 1987. Distribution and ultrastructure of neurons in opossum displaying immunoreactivity to GABA and GAD and high-affinity tritiated GABA uptake. J. Comp. Neurol. 266: 269–290.

Haberly, L.B., and Presto, S. 1986. Ultrastructural analysis of synaptic relationships of intracellularly stained pyramidal cell axons in piriform cortex. J. Comp. Neurol. 248: 464–474.

Haberly, L.B., and Price, J.L. 1977. The axonal projection patterns of the mitral and tufted cells of the olfactory bulb in the rat. Brain Res. 129: 152–157.

Haberly, L.B., and Price, J.L. 1978a. Association and commissural fiber systems of the olfactory cortex of the rat. I. Systems originating in the piriform cortex and adjacent areas. J. Comp. Neurol. 178: 711–740.

Haberly, L.B., and Price, J.L. 1978b. Association and commissural fiber systems of the olfactory cortex of the rat. II. Systems originating in the olfactory peduncle. J. Comp. Neurol. 181: 781–808.

Haberly, L.B., and Shepherd, G.M. 1973. Current density analysis of opossum prepyriform cortex. J. Neurophysiol. 36: 789–802.

Hablitz, J.J., and Langmoen, I.A. 1982. Excitation of hippocampal pyramidal cells by glutamate in the guinea-pig and rat. J. Physiol. (Lond.) 325: 317–331.

Hackett, J.T., Hou, S.M., and Cochran, S.L. 1979. Glutamate and synaptic depolarization of Purkinje cells evoked by climbing fibers. Brain Res. 170: 377–380.

Hagihara, K., Tsumoto, T., Sato, H., and Hata, Y. 1988. Actions of excitory amino acid antagonists on geniculo-cortical transmission in the cat's visual cortex. Exp. Brain Res. 60: 407–416.

Halasz, N., Ljungdahl, A., and Hökfelt, T. 1978. Transmitter histochemistry of the rat olfactory bulb. II. Fluorescence histochemical, autoradiographic and electron microscopic localization of monoamines. Brain Res. 154: 253–271.

Halasz, N., Ljungdahl, A., Hökfelt, T., Johannsson, O., Goldstein, M., Park, D., and Biberfeld, P. 1977. Transmitter histochemistry of the rat olfactory bulb. I. Immunohistochemical localization of monoamine-synthesizing enzymes. Support for intrabulbar, periglomerular dopamine neurons. Brain Res. 126: 455–474.

Halasz, N., and Shepherd, G.M. 1983. Neurochemistry of the vertebrate olfactory bulb. Neuroscience 10: 579–619.

Hall, W.C., and Ebner, F.F. 1970. Thalamotelencephalic projections in the turtle *(Pseudemys scripta)*. J. Comp. Neurol. 140: 101–122.

Halliwell, J.V., and Adams, P.R. 1982. Voltage-clamp analysis of muscarinic excitation in hippocampal neurons. Brain Res. 250: 71–92.

Halliwell, J.V., and Scholfield, C.N. 1984. Somatically recorded Ca-currents in guinea-pig hippocampal and olfactory cortex neurones are resistant to adenosine action. Neurosci. Lett. 50: 13–18.

Hamilton, K.A., and Kauer, J.S. 1985. Intracellular potentials of salamander mitral/tufted neurons in response to odor stimulation. Brain Res. 338: 181–185.

Hamlyn, L.H. 1962. The fine structure of the mossy fiber endings in the hippocampus of the rabbit. J. Anat. 96: 112–120.

Hammill, O.P., Bormann, J., and Sakmann, B. 1983. Activation of multiple-conductance state chloride channels in spinal neurones by glycine and GABA. Nature 305: 805–808.

Hámori, J., and Mezey, E. 1977. Serial and triadic synapses in the cerebellar nuclei of the cat. Exp. Brain Res. 30: 259–273.

Hámori, J., and Szentágothai, J. 1966. Identification under the electron microscope of climbing fibers and their synaptic contacts. Exp. Brain Res. 1: 65–81.

Hámori, J., and Szentágothai, J. 1966. Participation of Golgi neuron processes in the cerebellar glomeruli: An electron microscope study. Exp. Brain Res. 2: 35–48.

Hamos, J.E., Van Horn, S.C., Raczkowski, D., Uhlrich, D.J., and Sherman, S.M. 1985. Synaptic connectivity of a local circuit neurone in lateral geniculate nucleus of the cat. Nature 317: 618–621.

Hara, Y., Shiosaka, S., Senba, E., Sakanaka, M., Inagaki, S., Takagi, H., Kawai, Y., Takatsuki, K., Matsuzaki, T., and Tohyama, M. 1982. Ontogeny of the neurotensin-containing neuron system of the rat: Immunohistochemical analysis. I. Forebrain and diencephalon. J. Comp. Neurol. 208: 177–195.

Harlan, R.E., Shivers, B.D., Romano, G.J., Howells, R.D., and Pfaff, D.W. 1987. Localization of preproenkephalin mRNA in the rat brain and spinal cord by in situ hybridization. J. Comp. Neurol. 258: 159–184.

Harris, E.W., and Cotman, C.W. 1986. Long-term potentiation of guinea pig mossy fiber responses is not blocked by N-methyl-D-aspartate antagonists. Neurosci. Lett. 70: 132–137.

Harris, E.W., Ganong, A.H., and Cotman, C.W. 1984. Long-term potentiation in the hippocampus involves activation of N-methyl-D-aspartate receptors. Brain Res. 323: 132–137.

Harris, K.M., and Stevens, J.K. 1988. Dendritic spines of rat cerebellar Purkinje cells: Serial electron microscopy with reference to their biophysical characteristics. J. Neurosci. 8: 4455–4469.

Harting, J.K., Hashikawa, T., and Van Lieshout, D. 1986. Laminar distribution of tectal, parabigeminal and pretectal inputs to the primate superior colliculus: Connectional studies in *Galago crassicaudatus*. Brain Res. 366: 358–363.

Hartzell, H.C. 1981. Mechanisms of slow postsynaptic potentials. Nature. 291: 539–544.

Hasan, Z., and Stuart, D.G. 1984. Mammalian muscle receptors. In: Handbook of the Spinal Cord, Vol. I: Physiology (Davidoff, R.E., ed.). New York: Marcel Dekker, pp. 559–608.

Hattori, T., McGreer, P.L., Fibiger, H.C., and McGeer, E.G., 1973. On the source of GABA-containing terminals in the substantia nigra. Electron micrographic, autoradiographic, and biochemical studies. Brain Res. 54: 103–114.

Hearn, T.J., Ganong, A.H., and Cotman, C.W. 1986. Antagonism of lateral olfactory tract synaptic potentials in rat prepyriform cortex slices. Brain Res. 379: 372–376.

Hebb, D.O. 1949. The Organization of Behavior. New York: John Wiley.

Heggelund, P. 1981. Receptive field organization of simple cells in cat striate cortex. Exp. Brain Res. 42: 89–98.

Heimer, L. 1968. Synaptic distribution of centripetal and centrifugal nerve fibers in the olfactory system of the rat: An experimental anatomical study. J. Anat. 103: 413–432.

Heimer, L., Alheid, G.F., and Zaborszky, L. 1985. Basal ganglia. In: The Rat Nervous System: Forebrain and Midbrain (Paxinos, G., ed.). Sydney: Academic Press, pp. 37–86.

Heit, G., Smith, M.E., and Halgren, E. 1988. Neural encoding of individual words and faces by the human hippocampus and amygdala. Nature 333: 773–775.

Heizman, C.W. 1984. Parvalbumin, an intracellular calcium-binding protein: Distribution, properties and possible roles in mammalian cells. Experientia 40: 910–921.

Hendrickson, A.E., Hunt, S.P., and Wu, J-Y. 1981. Immunocytochemical localization of glutamic acid decarboxylase in monkey striate cortex. Nature 292: 605–607.

Hendry, A.H.C., Jones, E.G., De Felipe, J., Schmechel, D., Brandon, C., and Emson, P.C. 1984. Neuropeptide-containing neurons of the cerebral cortex are also GABA-ergic. Proc. Natl. Acad. Sci. USA 81: 6526–6530.

Henneman, E., and Mendell, L.M. 1981. Functional organization of motoneuron pool and its inputs. In: Handbook of Physiology, Section 1: The Nervous System, Vol. II, Chapter 11, Part 1 (Brooks, V.B., ed.). Bethesda: American Physiological Society, pp. 423–507.

Henneman, E., Lüscher, H.-R., and Mathis, J. 1984. Simultaneously active and inactive synapses of single Ia fibres on cat spinal motoneurones. J. Physiol. (Lond.) 352: 147–161.

Henneman, E., and Olson, C.B. 1965. Relations between structure and function in the design of skeletal muscles. J. Neurophysiol. 28: 581–598.

Henneman, E., Somjen, G., and Carpenter, D.O. 1965. Excitability and inhibitibility of motoneurons of different sizes. J. Neurophysiol. 28: 599–620.

Herkenham, M., and Pert, C.B. 1981. Mosaic distribution of opiate receptors, parafascicular projections and acetylcholinesterase in the rat striatum. Nature 291: 415–418

Herkenham, M., Moon Edley, S., and Stuart, J. 1984. Cell clusters in the nucleus accumbens of the rat, and the mosaic relationship of opiate receptors, acetylcholinesterase and subcortical afferent terminations. Neuroscience 11: 561–593.

Herrling, P.L., and Hull, C.D. 1980. Iontophoretically applied dopamine depolarizes and hyperpolarizes the membrane of cat caudate neurons. Brain Res. 192: 441–462.

Herron, C.E., Lester, R.A.J., Coan, E.J., and Collingridge, G.L. 1985. Intracellular demonstration of an N-methyl-D-aspartate receptor mediated component of synaptic transmission in the rat hippocampus. Neurosci. Lett. 60: 19–23.

Hersch, S.M., and White, E.L. 1981. Quantification of synapses formed with apical dendrites of Golgi-impregnated pyramidal cells: Variability in thalamocortical inputs, but consistence in the ratios of asymmetrical to symmetrical synapses. J. Neurosci. 6: 1043–1051.

Hicks, T.P. 1987. Excitatory amino acid pathways in cerebral cortex. Excitatory Amino Acid Transmission (Hicks, T.P., Lodge, D., and McLennan, H., eds.). New York: Alan R. Liss, pp. 373–380.

Hicks, T.P., Guedes, R.C.A., Veale, W.L., and Veenhuizen, J. 1985. Aspartate and glutamate as synaptic transmitters of parallel visual cortical pathways. Exp. Brain Res. 58: 421–425.

Hikosaka, O., and Wurtz, R.H. 1983. Visual and oculomotor functions of monkey substantia nigra pars reticulata. III. Memory-contingent visual and saccade responses. J. Neurophysiol. 49: 1268–1284.

Hille, B. 1984. Ionic Channels of Excitable Membranes. Sunderland, MA: Sinauer Associates.

Hille, B. 1987. Evolutionary origins of voltage-gated channels and synaptic transmission. In: Synaptic Function (Edelman, G.M., Gall, W.E., and Cowan, W.M., eds.). New York: John Wiley, pp. 163–176.

Hillman, D.E., and Chen, S.C. 1984. Reciprocal relationship between size of postsynaptic densities and their number: Constancy in contact area. Brain Res. 295: 325–343.

Hillman, D.E., and Chen, S.C. 1985a. Compensation in the number of presynaptic dense projections and synaptic vesicles in remaining parallel fibres following cerebellar lesions. J. Neurocytol. 14: 673–687.

Hillman, D.E., and Chen, S.C. 1985b. Plasticity in the size of presynaptic and postsynaptic membrane specializations. In: Synaptic Plasticity (Cotman, C.W., ed.). New York: Guilford Press, pp. 39–76.

Hinds, J.W. 1970. Reciprocal and serial dendrodendritic synapses in the glomerular layer of the rat olfactory bulb. Brain Res. 17: 530–534.

Hinton, G., and Anderson, J.A. 1981. Parallel Models of Associative Memory. Hillsdale, NJ: Lawrence Erlbaum Associates.

Hirata, Y. 1964. Some observations of the fine structure of synapses in the olfactory bulb of the mouse, with particular reference to the atypical synaptic configurations. Arch. Histol. Jpn. 24: 303–317.

Hirsch, J.C., Fourment, A., and Marc, M.E. 1983. Sleep-related variations of membrane potential in the lateral geniculate body relay neurons of the cat. Brain Res. 259: 308–312.

Hirsh, R. 1980. The hippocampus, conditional operations, and cognition. Physiol. Psychol. 8: 175–182.

Hirst, G.D.S., Redman, S.J., and Wong, K. 1981. Post-tetanic potentiation and facilitation of synaptic potentials evoked in cat spinal motoneurones. J. Physiol. (Lond.) 321: 97–109.

Hodgkin, A.L., and Huxley, A.F. 1952a. Currents carried by sodium and potassium ions through the membrane of the giant axon of *Loligo*. J. Physiol. (Lond.) 116: 449–472.

Hodgkin, A.L., and Huxley, A.F. 1952b. The components of membrane conductance in the giant axon of *Loligo*. J. Physiol. (Lond.) 116: 473–496.

Hodgkin, A.L., and Huxley, A.F. 1952c. The dual effect of membrane potential and sodium conductance in the giant axon of *Loligo*. J. Physiol. (Lond.) 116: 497–506.

Hodgkin, A.L., and Huxley, A.F. 1952d. A quantitative description of membrane current and its application to conduction and excitation in nerve. J. Physiol. (Lond.) 117: 500–544

Hodgkin, A.L., and Katz, B. 1949. The effect of sodium ions on the electrical activity of the giant axon of the squid. J. Physiol. (Lond.) 108: 37–77.

Hoffer, B.J., Siggins, G.R., Oliver, A.P., and Bloom, F.E. 1973. Activation of the pathway from locus coeruleus to rat cerebellar Purkinje neurons: Pharmacological evidence of noradrenergic central inhibition. J. Pharmacol. Exp. Ther. 184: 553–569.

Hoffman, W.H., and Haberly, L.B. 1989. Bursting induces persistent all-or-none EPSPs by an NMDA-dependent process in piriform cortex. J. Neurosci. 9: 206–215.

Hoffmann, K-P., and Stone, J. 1971. Conduction velocity of afferents to cat visual cortex: A correlation with cortical receptive field properties. Brain Res. 32: 460–466.

Hökfelt, T., Johansson, O., and Goldstein, M. 1984. Chemical anatomy of the brain. Science 225; 1326-1334.

Hollingworth, H.L. 1928. Psychology: Its Facts and Principles. New York: D. Appleton

Holsheimer, J. 1982. Generation of theta activity (RSA) in the cingulate cortex of the rat. Exp. Brain Res. 47: 309–312.

Honig, M.G., Collins, W.F., and Mendell, L.M. 1873. α-Motoneuron EPSPs exhibit different frequency sensitivites to single Ia-afferent fiber stimulation. J. Neurophysiol. 49: 886–901.

Honore, T., Davies, S.N., Drejer, J., Fletcher, E.J., Jacobsen, P., Lodge, D., and Nielsen, F.E. 1988. Quinoxalinediones: Potent competitive non-NMDA glutamate receptor antagonists. Science 241: 701–703,

Hopfield, J.J. 1982. Neural networks and physical systems with emergent collective computational abilities. Proc. Natl. Acad. Sci. USA 79: 2554–2558.

Hopfield, J.J. 1984. Neurons with graded response have collective computational properties like those of two-state neurons. Proc. Natl. Acad. Sci. USA 81: 3088–3092.

Hopkins, W.F., and Johnston, D. 1988. Noradrenergic enhancement of long-term potentiation at mossy fiber synapses in the hippocampus. J. Neurophysiol. 59: 667–687.

Hori, N., Auker, C.R., Braitman, D.J., and Carpenter, D.O. 1981. Lateral olfactory tract transmitter: Glutamate, aspartate, or neither? Cell. Neurobiol. 1: 115–120.

Hori, N., Auker, C.R., Braitman, D.J., and Carpenter, D.O. 1982. Pharmacologic sensitivity of amino acid responses and synaptic activation of *in vitro* prepiriform neurons. J. Neurophysiol. 48: 1289–1301.

Horn, J.P., and Stofer, W.D. 1988. Double labeling of the paravertebral sympathetic C system in the bullfrog and antiserum to LHRH and NPY. J. Auton. Nerv. Syst. 23: 17–24.

Hornung, J.P., and Garey, L.J. 1981. The thalamic projection to cat visual cortex: Ultrastructure of neurons identified by Golgi impregnation or retrograde horseradish peroxidase transport. Neuroscience 6: 1053–1068.

Horton, J.C., and Hubel, D.H. 1981. A regular patchy distribution of cytochrome oxidase staining in primary visual cortex of the macaque monkey. Nature 292: 762–764.

Hotson, J.R., and Prince, D.A. 1980. A calcium-activated hyperpolarization follows repetitive firing in hippocampal neurons. J. Neurophysiol. 43: 409–419.

Hotson, J.R., Prince, D.A., and Schwartzkroin, P.A. 1979. Anomalous rectification in hippocampal neurons. J. Neurophysiol. 42: 889–895.

Houk, J.C., and Rymer, W.A. 1981. Neural control of muscle length and tension. In: Handbook of Physiology, Section 1: The Nervous System, Vol. II, Part 1 (Brooks, V.B., ed). Bethesda: American Physiology Society, pp. 257–323.

Hounsgaard, J., Hultborn, H., Jespersen, B., and Kiehn, O. 1984. Intrinsic membrane properties causing a bistable behaviour of alpha-motoneurones. Exp. Brain Res. 55: 391–394.

Hounsgaard, J., Hultborn, H., Jespersen, B., and Kiehn, O. 1988. Bistability of α-motoneurones in the decerebrate cat and in the acute spinal cat after intravenous 5-hydroxytryptophan. J. Physiol. (Lond.) 405: 345–367.

Hounsgaard, J., and Kiehn, O. 1985. Ca^{++}-dependent bistability induced by serotonin in spinal motoneurones. Exp. Brain Res. 57: 422–425.

Houser, C.R., Crawford, G.D., Salvaterra, P.M., and Vaughn, J.E. 1985. Immunocytochemical localization of choline acetyltransferase in rat cerebral cortex: A study of cholinergic neurons and synapses. J. Comp. Neurol. 234: 17–34.

Hökfelt, T., and Ljungdahl, A. 1972. Autoradiographic identification of cerebral and cerebellar cortical neurons accumulating labeled gamma-aminobutyric acid (^{3}H-GABA). Exp. Brain Res. 14: 354–362.

Hubel, D.H., and Wiesel, T.N. 1962. Receptive fields, binocular interaction and functional architecture in the cat's visual cortex. J. Physiol. (Lond.) 160: 106–154.

Hubel, D.H., and Wiesel, T.N. 1977. Functional architecture of macaque monkey visual cortex (Ferrier lecture). Proc. R. Soc. Lond. [B] 198: 1–59.

Hudson, D.B., Valcana, T., Bean, G., and Timiras, P.S. 1976. Glutamic acid: A strong candidate as the neurotransmitter of the cerebellar granule cell. Neurochem. Res. 1: 73–81,

Huettner, J.E., and Baughman, R.W. 1986. Primary culture of identified neurons from the visual cortex of post-natal rats. J. Neurosci. 6: 3044–3060.

Huettner, J.E., and Baughman, R.W. 1988. The pharmacology of synapses formed by identified corticocollicular neurons in primary cultures of rat visual cortex. J. Neurosci. 8: 160–175.

Hughes, A. 1981. Cat retina and the sampling theorem; the relation of transient and sustained brisk-unit cut-off frequency to alpha- and beta-mode cell density. Exp. Brain Res. 42: 196–202.

Hull, C.L. 1929. A functional interpretation of the conditioned reflex. Psychol. Rev. 36: 498–511.

Hultborn, H., Jankowska, E., and Lindström, S. 1971a. Recurrent inhibition from motor axon collaterals of transmission in the Ia inhibitory pathway to motoneurones. J. Physiol. (Lond.) 215: 591–612.

Hultborn, H., Jankowska, E., and Lindström, S. 1971b. Recurrent inhibition of interneurones monosynaptically activated from group Ia afferents. J. Physiol. (Lond.) 215: 613–636.

Humphrey, A.L., Sur, M., Uhlrich, D.J., and Sherman, S.M. 1985. Projection patterns of individual X- and Y-cell axons from the lateral geniculate nucleus to cortical area 17 in the cat. J. Comp. Neurol. 233: 159–189.

Illert, M., Lundberg, A., and Tanaka, R. 1976. Integration in descending motor pathways controlling the forelimb in the cat. 2. Convergence on neurones mediating disynaptic cortico-motoneuronal excitation. Exp. Brain Res. 26: 521–540.

Illert, M., Jankowska, E., Lundberg, A., and Odutola, A. 1981. Integration in descending motor pathways controlling the forelimb in the cat. 7. Effects from the reticular formation on C3–C4 propriospinal neurons. Exp. Brain Res. 42: 269–281.

Illes, P. 1986. Mechanisms of receptor-mediated modulation of transmitter release in noradrenergic, cholinergic, and sensory neurones. Neuroscience 17: 909–928.

Inagaki, N. Yamatodani, A., Ando-Yamamoto, M., Tohyama, M., Watanabe, T., and Wada, H. 1988. Organization of histaminergic fibers in the rat brain. J. Comp. Neurol. 273: 283–300.

Inagaki, S., Kubota, Y., Shinoda, K., Kawai, Y., and Tohyama, M. 1983a. Neurotensin-containing pathway from the endopiriform nucleus and the adjacent prepiriform cortex to the dorsomedial nucleus in the rat. Brain Res. 260: 143–146.

Inagaki, S., Shinoda, K., Kubota, Y., Shiosaka, S., Matsuzaki, T., and Tohyama, M. 1983b. Evidence for the existence of a neurotensin-containing pathway from the endopiriform nucleus and the adjacent prepiriform cortex to the anterior olfactory nucleus and nucleus of diagonal band Broca of the rat. Neuroscience 8: 487–493.

Inagaki, S., Shiosaka, S., Takatsuki, K., Iida, H., Sakanaka, M., Senba, E., Hara, Y., Matsuzuki, T., Kawai, Y., and Tohyama, M. 1982. Ontogeny of somatostatin-containing neuron system of the rat cerebellum including its fiber connections: An experimental and immunohistochemical analysis. Dev. Brain Res. 3: 509–527.

Inoue, M., Oomara, Y., Yakushiji, T., and Akaike, N. 1986. Intracellular calcium ions decrease the affinity of the GABA receptor. Nature 324: 156–158.

Ishizuka, N., Mannen, H., Hongo, T., and Sasaki, S. 1979. Trajectory of group Ia afferent fibers stained with horseradish peroxidase in the lumbosacral spinal cord of the cat: Three dimensional reconstructions from serial sections. J. Comp. Neurol. 186: 189–211.

Israël, M., and Whittaker, V.P. 1965. The isolation of mossy fiber endings from the granular layer of the cerebellar cortex. Experientia 21: 325–326.

Ito, M., Yoshida, M., and Obata, K. 1964. Monosynaptic inhibition of the intracerebellar nuclei induced from the cerebellar cortex. Experientia 20: 575–576.

Jack, J.J.B. 1979. Introduction to linear cable theory. In: The Neurosciences: Fourth Study Program (Schmitt, F.O., and Worden, F.G., eds.). Cambridge, MA: MIT, pp. 423–437.

Jack, J.J.B., Miller, S., Porter, R., and Redman, S.J. 1971. The time course of minimal excitatory postsynaptic potentials evoked in spinal motoneurons by group Ia afferent fibres. J. Physiol. (Lond.) 215: 353–380.

Jack, J.J.B., Noble, D., and Tsien, R.W. 1975. Electric Current Flow in Excitable Cells. Oxford: Oxford University Press.

Jack, J.J.B., Redman, S.J., and Wong, K. 1981a. The components of synaptic potentials evoked in cat spinal motoneurones by impulses in single group Ia afferents. J. Physiol. (Lond.) 321: 65–96.

Jack, J.J.B., Redman, S.J., and Wong, K. 1981b. Modifications to synaptic transmission at group Ia synapses on cat spinal motoneurones by 4-aminopyridine. J. Physiol. (Lond.) 321: 111–126.

Jackowski, A., Parnevalas, J.G., and Lieberman, A.R. 1978. The reciprocal synapse in the external plexiform layer of the mammalian olfactory bulb. Brain Res. 195: 17–28.

Jacobson, M. 1978. Developmental Neurobiology, 2nd ed. New York: Plenum Press.

Jahnsen, H. 1986a. Responses of neurons in isolated preparations of the mammalian central nervous system. Prog. Neurobiol. 27: 351–372.

Jahnsen, H. 1986b. Electrophysiological characteristics of neurones in the guinea-pig deep cerebellar nuclei *in vitro*. J. Physiol. (Lond.) 372: 129–147.

Jahnsen, H. 1986c. Extracellular activation and membrane conductances of neurones in the guinea-pig deep cerebellar nuclei *in vitro*. J. Physiol. (Lond.) 372: 149–168.

Jahnsen, H., and Llinás, R. 1984a. Electrophysiological properties of guinea-pig thalamic neurones: An in vitro study. J. Physiol. (Lond.) 349: 205–226

Jahnsen, H., and Llinás, R. 1984b. Ionic basis for the electroresponsiveness and oscillatory properties of guinea-pig thalamic neurones in vitro. J. Physiol. (Lond.) 349: 227–247.

Jahr, C.E., and Nicoll, R.A. 1981. Primary afferent depolarization in the in vitro frog olfactory bulb. J. Physiol. (Lond.) 318: 375–384.

Jahr, C.E., and Nicoll, R.A. 1982. An intracellular analysis of dendrodendritic inhibition in the turtle in vitro olfactory bulb. J. Physiol. (Lond.) 326: 213–234.

Jahr, C.E., and Stevens, C.F. 1987. Glutamate activates multiple single channel conductances in hippocampal neurons. Nature 325: 522–525.

Jahr, C.E., and Yoshioka, K. 1986. Ia afferent excitation of motoneurones in the in vitro newborn rat spinal cord is selectively antagonized by kynurenate. J. Physiol. (Lond.) 370: 515–530.

Jameson, D., and Hurvich, L.M. 1959. Perceived color and its dependence on focal, surrounding and preceding stimulus variables. J. Opt. Soc. Am. 49: 890–898.

Jami, L., Murthy, K.S.K., and Petit, J. 1982. A quantitative study of skeletofusimotor innervation on the cat peroneus tertius muscle. J. Physiol. (Lond.) 325: 125–144.

Jan, L.Y., and Jan, Y.N. 1982. Peptidergic transmission in sympathetic ganglia of the frog. J. Physiol. (Lond.) 327: 219–246.

Janigro, D., and Schwartzkroin, P.A. 1988. Effects of GABA on CA3 pyramidal cell dendrites in rabbit hippocampal slices. Brain Res. 453: 265–274.

Jankowska, E., and Lindström, S. 1972. Morphology of interneurones mediating Ia reciprocal inhibition of motoneurones in the spinal cord of the cat. J. Physiol. (Lond.) 226: 805–823.

Jankowska, E., and Lundberg, A. 1981. Interneurones in the spinal cord. Trends Neurosci. 4: 230–233.

Jankowska, E., Lundberg, A., Rudomin, P., and Sykova, E. 1977. Effects of 4-aminopyridine on transmission in excitatory and inhibitory synapses in the spinal cord. Brain Res. 136: 387–392.

Jankowska, E., and Roberts, W.J. 1972a. An electrophysiological demonstration of the axonal projections of single spinal interneurones in the cat. J. Physiol. (Lond.) 222: 597–622.

Jankowska, E., and Roberts, W.J. 1972b. Synaptic actions of single interneurons mediating reciprocal Ia inhibition of motoneurones. J. Physiol. (Lond.) 222: 623–642.

Jankowska, E., and Smith, D.O. 1973. Antidromic activation of Renshaw cells and their axonal projections. Acta Physiol. Scand. 88: 198–214.

Jansen, J. 1969. Vertebrate history with special reference to factors related to cerebellar evolution. In: On Cerebellar Evolution and Organization from the Point of View of a Morphologist. Chicago: American Medical Association, pp. 881–893.

Jayaraman, A. 1980. Anatomical evidence for cortical projections from the striatum in the cat. Brain Res. 195: 29–36.

Jensen, R., and Devoe, R. 1983. Comparisons of directionally selective with other ganglion cells of the turtle retina: Intracellular recording and staining. J. Comp. Neurol. 217: 271–287.

Johnson, J.W., and Ascher, P. 1987. Glycine potentiates the NMDA response in cultured mouse brain neurons. Nature 325: 529–531.

Johnston, D. 1981. Passive cable properties of hippocampal CA3 pyramidal neurons. Cell. Mol. Neurobiol. 1: 41–55.

Johnston, D., and Brown, T.H. 1981. Giant synaptic potential hypothesis for epileptiform activity. Science 211: 294–297.

Johnston, D., and Brown, T.H. 1983. Interpretation of voltage-clamp measurements in hippocampal neurons. J. Neurophysiol. 50: 464–486.

Johnston, D., and Brown, T.H. 1984a. Mechanisms of neuronal burst generation. In: Electrophysiology of Epilepsy (Schwartzkroin, P.A., and Wheal, H.V., eds.). New York: Academic Press, pp. 277–301.

Johnston, D., and Brown, T.H. 1984b. Biophysics and microphysiology of synaptic transmission in hippocampus. In: Brain Slices (Dingledine, R., ed.). New York: Plenum Press, pp. 51–86.

Johnston, D., and Brown, T.H. 1986. Control theory applied to neural networks illuminates synaptic basis of interictal epileptiform activity. In: Advances in Neurology, Vol. 44: Basic Mechanisms of the Epilepsies. Molecular and Cellular Approaches (Delgado-Escueta, A.V., Ward, A.A., Jr., Woodbury, D.M., and Porter, R.J., eds.). New York: Raven Press, pp. 263–274.

Johnston, D., Hopkins, W.F., and Gray, R. 1988. Noradrenergic enhancement of long-term synaptic enhancement. In: Long-Term Potentiation: From Biophysics to Behavior (Landfield, P.W., and Deadwyler, S.A., eds.). New York: Alan R. Liss, pp. 355–376.

Johnston, D., Hopkins, W.F., and Gray, R. 1989. The role of norepinephrine in long-term potentiation at mossy fiber synapses in the hippocampus. In: Neural Models of Plasticity: Experimental and Theoretical Approaches (Byrne, J.H., and Berry, W.O., eds.). San Diego: Academic Press, pp. 307–328.

Jones, E.G. 1985. The Thalamus. New York: Plenum Press.

Jones, E.G., Coulter, J.D., Burton, H., and Porter, R. 1977. Cells of origin and terminal distribution of corticostriatal fibers arising in the sensory-motor cortex of monkeys. J. Comp. Neurol. 173: 53–80.

Jones, E.G., and Powell, T.P.S. 1969. Morphological variations in the dendritic spines of the neocortex. J. Cell. Sci. 5: 509–529.

Jones, K.A., and Baughman, R.W. 1988. NMDA- and non-NMDA-receptor components of excitatory synaptic potentials recorded from cells in layer V of rat visual cortex. J. Neurosci. 8: 3522–3534.

Jones, S.W. 1985. Muscarinic and peptidergic excitation of bullfrog sympathetic neurones. J. Physiol. (Lond.) 366: 63–87.

Jones, S.W. 1987. Luteinizing hormone-releasing hormone as a neurotransmitter in bullfrog sympathetic ganglia. Ann. N.Y. Acad. Sci. 519: 310–322.

Jones, S.W., and Adams, P.R. 1987. The M-current and other potassium currents of vertebrate neurons. In: Neuromodulation (Kaczmarek, L., and Levitan, I.B., eds.). New York: Oxford University Press, pp. 159–186.

Jourdan, F. 1975. Ultrastructure de l'épithelium olfatif du rat: Polymorphisme des récepteurs. C.R. Séances Acad. Sci. [III] 280: 443–446.

Kaczmarck, L.K., and Levitan, I.B. 1986. Neuromodulation. New York: Oxford University Press.

Kan, K.S.K., Chao, L.P., and Eng, L.F. 1978. Immunohistochemical localization of choline acetyltransferase in rabbit spinal cord and cerebellum. Brain Res. 146: 221–229.

Kan, K.S.K., Chao, L.P., and Forno, L.S. 1980. Immuno-histochemical localization of choline acetyltransferase in the human cerebellum. Brain Res. 193: 165–171.

Kanda, K., Burke, R.E., and Walmsley, B. 1977. Differential control of fast and slow twitch motor units in the decerebrate cat. Exp. Brain Res. 29: 57–74.

Kandel, E.R., Spencer, W.A., and Brinley, F.J. 1961. Electrophysiology of hippocampal neurons. I. Sequential invasion and synaptic organization. J. Neurophysiol. 24: 225–242.

Kaneko, A. 1970. Physiological and morphological identification of horizontal, bipolar, and amacrine cells in the goldfish retina. J. Physiol. (Lond.) 207: 623–633.

Karlsson, G., Pozza, M., and Olpe, H.-R. 1988. Phaclofen: A $GABA_B$ blocker reduces long-duration inhibition in the neocortex. Eur. J. Pharmacol. 148: 485–486.

Katayama, Y., Tsubokawa, T., and Moriyasu, N. 1980. Slow rhythmic activity of caudate neurons in the cat: Statistical analysis of caudate neuronal spike trains. Exp. Neurol. 68: 310–321.

Kauer, J.S. 1974. Response patterns of amphibian olfactory bulb neurons to odour stimulation. J. Physiol. (Lond.) 243: 695–715.

Kauer, J.S., and Shepherd, G.M. 1977. Analysis of the onset phase of olfactory bulb unit responses to odour pulses in the salamander. J. Physiol. (Lond.) 272: 495–516.

Kawamura, H., and Provini, L. 1970. Depression of cerebellar Purkinje cells by micro-iontophoretic application of GABA and related amino acids. Brain Res. 24: 293–304.

Kay, A.R., and Wong, R.K.S. 1987. Calcium current activation kinetics in isolated pyramidal neurones of the CA1 region of the mature guinea-pig hippocampus. J. Physiol. (Lond.) 392: 603–616.

Keenan, C.L., Chapman, P.F., Chang, V.C., and Brown, T.H. 1988. Videomicroscopy of acute brain slices from amygdala and hippocampus. Brain Res. Bull. 21: 373–383.

Keenan, C.L., Kairiss, E.W., Greenwood, A., Nobre, C.A., Rihm, L., and Brown, T.H. 1989. Confocal scanning laser microscopy of injected neurons in hippocampal slices. Soc. Neurosci. Abstr. 15: 980.

Kelso, S.R., and Brown, T.H. 1986. Differential conditioning of associative synaptic enhancement in hippocampal brain slices. Science 232: 85–87.

Kelso, S.R., Ganong, A.H., and Brown, T.H. 1986. Hebbian synapses in the hippocampus. Proc. Natl. Acad. Sci. USA 83: 5326–5331.

Kemp, J.A., and Sillito, A.M. 1982. The nature of the excitatory transmitter mediating X and Y cell inputs to the cat dorsal lateral geniculate nucleus. J. Physiol. (Lond.) 323: 377–391.

Kemp, J.M., and Powell, T.P.S. 1971a. The synaptic organization of the caudate nucleus. Philos. Trans. Soc. Lond. [B] 262: 403–412.

Kemp, J.M., and Powell, T.P.S. 1971b. The site of termination of afferent fibers in the caudate nucleus. Philos. Trans. Soc. Lond. [B] 262: 413–427.

Kerr, D.I.B., and Dennis, B.J. 1972. Collateral projection of the lateral olfactory tract to entorhinal cortical areas in the cat. Brain Res. 36: 399–403.

Kerr, D.I.B., Ong, J., Prager, R.H., Gynther, B.D., and Curtis, D.R. 1987. Phaclophen: A peripheral and central baclophen antagonist. Brain Res. 405: 150–154.

Ketchum, K.L., and Haberly, L.B. 1988. CSD analysis of oscillatory responses in rat piriform cortex reveals stereotyped cyclical components mediated by afferent and intrinsic association fibers. Soc. Neurosci. Abstr. 14: 278.

Kimura, M., Raijkowski, J., and Evarts, E. 1984. Tonically discharging putamen neurons exhibit set-dependent responses. Proc. Natl. Acad. Sci. USA 81: 4998–5001.

Kishi, K., Mori, K., and Ojima, H. 1984. Distribution of local axon collaterals of mitral, displaced mitral and tufted cells in the rabbit olfactory bulb. J. Comp. Neurol. 225: 511–526.

Kisvárday, Z.F., Martin, K.A.C., Whitteridge, D., and Somogyi, P. 1985. Synaptic connections of intracellularly filled clutch cells: A type of small basket cell in the visual cortex of the cat. J. Comp. Neurol. 241: 111–137.

Kita, H., and Kitai, S.T. 1987. Efferent projections of the subthalamic nucleus in the rat: Light and electron microscopic analysis with the PHA-L method. J. Comp. Neurol. 260: 435–452.

Kita, H., Kita, T., and Kitai, S.T. 1985a. Active membrane properties of rat neostriatal neurons in an in vitro slice preparation. Exp. Brain Res. 60: 54–62.

Kita, T., Kita, H., and Kitai, S.T. 1984. Passive electrical membrane properties of rat neostriatal neurons in an in vitro slice preparation. Brain Res. 300: 129–139.

Kita, T., Kita, H., and Kitai, S.T. 1985b. Local stimulation induced GABAergic response in rat striatal slice preparations: Intracellular recordings on QX-314 injected neurons. Brain Res. 360: 304–310.

Kita, T., Kita, H., and Kitai, S.T. 1985c. Effects of [D-Ala-2]methionine enkephalinamide and carbachol recorded intracellularly from rat neostriatum in slice preparation. Soc. Neurosci. Abstr. 11: 362.

Kitai, S.T. 1981. Electrophysiology of the corpus striatum and brain stem integrating systems. In: Handbook of Physiology, Section 1: The Nervous System, Vol. II: Motor Systems, Part 2 (Brooks, V.B., ed.). Bethesda: American Physiological Society, pp. 997–1015.

Kitai, S.T., Kocsis, J.D., Preston, R.J., and Sugimori, M. 1976a. Monosynaptic inputs to caudate neurons identified by intracellular injection of horseradish peroxidase. Brain Res. 109: 601–606.

Kitai, S.T., Sugimori, M., and Kocsis, J. 1976b. Excitatory nature of dopamine in the nigro-caudate pathway. Exp. Brain Res. 24: 351–363.

Knowles, W.D., Funch, P.G., and Schwartzkroin, P.A. 1982. Electrotonic and dye coupling in hippocampal CA1 pyramidal cells in vitro. Neuroscience 7: 1713–1722.

Knowles, W.D., and Schwartzkroin, P.A. 1981. Local circuit synaptic interactions in hippocampal brain slices. J. Neurosci. 1: 318–322.

Koch, C. 1985. Understanding the intrinsic circuitry of the cat's dLGN: Electrical properties of the spine–triad arrangement. Proc. R. Soc. Lond. [B] 225: 365–390.

Koch, C. 1987. The action of the corticofugal pathway on sensory thalamic nuclei: A hypothesis. Neuroscience 23: 399–406.

508 *References*

Koch, C., and Poggio, T. 1983. A theoretical analysis of the electrical properties of spines. Proc. R. Soc. Lond. [B] 218: 455–477.

Koch, C., and Poggio, T. 1987. Biophysics of computation: Neurons, synapses and membranes. In: Synaptic Function (Edelman, G.M., Gall, W.E., and Cowan, W.M., eds.). New York: John Wiley, pp. 637–697.

Koch, C., Poggio, T., and Torre, V. 1982. Retinal ganglion cells: A functional interpretation of dendritic morphology. Philos. Trans. R. Soc. Lond. [B] 298: 227–263.

Koch, C., Poggio, T., and Torre, V. 1983. Nonlinear interactions in a dendritic tree: Localization, timing and role in information processing. Proc. Natl. Acad. Sci. USA 80: 2799–2802.

Kocsis, J.D., Sugimori, M., and Kitai, S.T. 1976. Convergence of excitatory synaptic inputs to caudate spiny neurons. Brain Res. 124: 403–413.

Koerber, H.R., Druzinsky, R.E., and Mendell, L.M. 1988. Properties of somata of spinal dorsal root ganglion cells differ according to peripheral receptor innervated. J. Neurophysiol. 60: 1584–1596.

Kohler, C. 1986. Intrinsic connections of the retrohippocampal region in the rat brain. II. The medial entorhinal area. J. Comp. Neurol. 236: 504–504.

Kohonen, T. 1977. A principal of neural associative memory. Neuroscience 2: 1065–1076.

Kohonen, T. 1984. Self-Organization and Associative Memory. New York: Springer-Verlag.

Kolb, H. 1974. The connections between horizontal cells and photoreceptors in the retina of the cat: Electron microscopy of Golgi preparations. J. Comp. Neurol. 155: 1–14.

Kolb, H. 1977. The organization of the outer plexiform layer in the retina of the cat: Electron microscopic observations. J. Neurocytol. 6: 131–153.

Kolb, H. 1979. The inner plexiform layer in the retina of the cat: Electron microscopic observations. J. Neurocytol. 8: 295–329.

Kolb, H., and Nelson, R. 1983. Rod pathways in the retina of the cat. Vision Res. 23: 301–312.

Kolb, H., Boycott, B.B., and Dowling, J.E. 1969. A second type of midget bipolar cell in the primate retina. (Appendix.) Philos. Trans. R. Soc. Lond. [B] 225: 177–184.

Kolb, H., Nelson, R., and Mariani, A. 1981. Amacrine cells, bipolar cells, and ganglion cells of the cat retina: A Golgi study. Vision Res. 21: 1081–1114.

Kolb, H., and West, R.W. 1977. Synaptic connections of the interplexiform cell in the retina of the cat. J. Neurocytol. 6: 155–170.

Komatsu, Y., Nakajima, S., Toyama, K., and Fetz, E. 1988. Intracortical connectivity revealed by spike-triggered averaging in slice preparations of cat visual cortex. Brain Res. 422: 359–362.

Komatsu, Y., Toyama, K., Maeda, J., and Sakaguchi, H. 1981. Long-term potentiation investigated in a slice preparation of striate cortex of young kittens. Neurosci. Lett. 26: 269–274.

Korn, H., and Faber, D.S. 1987. Regulation and significance of probabilistic release mechanisms at central synapses. In: Synaptic Function (Edelman, G.M., Gall, W.E., and Cowan, W.M., eds.). New York: John Wiley, pp. 57–108.

Korn, H., Mallet, A., Triller, A., and Faber, D.S. 1982. Transmission at a central inhibitory synapse II. Quantal description of release, with a physical correlate for binomial neuron. J. Neurophysiol. 48: 679–707.

Kosaka, T., Katsumaru, H., Hama, K., Wu, J.-Y, and Heizman, C.W. 1987. GABAergic neurons containing the Ca^{2+}-binding protein parvalbumin in the rat hippocampus and dentate gyrus. Brain Res. 419: 119–130.

Kriegstein, A.R., and Connors, B.W. 1986. Cellular physiology of the turtle visual cortex: Synaptic properties and intrinsic circuitry. J. Neurosci. 6: 178–191.

Krnjević, K. 1974. Chemical nature of synaptic transmission in vertebrates. Physiol. Rev. 54: 418–540.

Krnjević, K. 1981. Transmitters in motor systems. In: Handbook of Physiology, Section 1: The Nervous System, Vol. II: Motor Systems, Part 1 (Brooks, V.B., ed.). Bethesda: American Physiological Society, pp. 107–154.

Krnjević, K., and Phillis, J.W. 1963a. Acetylcholine-sensitive cells in the cerebellar cortex. J. Physiol. (Lond.) 166: 296–327.

Krnjević, K., and Phillis, J.W. 1963b. Actions of certain amines on cerebral cortical neurones. Br. J. Pharmacol. 20: 471–490.

Krnjević, K., Pumain, R., and Renaud, L. 1971. The mechanism of excitation by acetylcholine in the cerebral cortex. J. Physiol. (Lond.) 215: 247–268.

Krnjević, K., Ropert, N., and Casullo, J. 1988. Septohippocampal disinhibition. Brain Res. 438: 182–192.

Krnjević, K., and Schwartz, S. 1967. The action of γ-aminobutyric acid on cortical neurones. Exp. Brain Res. 3: 320–336.

Kröller, J., and Grüsser, O.-J. 1982. Convergence of muscle spindle afferents on single neurons of the cat dorsal spino-cerebellar tract and their synaptic efficacy. Brain Res. 253: 65–80.

Kuba, K., and Minota, S. 1986. Presynaptic modulation: The mechanism and regulation of transmitter liberation in sympathetic ganglia. In: Autonomic and Enteric Ganglia, Transmission and Pharmacology (Kaczmarek, A.G., Koketsu, S., and Nishi, S., eds.). New York: Plenum Press, pp. 225–251.

Kuba, K., and Nishi, S. 1979. Characteristics of fast excitatory postsynaptic currents in bullfrog sympathetic ganglion cells. Pflugers Arch. 378: 205–212.

Kubo, T., Fukada, K., Mikami, A., Maeda, A., Takashashi, H., Mishina, M., Haga, K., Ichiyama, A., Kangawa, K., Kojima, M., Matsuo, H., Hirose, T., and Numa, S. 1986. Cloning, sequencing and expression of complementary DNA encoding the muscarinic acetylcholine receptor. Nature 323: 411–416.

Kuffler, S.W. 1953. Discharge patterns and functional organization of mammalian retina. J. Neurophysiol. 16: 37–68.

Kuffler, S.W., Nicolls, J.G., and Martin, A.R. 1984. From Neuron to Brain. Sunderland, MA: Sinauer Associates.

Kuffler, S.W., and Sejnowski, T.J. 1983. Peptidergic and muscarinic excitation at amphibian sympathetic synapses. J. Physiol. (Lond.) 341: 257–278.

Kuno, M. 1964. Quantal components of excitatory synaptic potentials in spinal motoneurones. J. Physiol. (Lond.) 175: 81–99.

Kuno, M., and Llinás, R. 1970. Enhancement of synaptic transmission by dendritic potentials in chromatolyzed motoneurones of the cat. J. Physiol. (Lond.) 210: 807–821.

Künzle, H. 1975. Bilateral projections from precentral motor cortex to the putamen and other parts of the basal ganglia. An autoradiographic study in *Macaca fascicularis*. Brain Res. 88: 195–209.

Lacaille, J.-C., Mueller, A.L., Kunkel, D.D., and Schwartzkroin, P.A. 1987. Local circuit interactions between oriens/alveus interneurons and CA1 pyramidal cells in hippocampal slices: Electrophysiology and morphology. J. Neurosci. 7: 1979–1993.

Lacaille, J.-C., and Schwartzkroin, P.A. 1988. Stratum lacunosum-moleculare inter-neurons of hippocampal Ca1 region. I. Intracellular response characteristics, synaptic responses, morphology. J. Neurosci. 8: 1400–1410.

Lacey, M.G., Mercuri, N.B., and North, R.A. 1987. Dopamine acts on D2 receptors to increase potassium conductance in neurons of the rat substantia nigra zona compacta. J. Physiol. (Lond.) 392: 397–416.

Lägerbäck, P.-Å., and Kellerth, J.-O. 1985a. Light microscopic observations on cat Renshaw cells after intracellular staining with horseradish peroxidase. I. The axonal systems. J. Comp. Neurol. 240: 359–367.

Lägerbäck, P.-Å., and Kellerth, J.-O. 1985b. Light microscopic observations on cat Renshaw cells after intracellular staining with horseradish peroxidase. II. The cell bodies and dendrites. J. Comp. Neurol. 240: 368–376.

Lägerbäck, P.-Å., Ronnevi, L.-O., Cullheim, S., and Kellerth, J.-O. 1981. An ultrastructural study of the synaptic contacts of alpha-motoneurone axon collaterals. I. Contacts in lamina IX and with identified alpha-motoneurone dendrites in lamina VII. Brain Res. 207: 247–266.

Lancaster, B., and Adams, P.R. 1986. Calcium-dependent current generating the after-hyperpolarization of hippocampal neurons. J. Neurophysiol. 55: 1268–1282.

Lancet, D. 1986. Vertebrate olfactory reception. Annu. Rev. Neurosci. 9: 329–355.

Land, E.H. 1959. Color vision and the natural image. Part 1. Proc. Natl. Acad. Sci. USA 45: 115–129.

Land, E.H. 1983. Recent advances in retinex theory and some implications for cortical computations: Color vision and the natural image. Proc. Natl. Acad. Sci. USA 80: 5163–5169.

Land, L.J., and Shepherd, G.M. 1974. Autoradiographic analysis of olfactory receptor projections in the rabbit. Brain Res. 70: 506–510.

Landis, D.M.D., and Reese, T.S. 1983. Cytoplasmic organization in cerebellar dendritic spines. J. Cell Biol. 97: 1169–1178.

Landmesser, L. 1978. The distribution of motoneurones supplying chick hind limb muscles. J. Physiol. (Lond.) 264: 371–389.

Laporte, Y., and Lloyd, D.P.C. 1952. Nature and significance of the reflex connections established by large afferent fibers of muscular origin. Am. J. Physiol. 169: 609–621.

Laughlin, S. 1987. Form and function in retinal processing. Trends Neurosci. 10: 478–483.

Laurberg, S., and Sorensen, K.E. 1981. Associational and commissural collaterals of neurons in the hippocampal formation (hilus fasciae dentatae and subfield CA3). Brain Res. 212: 287–300.

Lawrence, D.G., Porter, R., and Redman, S.J. 1985. Cortico-motoneuronal synapses in the monkey: Light microscopic localization upon motoneurons of intrinsic muscles of the hand. J. Comp. Neurol. 232: 499–510.

Legge, K.F., Randic, M., and Straughan, D.W. 1966. The pharmacology of neurones in the piriform cortex. Br. J. Pharmacol. 26: 87–107.

Lehky, S.R., and Sejnowski, T.J. 1988. Network model of shape-from-shading: Neural function arises from both receptive and projective fields. Nature 333: 452–454.

Lev-Tov, A., Meyers, D.E.R., and Burke, R.E. 1988. Activation of type B γ-aminobutyric acid receptors in the intact mammalian spinal cord mimics the effects of reduced presynaptic Ca^{2+} influx. Proc. Natl. Acad. Sci. USA 85: 5330–5334.

Lev-Tov, A., Pinter, M.J., and Burke, R.E. 1983. Post-tetanic potentiation of group Ia EPSPs: Possible mechanisms for differential distribution in the MG motor nucleus. J. Neurophysiol. 50: 379–398.

LeVay, S. 1973. Synaptic patterns in the visual cortex of the cat and monkey. Electron microscopy of Golgi preparations. J. Comp. Neurol. 150: 53–86.

LeVay, S., and Gilbert, C.D. 1976. Laminar patterns of geniculocortical projection in the cat. Brain Res. 113: 1–19.

Leveteau, J., and MacLeod, P. 1966. Olfactory discrimination in the rabbit olfactory glomerulus. Science 153: 175–176.

Levey, A.I., Hallanger, A.E., and Wainer, B.H. 1987. Choline acetyltransferase immunoreactivity in the rat thalamus. J. Comp. Neurol. 257: 317–332.

Liddell, E.G.T., and Sherrington, C.S. 1925. Recruitment and some other factors of reflex inhibition. Proc. R. Soc. Lond. [B] 97: 488–518.

Liesi, P. 1985. Laminin-immunoreactive glia distinguish regenerative adult CNS systems from non-regenerative ones. EMBO J. 4: 2505–2511.

Linsenmeier, R.A., Frishman, L.A., Jakiela, H.G., and Enroth-Cugell, C. 1982. Receptive field properties of X and Y cells in the cat retina derived from contrast sensitivity measurements. Vision Res. 22: 1173–1183.

Linsker, R. 1986. From basic network principles to neural architecture: Emergence of orientation columns. Proc. Natl. Acad. Sci. USA 83: 8779–8783.

Livingstone, M.S., and Hubel, D.H. 1987. Connections between layer 4B of area 17 and thick cytochrome oxidase stripes of area 18 in the squirrel monkey. J.Neurosci. 7: 3371–3377.

Livingstone, M.S., and Hubel, D.H. 1988. Segregation of form, color, movement, and depth: Anatomy, physiology, and perception. Science 240: 740–749.

Ljungdahl, A., Hökfelt, T., and Nilsson, G. 1978. Distribution of substance P-like immunoreactivity in the central nervous system of the rat. I. Cell bodies and nerve terminals. Neuroscience 3: 861–943.

Llinás, R.R. 1981. Electrophysiology of the cerebellar networks. In: Handbook of Physiology, Section 1: The Nervous System, Vol. II: Motor control, Part 2 (Brooks, V.B., ed.). Bethesda: American Physiological Society, pp. 831–876.

Llinás, R. 1985. Electrotonic transmission in the mammalian central nervous system. In: Gap Junctions (Bennett, M.V.L., and Spray, D.C., eds.). Cold Spring Harbor, NY: Cold Spring Harbor Laboratory, pp. 337–353.

Llinás, R. 1988. The intrinsic electrophysiological properties of mammalian neurons: Insights into central nervous system function. Science 242: 1654–1664.

Llinás, R., and Hess, R. 1976. Tetrodotoxin-resistant dendritic spikes in avian Purkinje cells. Soc. Neurosci. Abstr. 2: 112.

Llinás, R., and Mühlethaler, M. 1988a. An electrophysiological study of the in vitro, perfused brainstem–cerebellum of adult guinea pig. J. Physiol. (Lond.) 404: 215–240.

Llinás, R., and Mühlethaler, M. 1988b. An electrophysiological study of the in vitro, perfused brainstem–cerebellum of adult guinea pig. J. Physiol. (Lond.) 404: 241–268.

Llinás, R., and Nicholson, C. 1971. Electrophysiological properties of dendrites and somata in alligator Purkinje cells. J. Neurophysiol. 34: 532–551.

Llinás, R., and Nicholson, C. 1976. Reversal properties of climbing fiber potential in cat Purkinje cells: An example of a distributed synapse. J. Neurophysiol. 39: 311–323.

Llinás, R., and Sugimori, M. 1978. Dendritic calcium spiking in mammalian Purkinje cells: In vitro study of its function and development. Soc. Neurosci. Abstr. 4: 66.

Llinás, R., and Sugimori, M. 1980a. Electrophysiological properties of in vitro Purkinje cell somata in mammalian cerebellar slices. J. Physiol. (Lond.) 305: 171–195.

Llinás, R., and Sugimori, M. 1980b. Electrophysiological properties of in vitro Purkinje cell dendrites in mammalian cerebellar slices. J. Physiol. (Lond.) 305: 197–213.

Llinás, R., and Volkind, R.A. 1973. The olivo-cerebellar system: Functional properties as revealed by harmaline-induced tremor. Exp. Brain Res. 18: 69–87.

Llinás, R., and Yarom, Y. 1981a. Electrophysiological properties of mammalian inferior olivary cells in vitro: Different types of voltage-dependent conductances. J. Physiol. (Lond.) 315: 549–567

Llinás, R., and Yarom, Y. 1981b. Properties and distribution of ionic conductances generating electroresponsiveness of mammalian inferior olivary neurones in vitro. J. Physiol. (Lond.) 315: 569–584.

Llinás, R., and Yarom, Y. 1986. Oscillatory properties of guinea pig inferior olivary neurons and their pharmacological modulation: An in vitro study. J. Physiol. (Lond.) 376: 163–182.

Lloyd, D.P.C. 1960. Spinal mechanisms involved in somatic activities. In: Handbook of Physiology, Section 1: Neurophysiology, Vol. II (Magoun, H.W., ed.). Washington, D.C.: American Physiology Society, pp. 929–949.

Lo, F.-S., and Sherman, S.M. 1989. Dependence of retinogeniculate transmission on membrane voltage in the cat: Differences between X and Y cells. Eur. J. Neurosci.

Lopez, H.S., and Adams, P.R. 1989. A G protein mediates the inhibition of the voltage-dependent potassium M current by muscarine, LHRH, substance P and UTP in bullfrog sympathetic neurons. Eur. J. Neurosci. 1: 529–542.

Loren, I., Emson, P.C., Fahrenkrug, A., Björklund, J., Alumets, J., Hakanson, R., and Sundler, F. 1979. Distribution of vasoactive intestinal polypeptide in the rat and mouse brain. Neuroscience 4: 1953–1976.

Loy, R., Koziell, D.A., Lindsey, J.D., and Moore, R.Y. 1980. Noradrenergic innervation of the adult rat hippocampal formation. J. Comp. Neurol. 189: 699–710.

Lund, J.S. 1973. Organization of neurons in the visual cortex, area 17, of the monkey *(Macaca mulatta)*. J. Comp. Neurol. 147: 455–495.

Lund, J.S. 1987. Local circuit neurons of macaque monkey striate cortex. I. Neurons of laminae 4C and 5A. J. Comp. Neurol. 257: 60–92.

Lundberg, A. 1969. Convergence of excitatory and inhibitory action on interneurones in the spinal cord. In: The Interneuron, Vol. 2 (Brazier, M.A.B., ed.). UCLA Forum in Medical Sciences. Los Angeles: Univ. of Calif. Press, pp. 231–265.

Lundberg, A. 1971. Function of the ventral spinocerebellar tract, a new hypothesis. Exp. Brain Res. 12: 317–330.

Lundberg, A. 1975. Control of spinal mechanisms from the brain. In: The Nervous System: the Basic Neurosciences, Vol. I (Brady, R.O., ed.). New York: Raven Press, pp. 253–265.

Lundberg, A. 1979. Multisensory control of spinal reflex pathways. In: Reflex Control of Posture and Movement, Progress in Brain Research, Vol. 50 (Granit, R., and Pompeiano, O., eds.). Amsterdam: Elsevier, pp. 11–28.

Lundberg, A., Malmgren, K., and Schomburg, E.D. 1987. Reflex pathways from group II muscle afferents. 3. Secondary spindle afferents and the FRA; a new hypothesis. Exp. Brain Res. 65: 294–306.

Lundberg, A., and Weight, F.F. 1971. Functional organization of connections to the ventral spinocerebellar tracts. Exp. Brain Res. 12: 295–316.

Lundberg, J.M., and Hökfelt, T. 1983. Co-existence of peptides and classical neurotransmitters. Trends Neurosci. 6: 325–333.

Lüscher, H.-R., Ruenzel, P., and Henneman, E. 1983. Composite EPSPs in motoneurons of different sizes before and during PTP: Implications for transmission failure and its relief in Ia projections. J. Neurophysiol. 49: 269–289.

Luskin, M.B., and Price, J.L. 1982. The distribution of axon collaterals from the olfactory bulb and the nucleus of the horizontal limb of the diagonal band to the olfactory cortex, demonstrated by double retrograde labeling techniques. J. Comp. Neurol. 209: 249–263.

Luskin, M.B., and Price, J.L. 1983a. The topographic organization of associational fibers to the olfactory bulb. J. Comp. Neurol. 216: 264–291.

Luskin, M.B., and Price, J.L. 1983b. The laminar distribution of intracortical fibers originating in the olfactory cortex of the rat. J. Comp.Neurol. 216: 292–302.

Lynch, G., and Baudry, M. 1988. Structure–function relationships in the organization of memory. In: Perspectives in Memory Research (Gazzaniga, M.S., ed.). Cambridge, MA: MIT Press, pp. 23–92.

MacDermott, A.B., and Dale, N. 1987. Receptors, ion channels and synaptic potentials underlying the integrative actions of excitatory amino acids. Trends Neurosci. 10: 280–284.

Macrides, F., and Davis, B.J. 1983. The olfactory bulb. In: Chemical Neuroanatomy (Emson, P.C., ed.). New York: Raven Press, pp. 391–426.

Macrides, F., and Schneider, S.P. 1982. Laminar organization of mitral and tufted cells in the main olfactory bulb of the adult hamster. J. Comp. Neurol. 208: 419–430.

Macrides, F., Schoenfeld, T., Marchland, J., and Clancy, A. 1985. Evidence for morphologically, neurochemically and functionally heterogeneous classes of mitral and tufted cells in the olfactory bulb. Chem. Senses 10: 175–202.

MacVicar, V.A., and Dudek, F.E. 1981. Electrotonic coupling between pyramidal cells. A direct demonstration in rat hippocampal slices. Science 213: 782–785.

Madison, D.V., Lancaster, B., and Nicoll, R.A. 1987. Voltage clamp analysis of cholinergic action in the hippocampus. J. Neurosci. 7: 733–741.

Madison, D.V., Malenka, R.C., and Nicoll, R.A. 1986. Phorbol esters block a voltage-sensitive chloride current in hippocampal pyramidal cells. Nature 321: 695–697.

Madison, D.V., and Nicoll, R.A. 1982. Noradrenaline blocks accommodation of pyramidal cell discharge in hippocampus. Nature 291: 554–561.

Madison, D.V., and Nicoll, R.A. 1984. Control of the repetitive discharge of rat CA1 pyramidal neurones in vitro. J. Physiol. (Lond.) 354: 319–331.

Madison, D.V., and Nicoll, R.A. 1986a. Actions of noradrenaline recorded intracellularly in rat hippocampal CA1 pyramidal neurones, in vitro. J. Physiol. (Lond.) 372: 221–244.

Madison, D.V., and Nicoll, R.A. 1986b. Cyclic adenosine 3',5'-monophosphate mediates beta-receptor actions of noradrenaline in rat hippocampal pyramidal cells. J. Physiol. (Lond.) 372: 245–259.

Madison, D.V., and Nicoll, R.A. 1988. Norepinephrine decreases synaptic inhibition in the rat hippocampus. Brain Res. 442: 131–138.

Magistretti, P.J., Dietl, M.M., Hof, P.R., Martin, J.-L., Palacios, J.M., Schaad, N., and Schorderet, M. 1988. Vasoactive intestinal peptide as a mediator of intracellular communication in the cerebral cortex: Release, receptors, actions, and interactions with norepinephrine. Ann. N.Y. Acad. Sci. 527: 110–129.

Makowski, L., Casper, D.L.D., Phillips, W.C., and Goodenough, D.A. 1977. Gap junction structure. II. Analysis of x-ray diffraction data. J. Cell. Biol. 74: 629–645.

Malach, R., and Graybiel, A.M. 1986. Mosaic architecture of the somatic sensory-recipient sector of the cat's striatum. J. Neurosci. 6: 3436–3458.

Malenka, R.C., Kauer, J.A., Zucker, R.S., and Nicoll, R.A. 1988. Postsynaptic calcium is sufficient for potentiation of hippocampal synaptic transmission. Science 242: 81–84.

Malinow, R., and Miller, J.P. 1986. Postsynaptic hyperpolarization during conditioning reversibly blocks induction of long-term potentiation. Nature 320: 529–530.

Mancillas, J.R., Siggins, G.R., and Bloom, F.E. 1986. Somatostatin selectively enhances acetylcholine-induced excitations in rat hippocampus and cortex. Proc. Natl. Acad. Sci. USA 83: 7518–7521.

Margolis, F.L. 1988. Molecular cloning of olfactory-specific gene products. In: Molecular Neurobiology of the Olfactory System (Margolis, F.L., and Getchell, T.V., eds.). New York: Plenum Press, pp. 237–265.

Mariani, A.P. 1982. Biplexiform cells: Ganglion cells of the primate retina that contact photoreceptors. Science 216: 1134–1136.

Mariani, A.P. 1984. Bipolar cells in monkey retina selective for the cones likely to be blue-sensitive. Nature 308: 184–186.

Mariani, J., Crepel, F., Mikoshiba, K., Changeux, J.P., and Sotelo, C. 1977. Anatomical, physiological and biochemical studies of the cerebellum from Reeler mutant mouse. Philos. Trans. R. Soc. Lond. [B] 281: 1–28.

Markham, J.A., and Fifkova, E. 1986. Actin filament organization within dendrites and dendritic spines during development. Brain Res. 392: 263–269.

Marr, D. 1971. Simple memory: A theory for archicortex. Philos. Trans. R. Soc. Lond. [B] 262: 23–81.

Marr, D., and Poggio, T. 1979. A computational theory of human stereo vision. Proc. R. Soc. [B] 204: 301–328.

Marshall, L.M. 1981. Synaptic localization of a bungarotoxin binding which blocks nicotinic transmission at frog sympathetic neurones. Proc. Natl. Acad. Sci. USA 78: 1948–1952.

Martin, A.R. 1977. Functional transmission. II. Presynaptic mechanisms. In: Handbook of Physiology, Section 1: The Nervous System, Vol. I: The Cellular Biology of Neurons (Kandel, E.R., ed.). Bethesda: American Physiological Society, pp. 329–355.

Martin, K.A.C. 1984. Neuronal circuits in cat striate cortex. In: Cerebral Cortex, Vol. 2: Functional Properties of Cortical Cells (Jones, E.G., and Peters, A., eds.). New York, Plenum Press, pp. 241–284.

Martin, K.A.C. 1988a. The Wellcome Prize Lecture: From single cells to simple circuits in the cerebral cortex. Q. J. Exp. Physiol. 73: 637–702.

Martin, K.A.C. 1988b. From enzymes to visual perception: A bridge too far? Trends Neurosci. 11: 380–387.

Masland, R.H., Mills, J.W., and Cassidy, C. 1984. The functions of acetylcholine in the rabbit retina. Proc. R. Soc. Lond. [B] 223: 121–139.

Masland, R.H., and Tauchi, M. 1986. The cholinergic amacrine cell. Trends Neurosci. 9: 218–223.

Massey, S.C., and Miller, R. 1988. Glutamate receptors of ganglion cells in the rabbit retina: Evidence for glutamate as a bipolar cell transmitter. J. Physiol. (Lond.) 405: 635–655.

Massey, S.C., and Redburn, D. 1987. Transmitter circuits in the vertebrate retina. Prog. Neurobiol. 28: 55–96.

Mastronarde, D.N. 1983. Correlated firing of cat retinal ganglion cells. II. Responses of X- and Y-cells to single quantal events. J. Neurophysiol. 49: 325–349.

Masukawa, L.M., and Prince, D.A. 1982. Enkephalin inhibition of inhibitory input to CA1 and CA3 pyramidal neurons in the hippocampus. Brain Res. 249: 271–280.

Masukawa, L.M., and Prince, D.A. 1984. Synaptic control of excitability in isolated dendrites of hippocampal neurons. J. Neurosci. 4: 217–227.

Mates, S.L., and Lund, J.S. 1983. Neuronal composition and development of lamina 4C of monkey striate cortex. J. Comp. Neurol. 221: 60–90.

Matsushita, M., and Hosoya, Y. 1979. Cells of origin of the spinocerebellar tract in the rat, studied with the method of retrograde transport of horseradish peroxidase. Brain Res. 173: 185–200.

Matthews, D.A., Cotman, C., and Lynch, G. 1976. An electron microscopic study of lesion-induced synaptogenesis in the dentate gyrus of the adult rat I. Magnitude and time course of degeneration. Brain Res. 115: 1–21.

Matthews, M.A., Willis, W.D., and Williams, V. 1971. Dendrite bundles in lamina IX of cat spinal cord: A possible source for electrical interaction between motoneurons. Anat. Rec. 171: 313–327.

Matthews, P.B.C. 1972. Mammalian Muscle Receptors and Their Central Axons. Physiol. Soc. Monographs. London: Arnold.

Matthews, P.B.C. 1981. Muscle spindles: Their messages and their fusimotor supply. In: Handbook of Physiology, Section 1: The Nervous System, Vol. II: Motor Control, Part 1. Bethesda: American Physiological Society, pp. 189–228.

Maturana, H.R., Lettvin, J.Y., McCulloch, W.S., and Pitts, W.H. 1960. Anatomy and physiology of vision in the frog *(Rana pipiens)*. J. Gen. Physiol. 43: 129–175.

Mayer, M.L., and Westbrook, G.L. 1987. The physiology of excitatory amino acids in the vertebrate nervous system. Prog. Neurobiol. 28: 197–276.

Mayer, M.L., Westbrook, G.L., and Guthrie, P.B. 1984. Voltage-dependent block by Mg^{2+} of NMDA responses in spinal cord neurones. Nature 309: 261–263.

McArdle, C.B., Dowling, J.E., and Masland, R.H. 1977. Development of outer segments and synapses in the rabbit retina. J. Comp. Neurol. 175: 253–274.

McBride, W.J., Aprison, M.H., and Kusano, K. 1976. Contents of several amino acids in the cerebellum, brainstem and cerebrum of the "staggerer," "weaver" and "nervous" neurologically mutant mice. J. Neurochem. 26: 867–870.

McBurney, R.N., and Neering, I.R. 1985. The measurement of changes in intracellular free calcium during action potentials in mammalian neurones. J. Neurosci. Methods 13: 65–76.

McBurney, R.N., and Neering, I.R. 1987. Neuronal calcium homeostasis. Trends Neurosci. 10: 164–169.

McCormick, D.A., Connors, B.W., Lighthall, J.W., and Prince, D.A. 1985. Comparative electrophysiology of pyramidal and sparsely spiny stellate neurons of the neocortex. J. Neurophysiol. 59: 782–806.

McCormick, D.A., and Pape, H.-C. 1988. Acetylcholine inhibits identified interneurons in the cat lateral geniculate nucleus. Nature 334: 246–248.

McCormick, D.A., and Prince, D.A. 1985. Two types of muscarinic response to acetylcholine in mammalian cortical neurons. Proc. Natl. Acad. Sci. USA 82: 6344–6348.

McCormick, D.A., and Prince, D.A. 1986a. Acetylcholine induces burst firing in thalamic reticular neurones by activating a potassium conductance. Nature 319: 402–405.

McCormick, D.A., and Prince, D.A. 1986b. Mechanisms of action of acetylcholine in the guinea-pig cerebral cortex in vitro. J. Physiol. (Lond.) 375: 169–194.

McCormick, D.A., and Prince, D.A. 1987a. Actions of acetylcholine in the guinea-pig and cat medial and lateral geniculate nuclei, in vitro. J. Physiol. (Lond.) 392: 147–165.

McCormick, D.A., and Prince, D.A. 1987b. Acetylcholine causes rapid nicotinic excitation in the medial habenular nucleus of guinea-pig, in vitro. J. Neurosci. 7: 742–752.

McCormick, D.A., and Prince, D.A. 1988. Noradrenergic modulation of firing pattern in guinea-pig and cat thalamic neurons, in vitro. J. Neurophysiol. 59: 978–996.

McCormick, D.A., and Williamson, A. 1989. Convergence and divergence of neurotransmitter action in the human cerebral cortex. Proc. Natl. Acad. Sci. USA. 86: 8098–8102.

McGeer, P.L., McGeer, E.G., Scherer, U., and Singh, K. 1977. A glutamatergic corticostriatal path? Brain Res. 128: 369–373.

McGuire, B.A., Stevens, J.K., and Sterling, P. 1984. Microcircuitry of bipolar cells in cat retina. J. Neurosci. 4: 2920–2938.

McGuire, B.A., Stevens, J.K., and Sterling, P. 1986. Microcircuitry of beta ganglion cells in cat retina. J. Neurosci. 6: 907–918.

McIntyre, D.C., and Wong, R.K.S. 1986. Cellular and synaptic properties of amygdala kindled pyriform cortex in vitro. J. Neurophysiol. 55: 1295–1307.

McLachlan, E. 1978. The statistics of transmitter release at chemical synapses. In: International Review of Physiology: Neurophysiology III, Vol. 17 (Porter, R., ed.). Baltimore: University Park Press, pp. 49–117.

McLaughlin, B.J., Barber, R., Saito, K., Roberts, E., and Wu, J.Y. 1975. Immunocytochemical localization of glutamine decarboxylase in rat spinal cord. J. Comp. Neurol. 164: 305–322.

McLaughlin, B.J., Woods, J.G., Saito, K., Barber, R., Vaughn, J.E., Roberts, E., and Wu, J. 1974. The fine structural localization of glutamate decarboxylase in synaptic terminals of rodent cerebellum. Brain Res. 76: 377–391.

McLennan, H. 1971. The pharmacology of inhibition of mitral cells in the olfactory bulb. Brain Res. 29: 177–187.

McLennan, H. 1983. Receptors for the excitatory amino acids in the mammalian central nervous system. Prog. Neurobiol. 20: 251–271.

McWilliams, J.R., and Lynch, G.S. 1979. Terminal proliferation in the partially deafferented dentate gyrus: Time courses for the appearance and removal of degeneration and the replacement of lost terminals. J. Comp. Neurol. 187: 191–198.

Mead, C., and Conway, L. 1980. Introduction to VLSI Systems. Reading, MA: Addison-Wesley.

Mendell, L.M. 1984. Modifiability of spinal synapses. Physiol. Rev. 64: 260–324.

Mendell, L.M., and Henneman, E. 1971. Terminals of single Ia fibers: Location, density and distribution within a pool of 300 homonymous motoneurons. J. Neurophysiol. 34: 171–187.

Merzenich, M.M., Nelson, R.J., Stryker, M.P., Cynader, M.S., Shoppmann, A., and Zook, J.M. 1984. Somatosensory cortical map changes following digital amputation in adult monkey. J. Comp. Neurol. 224: 591–605.

Mesulam, M.M., Mufson, E.J., Levey, A.I., and Wainer, B.H. 1983. Cholinergic innervation of cortex by the basal forebrain: Cytochemistry and cortical connections of the septal area, diagonal band nucleus, nucleus basalis (substantia innominata), and hypothalamus in the rhesus monkey. J. Comp. Neurol. 214: 170–197.

Miles, R., and Wong, R.K.S. 1984. Unitary inhibitory synaptic potentials in the guinea-pig hippocampus in vitro. J. Physiol (Lond.) 356: 97–113.

Miles, R., and Wong, R.K.S. 1986. Excitatory synaptic interactions between CA3 neurones in the guinea pig hippocampus. J. Physiol. (Lond.) 373: 397–418.

Miles, R., and Wong, R.K.S. 1987. Inhibitory control of local excitatory circuits in the guinea-pig hippocampus. J. Physiol. (Lond.) 388: 611–629.

Miller, J.P., Rall, W., and Rinzel, J. 1985. Synaptic amplification by active membrane in dendritic spines. Brain Res. 325: 325–330.

Miller, R.F. 1988. Are single retinal neurons both excitatory and inhibitory? Nature 336: 517–519.

Miller, R.F., and Bloomfield, S.A. 1983. Electroanatomy of a unique amacrine cell in the rabbit retina. Proc. Natl. Acad. Sci. USA 80: 3069–3073.

Miller, R.F., and Slaughter, M. 1986. Excitatory amino acid receptors of the retina: Diversity of subtypes and conductance mechanisms. Trends Neurosci. 9: 211–218.

Mishkin, M., and Petri, H.L. 1984. Memories and habits: Some implications for the analysis of learning and retention. In: Neuropsychology of Memory (Squire, L.R., and Butters, N., eds.). New York: Guilford Press, pp. 287–296.

Mitchell, S.J., and Ranck, J.B. 1980. Generation of theta rhythm in medial entorhinal cortex of freely moving rats. Brain Res. 189: 49–66.

Mitzdorf, U., and Singer, W. 1978. Prominent excitatory pathways in the cat visual cortex (A17 and A18): A current source density analysis of electrically evoked potentials. Exp. Brain Res. 33: 371–394.

Moczydlowski, E., and Latorre, R. 1983. Gating kinetics of Ca^{2+}-activated K^+ channels from rat muscle incorporated into planar lipid bilayers: Evidence for two voltage-dependent Ca^{2+} binding reactions. J. Gen. Physiol. 82: 511–542.

Monaghan, D.T., and Cotman, C.W. 1982. The distribution of [³H]kainic acid binding sites in rat CNS as determined by autoradiography. Brain Res. 252: 91–100.

Monaghan, D.T., and Cotman, C.W. 1985. Distribution of N-methyl-D-aspartate-sensitive L-[³H]glutamate-binding sites in rat brain. J. Neurosci. 5: 2909–2919.

Monaghan, D.T., Yao, D., and Cotman, C.W. 1984. Distribution of [³H]AMPA binding sites in rat brain as determined by quantitative autoradiography. Brain Res. 324: 160–164.

Montero, V.M. 1987. Ultrastructural identification of synaptic terminals from the axon of type 3 interneurons in the cat lateral geniculate nucleus. J. Comp. Neurol. 264: 268–283.

Monti-Graziadei, G., Stanley, R., and Graziadei, P. 1980. The olfactory marker protein in the olfactory system of the mouse during development. Neuroscience 5: 1239–1252.

Moody, C.I., and Sillito, A.M. 1988. The role of n-methyl-D-aspartate (NMDA) receptor in the transmission of visual information in the feline dorsal lateral geniculate nucleus (dLGN). J. Physiol. (Lond.) 396: 62P.

Moore, R.Y., and Bloom, F.E. 1979. Central catecholamine neuron systems: Anatomy and physiology of the norepinephrine and epinephrine systems. Annu. Rev. Neurosci. 2: 113–168.

Moore, R.Y., and Card, J.P. 1984. Noradrenaline-containing neuron systems. In: Handbook of Chemical Neuroanatomy, Vol. 2: Classical Transmitters in the CNS, Part I (Björklund, A., and Hökfelt, T., eds.). Amsterdam: Elsevier, pp. 123–156.

Moran, D., Rowles, J., and Jafek, B. 1982. Electronmicroscopy of human olfactory epithelium reveals a new cell type: The microvillar cell. Brain Res. 253: 39–46.

Morgan, J.I. 1988. Monoclonal antibody mapping of the rat olfactory tract. In: Molecular Neurobiology of the Olfactory System (Margolis, F.L., and Getchell, T.V., eds.). New York: Plenum Press, pp. 269–296.

Mori, K. 1987. Membrane and synaptic properties of identified neurons in the olfactory bulb. Prog. Neurobiol. 29: 275–320.

Mori, K., Kishi, K., and Ojima, H. 1983. Distribution of dendrites of mitral, displaced mitral, tufted and granule cells in the rabbit olfactory bulb. J. Comp. Neurol. 219: 339–355.

Mori, K., Nowycky, M.C., and Shepherd, G.M. 1981a. Electrophysiological analysis of mitral cells in isolated turtle olfactory bulb. J. Physiol. (Lond.) 314: 281–294.

Mori, K., Nowycky, M.C., and Shepherd, G.M. 1981b. Analysis of synaptic potentials in mitral cells of the isolated turtle olfactory bulb. J. Physiol. (Lond.) 314: 295–309.

Mori, K., Nowycky, M.C., and Shepherd, G.M. 1981c. Analysis of long duration inhibitory potential in mitral cells in the isolated turtle olfactory bulb. J. Physiol. (Lond.) 314: 311–320.

Moruzzi, G., and Magoun, H.W. 1949. Brain stem reticular formation and activation of the EEG. Electroencephal. Clin. Neurophysiol. 1: 455–473.

Mugnaini, E., and Oertel, W.H. 1981. Distribution of glutamate decarboxylase positive neurons in the rat cerebellar nuclei. Soc. Neurosci. Abstr. 7: 122.

Mugnaini, E., and Oertel, W.H. 1985. An atlas of the distribution of GABAergic neurons and terminals in the rat CNS as revealed by GAD immunohistochemistry. In: Handbook of Chemical Neuroanatomy, Vol. 4: GABA and Neuropeptides in the CNS, Part I (Bjorklund, A., and Hökfelt, T., eds.). Amsterdam: Elsevier, pp. 436–622.

Mugnaini, E., Oertel, W.H., and Wouterlood, F.F. 1984. Immunocytochemical localization of GABA neurons and dopamine neurons in the rat main and accessory olfactory bulbs. Neurosci. Lett. 47: 221–226.

Muir, R.B., and Porter, R. 1973. The effect of a preceding stimulus on temporal facilitation at corticomotoneuronal synapses. J. Physiol. (Lond.) 228: 749–763.

Murthy, K.S.K., Ledbetter, W.D., Eidelberg, E., Cameron, W.E., and Petit, J. 1982. Histochemical evidence for the existence of skeletofusimotor (β) innervation in the primate. Exp. Brain Res. 46: 186–190.

Nadel, L., Willner, J., and Kurz, E.M. 1985. Cognitive maps and environmental context. In: Context and Learning (Balsam, P.D., and Tomie, A., eds.). Hillsdale, NJ: Lawrence Erlbaum Associates, pp. 385–406.

Nakamura, Y., McGuire, B.A., and Sterling, P. 1980. Interplexiform cell in cat retina: Identification by uptake of γ-[3H]aminobutyric acid and serial reconstruction. Proc. Natl. Acad. Sci. USA 77: 658–661.

Nakanishi, H., Kita, H., and Kitai, S.T. 1987. Intracellular study of rat substantia nigra pars reticulata neurons in an in vitro slice preparation: Electrical membrane properties and response characteristics to subthalamic stimulation. Brain Res. 437: 45–55.

Nathans, J., Thomas, D., and Hogness, C. 1986. Molecular genetics of human color vision: The genes encoding blue, green and red pigments. Science 232: 193–210.

Nawy, S., and Copenhagen, D.R. 1987. Multiple classes of glutamate receptor on depolarizing bipolar cells in retina. Nature 325: 56–58.

Neer, E.J., and Clapman, D.E. 1988. Roles of G protein subunits in transmembrane signalling. Nature 333: 129–134.

Nelson, P.G. 1966. Interaction between spinal motoneurons of the cat. J. Neurophysiol. 29: 275–287.

Nelson, R. 1977. Cat cones have rod input: A comparison of the response properties of cones and horizontal cell bodies in the retina of the cat. J. Comp. Neurol. 172: 109–136.

Nelson, R., Famiglietti, E.V., Jr., and Kolb, H. 1978. Intracellular staining reveals different levels of stratification for on-center and off-center ganglion cells in cat retina. J. Neurophysiol. 41: 472–483.

Nelson, R., and Kolb, H. 1983. Synaptic patterns and response properties of bipolar and ganglion cells in the cat retina. Vision Res. 23: 1183–1195.

Newberry, N.R., and Nicoll, R.A. 1984. A bicuculline resistant inhibitory postsynaptic potential in rat hippocampal pyramidal cells in vitro. J. Physiol. (Lond.) 348: 239–254.

Newberry, N.R., and Nicoll, R.A. 1985. Comparison of the action of baclofen with gamma-aminobutyric acid on rat hippocampal pyramidal cells in vitro. J. Physiol. (Lond.) 360; 161–185.

Nicoll, R.A. 1971. Pharmacological evidence for GABA as the transmitter in granule cell inhibition in the olfactory bulb. Brain Res. 35: 137–149.

Nicoll, R.A. 1982. Neurotransmitters say more than "yes" or "no." Trends Neurosci. 5: 369.

Nicoll, R.A. 1988. The coupling of neurotransmitter receptors to ion channels in the brain. Science 241: 545–551.

Nicoll, R.A., Alger, B.E., and Jahr, C.E. 1980. Peptides as putative excitatory neurotransmitters: Carnosine, enkephalin, substance P and TRH. Proc. R. Soc. Lond. [B] 210: 133–149.

Nicoll, R.A., Kauer, J.A., and Malenka, R.C. 1988. The current excitement in long-term potentiation. Neuron 1: 97–103.

Nishi, S., Soeda, H., and Koketsu, K. 1965. Studies in sympathetic B and C neurons and patterns of preganglionic innervation. J. Cell. Comp. Physiol. 66: 19–32.

Nishimura, Y., and Rakic, P. 1987a. Development of the rhesus monkey retina II. A three-dimensional analysis of the sequences of synaptic combinations in the inner plexiform layer. J. Comp. Neurol. 262: 290–313.

Nishimura, Y., and Rakic, P. 1987b. Synaptogenesis in primate retina proceeds from the ganglion cells towards the photoreceptors. Neurosci. Res. [Suppl.] 6: S253–S268.

Nistri, A. 1983. Spinal cord pharmacology of GABA and chemically related amino acids. In: Handbook of the Spinal Cord, Vol. I: Pharmacology (Davidoff, R.E., ed.). New York: Marcel Dekker, pp. 45–104.

North, R.A. 1987. Receptors of individual neurones. Neuroscience 17: 899–907.

Nowak, L., Bregestovski, P., Ascher, P., Herbet, A., and Prochiantz, A. 1984. Magnesium gates glutamate-activated channels in mouse central neurones. Nature 307: 462–465.

Nowycky, M.C., Fox, A.P., and Tsien, R.W. 1985. Three types of neuronal calcium channel with different calcium agonist sensitivity. Nature 316: 440–443.

Nowycky, M.C., Halasz, N., and Shepherd, G.M. 1981a. Studies of dopamine as a neurotransmitter in the turtle olfactory bulb. Soc. Neurosci. Abstr. 7: 573.

Nowycky, M.C., Mori, C., and Shepherd, G.M. 1981b. Blockade of synaptic inhibition reveals long-lasting synaptic excitation in isolated turtle olfactory bulb. J. Neurophysiol. 47: 649–658.

Nowycky, M.C., Mori, K., and Shepherd, G.M. 1981c. GABAergic mechanisms of dendrodendritic synapses in isolated turtle olfactory bulb. J. Neurophysiol. 46: 639–648.

Nowycky, M.C., Waldow, U., and Shepherd, G.M. 1978. Electrophysiological studies in the isolated turtle brain. Soc. Neurosci. Abstr. 4: 583.

Numann, R.E., Wadman, W.J., and Wong, R.K.S. 1987. Outward currents of single hippocampal cells obtained from the adult guinea pig. J. Physiol. (Lond.) 393: 331–354.

Obata, K., Ito, M., Ochi, R., and Sato, N. 1967. Pharmacological properties of the postsynaptic inhibition by Purkinje cell axons and the action of γ-aminobutyric acid on Dieters neurons. Exp. Brain Res. 4: 43–57.

Obata, K.T., and Shinozaki, H. 1970. Further study on pharmacological properties of the cerebellar-induced inhibition of Dieters neurones. Exp. Brain Res. 11: 327–342.

Oertel, W.H., and Mugnaini, E. 1984. Immunocytochemical studies of GABAergic neurons in rat basal ganglia and their relations to other neuronal systems. Neurosci. Lett. 47: 233–238.

Oertel, W.H., Schmechel, D.E., Mugnaini, E., Tappaz, M.L., and Kopin, I.J. 1981. Immunocytochemical localization of glutamate decarboxylase in rat cerebellum with a new antiserum. Neuroscience 6: 2715–2735.

Oertel, W.H., Tappaz, M.L., Berod, A., and Mugnaini, E. 1982. Two-color immunohistochemistry for dopamine and GABA neurons in rat substantia nigra and zona incerta. Brain Res. Bul. 9: 463–474.

Ogawa, T.S., Ito, S., and Kato, H. 1981. Membrane characteristics of visual cortical neurons in in vitro slices. Brain Res. 226: 315–319.

Ohara, P.T., and Lieberman, A.R. 1985. The thalamic reticular nucleus of the adult rat: Experimental anatomical studies. J. Neurocytol. 14: 365–411.

Ohara, P.T., Lieberman, A.R., Hunt, S.P., and Wu, J.Y. 1983. Neural elements containing glutamic acid decarboxylase (GAD) in the dorsal lateral geniculate nucleus of the rat: Immunohistochemical studies by light and electron microscopy. Neuroscience 8: 189–211.

Ohtsuka, T. 1985. Spectral sensitivities of seven morphological types of photoreceptors in the retina of the turtle, *Geoclemys reevesii*. J. Comp. Neurol. 237: 145–154.

Ojima, H., Mori, K., and Kishi, K. 1984. The trajectory of mitral cell axons in the rabbit olfactory cortex revealed by intracellular HRP injection. J. Comp. Neurol. 230: 77–87.

Okamoto, K., Quastel, D.M.J., and Quastel, J.H. 1976. Action of amino acids and convulsants on cerebellar spontaneous action potentials in vitro: effects of deprivation of Cl^-, K^+ or Na^+. Brain Res. 113: 147–158.

Okamoto, K., and Sakai, Y. 1981. Inhibitory actions of taurocyamine, hypotaurine, homotaurine, taurine and GABA on spike discharges of Purkinje cells, and localization of sensitive site, in guinea pig cerebellar slices. Brain Res. 206: 371–386.

O'Keefe, J. 1989. Computations the hippocampus might perform. In: Neural Connections, Mental Computation (Nadel, L., Cooper, L.A., Culicover, P., and Harnish, R.M., eds.). Cambridge, MA: MIT Press, pp. 225–284.

O'Keefe, J., and Nadel, L. 1978. The Hippocampus as a Cognitive Map. Oxford: Clarendon Press.

O'Leary, J.L. 1937. Structure of the primary olfactory cortex of the mouse. J. Comp. Neurol. 67: 1–31.

Olpe, H.R. 1981. The cortical projection of the dorsal raphe nucleus: Some electrophysiological and pharmacological properties. Brain Res. 216: 61–71.

Olton, D.S. 1983. Memory functions and the hippocampus. In: Neurobiology of the Hippocampus (Seifert, W., ed.). New York: Academic Press, pp. 335–373.

Orona, E., Rainer, E., and Scott, J. 1984. Dendritic and axonal organization of mitral and tufted cells in the rat olfactory bulb. J. Comp. Neurol. 226: 346–356.

Orona, E., Scott, J., and Rainer, E. 1983. Different granule cell populations innervate superficial and deep regions of the external plexiform layer in rat olfactory bulb. J. Comp. Neurol. 217: 227–237.

Ottersen, O.P., Fisher, B.O., and Storm-Mathisen, J. 1983. Retrograde transport of D-(^{3}H)aspartate in thalamocortical neurones. Neurosci. Lett. 42: 19–24.

Palay, S.L., and Chan-Palay, V. 1974. Cerebellar Cortex: Cytology and Organization. New York: Springer-Verlag.

Palkovits, M., and Brownstein, M.J. 1985. Distribution of neuropeptides in the central nervous system using biochemical micromethods. In: Handbook of Chemical Neuroanatomy, Vol. 4: GABA and Neuropeptides in the CNS, Part I (Björklund, A., and Hökfelt, T., eds.). Amsterdam: Elsevier, pp. 1–71.

Palkovits, M., Mezey, E., Hámori, J., and Szentágothai, J. 1977. Quantitative histological analysis of the cerebellar nuclei in the cat. I. Numerical data on cells and on synapses. Exp. Brain Res. 28: 189–209.

Passingham, R. 1982. The Human Primate. Oxford: W.H. Freeman.

Pasternak, T., and Merigan, W.H. 1981. The luminance dependence of spatial vision in the cat. Vision Res. 21: 1333–1339.

Pedersen, P.E., Greer, C.A., and Shepherd, G.M. 1985. Early development of olfactory function. In: Handbook of Behavioral Neurobiology, Vol. 8 (Blass, E.M., ed.). New York: Plenum Press, pp. 163–203.

Pedersen, P.E., Shepherd, G.M., and Greer, C.A. 1987. Cytochrome oxidase staining in the olfactory epithelium and bulb of normal and odor-deprived neonatal rats. In: Olfaction and Taste IX (Roper, S., and Atema, J., eds.). Annu. N.Y. Acad. Sci. 510; 544–546.

Penny, G.R., Afsharpo, S., and Kitai, S.T. 1986a. Substance P-immunoreactive neurons in the neocortex of the rat. Subset of the glutamic acid decarboxylase-immunoreactive neurons. Neurosci. Lett. 65: 53–59.

Penny, G.R., Afsharpo, S., and Kitai, S.T. 1986b. The glutamate decarboxylase-immunoreactive, met-enkephalin-immunoractive, and substance P-immunoreactive neurons in the neostriatum of the rat and cat. Evidence for partial population overlap. Neuroscience 17: 1011–1045.

Penny, G.R., Wilson, C.J., and Kitai, S.T. 1988. Relationship of the axonal and dendritic geometry of spiny projection neurons to the compartmental organization of the neostriatum. J. Comp. Neurol. 269: 275–289.

Perkins, M.N., and Stone, T.W. 1985. Actions of kynurenic acid and quinolinic acid in the rat hippocampus in vivo. Exp. Neurol. 88: 570–579.

Peters, A., 1984. Chandelier cells. In: Cerebral Cortex, Vol. 1: Cellular Components of the Cerebral Cortex (Peters, A., and Jones, E.G., eds.). New York: Plenum Press, pp. 361–380.

Peters, A., 1987. Number of neurons and synapses in the primary visual cortex. In: Cerebral Cortex, Vol. 6: Further Aspects of Cortical Functions Including Hippocampus (Jones, E.G., and Peters, A., eds.). New York: Plenum Press, pp. 267–294.

Peters, A., and Kaiserman-Abramof, I.R. 1970. The small pyramidal neuron of the rat cerebral cortex: The perikaryon, dendrites and spines. Am. J. Anat. 127: 321–356.

Peters, A., and Regidor, J. 1981. A reassessment of the forms of nonpyramidal neurons in area 17 of cat visual cortex. J. Comp. Neurol. 203: 685–716.

Pfaffinger, P.J. 1988. Muscarine and t-LHRH suppress M-current by activating an IAP-insensitive G protein. J. Neurosci. 8: 3343–3353.

Phelps, P.E., Houser, C.R., and Vaughn, J.E. 1985. Immunocytochemical localization of choline acetyltransferase within the rat neostriatum: A correlated light and electron microscopic study of cholinergic neurons and synapses. J. Comp. Neurol. 238: 286–307.

Phillips, C.G. 1959. Actions of antidromic pyramidal volleys on single Betz cells in the cat. Q. J. Exp. Physiol. 44: 1–25.

Phillips, C.G. 1969. Motor apparatus of the baboon's hand. Proc. R. Soc. Lond. [B] 173: 141–174.

Phillips, C.G. 1973. Pyramidal apparatus for control of the baboon's hand. In: New Developments in Electromyography and Clinical Neurophysiology, Vol. 3 (Desmedt, J.E., ed.). Basel: Karger, pp. 136–144.

Phillips, C.G., Powell, T.P.S., and Shepherd, G.M. 1963. Responses of mitral cells to stimulation of the lateral olfactory tract in the rabbit. J. Physiol. (Lond.) 168: 65–88.

Phillis, J.W. 1968. Acetylcholinesterase in the feline cerebellum. J. Neurochem. 15: 691–698.

Phillis, J.W., and Wu, P.H. 1981. Catecholamines and the sodium pump in excitable cells. Prog. Neurobiol. 17: 141–184.

Piccolino, M. 1988. Cajal and the retina: A 100-year retrospective. Trends Neurosci. 11: 521–525.

Piccolino, M., and Gerschenfeld, H.M. 1980. Characteristics and ionic processes involved in feedback spikes of turtle cones. Proc. R. Soc. Lond. [B] 206: 439–463.

Pikel, V.M., Segal, M., and Bloom, F.E. 1974. A radioautographic study of the efferent pathways of the nucleus locus coeruleus. J. Comp. Neurol. 155: 15–42.

Pinching, A.J. 1971. Myelinated dendritic segments in the monkey olfactory bulb. Brain Res. 29: 133–138.

Pinching, A.J., and Powell, T.P.S. 1971a. The neuron types of the glomerular layer of the olfactory bulb. J. Cell. Sci. 9: 305–345.

Pinching, A.J., and Powell, T.P.S. 1971b. The neuropil of the glomeruli of the olfactory bulb. J. Cell. Sci. 9: 347–377.

Pinter, M.J., Burke, R.E., O'Donovan, M.J., and Dum, R.P. 1982. Supraspinal facilitation of cutaneous polysynaptic EPSPs in cat medial gastrocnemius motoneurons. Exp. Brain Res. 45: 133–143.

Piredda, S., and Gale, K. 1985. A crucial epileptogenic site in the deep prepiriform cortex. Nature 317: 623–625.

Porter, R. 1973. Functions of the mammalian cerebral cortex in movement. In: Progress in Neurobiology, Vol. I (Kerkut, G.A., and Phillis, J.W., eds.). New York: Pergamon Press, pp. 1–51.

Porter, R. 1987. Corticomotoneuronal projections: Synaptic events related to skilled movement. Proc. R. Soc. Lond. [B] 231: 147–168.

Potter, D.D., Landis, S.C., and Furshpan, E.J. 1981. Adrenergic–cholinergic dual function in cultured sympathetic neurons of the rat. Ciba Found. Symp. 83: 135–150.

Pourcho, R.G. 1982. Dopaminergic amacrine cells in the cat retina. Brain Res. 252: 101–109.

Pourcho, R.G., and Goebel, D.J. 1985. A combined Golgi and autoradiographic study of (^{3}H)glycine-accumulating amacrine cells in the cat retina. J. Comp. Neurol. 233: 473–480.

Pourcho, R.G., and Goebel, D.J. 1987. A combined Golgi and autoradiographic study of ^{3}H-glycine-accumulating cone bipolar cells in cat retina. J. Neurosci. 7: 1178–1188.

Powell, T.P.S. 1981. Certain aspects of the intrinsic organization of the cerebral cortex. In: Brain Mechanisms and Perceptual Awareness (Pompeiano, O., and Ajmone Marsan, C., eds.). New York: Raven Press, pp. 1–19.

Powell, T.P.S., Cowan, W.M., and Raismann, G. 1965. The central olfactory connections. J. Anat. 99: 791–813.

Powell, T.P.S., and Mountcastle, V.B. 1959. Some aspects of the functional organization of the cortex of the postcentral gyrus of the monkey: A correlation of findings obtained in a single unit analysis with cytoarchitecture. Bull. Johns Hopkins Hospital 105: 133–162.

Precht, W., and Yoshinda, M. 1971. Blockage of caudate-evoked inhibition of neurons in the substantia nigra by picrotoxin. Brain Res. 32: 229–233.

Preston, R.J., Bishop, G.A., and Kitai, S.T. 1980. Medium spiny neurons from rat striatum: An intracellular horseradish peroxidase study. Brain Res. 183: 253–263.

Pribram, K.H., and Kruger, L. 1954. Functions of the "olfactory brain." Ann. N.Y. Acad. Sci. 58: 109–138.

Price, J.L. 1973. An autoradiographic study of complementary laminar patterns of termination of afferent fibers to the olfactory cortex. J. Comp. Neurol. 150: 87–108.

Price, J.L. 1985. Beyond the primary olfactory cortex: Olfactory-related areas in the neocortex, thalamus and hypothalamus. Chem. Senses 10: 239–258.

Price, J.L. 1987. The central olfactory and accessory olfactory systems. In: Neurobiology of Taste and Smell (Finger, T.E., and Silver, W.L., eds.). New York: John Wiley, pp. 179–203.

Price, J.L., and Powell, T.P.S. 1970a. The morphology of granule cells of the olfactory bulb. J. Cell Sci. 7: 91–123.

Price, J.L., and Powell, T.P.S. 1970b. The synaptology of the granule cells of the olfactory bulb. J. Cell Sci. 7: 125–155.

Price, J.L., and Sprich, W.W. 1975. Observations on the lateral olfactory tract of the rat. J. Comp. Neurol. 162: 321–336.

Prince, D.A., and Huguenard, J.R. 1988. Functional properties of neocortical neurons. In: Neurobiology of Neocortex (Rakic, P., and Singer, W., eds.). Chichester: John Wiley, pp. 153–176.

Puil, E. 1983. Actions and interactions of S-glutamate in the spinal cord. In: Handbook of the Spinal Cord, Vol. 1: Pharmacology (Davidoff, R.A., ed.). New York: Marcel Dekker, pp. 105–169.

Purkinje, J.E. 1837. Uber die Gangliösen Körperchen in Verschiedenen Theilen des Gehirns. Ber Versamm. dt., Naturf. S. 179.

Purves, D., and Lichtman, J.W. 1985. Principles of Neural Development. Sunderland, MA: Sinauer Associates.

Purves, D., Hadley, R.D., and Voydovic, J.T. 1986a. Dynamic changes in the dendritic geometry of individual neurons visualized over periods of up to three months in the superior cervical ganglion of living mice. J. Neurosci. 6: 1051–1060.

Purves, D., Rubin, E., Snider, W.D., and Lichtman, J.W. 1986b. Relation of animal size to convergence, divergence and neuronal number in peripheral sympathetic pathways. J. Neurosci. 6: 158–163.

Qian, N., and Sejnowski, T.J. 1989. An electro-diffusion model for computing membrane potentials and ionic concentrations in branching dendrites, spines and axons. Biol. Cybern.

Racine, R.J., Mosher, M., and Kairiss, E.W. 1988. The role of the pyriform cortex in the generation of interictal spikes in the kindled preparation. Brain Res. 454: 251–263.

Rack, P.M.H. 1981. Limitations of somatosensory feedback in control of posture and movement. In: Handbook of Physiology, Section 1: The Nervous System, Vol. II: Motor Control, Part 1 (Brooks, V.B., ed.). Bethesda: American Physiological Society, pp. 229–256.

Ragsdale, C.W., Jr., and Graybiel, A.M. 1981. The fronto-striatal projection in the cat and monkey and its relationship to inhomogeneities established by acetylcholinesterase histochemistry. Brain Res. 208: 259–266.

Rainbow, T.C., Wieczorek, C.M., and Halpain, S. 1984. Quantitative autoradiography of binding sites for [^{3}H]AMPA, a structural analogue of glutamic acid. Brain Res. 309: 173–177.

Rakic, P. 1976. Prenatal genesis of connections subserving ocular dominance in the rhesus monkey. Nature 261: 467–471.

Rakic, P. 1986. Mechanisms of ocular dominance segregation of the lateral geniculate nucleus: Competitive elimination hypothesis. Trends Neurosci. 9: 11–15.

Rakic, P., and Riley, K.P. 1983. Overproduction and elimination of retinal axons in the fetal rhesus monkey. Science 219: 1441–1444.

Rall, W. 1957. Membrane time constant of motoneurons. Science 126: 454–455.

Rall, W. 1959a. Dendritic current distribution and whole neuron properties. Naval Med. Res. Inst., Research Report NM 01-05-00.01.01.

Rall, W. 1959b. Branching dendritic trees and motoneuron membrane resistivity. Exp. Neurol. 1: 491–527.

Rall, W. 1962. Electrophysiology of a dendritic neuron model. Biophys. J. 2: 145–167.

Rall, W. 1964. Theoretical significance of dendritic trees for neuronal input–output relations. In: Neural Theory and Modelling (Reiss, R.F., ed.). Stanford: Stanford University Press, pp. 73–97.

Rall, W. 1967. Distinguishing theoretical synaptic potentials computed for different soma–dendritic distributions of synaptic input. J. Neurophysiol. 30: 1138–1168.

Rall, W. 1969. Time constants and electrotonic length of membrane cylinders and neurons. Biophys. J. 9: 1483–1508.

Rall, W. 1970. Cable properties of dendrites and effects of synaptic location. In: Excitatory Synaptic Mechanisms (Andersen, P., and Jansen, J.K.S., eds.). Oslo: Universitetsforlag, pp. 175–187.

Rall, W. 1974a. Dendritic spines and synaptic potency. In: Studies in Neurophysiology (Porter, R., ed.). Cambridge: Cambridge University Press, pp. 203–209.

Rall, W. 1974b. Dendritic spines, synaptic potency in neuronal plasticity. In: Cellular Mechanisms Subserving Changes in Neuronal Activity (Woody, C.D., Brown, K.A., Crow, T.J., and Knispel, J.D., eds.). Los Angeles: Brain Information Service, pp. 13–21.

Rall, W. 1977. Core conductor theory and cable properties of neurons. In: The Nervous

System, Vol. I: Cellular Biology of Neurons, Part 1 (Kandel, E.R., ed.). Bethesda: American Physiological Society, pp. 39–97.

Rall, W. 1978. Dendritic spines and synaptic potency. In: Studies in Neurophysiology (Porter, R., ed.). Cambridge: Cambridge University Press, pp. 203–209.

Rall, W. 1981. Functional aspects of neuronal geometry. In: Neurones Without Impulses (Bush, B.M.H., and Roberts, A., eds.). Cambridge: Cambridge University Press, pp. 223–254.

Rall, W. 1985. Cable theory for dendritic neurons. In: Methods in Neuronal Modeling (Koch, C., and Segev, I., eds.). Cambridge, MA: MIT Press, pp. 9–62.

Rall, W., Burke, R.E., Smith, T.G., Nelson, P.G., and Frank, K. 1967. Dendritic location of synapses and possible mechanisms for the monosynaptic EPSP in motoneurons. J. Neurophysiol. 30: 1169–1193.

Rall, W., and Hunt, C.C. 1956. Analysis of reflex variability in terms of partially correlated excitability fluctuations in a population of motoneurons. J. Gen. Physiol. 39: 397–422.

Rall, W., and Rinzel, J. 1973. Branch input resistance and steady attenuation for input to one branch of a dendritic neuron model. Biophys. J. 13: 648–688.

Rall, W., and Segev, I. 1987. Functional possibilities for synapses on dendrites and on dendritic spines. In: Synaptic Function (Edelman, G.M., Gall, W.F., and Cowan, W.M., eds.). Neurosci. Res. Found. New York: John Wiley, pp. 605–636.

Rall, W., and Shepherd, G.M. 1968. Theoretical reconstruction of field potentials and dendro-dendritic synaptic interactions in olfactory bulb. J. Neurophysiol. 31: 884–915.

Rall, W., Shepherd, G.M., Reese, T.S., and Brightman, M.W. 1966. Dendrodendritic synaptic pathway for inhibition in the olfactory bulb. Exp. Neurol. 14: 44–56.

Ralston, H.J. 1971. Evidence for presynaptic dendrites and a proposal for their mechanism of action. Nature 230: 585–587.

Ralston, H.J. 1983. The synaptic organization of the ventrobasal thalamus in the rat, cat and monkey. In: Somatosensory Integration in the Thalamus (Macchi, G., Rustioni, A., and Spreafico, R., eds.). New York: Elsevier, pp. 241–250.

Ralston, H.J., Ohara, P.T., Ralston, D.D., and Chazal, G. 1988. The neuronal and synaptic organization of the cat and primate somatosensory thalamus. In: Cellular Thalamic Mechanisms (Macchi, G., Bentivoglio, M., and Spreafico, R., eds.). New York: Elsevier, pp. 127–141.

Ramoa, A., Campbell, G., and Shatz, C.J. 1988. Dendritic growth and remodeling of cat retinal ganglion cells during fetal and postnatal development. J. Neurosci. 8: 4239–4261.

Ramón-Moliner, E. 1977. The reciprocal synapses of the olfactory bulb: Questioning the evidence. Brain Res. 128: 1–20.

Ramón y Cajal, S. 1911. Histologie du Système Nerveux de l'Homme et des Vertébrés, II, trans. L. Azoulay. Paris: Maloine.

Ratliff, F. 1965. Mach bands: Quantitative studies on neural networks in the retina. San Francisco: Holden-Day.

Reader, T.A. 1978. The effects of dopamine, noradrenaline and serotonin in the visual cortex of the cat. Experientia 34: 1586–1588.

Reader, T.A., Ferron, A., Descarries, L., and Jasper, H.H. 1979. Modulatory role for biogenic amines in the cerebral cortex. Microiontophoretic studies. Brain Res. 160: 217–229.

Redman, S.J. 1979. Junctional mechanisms at group Ia synapses. Prog. Neurobiol. 12: 33–83.

Redman, S.J., and Walmsley, B. 1983a. Amplitude fluctuations in synaptic potentials evoked in cat spinal motoneurons at identified group Ia synapses. J. Physiol. (Lond.) 343: 135–145.

Redman, S.J., and Walmsley, B. 1983b. The time course of synaptic potentials evoked in cat spinal·motoneurones at identified group Ia synapses. J. Physiol. (Lond.) 343: 117–133.

Reese, T.S., and Brightman, M.W. 1970. Olfactory surface and central olfactory connections in some vertebrates. In: Taste and Smell in Vertebrates (Wolstenholme, G.E.W., and Knight, J., eds.). London: J. & A. Churchill, pp. 115–149.

Reese, T.S., and Shepherd, G.M. 1972. Dendro-dendritic synapses in the central nervous system. In: Structure and Function of Synapses (Pappas, G.D., and Purpura, D.P., eds.). New York: Raven Press, pp. 121–136.

Reh, T., and Constantine-Paton, M. 1984. Retinal ganglion cell terminals change their projection sites during larval development of *Rana pipiens*. J. Neurosci. 4: 442–457.

Reubi, J.C., and Maurer, R. 1985. Autoradiographic mapping of somatostatin receptors in the rat central nervous system and pituitary. Neuroscience 15: 1183–1193.

Rexed, B. 1952. The cytoarchitectonic organization of the spinal cord in the cat. J. Comp. Neurol. 96: 415–496.

Reymond, L. 1985. Spatial visual acuity of the eagle *Aquila audax:* A behavioural, optical and anatomical investigation. Vision Res. 25: 1477–1491.

Ribak, C.E. 1978. Aspinous and sparsely-spinous stellate neurons in the visual cortex of rats contain glutamic acid decarboxylase. J. Neurocytol. 7: 461–478.

Ribak, C.E., and Seress, L. 1983. Five types of basket cell in the hippocampal dentate gyrus: A combined Golgi and electron microscopic study. J. Neurocytol. 12: 577–597.

Ribak, C.E., Seress, L., and Amaral, D.G. 1985. The development, ultrastructure and synaptic connections of the mossy cells of the dentate gyrus. J. Neurocytol. 14: 835–857.

Ribak, C.E., Vaughn, J.E., Saito, K., Barber, R., and Roberts, E. 1977. Glutamate decarboxylase localization in neurons in the olfactory bulb. Brain Res. 126: 1–18.

Richmond, B., Wurtz, R., and Sato, T. 1983. Visual responses of inferior temporal neurons in the awake rhesus monkey. J. Neurophysiol. 50: 1415–1432.

Rinzel, J., and Rall, W. 1974. Transient response in a dendritic neuronal model for current injected at one branch. Biophys. J. 14: 759–790.

Robbins, S.L., and Kumar, V. 1987. Basic Pathology. Philadelphia: W. B. Saunders.

Roberts, E. 1986. What do GABA neurons really do? They make possible variability generation in relation to demand? Exp. Neurol. 93: 279–290.

Roberts, G.W., Woodhams, P.L., Polak, J.M., and Crow, T.J. 1982. Distribution of neuropeptides in the limbic system of the rat: The amygdaloid complex. Neuroscience 7: 99–131.

Rockel, A.J., Hiorns, R.W., and Powell, T.P.S. 1980. The basic uniformity in structure of the neocortex. Brain 103: 221–244.

Rodieck, R.W. 1965. Quantitative analysis of cat retinal ganglion cell response to visual stimuli. Vision Res. 5: 583–601.

Rodieck, R.W. 1979. Visual pathways. Annu. Rev. Neurosci. 2: 193–225.

Rodieck, R.W. 1988. The primate retina. Comp. Primate Biol. 4: 203–278.

Rodieck, R.W., and Brening, R.K. 1983. Retinal ganglion cells: Properties, types, genera, pathways, and trans-species comparisons. Brain Behav. Evol. 23: 121–164.

Rodriguez, R., and Haberly, L.B. 1989. Analysis of synaptic events in the opossum piriform cortex with improved current source density techniques. J. Neurophysiol. 61: 702–718.

Roffler-Tarlov, S., and Turey, M. 1982. The content of amino acids in the developing cerebellar cortex and deep cerebellar nuclei of granule cell deficient mutant mice. Brain Res. 247: 65–73.

Rohde, B.H., Rea, M.A., Simon, J.R., and McBride, W.J. 1979. Effects of X-irradiation induced loss of cerebellar granule cells on the synaptosomal levels and the high affinity uptake of amino acids. J. Neurochem. 32: 1431–1435.

Rolls, E.T. 1988. Visual information processing in the primate temporal lobe. In: Models of Visual Perception: From Natural to Artificial (Imbert, M., ed.). Oxford: Oxford University Press.

Rolls, E.T. 1989. Functions of neuronal networks in the hippocampus and neocortex in memory. In: Neural Models of Plasticity: Experimental and Theoretical Approaches (Byrne, J.H. and Berry, W.O., eds.). San Diego: Academic Press, pp. 240–265.

Roman, F., Staubli, U., and Lynch, G. 1987. Evidence for synaptic potentiation in a cortical network during learning. Brain Res. 418: 221–226.

Romanes, G.J. 1951. The motor cell columns of the lumbo-sacral spinal cord of the cat. J. Comp. Neurol. 94: 313–363.

Romer, A.S. 1969. Vertebrate history with special reference to factors related to cerebellar evolution. In: Neurobiology of Cerebellar Evolution and Development (Llinás, R., ed.). Chicago: American Medical Association, pp. 1–18.

Ronnevi, L.-O. 1979. Spontaneous phagocytosis of C-type synaptic terminals by spinal α-motoneurons in newborn kittens. An electron microscopic study. Brain Res. 162: 189–199.

Rose, A. 1973. Vision: Human and Electronic. New York: Plenum Press.

Rose, D., and Dobson, V.G. 1985. Models of the Visual Cortex. New York: John Wiley.

Rose, P.K., and Richmond, F.J.R. 1981. White-matter dendrites in the upper cervical spinal cord of the adult cat: A light and electron microscopic study. J. Comp. Neurol. 199: 191–203.

Ross, W.N., and Werman, J.R. 1986. Mapping calcium transients in the dendrites of Purkinje cells from the guinea-pig cerebellum in vitro. J. Physiol. (Lond.) 389: 319–336.

Rotter, A., Birdsall, N.J.M., Burgen, A.S.V., Field, P.M., Hulme, E.C., Raisman, G. 1979. Muscarinic receptors in the central nervous system of the rat. I. Technique for autoradiographic localization of the binding of [³H]propylbenzilylcholine mustard and its distribution in the forebrain. Brain Res. Rev. 1: 141–165.

Rovira, C., Ben Ari, Y., and Cherubini, E. 1983. Dual cholinergic modulation of hippocampal somatic and dendritic field potentials by the septohippocampal pathway. Exp. Brain Res. 49: 151–155.

Rudy, B. 1988. Diversity and ubiquity of K+ channels. Neuroscience 25: 729–749.

Rumelhart, D.E., and McClelland, J.L. 1986. Parallel Distributed Processing: Explorations in the Microstructure of Cognition, Vol. 1: Foundations. Cambridge, MA: MIT Press.

Ryall, R.W., Piercey, M.F., and Polosa, C. 1971. Intersegmental and intrasegmental distribution of mutual inhibition of Renshaw cells. J. Neurophysiol. 34: 700–707.

Rymer, W.Z. 1984. Spinal mechanisms for control of muscle length and tension. In: Handbook of the Spinal Cord, Vol. I: Physiology (Davidoff, R.E., ed.). New York: Marcel Dekker, pp. 609–646.

Saito, H. 1983a. Morphology of physiologically identified X-, Y-, and W-type retina ganglion cells of the cat. J. Comp. Neurol. 221: 279–288.

Saito, H. 1983b. Pharmacological and morphological differences between X-type and Y-type ganglion cells in the cat's retina. Vision Res. 23: 1299–1308.

Sakai, H., and Naka, K. 1985. Novel pathway connecting the outer and inner vertebrate retina. Nature 315: 570–571.

Sakanaka, M., Shibsdski, T., and Lederis, K. 1987. Corticotropin releasing factor-like immunoreactivity in the rat brain as revealed by a modified cobalt–glucose oxidase–diaminobenzidine method. J. Comp. Neurol. 260: 256–298.

Salt, T.E. 1988. Electrophysiological studies of excitatory amino acid neurotransmission in the ventrobasal thalamus. In: Cellular Thalamic mechanisms (Macchi, G., Bentivoglio, M., and Spreafico, R., eds.). Amsterdam: Elsevier, pp. 297–310.

Salt, T.E., and Sillito, A.M. 1984. The action of somatostatin (SSt) on the response properties of cells in the cat's visual cortex. J. Physiol. (Lond.) 350: 28P.

Sandell, J., and Masland, R. 1986. A system of indoleamine-accumulating neurons in the rabbit retina. J. Neurosci. 6: 3331–3347.

Sanderson, K.J. 1971. Visual field projection columns and magnification factors in the lateral geniculate nucleus of the cat. Exp. Brain Res. 13: 159–177.

Sandoval, M.E., and Cotman, C.W. 1978. Evaluation of glutamate as a neurotransmitter of cerebellar parallel fibers. Neuroscience 3: 199–206.

Sanides, F., and Sanides, D. 1972. The "extraverted neurons" of the mammalian cerebral cortex. Z. Anat. Entwickl-Gesch. 136: 272–293.

Sasaki, K., and Llinas, R. 1985. Evidence for dynamic electrotonic coupling in mammalian inferior olive *in vivo. Soc. Neurosci. Abst.* 11: 181.

Satou, M., Mori, K., Tazawa, Y., and Takagi, S.F. 1982. Two types of postsynaptic inhibition in pyriform cortex of the rabbit: Fast and slow inhibitory postsynaptic potentials. J. Neurophysiol. 48: 1142–1156.

Satou, M., Mori, K., Tazawa, Y., and Takagi, S.F. 1983. Interneurons mediating fast postsynaptic inhibition in pyriform cortex of the rabbit. J. Neurophysiol. 50: 89–101.

Scharfman, H.E., and Sarvey, J.M. 1987. Responses to GABA recorded from identified rat visual cortical neurons. Neuroscience 23: 407–422.

Scharfman, H.E., and Schwartzkroin, P.A. 1988. Electrophysiology of morphologically identified mossy cells of the dentate hilus in guinea pig hippocampal slices. J. Neurosci. 8: 3812–3821.

Schein, S.J. 1988. Anatomy of macaque fovea and spatial densities of neurons in foveal representation. J. Comp. Neurol. 269: 479–505.

Schiller, P.H., Sandell, J.H., and Maunsell, J.H.R. 1986. Functions of the ON and OFF channels of the visual system. Nature 322: 824–825.

Schmechel, D.E., Vickrey, B.G., Fitzpatrick, D., and Elde, R.P. 1984. GABAergic neurons of mammalian cerebral cortex: Widespread subclass defined by somatostatin content. Neurosci. Lett. 47: 227–232.

Schmidt, M., Humphrey, M., and Wässle, H. 1987. Action and localization of acetylcholine in the cat retina. J. Neurophysiol. 58: 997–1015.

Schneider, S.P., and Macrides, F. 1978. Laminar distribution of interneurons in the main olfactory bulb of adult hamster. Brain Res. Bull. 3: 73–82.

Schoenfeld, T.A., Marchand, J.E., and Macrides, F. 1985. Topographic organization of tufted cell axonal projections in the hamster main olfactory bulb: An intrabulbar associational system. J. Comp. Neurol. 235: 503–518.

Scholfield, C.N. 1978a. Electrical properties of neurones in the olfactory cortex slice in vitro. J. Physiol. (Lond.) 275: 535–546.

Scholfield, C.N. 1978b. A depolarizing inhibitory potential in neurones of the olfactory cortex in vitro. J. Physiol. (Lond.) 275: 547–557.

Scholfield, C.N. 1978c. A barbiturate induced intensification of the inhibitory potential in slices of guinea-pig olfactory cortex. J. Physiol. (Lond.) 275: 559–566.

Scholfield, C.N. 1978d. The action of depressant amino-acids on neurones in the isolated olfactory cortex. In: Iontophoresis and Transmitter Mechanisms in the Mammalian Central Nervous System (Ryall, R.W., and Kelly, J.S., eds.). Amsterdam: Elsevier, pp. 188–190.

Scholfield, C.N. 1980. Convulsants antagonise inhibition in the olfactory cortex slice. Arch. Pharmacol. 314: 29–36.

Schwartz, E. 1987. Depolarization without calcium can release gamma-aminobutyric acid from a retinal neuron. Science 238: 350–355.

Schwartzkroin, P.A. 1975. Characteristics of CA1 neurons recorded intracellularly in the hippocampal in vitro slice preparation. Brain Res. 85: 423–436.

Schwartzkroin, P.A., and Kunkel, D.D. 1985. Morphology of identified interneurons in the CA1 regions of guinea pig hippocampus. J. Comp. Neurol. 232: 205–218.

Schwartzkroin, P.A., and Mathers, L.H. 1978. Physiological and morphological identification of a nonpyramidal hippocampal cell type. Brain Res. 157: 1–10.

Schwartzkroin, P.A., and Slawsky, M. 1977. Probable calcium spikes in hippocampal neurons. Brain Res. 135: 157–161.

Schwindt, P., and Crill, W.E. 1977. A persistent negative resistance in cat lumbar motoneurons. Brain Res. 120: 173–178.

Schwindt, P.C., and Crill, W.E. 1980. Properties of a persistent inward current in normal and TEA-injected motoneurons. J. Neurophysiol. 43: 1700–1724.

Schwindt, P.C., Spain, W.J., Foehring, R.C., Stafstrom, C.E., Chubb, M.C., and Crill, W.E. 1988a. Multiple potassium conductances and their functions in neurons from cat sensorimotor cortex in vitro. J. Neurophysiol. 59: 424–449.

Schwindt, P.C., Spain, W.J., Foehring, R.C., Chubb, M.C., and Crill, W.E. 1988b. Slow conductances in neurons from cat sensorimotor cortex in vitro and their role in slow excitability changes. J. Neurophysiol. 59: 450–467.

Schwob, J.E., and Price, J.L. 1984. The development of lamination of afferent fibers to the olfactory cortex in rats, with additional observations in the adult. J. Comp. Neurol. 223: 203–222.

Scott, J.W., and Harrison, T.A. 1987. The olfactory bulb: Anatomy and physiology. In: Neurobiology of Taste and Smell (Finger, T.E., and Silver, W.L., eds.). New York: John Wiley, pp. 151–178.

Scott, J.W., McBride, R.L., and Schneider, S.P. 1980. The organization of projections from the olfactory bulb to the piriform cortex and olfactory tubercle in the rat. J. Comp. Neurol. 194: 519–534.

Segal, M., and Barker, J.L. 1984. Rat hippocampal neurons in culture: Properties of GABA-activated Cl^- ion conductance. J. Neurophysiol. 51: 500–515.

Segev, I., Fleshman, J.W., and Burke, R.E. 1989. Compartmental models of complex neurons. In: Methods in Neuronal Modeling: From Synapse to Network (Koch, C., and Segev, I., eds.). Cambridge, MA: MIT Press, pp. 63–96.

Segev, I., Fleshman, J.W., Miller, J.P., and Bunow, B. 1985. Modeling the electrical behavior of anatomically complex neurons using a network analysis program: Passive membrane. Biol. Cybern. 53: 27–40.

Segev, I., and Rall, W. 1988. Computational study of an excitable dendritic spine. J. Neurophysiol. 60: 499–523.

Sejnowski, T., Koch, C., and Churchland, P. 1988. Computational neuroscience. Science 241: 1299–1306.

Selemon, L.D., and Goldman-Rakic, P.S. 1986. Longitudinal topography and interdigitation of corticostriatal projections in the rhesus monkey. J. Neurosci. 5: 776–794.

Seress, L., and Ribak, C.E. 1983. GABAergic cells in the dentate gyrus appear to be local circuit and projection neurons. Exp. Brain Res. 50: 173–182.

Shannon, C.E., and Weaver, W. 1949. The Mathematical Theory of Communication. Urbana: University of Illinois Press.

Shapley, R., and Enroth-Cugell, C. 1984. Visual adaptation and retinal gain controls. Prog. Retinal Res. 3: 263–346.

Shapley, R., and Lennie, P. 1985. Spatial frequency analysis in the visual system. Annu. Rev. Neurosci. 8: 547–583.

Shapovalov, A.I., and Shiriaev, B.I. 1980. Dual mode of junctional transmission at synapses between single primary afferent fibers and motoneurons in amphibian. J. Physiol. (Lond.) 306: 1–15.

Shapovalov, A.I., and Shiraev, B.I. 1982. Selective modulation of chemical transmission at a dual-action synapse (with special reference to baclofen). Gen. Physiol. Biophys. 1: 423–433.

Sharp, F.R., Kauer, J.S., and Shepherd, G.M. 1977. Laminar analysis of 2-deoxyglucose uptake in olfactory bulb and olfactory cortex of rabbit and rat. J. Neurophysiol. 40: 800–813.

Sharrard, W.J.W. 1955. The distribution of the permanent paralysis in the lower limb in poliomyelitis. J. Bone Jt. Surg. 37: 540–558.

Shatz, C.J., and Kirkwood, P.A. 1984. Prenatal development of functional connections in the cat's retinogeniculate pathway. J. Neurosci. 4: 1378–1397.

Shatz, C., and Stryker, M. 1988. Prenatal tetrodotoxin infusion blocks segregation of retinogeniculate afferents. Science 242: 87–89.

Shepherd, G.M. 1963. Responses of mitral cells to olfactory nerve volleys in the rabbit. J. Physiol. (Lond.) 168: 89–100.

Shepherd, G.M. 1972. The neuron doctrine: A revision of functional concepts. Yale J. Biol. Med. 45: 584–599.

Shepherd, G.M. 1974. The Synaptic Organization of the Brain. New York: Oxford University Press.

Shepherd, G.M. 1978. Microcircuits in the nervous system. Sci. Am. 238: 93–103.

Shepherd, G.M. 1979. The Synaptic Organization of the Brain, 2nd ed. New York: Oxford University Press.

Shepherd, G.M. 1988a. Neurobiology, 2nd ed. New York: Oxford University Press.

Shepherd, G.M. 1988b. A basic circuit for cortical organization. In: Perspectives on Memory Research (Gazzaniga, M.C., ed.). Cambridge, MA: MIT Press, pp. 93–134.

Shepherd, G.M. 1990. The significance of real neuron architectures for neural network simulations. In: Computational Neuroscience (Schwartz, E., ed.). Cambridge, MA: MIT Press.

Shepherd, G.M., and Brayton, R.K. 1979. Computer simulation of a dendrodendritic synaptic circuit for self- and lateral inhibition in the olfactory bulb. Brain Res. 175: 377–382.

Shepherd, G.M., and Brayton, R.K. 1987. Logic operations are properties of computer-simulated interactions between excitable dendritic spines. Neuroscience 21: 151–166.

Shepherd, G.M., Brayton, R.K., Miller, J.F., Segev, I., Rinzel, J., and Rall, W. 1985. Signal enhancement in distal cortical dendrites by means of interactions between active dendritic spines. Proc. Natl. Acad. Sci. USA 82: 2192–2195.

Shepherd, G.M., Carnevale, N.T., and Woolf, T.B. 1989. Comparisons between computational operations generated by active responses in dendritic branches and spines. J. Cognit. Neurosci. 1: 273–286.

Shepherd, G.M., and Greer, C.A. 1988. The dendritic spine: Adaptations of structure and function for different types of synaptic integration. In: Intrinsic Determinants of Neuronal Form and Function (Lasek, R., and Black, M., eds.). New York: Alan R. Liss, pp. 245–262.

Sherman, S.M. 1985. Functional organization of the W-, X-, and Y-cell pathways: A review and hypothesis. In: Progress in Psychobiology and Physiological Psychobiology, Vol. 11 (Sprague, J.M., and Epstein, A.N., eds.). Amsterdam: Elsevier, pp. 233–314.

Sherman, S.M. 1988. Functional organization of the cat's lateral geniculate nucleus. In: Cellular Thalamic Mechanisms (Macchi, G., Bentivoglio, M., and Spreafico, E., eds.). Amsterdam: Elsevier, pp. 163–183.

Sherman, S.M., and Friedlander, M.J. 1988. Identification of X versus Y properties for interneurons in the A-laminae of the cat's lateral geniculate nucleus. Exp. Brain Res. 73: 384–392.

Sherman, S.M., and Koch, C. 1986. The control of retinogeniculate transmission in the mammalian lateral geniculate nucleus. Exp. Brain Res. 63: 1–20.

Sherman, S.M., and Spear, P.D. 1982. Organization of the visual pathways in normal and visually deprived cats. Physiol. Rev. 62: 738–855.

Sherrington, C.S. 1941. Man on His Nature. Cambridge: Cambridge University Press.

Shinoda, Y., Ohgaki, T., and Futami, T. 1986. The morphology of single lateral vestibulospinal tract axons in the lower cervical spinal cord of the cat. J. Comp. Neurol. 249: 226–241.

Shipley, M.T., and Geinisman, Y. 1984. Anatomical evidence for convergence of olfactory, gustatory, and visual afferent pathways in mouse cerebral cortex. Brain Res. Bull. 12: 221–226.

Siggins, G.R., and Bloom, F.E. 1971. Activation of a central noradrenergic projection to the cerebellum. Nature 233: 481–483.

Siggins, G.R., Hoffer, B.J., Oliver, A.P., and Bloom, F.E. 1971. Activation of a central noradrenergic projection to cerebellum. Nature 233: 481–483.

Siggins, G.R., Hoffer, B.J., and Bloom, F.E. 1971. Studies on norepinephrine-containing afferents to Purkinje cells of rat cerebellum. III. Evidence for mediation of norepinephrine effects by cyclic 3′,5′-adenosine monophosphate. Brain Res. 25: 535–553.

Siggins, G.R., Oliver, A.P., Hoffer, B.J., and Bloom, F.E. 1971. Cyclic adenosine monophosphate and norepinephrine: Effects on transmembrane properties of cerebellar Purkinje cells. Science 171: 192–194.

Sillito, A.M. 1975. The effectiveness of bicuculline as an antagonist of GABA and visually evoked inhibition in the cat's striate cortex. J. Physiol. (Lond.) 250: 287–304.

Sillito, A.M. 1984. Functional considerations of the operation of GABAergic inhibitory processes in the visual cortex. In: The Cerebral Cortex, Vol. II (Peters, A., and Jones, E.G., eds.). New York: Plenum Press, pp. 91–118.

Sillito, A.M., and Kemp, J.A. (1983. Cholinergic modulation of the functional organization of the cat visual cortex. Brain Res. 289: 143–155.

Sillito, A.M., Kemp, J.A., Wilson, J.A., and Berardi, N. 1980. A re-evaluation of the mechanisms underlying simple cell orientation selectivity. Brain Res. 194: 517–520.

Sims, S.M., Singer, J.J., and Walsh, J.V., Jr. 1988. Antagonistic adrenergic–muscarinic regulation of m-current in smooth muscle cells. Science 239: 190–193.

Simon, H., Moal, M.L., and Calas, A. 1979. Efferents and afferents of the ventral tegmental-A-10 region studied after local injection of (^{3}H)leucine and horseradish peroxidase. Brain Res. 175: 1–23.

Singer, W. 1977. Control of thalamic transmission by corticofugal and ascending reticular pathways in the visual system. Physiol. Rev. 57: 386–420.

Skeen, L.C., and Hall, W.C. 1977. Efferent projections of the main and accessory olfactory bulb in the tree shrew *(Tupaia glis)*. J. Comp. Neurol. 172: 1–36.

Skrede, K.K., and Westgaard, R.H. 1971. The transverse hippocampal slice: A well-defined cortical structure maintained in vitro. Brain Res. 35: 589–593.

Slaughter, M.M., and Miller, R.F. 1981. 2-Amino-4-phosphonobutyric acid: A new pharmacological tool for retinal research. Science 211: 182–184.

Slaughter, M.M., and Miller, R.F. 1983. An excitatory amino acid antagonist blocks cone input to sign-conserving second-order retinal neurons. Science 219: 1230–1232.

Sloper, J.J., Hiorns, R.W., and Powell, T.P.S. 1979. A qualitative and quantitative electron microscopic study of the neurons in the primate motor and somatic sensory cortices. Philos. Trans. R. Soc. Lond. [B] 285: 141–171.

Sloper, J.J., and Powell, T.P.S. 1979a. Ultrastructural features of the sensorimotor cortex of the primate. Philos. Trans. R. Soc. Lond. [B] 285: 123–139.

Sloper, J.J., and Powell, T.P.S. 1979b. A study of the axon initial segment and proximal axon of neurons in the primate motor and somatic sensory cortices. Philos. Trans. R. Soc. Lond. [B] 285: 173–197.

Sloper, J.J., and Powell, T.P.S. 1979c. An experimental electron microscopic study of afferent connections to the primate motor and somatic sensory cortices. Philos. Trans. R. Soc. Lond. [B] 285: 199–266.

Smith, J.L., Betts, B., Edgerton, V.R., and Zernicke, R.F. 1980. Rapid ankle extension during paw shakes: Selective recruitment of fast ankle extensors. J. Neurophysiol. 43: 612–620.

Smith, R.G., Freed, M., and Sterling, P. 1986. Microcircuitry of the dark adapted cat retina: Functional architecture of the rod–cone network. J. Neurosci. 6: 3505–3517.

Smith, R.G., and Sterling, P. 1990a. Cone receptive field in cat retina computed from microcircuitry and ganglion cell receptive fields. In preparation.

Smith, R.G., and Sterling, P. 1990b. Cone receptive field of cat retina derived from computational model of the cone-horizontal cell network. In preparation.

Smith, S.J., MacDermott, A.B., and Weight, F.F. 1983. Detection of intracellular Ca2+ transients in sympathetic neurones using arsenazo III. Nature 304: 350–352.

Smith, T.G., Wuerker, R.B., and Frank, K. 1967. Membrane impedance changes during synaptic transmission in cat spinal motoneurons. J. Neurophysiol. 30: 1072–1096.

Somogyi, P. 1977. A specific 'axo-axonal' interneuron in the visual cortex of the rat. Brain Res. 136: 345–350.

Somogyi, P., and Cowey, A. 1981. Combined Golgi and electron microscopic study on the synapses formed by double bouquet cells in the visual cortex of the cat and monkey. J. Comp. Neurol. 195: 547–566.

Somogyi, P., Freund, T.F., and Cowey, A. 1982. The axo-axonic interneuron in the cerebral cortex of the rat, cat and monkey. Neuroscience 7: 2577–2607.

Somogyi, P., Halasy, K., Somogyi, J., Storm-Mathisen, J., and Ottersen, O.P. 1986. Quantification of immunogold labelling reveals enrichment of glutamate in mossy and parallel fiber terminals in cat cerebellum. Neuroscience 19: 1045–1050.

Somogyi, P., Hodgson, A.J., Smith, A.D., Nunzi, M.G., Gorio, A., and Wu, J-Y. 1984. Different populations of GABAergic neurons in the visual cortex and hippocampus of the cat contain somatostatin- or cholecystokinin-immunoreactive material. J. Neurosci. 4: 2590–2603.

Somogyi, P., Kisvarday, Z.F., Martin, K.A.C., and Whitteridge, D. 1983. Synaptic connections of morphologically identified and physiologically characterized large basket cells in the striate cortex of the cat. Neuroscience 10: 261–294.

Somogyi, P., and Soltész, I. 1986. Immunogold demonstration of GABA in synaptic terminals of intracellularly recorded, horseradish peroxidase-filled basket cells and clutch cells in the cat's visual cortex. Neurosci. 19: 1051–1065.

Sonnhof, U., Richter, D.W., and Taugner, R. 1977. Electrotonic coupling between frog spinal motoneurons. An electro-physiological and morphological study. Brain Res. 138: 197–215.

Sotelo, C., Gotow, T., and Wassef, M. 1986. Localization of glutamate acid-decarboxyl-ase immunoreactive axon terminals in the inferior olive of the rat with special emphasis on anatomical relations between GABAergic synapses and dendro-dendritic gap junctions. J. Comp. Neurol. 252: 32–50.

Sotelo, C., Hillman, D.E., Zamora, A.J., and Llinás, R. 1975. Climbing fiber deafferen-tation: Its action on Purkinje cell dendritic spines. Brain Res. 98: 574–581.

Sotelo, C., Privat, A., and Drian, M. 1972. Localization of [^{3}H]GABA in tissue culture of rat cerebellum using electron microscopy radioautography. Brain Res. 45: 302–308.

Spreafico, R., Schmechel, D.E., Ellis, L.C., and Rustioni, A. 1983. Cortical relay neurons and interneurons in the n. ventralis posterlateralis of cats: A horseradish peroxidase electron microscopic Golgi and immunocytochemical study. Neuro-science 9: 491–509.

Squire, L.R. 1987. Memory and Brain. New York: Oxford University Press.

Squire, L.R., Shimamura, A.P., and Amaral, D.G. 1989. Memory and the hippocampus. In: Neural Models of Plasticity: Experimental and Theoretical Approaches (Byrne, J.H., and Berry, W.O., eds.). San Diego: Academic Press, pp. 208–239.

Srinivasan, M.V., Laughlin, S.B., and Dubs, A. 1982. Predictive coding: A fresh view of inhibition in the retina. Proc. R. Soc. Lond. [B] 216: 427–459.

Stafstrom, C.E., Schwindt, P.E., Chubb, M.C., and Crill, W.E. 1985. Properties of persistent sodium conductance and calcium conductance of layer V neurons from cat sensorimotor cortex in vitro. J. Neurophysiol. 53: 153–170.

Stafstrom, C.E., Schwindt, P.C., and Crill, W.E. 1982. Negative slope conductance due to a persistent subthreshold sodium current in cat neocortical neurons in vitro. Brain Res. 236: 221–226.

Stafstrom, C.E., Schwindt, P.C., and Crill, W.E. 1984. Cable properties of layer V neurons from cat sensorimotor cortex in vitro. J. Neurophysiol. 52: 278–289.

Stanford, L.R., and Sherman, S.M. 1984. Structure/function relationships of retinal ganglion cells in the cat. Brain Res. 297: 381–386.

Stein, P.S.G. 1984. Central pattern generators in the spinal cord. In: Handbook of the Spinal Cord, Vol. I: Physiology (Davidoff, R.E., ed.). New York: Marcel Dekker, pp. 647–672.

Steinberg, R.H., Reid, M., and Lacey, P.L. 1973. The distribution of rods and cones in the retina of the cat *(Felis domesticus)*. J. Comp. Neurol. 148: 229–248.

Steinbusch, H.W.M. 1984. Serotonin-immunoreactive neurons and their projections in the CNS. In: Handbook of Chemical Neuroanatomy, Vol. 3: Classical Transmitters and Transmitter Receptors in the CNS, Part II (Björklund, A., Hökfelt, T., and Kuhar, M.J., eds.). Amsterdam: Elsevier, pp. 68–125.

Stelzer, A., and Wong, R.K.S. 1989. GABA$_A$ responses in hippocampal neurons are potentiated by glutamate. Nature 337: 170–173.

Stent, G.S. 1973. A physiological mechanism for Hebb's postulate of learning. Proc. Natl. Acad. Sci. USA 73: 997–1001.

Steriade, M., and Deschênes, M. 1984. The thalamus as a neuronal oscillator. Brain Res. Rev. 8: 1–63.

Steriade, M., Domich, L., Oakson, G., and Deschênes, M. 1987. The deafferented reticular thalamic nucleus generates spindle rhythmicity. J. Neurophysiol. 57: 260–273.

Steriade, M., and Llinás, R.R. 1988. The functional states of the thalamus and the associated neuronal interplay. Physiol. Rev. 68: 649–742.

Sterling, P. 1983. Microcircuitry of the cat retina. Annu. Rev. Neurosci. 6: 149–185.

Sterling, P., Cohen, E., Freed, M., and Smith, R.G. 1987. Microcircuitry of the on-beta ganglion cell in daylight, twilight and starlight. Neurosci. Res. [Suppl.] 6: 269–285.

Sterling, P., Freed, M., and Smith, R. 1986. Microcircuitry and functional architecture of the cat retina. Trends Neurosci. 9: 186–192.

Sterling, P., Freed, M., and Smith, R.G. 1988. Architecture of rod and cone circuits to the on-beta ganglion cell. J. Neurosci. 8: 623–642.

Sterling, P., and Lampson, L. 1986. Molecular specificity of defined types of amacrine synapse in cat retina. J. Neurosci. 6: 1314–1324.

Stevens, C.F. 1969. Structure of the cat frontal olfactory cortex. J. Neurophysiol. 32: 184–192.

Steward, O. 1976. Topographic organization of the projections from the entorhinal area to the hippocampal formation of the rat. J. Comp. Neurol. 167: 347–370.

Stone, J. 1983. Parallel Processing in the Visual System. New York: Plenum Press.

Stone, T.W., and Taylor, D.A. 1977. Microiontophoretic studies of the effects of cyclic nucleotides on excitability of neurones in the rat cerebral cortex. J. Physiol. (Lond.) 266: 523–543.

Storm-Mathisen, J. 1977. Localization of transmitter candidates in the brain: The hippocampal formation as a model. Prog. Neurobiol. 8: 119–181.

Stripling, J.S., Patneau, D.K., and Gramlich, C.A. 1988. Selective long-term potentiation in the pyriform cortex. Brain Res. 441: 281–291.

Struble, R.G., Desmond, N.L., and Levy, L.B. 1978. Anatomical evidence for interlamellar inhibition in the fascia dentata. Brain Res. 152: 580–585.

Struble, R.G., and Walters, C.P. 1982. Light microscope differentiation of two populations of rat olfactory bulb granule cells. Brain Res. 236: 237–251.

Sugimori, M., and Llinás, R. 1981. Localization of ionic conductances in soma–dendritic regions of Purkinje cells: An in vitro study of guinea pig cerebellar slices. Soc. Neurosci. Abstr. 7: 76.

Sugimori, M., Llinás, R., and Angelides, A. 1986. Fluorescence localization of tetrodotoxin receptors in mammalian cerebellar cortex in vitro. Soc. Neurosci. Abstr. 12: 463.

Sugimori, M., Preston, R.J., and Kitai, S.T. 1978. Response properties and electrical constants of caudate nucleus neurons in the cat. J. Neurophysiol. 41: 1662–1675.

Sur, M., Esguerra, M., Garraghty, P.E., Kritzer, M.F., and Sherman, S.M. 1987. Morphology of physiologically identified retinogeniculate X and Y axons in the cat. J. Neurophysiol. 58: 1–32.

Swanson, L.W. 1983. The hippocampus and the concept of the limbic system. In: Neurobiology of the Hippocampus (Seifert, W., ed.). London: Academic Press, pp. 3–19.

Swanson, L.W., and Cowan, W.M. 1975. Hippocampo-hypothalamic connections: Origin in subicular cortex, not Ammon's horn. Science 189: 303–304.

Swanson, L.W., Wyss, J.M., and Cowan, W.M. 1978. An autoradiographic study of the organization of intrahippocampal association pathways in the rat. J. Comp. Neurol. 181: 681–716.

Swindale, N. 1981. Dendritic spines only connect. Trends Neurosci. 4: 240–241.

Switzer, R.C., de Olmos, J., and Heimer, L. 1985. Olfactory system. In: The Rat Nervous System: Forebrain and Midbrain (Paxinos, G., ed.). Sydney: Academic Press, pp. 1–36.

Sypert, G.W., and Munson, J.B. 1984. Excitatory synapses. In: Handbook of the Spinal Cord, Vol. I: Physiology (Davidoff, R.E., ed.). New York: Marcel Dekker, pp. 315–384.

Szentagothai, J. 1978. The neuron network of the cerebral cortex: A functional interpretation. Proc. R. Soc. Lond. [B] 201: 219–248.

Tachibana, M., and Kaneko, A. 1984. Gamma-aminobutyric acid acts at axon terminals of turtle photoreceptors: Difference in sensitivity among cell types. Proc. Natl. Acad. Sci. USA 81: 7961–7964.

Takagi, H., Somogyi, P., Somogyi, J., and Smith, A.D. 1983. Fine structural studies of a type of somatostatin-immunoreactive neuron and its synaptic connections in the rat neostriatum: A correlated light and electron microscopic study. J. Comp. Neurol. 214: 1–16.

Takagi, S.F. 1986. Studies on the olfactory nervous system in the Old World monkey. Prog. Neurobiol. 27: 195–250.

Takeuchi, Y., Kimura, H., and Sano, Y. 1982. Immunohistochemical demonstration of serotonin-containing nerve fibers in the cerebellum. Cell Tissue Res. 226: 1–12.

Tanabe, Y., Iino, M., and Takagi, S.F. 1975. Discrimination of odors on olfactory bulb pyriform–amygdaloid areas, and orbitofrontal cortex of the monkey. J. Neurophysiol. 38: 1284–1296.

Tank, D.S., Sugimori, M., Connor, J.A., and Llinás, R. 1988. Spatially resolved calcium dynamics of mammalian Purkinje cells in cerebellar slices. Science 242: 241–245.

Tank, D.W., and Hopfield, J.J. 1987. Neural computation by concentrating information in time. Proc. Natl. Acad. Sci. USA 84: 1896–1900.

Tauchi, M., and Masland, R.H. 1984. The shape and arrangement of the cholinergic neurons in the rabbit retina. Proc. R. Soc. Lond. [B] 223: 101–119.

Taylor, C.P., and Dudek, F.E. 1984a. Excitation of hippocampal pyramidal cells by an electrical field effect. J. Neurophysiol. 52: 126–142.

Taylor, C.P., and Dudek, F.E. 1984b. Excitation of hippocampal pyramidal cells by electrical field effect. Trends Neurosci. 11: 431–438.

ten Bruggencate, G., and Engberg, I. 1971. Iontophoretic studies in Deiters' nucleus of the inhibitory actions of GABA and related amino acids and the interactions of strychnine and picrotoxin. Brain Res. 25: 431–448.

Thalman, R.H. 1988. Evidence that guanosine triphosphate (GTP)-binding proteins control a synaptic response in brain: Effect of pertussis toxin and GTP S on the late inhibitory postsynaptic potential of hippocampal CA3 neurons. J. Neurosci. 8: 4589–4602.

Thalmann, R.H., Peck, E.J., and Ayala, G.F. 1981. Biphasic response of hippocampal pyramidal neurons to GABA. Neurosci. Lett. 21: 319–324.

Thomson, A.M. 1986. A magnesium-sensitive post-synaptic potential in rat cerebral cortex resembles neuronal responses to N-methyl-aspartate. J. Physiol. (Lond.) 370: 531–549.

Thomson, A.M., Girdlestone, D., and West, D.C. 1988. Voltage-dependent currents prolong single axon postsynaptic potentials in layer III pyramidal neurons in rat neocortical slices. J. Neurophysiol. 6: 1896–1907.

Torre, V., and Poggio, T. 1981. A new approach to synaptic interaction. In: Theoretical Approaches in Neurobiology (Reichardt, W., and Poggio, T., eds.). Cambridge, MA: MIT Press, pp. 39–46.

Toyama, K., Kimura, M., and Tanaka, K. 1981. Cross-correlation analysis of inter-neuronal connectivity in cat visual cortex. J. Neurophysiol. 46: 191–201.

Toyama, K., Matsunami, K., Ohno, T., and Tokashiki, S. 1974. An intracellular study of neuronal organization in the visual cortex. Exp. Brain Res. 21: 45–66.

Traub, R.D., Knowles, W.D., Miles, R., and Wong, R.K.S. 1987. Models of the cellular mechanism underlying propagation of epileptiform activity in the CA2–CA3 region of the hippocampal slice. Neuroscience 21: 457–470.

Traub, R.D., and Llinás, R. 1977. The spatial distribution of ionic conductances in normal and axotomized motoneurons. Neuroscience 2: 829–849.

Traub, R.D., and Llinás, R. 1979. Hippocampal pyramidal cells: Significance of dendritic ionic conductances for neuronal function and epileptogenesis. J. Neurophysiol. 42: 476–496.

Traub, R.D., Miles, R., and Wong, R.K.S. 1989. Models of the origin of rhythmic population oscillations in the hippocampal slice. Science 243: 1319–1325.

Tseng, G.-F., and Haberly, L.B. 1988. Characterization of synaptically mediated fast and slow inhibitory processes in piriform cortex in an in vitro slice preparation. J. Neurophysiol. 59: 1352–1376.

Tseng, G.-F., and Haberly, L.B. 1989a. Deep neurons in piriform cortex. I. Morphology and synaptically evoked responses including a unique high amplitude paired shock facilitation. J. Neurophysiol. 62: 369–385.

Tseng, G.-F., and Haberly, L.B. 1989b. Deep neurons in piriform cortex. II. Membrane properties that underlie unusual synaptic responses. J. Neurophysiol. 62: 386–400.

Tsien, R.W., Lipscombe, D., Madison, D.V., Bley, K.R., and Fox, A.P. 1988. Multiple types of neuronal calcium channels and their selective modulation. Trends Neurosci. 11: 431–438.

Ts'o, D.Y., and Gilbert, C.D. 1988. The organization of chromatic and spatial interactions in the primate striate cortex. J. Neurosci. 8: 1712–1727.

Ts'o, D.Y., Gilbert, C.D., and Wiesel, T.N. 1986. Relationships between horizontal interactions and functional architecture as revealed by cross-correlation analysis in cat striate cortex. J. Neurosci. 6: 1160–1170.

Tsukamoto, Y., Smith, R.G., and Sterling, P. 1990. 'Collective coding' of correlated cone signals in the retinal ganglion cell. Proc. Natl. Acad. Sci.

Tsumoto, T., Eckart, W., and Creutzfeldt, O.D. 1979. Modification of orientation sensitivity of cat visual cortex neurons by removal of GABA-mediated inhibition. Exp. Brain Res. 46: 157–169.

Tsumoto, T., Masui, H., and Sato, H. 1986. Excitatory amino acid transmitters in neuronal circuits of the cat visual cortex. J. Neurophysiol. 55: 469–483.

Tsumoto, T., and Suda, K. 1980. Three groups of cortico-geniculate neurons and their distribution in binocular and monocular segments of cat striate cortex. J. Comp. Neurol. 193: 223–236.

Turner, D.A., and Schwartzkroin, P.A. 1983. Electrical characteristics of dendrites and dendritic spines in intracellularly stained CA3 and dentate hippocampal neurons. J. Neurosci. 3: 2381–2394.

Ueki, A. 1983. The mode of nigro-thalamic transmission investigated with intracellular recording in the cat. Exp. Brain Res. 49: 116–124.

Ulfhake, B., and Cullheim, S. 1981. A quantitative light microscopic study of the dendrites of cat spinal γ-motoneurons after intracellular staining with horseradish peroxidase. J. Comp. Neurol. 202: 585–596.

Ulfhake, B., and Kellerth, J.-O. 1981. A quantitative light microscopic study of the dendrites of cat spinal α-motoneurons after intracellular staining with horseradish peroxidase. J. Comp. Neurol. 202: 571–584.

Valcana, T., Hudson, D., and Timiras, P.S. 1972. Effects of X-irradiation on the content of amino acids in the developing rat cerebellum. J. Neurochem. 19: 2229–2232.

Valentino, K.L., Tatemoto, K., Hunter, J., and Barchas, J.D. 1986. Distribution of neuropeptide K-immunoreactivity in the rat central nervous system. Peptides 7: 1043–1059.

Valverde, F. 1965. Studies on the Piriform Lobe. Cambridge, MA: Harvard University Press.

Van Essen, D. 1985. Functional organization of primate visual cortex. In: Cerebral Cortex, Vol. 3: Visual Cortex (Peters, A., and Jones, E.G., eds.). New York: Plenum Press, pp. 259–329.

Van Essen, D.C., and Maunsell, J.H.R. 1983. Hierarchical organization and functional streams in the visual cortex. Trends Neurosci. 6: 370–375.

Van Keulen, L. 1981. Autogenetic recurrent inhibition of individual spinal motoneurones of the cat. Neurosci. Lett. 21: 297–300.

Vandermaelen, C.P., and Aghajanian, G.K. 1983. Electrophysiological and pharmacological characterization of serotonergic dorsal raphe neurons recorded extracellularly and intracellularly in rat brain slices. Brain Res. 289: 109–119.

Vanderwolf, C.H., Kramis, R., Gillespie, L.A., and Bland, B.H. 1975. Hippocampal rhythmic slow activity and neocortical low voltage fast activity: Relations to behaviour. In: The Hippocampus, Vol. II (Isaacson, R.L., and Pribram, K.L., eds.). New York: Plenum Press, pp. 101–128.

Vardi, N., Masarachia, P., and Sterling, P. 1990. Structure of the starburst amacrine network in the cat retina and its association with alpha ganglion cells. J. Comp. Neurol. 288: 601–611.

Vertes, R.P. 1982. Brain stem generation of the hippocampal EEG. Prog. Neurobiol. 19: 159–186.

Vidyasagar, T.R., and Heide, W. 1986. The role of GABAergic inhibition in the response properties of neurones in cat area 18. Neuroscience 17: 49–55.

Vincent, S.R., Satoh, K., Armstrong, D.M., and Fibiger, H.C. 1983. Substance P in the ascending cholinergic reticular system. Nature 306: 688–691.

Vinogradova, O.S. 1975. Functional organization of the limbic system in the process of registration of information: Facts and hypotheses. In: The Hippocampus (Isaacson, R.L., and Pribram, K., eds.). New York: Plenum Press, pp. 3–70.

Vogel, M. 1978. Postnatal development of the cat's retina. Adv. Anat. Embryol. Cell Biol. 54: 7–64.

Voigt, T., and Wässle, H. 1987. Dopaminergic innervation of A II amacrine cells in mammalian retina. J. Neurosci. 7. 4115–4128.

Voogd, J., and Bigaré, F. 1980. Topographical distribution of olivary and cortico nuclear fibers in the cerebellum: A review. In: The Inferior Olivary Nucleus (Courville, J., de Montigny, C., and Lamarre, Y., eds.). New York: Raven Press, pp. 207–234.

Walker, J.E., and Fonnum, F. 1983. Effect of regional cortical ablations on high-affinity D-aspartate uptake in striatum, olfactory tubercle, and pyriform cortex of the rat. Brain Res. 278: 283–286.

Walls, G.L. 1942. The Vertebrate Eye and Its Adaptive Radiation. Cranbrook Institute of Science, Bloomfield Hills. (Reprinted in 1963.) New York: Hafner Publishing.

Walmsley, B., Edwards, F.R., and Tracey, D.J. 1987. The probabilistic nature of synaptic transmission at a mammalian excitatory central synapse. J. Neurosci. 7: 1027–1048.

Walmsley, B., Hodgson, J.A., and Burke, R.E. 1978. Forces produced by medial gastrocnemius and soleus muscles during locomotion in freely moving cats. J. Neurophysiol. 41: 1203–1216.

Walmsley, B., and Tracey, D.J. 1981. An intracellular study of Renshaw cells. Brain Res. 223: 170–175.

Wässle, H., Boycott, B.B., and Illing, R.B. 1981b. Morphology and mosaic of on- and off-beta cells in the cat retina and some functional considerations. Proc. R. Soc. Lond. [B] 212: 177–195.

Wässle, H., and Chun, M. 1988. Dopaminergic and indoleamine-accumulating amacrine cells express GABA-like immunoreactivity in the cat retina. J. Neurosci. 8: 2920–2938.

Wässle, H., Peichl, L., and Boycott, B.B. 1981a. Morphology and topography of on- and off-alpha cells in the cat retina. Proc. R. Soc. Lond. [B] 212: 157–175.

Wässle, H., and Riemann, H.J. 1978. The mosaic of nerve cells in the mammalian retina. Proc. R. Soc. Lond. [B] 200: 441–461.

Wässle, H., Schafer-Trenkler, I., and Voigt, T. 1986. Analysis of a glycinergic inhibitory pathway in the cat retina. Neuroscience 6: 594–604.

Watanabe, K., and Kawana, E. 1984. Selective retrograde transport of tritiated D-aspartate from the olfactory bulb to the anterior olfactory nucleus of the lateral tract in the rat. Brain Res. 296: 148–151.

Watanabe, K., Taguchi, Y., Shiosaka, S., Tanaka, J., Kubota, H., Terano, Y., Tohyama, M., and Wada, H. 1984. Distribution of the histaminergic neuron system in the central nervous system of rats: A flourescent immunohistochemical analysis with histidine decarboxylase as a marker. Brain Res. 295: 13–25.

Watkins, J.C., and Olverman, H.J. 1987. Agonists and antagonists for excitatory amino acid receptors. Trends Neurosci. 10: 265–272.

Weight, F.F. 1983. Synaptic mechanisms in amphibian sympathetic ganglia. In: Autonomic Ganglia (Elfvin, L.G., ed.). New York: John Wiley, pp. 309–344.

Weitsen, H.A., and Weight, F.F. 1977. Synaptic innervation of sympathetic ganglion cells in the bullfrog. Brain Res. 128: 197–211.

Wellis, D.P., and Scott, J.W. 1987. Intracellular recordings of odor-induced responses in the rat olfactory bulb. Chem. Senses 12: 707.

Werblin, F.S., and Dowling, J.E. 1969. Organization of the retina of the mudpuppy, *Necturus maculosus*. II. Intracellular recording. J. Neurophysiol. 32: 339–355.

Westenbroek, R.E., Westrum, L.E., Hendrickson, A.E., and Wu, J.-Y. 1987. Immunocytochemical localization of cholecystokinin and glutamic acid decarboxylase during normal development in the prepiriform cortex of rats. Dev. Brain Res. 34: 191–206.

Westenbroek, R.E., Westrum, L.E., Hendrickson, A.E., and Wu, J.-Y. 1988. Ultrastructural localization of immunoreactivity in the developing piriform cortex. J. Comp. Neurol. 274: 319–333.

Westrum, L.E. 1969. Electron microscopy of degeneration in the lateral olfactory tract and plexiform layer of the prepiriform cortex of the rat. Z. Zellforsch. 98: 157–187.

Westrum, L.E. 1970. Observations on initial segments of axons in the prepiriform cortex of the rat. J. Comp. Neurol. 139: 337–356.

White, E.L. 1972. Synaptic organization in the olfactory glomerulus of the mouse. Brain Res. 37: 69–80.

White, E.L., and Rock, M.P. 1980. Three-dimensional aspects and synaptic relationships of a Golgi-impregnated spiny stellate cell reconstructed from serial thin sections. J. Neurocytol. 9: 615–636.

Wigstrom, H., and Gustafsson, B. 1983. Facilitated induction of hippocampal long-lasting potentiation during blockade of inhibition. Nature 301: 603–604.

Wigstrom, H., Gustafsson, B., and Huang, Y.-Y. 1986. Hippocampal long-lasting potentiation is induced by pairing single afferent volleys with intracellularly injected depolarizing current pulses. Acta Physiol. Scand. 126: 317–319.

Williams, D.R. 1986. Seeing through the photoreceptor mosaic. Trends Neurosci. 9: 193–197.

Williams, J.T., North, R.A., Shefner, S.A., Nishi, S., and Egan, T.M. 1984. Membrane properties of rat locus coeruleus neurones. Neuroscience 13: 137–156.

Williams, R.W., Bastiani, M.J., and Chalupa, L.M. 1983. Loss of axons in the cat optic nerve following fetal unilateral enucleation: An electron microscope analysis. J. Neurosci. 3: 133–144.

Wilson, C.J. 1984. Passive cable properties of dendritic spines and spiny neurons. Neuroscience 4: 281–297.

Wilson, C.J., 1986a. Postsynaptic potentials evoked in spiny neostriatal neurons by stimulation of ipsilateral and contralateral neocortex. Brain Res. 367: 201–213.

Wilson, C.J. 1986b. Three dimensional analysis of dendritic spines by means of HVEM. J. Electron Microsc. 35 (Suppl.): 1151–1155.

Wilson, C.J. 1988. Cellular mechanisms controlling the strength of synapses. J. Electron Microscopic Tech. 10: 293–313.

Wilson, C.J., Chang, H.T., and Kitai, S.T. 1982. Origins of postsynaptic potentials evoked in identified rat neostriatal neurons by stimulation in substantia nigra. Exp. Brain Res. 45: 157–167.

Wilson, C.J., Chang, H.T., and Kitai, S.T. 1983a. Origins of postsynaptic potentials evoked in spiny neostriatal projection neurons by thalamic stimulation in the rat. Exp. Brain Res. 51: 217–226.

Wilson, C.J., Chang, H.T., and Kitai, S.T. 1983b. Disfacilitation and long-lasting inhibition of neostriatal neurons in the rat. Exp. Brain Res. 51: 227–235.

Wilson, C.J., Chang, H.T., and Kitai, S.T. 1990. Firing pattern and synaptic potentials of identified giant aspiny interneurons in the rat neostriatum. J. Neurosci. 10.

Wilson, C.J., and Groves, P.M. 1980. Fine structure and synaptic connections of the common spiny neuron of the rat neostriatum: A study employing intracellular injection of horseradish peroxidase. J. Comp. Neurol. 194: 599–616.

Wilson, C.J., and Groves, P.M. 1981. Spontaneous firing patterns of identified spiny neurons in the rat neostriatum. Brain Res. 220: 67–80.

Wilson, C.J., Groves, P.M., Kitai, S.T., and Linder, J.C. 1983c. Three-dimensional structure of dendritic spines in the rat neostriatum. J. Neurosci. 3: 383–398.

Wilson, J.R., Friedlander, M.J., and Sherman, S.M. 1984. Ultrastructural morphology of identified X- and Y-cells in the cat's lateral geniculate nucleus. Proc. R. Soc. Lond. [B] 221: 411–436.

Wilson, M.A., and Bower, J.M. 1988. A computer simulation of olfactory cortex with functional implications for storage and retrieval of olfactory information. In: Neural Information Processing Systems (Anderson, D.Z., ed.). New York: American Institute of Physics, pp. 114–126.

Winson, J. 1976. Hippocampal theta rhythm. II. Depth profiles in the freely moving rabbit. Brain Res. 103: 71–79.

Winson, J., and Abzug, C. 1978. Neuronal transmission through the hippocampal pathways dependent on behavior. J. Neurophysiol. 41: 716–732.

Wong, R.K.S., and Prince, D.A. 1978. Participation of calcium spikes during intrinsic burst firing in hippocampal neurons. Brain Res. 159: 385–390.

Wong, R.K.S., Prince, D.A., and Basbaum, A.I. 1979. Intradendritic recordings from hippocampal neurons. Proc. Natl. Acad. Sci. USA 76: 986–990.

Wong, R.K.S., Traub, R.D., and Miles, R. 1984. Epileptogenic mechanisms as revealed by studies of the hippocampal slice. In: Electrophysiology of Epilepsy (Schwartzkroin, P.A., and Wheal, H.V., eds.). New York: Academic Press, pp. 253–276.

Wong-Riley, M.T.T., and Carroll, E.W. 1984. Quantitative light and electron microscopic analysis of cytochrome oxidase-rich zones in VII prestriate cortex of the squirrel monkey. J. Comp. Neurol. 222: 18–37.

Woolf, N.J., and Butcher, L.L. 1986. Cholinergic systems in the rat brain. III. Projections from the pontomesencephalic tegmentum to the thalamus, tectum, basal ganglia and basal forebrain. Brain Res. Bull. 16: 603–637.

Woolf, T.B., Shepherd, G.M., and Greer, C.A. 1988. Models of local electrical interactions within spiny dendrites of granule cells in mouse olfactory bulb. Soc. Neurosci. Abstr. 14: 620.

Wouterlood, F.G., Mugnaini, E., and Nederlof, J. 1985. Projection of olfactory bulb efferents to layer I GABAergic neurons in the entorhinal area. Combination of antegrade degeneration and immunoelectron microscopy in rat. Brain Res. 343: 283–296.

Yamada, W., Koch, C., and Adams, P.R. 1989. Multiple channels and calcium dynamics. In: Methods in Neuronal Modelling: From Synapses to Networks (Koch, C., and Segev, I., eds.). Cambridge, MA: MIT Press, pp. 97–133.

Yamamoto, C., and McIlwain, H. 1966. Electrical activities in thin sections from the mammalian brain maintained in chemically defined media in vitro. J. Neurochem. 13: 1333–1343.

Yang, X.L., and Wu, S.M. 1989a. Effects of background illumination on the horizontal cell responses in the tiger salamander retina. J. Neurosci. 9: 815–827.

Yang, X.L., and Wu, S.M. 1989b. Modulation of rod-cone coupling by light. Science 244: 352–354.

Yeterian, E.H., and Van Hoesen, G.W. 1978. Cortico-striate projections in the rhesus monkey: The organization of certain cortico-caudate connections. Brain Res. 139: 43–63.

Young, A.B., and Macdonald, R.L. 1983. Glycine as a spinal cord neurotransmitter. In: Handbook of the Spinal Cord, Vol. I: Pharmacology (Davidoff, R.E., ed.). New York: Marcel Dekker, pp. 1–43.

Young, A.B., Oster-Granite, M.L., Herndon, R.M., and Snyder, S.H. 1974. Glutamic acid: Selective depletion by viral induced granule cell loss in hamster cerebellum. Brain Res. 73: 1–13.

Zaborszky, L., Carlsen, J., Brashear, H.R., and Heimer, L. 1986. Cholinergic and

GABAergic afferents to the olfactory bulb in the rat with special emphasis on the projection neurons in the nucleus of the horizontal limb of the diagonal band. J. Comp. Neurol. 243: 488–509.

Zador, A.M., Koch, C.K., and Brown, T.H. 1990. Biophysical model of a hebbian synapse. Proceedings of the International Joint Conference on Neural Networks. Washington, D.C.

Zarbin, M.A., Innis, R.B., Wamsley, J.K., Snyder, S.H., and Kuhar, M.J. 1983. Auto-radiographic localization of cholecystokinin receptors in rodent brain. J. Neurosci. 3: 877–906.

Zbicz, K.L., and Weight, F.F. 1985. Transient voltage and calcium-dependent outward currents in the hippocampal CA3 pyramidal neurons. J. Neurophysiol. 53: 1038–1058.

Zeki, S., and Shipp, S. 1988. The functional logic of cortical connections. Nature 335: 311–317.

Zengel, J.E., Reid, S.A., Sypert, G.W., and Munson, J.B. 1985. Membrane electrical properties and prediction of motor-unit type of cat medial gastrocnemius motoneurons in the cat. J. Neurophysiol. 53: 1323–1344.

Zucker, C.L., and Dowling, J.E. 1987. Centrifugal fibers synapse on dopaminergic interplexiform cells in the teleost retina. Nature 330: 166–168.

Zucker, R.S. 1989. Short-term synaptic plasticity. Annu. Rev. Neurosci. 12: 13–31.

AUTHOR INDEX

Abzug, C., 362, 378
Adams, P. R., 38, 44, 45, 57, 60, 76, 79, 82, 379, 382, 409
Adrian, E. D., 158
Aghanjanian, G. K., 41, 61
Airaksinen, M. S., 340
Akaike, A., 306
Akeson, R. A., 158
Alger, B. E., 421, 422
Allison, A. C., 135, 140
Amaral, D. G., 347, 349, 350, 352, 353, 354, 361, 363, 364
Amthor, R. R., 212
Andersen, P., 235, 353, 359, 415, 421, 422
Anderson, C. L., 426
Anderson, J. A., 345, 348
Andrade, R., 60, 379
Andres, K., 145
Andrews, S. B., 411
Anson, J., 337
Armstrong, D. M., 342
Armstrong-James, M., 423
Aronin, N., 293, 298
Artola, A., 411
Ascher, P., 54, 57, 375, 381, 414, 465
Ashmore, J. F., 194
Ashwood, T. J., 421
Attwell, D., 191, 202
Atwood, R. P., 214, 218
Ayala, G. F., 365
Ayoub, G. S., 194

Baker, H., 158, 159
Baldissera, F., 90, 98, 100, 102, 103, 105, 108
Banks, M. S., 211
Barber, P. C., 147
Bargas, J., 308, 309
Barlow, H. B., 172, 189, 190, 196, 198, 202, 203, 210, 212, 430, 432, 433
Barrett, E. F., 113, 126
Barrett, J. N., 118, 120, 126
Barrionueva, G., 354, 358, 364, 373
Baudry, M., 384
Baughman, R. W., 420, 421, 423
Baylor, D. A., 191, 198, 208, 211, 212
Beaulieu, C., 398, 418, 430
Behan, M., 325, 327, 328, 330
Bell, C. C., 243
Benardo, L. S., 383
Bennett, M. V. L., 110
Benson, T., 146, 149
Berger, T. W., 384
Berman, N. J., 413, 420
Biedenbach, M. A., 330, 332, 333
Bigaré, F., 221

Bindman, L. J., 411
Bishop, G. A., 293
Bisti, S., 238
Björklund, A., 340
Black, J. G., 74
Blackstad, T. W., 354
Bland, B. H., 378
Blaxter, T. J., 421, 422
Bliss, T. V. P., 56, 135, 371, 410
Bloedel, J. R., 106, 214
Bloom, F. E., 161, 162, 234, 236, 238, 380
Bloomfield, S. A., 18, 193, 261, 274, 416
Bogan, N., 161
Bolam, J. P., 286, 293, 298, 300
Boltz, J., 195, 196, 353
Bower, J. M., 24, 326, 328, 330, 332, 333, 334, 335, 345
Bowery, N. G., 51, 116
Bowling, D. B., 252
Boycott, B. B., 143, 175, 177, 184, 185, 186, 187
Boyd, I. A., 91
Braitenberg, V., 214, 218
Braitman, D. J., 337
Brayton, R. K., 16, 429, 430, 463, 465
Brazier, M. A. B., 96
Brening, R. K., 170
Brock, L. G., 96, 110
Brodmann, , 390
Brown, A. G., 90, 91, 94, 95, 110, 120
Brown, D. A., 18, 45, 60, 338, 347, 371, 374, 375, 378, 381, 382, 383
Brown, T. H., 79, 347, 354, 356, 358, 363, 364, 365, 366, 368, 369, 384, 385, 387, 422
Brownstein, M. J., 339
Bruce, C., 433
Buchsbaum, G., 203
Bührle, C. P., 115, 116
Bullier, J., 403
Burd, G., 147
Burgoyne, R. D., 409, 411
Burke, R. E., 91, 92, 94, 98, 106, 107, 110, 112, 114, 115, 116, 118, 120, 124, 126, 128, 129, 130, 131
Butcher, L. L., 422

Calabresi, P., 306
Caldwell, J. H., 212
Calleja, C., 323
Calvin, W. H., 126
Carbonne, E., 42
Card, M. P., 340
Carey, J., 288, 294
Carroll, E. W., 403
Cattarelli, M., 345

Caviness, V. S., 325
Celio, M. R., 298
Chang, F-LF, 382
Chang, H. T., 286, 288, 289, 294, 381, 382
Chan-Palay, V., 216, 217, 221, 238, 239
Chen, S. C., 224, 225
Cherubini, E., 310, 312
Chesselet, M. F., 302
Chevalier, G., 315
Chover, J., 345
Chujo, T., 237
Chun, M. H., 195
Churchland, P. S., 388
Claiborne, B. J., 352, 354, 359, 363
Clapman, D. E., 51
Clark, W. E. leGros, 149
Cleland, B. G., 198, 200
Cohen, E., 177, 186, 187, 195, 199
Cohen, N. H., 346
Cole, A. E., 379, 413
Cole, K. S., 118
Collingridge, G. L., 56, 374, 410
Collins, G. G. S., 335, 337, 338
Colonnier, M., 398, 418, 430
Connor, J. A., 45, 382
Connors, B. W., 23, 43, 46, 407, 408, 409,
   413, 420, 421
Conradi, S., 109
Constanti, A., 331–332
Constantine-Paton, M., 189
Conway, L., 20
Coombs, J. S., 110, 115
Copenhagen, D. R., 194
Corkin, S., 346
Cotman, C. W., 160, 238, 337, 371, 373,
   374
Cottrell, G. A., 421, 422
Courville, J., 106, 214
Cowan, W. M., 180, 352, 358
Cowey, A., 400
Crepel, F., 46, 237
Creutzfeldt, O. D., 390, 415
Crick, F., 426, 465
Crill, W. E., 118, 120, 127, 407
Crone, C., 127
Crunelli, V., 414
Cuello, A. C., 339
Cuénod, M., 337
Cull-Candy, S. G., 381
Cullheim, S., 92, 100, 110, 115, 117, 118
Curtis, D. R., 113, 117, 237, 238
Czarkowska, J., 91

Dacey, D., 179
Dahl, D., 380
Dale, H. H., 160
Dale, N., 54, 414
Daniel, P. M., 396
Dann, J. F., 188
Darian-Smith, I., 90, 95
Datta, A. K., 127

Davey, M. R., 91
Davis, B. J., 161
Davis, B. M., 137, 158
Daw, N., 193, 195
Deadwyler, S. A., 361
Dehay, C., 418
Delay, R., 426, 465
DeLong, M. R., 312
Demeter, S., 349
Demeulemeester, H., 423
De Monasterio, F. M., 173
De Montigny, C., 243
Deniau, M. M., 315
Dennis, B. J., 335
Denny-Brown, D., 126, 127
Dent, J. A., 354
De Quidt, M. E., 339
Derrington, A. M., 198, 207
Deschênes, M., 43
Desmedt, J. E., 129
Devoe, R., 212
Devor, M., 320
DiFiglia, M., 286, 288, 289, 290, 293, 294,
   298
DiFrancesco, D., 45–46
Dingledine, R., 417
Dobson, V. G., 29
Dodd, J., 69, 83
Dodge, F. A., 118, 119
Dodt, H. U., 307
Doerner, D., 393
Dolphin, A. C., 113, 117
Donoghue, J. P., 299
Douglas, R. J., 24, 417, 418, 420, 437
Dowling, J. E., 143, 177, 180, 185, 187,
   192, 198
Dreifuss, J. J., 418
Dubé, L., 291
Dubin, H., 172, 187
Dudek, F. E., 49, 359
Dunlap, K., 113
Dunwiddie, T. V., 377, 380
Dupont, J. L., 238
Durand, D., 118, 120
Dutar, P., 59, 368, 369, 419, 421
Dykes, R. W., 252

Easter, S. S., 189
Ebner, F. F., 320
Eccles, J. C., 91, 94, 98, 101, 113, 232,
   234, 235
Eccles, R. M., 90, 91
Eckenstein, F., 423
Edwards, R. F., 122
Emonet-Denand, F., 94
Emson, P. C., 339
Eng, D. L., 135
Engberg, I., 110, 112, 238
Enoch,   , 191
Enroth-Cugell, C., 197, 198, 207,
   213

Eränkö, O., 70
Ericson, H., 340

Faber, D. S., 9
Fahrenbach, W. H., 27
Fallon, J. H., 339
Fanglietti, E. V., 176, 190, 195, 196, 255
Farbman, A. I., 135
Fatt, P., 417
Feig, S., 323, 324, 325, 326, 329
Feldman, A. G., 103
Felix, D., 238
Ferreyra-Moyana, H., 334
Ferster, D., 23, 27
Fetcho, J. R., 92
ffrench-Mullen, J. M. H., 337, 344
Fifkova, E., 382, 409, 411, 426, 465
Finch, D. M., 332
Finkel, A. S., 120
Fischbach, G. D., 113
Fitzpatrick, D., 24, 253, 254–255, 275
Fleshman, J. W., 91, 105, 118, 119, 120, 126, 127
Fonnum, F., 238, 336–337
Foote, S. L., 238
Forehand, C. J., 70, 85
Forsythe, I. D., 369, 381
Fox, C. A., 284, 286, 290
Fox, K., 423
Fox, S., 117
Frank, B., 189
Frank, K., 116
Frederickson, R. C. A., 238
Freed, M. A., 177, 182, 186, 187, 188, 190, 191, 195, 205, 206, 212
Freeman, W. J., 25, 149, 158, 342
Freund, T. F., 293, 393, 396, 398, 399, 415, 417, 433
Friedlander, M. J., 256, 261
Fukuda, Y., 175
Fuller, T. A., 337
Fuortes, M. G. F., 116
Fyffe, R. E. W., 90, 91, 109, 110, 120

Gabbott, P. L. A., 393, 415
Galarraga, , 314
Gale, K., 317
Gall, C., 358
Galvan, M., 38, 44, 57, 76, 331–332
Gamble, E., 382, 426–427, 463, 464
Gardner-Medwin, A. R., 371
Garey, L. J., 393
Garnett, R., 127, 128, 129
Gerfen, C. R., 298, 299, 300, 302
Gerschenfeld, H. M., 192
Gerson, M. D., 67
Gesteland, R. C., 135
Getchell, T. V., 135, 149, 167
Ghosh, S., 323
Gilbert, C. D., 396, 397, 405, 420
Giuffrida, R., 253

Gobel, S., 143
Godaux, E., 129
Goebel, D. J., 177, 195
Gogan, P., 109, 110
Goldman, D. E., 35
Goldman, P. S., 299
Goldman-Rakic, P. S., 285, 299
Golgi, C., 141
Gottlieb, D. I., 358
Gottschalk, A., 203
Granger, R., 24, 345
Graveland, G. A., 290
Gray, E. G., 6, 398
Graybiel, A. M., 282, 285, 299, 300, 301, 302
Graziadei, P. P. C., 135, 158
Greenough, W. T., 382
Greenwood, R. S., 339
Greer, C. A., 140, 141, 145, 146, 165
Grieve, K. L., 424
Griffith, W. H., 263, 369, 383
Grillner, S., 103
Grimm, R. J., 243
Gross, C., 433
Groves, P. M., 286, 293, 312
Grüsser, O. -J., 106
Guillery, R. W., 255, 256
Gustafsson, B., 410, 411

Haas, H. L., 60
Haberly, L. B., 24, 317, 320, 321, 322, 323, 324, 325, 326, 327, 328, 329, 330, 331, 332, 333, 334, 335, 337, 338, 339, 340, 342, 345
Hablitz, J. J., 414
Hackett, J. T., 237
Hagihara, K., 420
Halasz, N., 146, 158, 160
Hall, W. C., 150, 320
Halliwell, J. V., 45, 332, 379, 382
Hamilton, K. A., 168
Hamlyn, L. H., 354
Hammill, O. P., 57
Hámori, J., 221, 225
Hamos, J. E., 255, 256, 260, 263
Hara, Y., 339
Harkonen, M., 70
Harland, R. E., 339
Harris, E. W., 373, 374
Harris, K. M., 465
Harrison, T. A., 137
Harting, J. K., 253
Hartzell, H. C., 57
Hasan, Z., 92
Hattori, T., 293
Hearn, T. J., 337
Hebb, D. O., 345, 347, 373, 374, 410
Heggelund, P., 30
Heimer, L., 321, 327
Heit, G., 346
Heizman, C. W., 298

Hendrickson, A. E., 403
Hendry, A. H. C., 423
Henneman, E., 91, 110, 113, 124, 126, 127, 129
Henry, G. H., 403
Herkenham, M., 299, 300, 301
Herrling, P. L., 306
Herron, C. E., 421
Hersch, S. M., 393
Hess, R., 239
Hicks, T. P., 420
Hikosaka, O., 315, 316
Hille, B., 32, 34, 38, 50, 382, 409
Hillman, D. E., 224, 225
Hinds, J. W., 46
Hinton, G., 384
Hirata, Y., 145
Hirsh, R., 384
Hodgkin, A. L., 35, 36, 457
Hoffer, B. J., 236
Hoffman, W. H., 337, 338
Hoffmann, K-P., 403
Hökfelt, T., 238
Hollingworth, H. L., 387
Hollyfield, J., 189
Holsheimer, J., 378
Honig, M. G., 103
Honore, T., 380
Hopfield, J. J., 345, 384
Hopkins, W. F., 373
Hori, N., 337
Horn, J. P., 69, 74, 83
Hornung, J. P., 393
Horton, J. C., 403
Hosoya, Y., 107
Hotson, J. R., 41, 407, 409
Houk, J. C., 98
Hounsgaard, J., 127
Houser, C. R., 423
Howlett, S. J., 337
Hubel, D. H., 29, 189, 397, 401, 403
Hudson, D. B., 237
Huettner, J. E., 421
Hughes, A., 184
Huguenard, J. R., 407, 413
Hull, C. D., 306
Hull, C. L., 387
Hultborn, H., 101, 102
Humphrey, A. L., 396
Hunt, C. C., 127
Hurvich, L. M., 434
Huxley, A. F., 36, 457

Illert, M., 103, 105
Illes, P., 116
Inagaki, S., 239, 339, 340
Inoue, M., 420
Ishizuka, N., 91
Israël, M., 239
Ito, M., 221, 237
Iversen, L. L., 371

Jack, J. J. B., 91, 110, 112, 113, 114, 117, 119, 120, 369, 414, 415, 429, 430, 440, 463
Jackowski, A., 146
Jacobson, M., 28, 146
Jahnsen, H., 38, 43, 62, 230
Jahr, C. E., 113, 150, 155, 160, 161, 167, 194, 374, 381, 414
Jameson, D., 434
Jami, L., 94
Jan, L. Y., 73, 74, 82
Jan, Y. N., 73, 74, 82
Janigro, D., 368, 378
Jankowska, E., 91, 100, 101, 102, 105, 113
Jansen, J., 214
Jensen, R., 212
Johnson, J. W., 57
Johnston, D., 347, 354, 356, 363, 364, 365, 366, 368, 369, 373, 380, 381
Johnston, G. A. R., 237
Jones, E. G., 246, 247, 248, 252, 253, 256, 259, 284, 430
Jones, K. A., 421
Jones, S. W., 74, 79
Jourdan, F., 135

Kaczmarck, L. K., 382
Kairiss, E. W., 384, 387
Kaiserman-Abramof, I. R., 323, 426, 430
Kan, K. S. K., 239
Kanazawa, I., 339
Kanda, K., 127
Kandel, E. R., 383
Kaneko, A., 195, 198
Karlsson, G., 421
Katayama, Y., 313
Katz, B., 35, 417
Kauer, J. S., 168
Kawamura, H., 238
Kawana, E., 337
Kay, A. R., 383
Keenan, C. L., 364, 378
Kellerth, J. -O., 92, 99, 100, 110, 115
Kelly, J. P., 405
Kelso, S. R., 354, 358, 373
Kemp, J. A., 252, 422
Kemp, J. M., 291
Kerr, D. I. B., 335, 421
Ketchum, K. L., 340, 342
Kimura, M., 312, 314
Kirkwood, P. A., 188
Kishi, K., 137, 139
Kisvárday, Z. F., 400
Kita, H., 294
Kita, T., 303, 307, 308, 310
Kitai, S. T., 288, 294, 305, 306
Kjaerheim, A., 354
Knowles, W. D., 359, 363
Koch, C., 3, 5, 7, 10, 12, 14, 18, 20, 23, 27, 83, 193, 246, 248, 253, 259, 264,

266, 270, 273, 274, 275, 383, 425, 426–427, 428, 430, 463, 464, 468, 471, 472
Kocsis, J. D., 135, 299
Koehn, O., 427
Koerber, H. R., 90
Kohler, C., 361
Kohonen, T., 345, 384, 387
Kolb, H., 170, 175, 176, 177, 180, 185, 186, 190
Kölliker, 288
Komatsu, Y., 399, 411, 415
Konnerth, A., 60
Korn, H., 9, 75
Kosaka, T., 298
Kriegstein, A. R., 23
Krnjević, K., 57, 113, 237, 378, 416, 420, 423
Kroller, J., 106
Kruger, L., 319
Kuba, K., 74, 75
Kubo, T., 73
Kuffler, S. W., 32, 34, 38, 73, 82, 196, 197
Kumar, V., 347
Kunkel, D. D., 363
Kuno, M., 110, 112
Künzle, H., 284

Lacaille, J. -C., 352, 353
Lacey, M. G., 306
Lägerbäck, P. -Å., 99
Lamarre, Y., 243
Lampson, L., 187
Lancaster, B., 409
Lancet, D., 135
Land, E. H., 434
Land, L. J., 149
Landis, D. M. D., 465
Landmesser, L., 92
Lanfumey, L., 310
Langmoen, I. A., 414, 417
Laporte, Y., 101
Latorre, R., 382–383
Laughlin, S., 203, 211
Laurberg, S., 350
Lawrence, D. G., 114
Legge, K. F., 340
Lehky, S. R., 435
Lennie, P., 198, 207, 252, 277
Leslie, F. M., 339
LeVay, S., 396, 399, 415, 417
Leveteau, J., 140
Levey, A. I., 275
Levick, W. R., 198
Levitan, I. B., 382
Lev-Tov, A., 112, 113, 116
Lichtman, J. W., 84
Liddell, E. G. T., 124
Lieberman, A. R., 145–146, 247, 253
Liesi, P., 147
Light, A. R., 109
Lindsay, R. M., 147

Lindström, S., 91, 102
Lindvall, O., 340
Linsenmeier, R. A., 200
Linsker, R., 436
Livingstone, M. S., 307, 403
Ljungdahl, A., 238, 339
Llinás, R., 25, 41, 43, 62, 88, 118, 228, 230, 232, 237, 239, 242, 243, 244, 267, 269, 270, 271, 273, 275, 381, 382, 383, 407, 429
Lloyd, D. P. C., 91, 96, 101
Lo, F. -S., 252, 274
Lomo, T., 371, 410
Lopez, H. S., 82
Loren, I., 339
Loy, R., 380
Lund, J. S., 400
Lundberg, A., 90, 91, 97, 101, 105, 106, 107, 108
Lüscher, H. -R., 112
Luskin, M. B., 321, 328, 329, 340
Lux, H. D., 42
Lynch, G., 377
Lynch, J., 384, 410

McArdle, C. B., 188
McBride, W. J., 237
McBurney, R. N., 409, 410
McClelland, J. L., 384
McCormick, D. A., 41, 46, 57, 62, 267, 269, 275, 276, 411, 413, 423
MacDermott, A. B., 54, 414
Macdonald, R. L., 115
McGreer, P. L., 298
McGuire, B. A., 177, 180, 186, 187, 190, 195
McIlwain, H., 364
McIntyre, D. C., 317
McLachlan, E., 75
McLaughlin, B. J., 116, 238
McLennan, H., 54, 160
MacLeod, P., 149
Macrides, F., 137, 139, 140, 141, 146, 150, 158
MacVicar, V. A., 359
Madison, D. V., 60, 380, 383, 408, 409, 413, 423
Magistretti, P. J., 339
Magleby, K. L., 113
Magoun, H. W., 62
Malach, R., 285
Malenka, R. C., 377
Malinow, R., 373
Malouf, 368
Mancillas, J. F., 422, 424
Margolis, F. L., 158
Mariani, A. P., 175, 177
Mariani, J., 224
Markham, J. A., 426
Marr, D., 384, 434
Marshall, K. C., 110, 112

Martin, A. R., 114
Martin, K. A. C., 24, 393, 394, 396, 397,
    398, 403, 418, 420, 430, 437
Masland, R. H., 195, 196, 206
Massey, S. C., 194, 196
Mastronarde, D. N., 198, 208, 210
Masukawa, L. M., 379, 383, 429
Mates, S. L., 400
Mathers, L. H., 352
Matsushita, M., 107
Matthews, D. A., 352
Matthews, M. A., 110
Matthews, P. B. C., 92
Maturana, H. R., 172
Maunsell, J. H. R., 397
Maurer, R., 339
Mayer, M. L., 273, 337, 374
Mead, C., 20
Mendell, L. M., 91, 110, 124, 129
Merigan, W. H., 172
Merzenich, M. M., 86
Mesulam, M. M., 422
Mezey, E., 225
Michael, C. R., 252
Miller, J. P., 373, 429, 465
Miller, R. F., 18, 193, 194, 196, 363
Minota, S., 74, 75
Misgeld, U., 307
Mishkin, M., 346, 384
Mitchell, S. J., 378
Mitzdorf, U., 403
Moczydlowski, E., 382–383
Monaghan, D. T., 337, 371
Montero, V. M., 256, 263
Monti-Graziadei, G. A., 135, 158
Moody, C. I., 252, 274
Moore, R. Y., 340, 380
Moran, D., 135
Morgan, J. L., 158
Mori, K., 137, 139, 140, 141, 145, 150,
    153, 160, 161
Moruzzi, G., 62
Mountcastle, V. B., 401
Mugnaini, E., 159, 244, 339
Mühlethaler, M., 230, 237
Muir, R. B., 114
Munson, J. B., 97
Murthy, K. S. K., 94

Nadel, L., 346, 378, 384
Naka, K., 177
Nakamura, Y., 180
Nakanishi, H., 314
Nathans, J., 192
Nauta, W. J., 299
Nawys, S., 194
Neer, E. J., 51
Neering, I. R., 409, 410
Nelson, C., 418
Nelson, P. G., 109, 126, 176
Nelson, R., 177, 179, 186, 190, 208

Newberry, N. R., 59, 379
Nicholson, C., 232, 239, 242
Nicoll, R. A., 56, 57, 59, 60, 151, 155,
    160, 161, 167, 368, 378, 379, 380, 408,
    409, 413, 419, 421, 422, 423
Nishi, S., 75
Nishimura, Y., 146, 188
Nistri, A., 69, 116
North, R. A., 57
Nowak, L., 54, 374, 375, 381, 414,
    465
Nowycky, M. C., 42, 150, 160, 383
Numann, R. E., 382

Obata, K. T., 237, 238
Oertel, W. H., 159, 237, 244, 298, 339
Ogawa, T. S., 417, 418
Ohara, P. T., 247, 253, 254–255
Ohtsuka, T., 174–175
Ojima, H., 325, 329
Okamoto, K., 238
O'Keefe, J., 246, 378, 384
O'Leary, J. L., 320, 323, 324
Olpe, H. R., 423
Olson, C. B., 126, 127
Olton, D. S., 384
Olverman, H. J., 54, 380
Orlovsky, G. N., 103
Orona, E., 137, 139, 140, 141, 150
Ottersen, O. P., 420

Palay, S. I., 216, 221
Palkovits, M., 221, 225, 339
Panula, P., 341
Pape, H. -C., 57, 275
Passingham, R., 389
Pasternak, T., 172
Pedersen, P. E., 150
Penit-Soria, J., 46
Penny, G. R., 296, 300
Perkins, M. N., 421
Pert, C. B., 299, 301
Peters, A., 255, 323, 392, 393, 399, 400,
    415, 417, 426, 430
Petri, H. L., 346, 384
Pfaffinger, P. J., 82
Phelps, P. E., 293, 298
- Phillips, C. G., 96, 106, 114, 154, 155, 165,
    415
Phillis, J. W., 237, 239, 420
Piccolino, M., 179, 192
Pikel, V. M., 234
Pinching, A. J., 140, 141, 143, 145, 147
Pinter, M. J., 128
Piredda, S., 317
Poggio, T., 3, 5, 20, 83, 428, 430, 434,
    463, 467, 468, 471
Porter, R., 106, 114
Potter, D. D., 73
Pourcho, R. G., 177, 195, 196
Powell, T. P. S., 140, 141, 143, 145, 146,

147, 154, 291, 349, 390, 393, 397, 398, 401, 430
Precht, W., 296
Presto, S., 325, 326, 327, 330
Preston, R. J., 286
Pribram, K. H., 319
Price, J. L., 141, 145, 146, 154, 318, 319, 320, 321, 323, 327, 328, 329, 337, 340
Prince, D. A., 41, 57, 62, 267, 269, 275, 276, 379, 407, 408, 409, 413, 423, 429
Probett, G. A., 335
Provini, L., 238
Pull, E., 112
Purkinje, J. E., 218
Purves, D., 83, 84, 85

Qian, N., 465

Racine, R. J., 317
Rack, P. M. H., 105
Rafols, J. A., 286, 290
Ragsdale, C. W., Jr., 282, 299
Rainbow, T. C., 337
Rakic, P., 7, 146, 188, 249
Rall, W., 10, 85, 91, 110, 112, 117, 118, 120, 122, 124, 127, 145, 154, 155, 157, 163, 164, 165, 167, 261, 422, 423, 426, 429, 430, 440, 451, 453, 455, 456, 459, 461, 463, 464, 465, 467, 471
Ralston, H. J., 254–255, 256, 259, 260, 261
Ramoa, A., 188
Ramón y Cajal, S., 133, 139, 140, 141, 172, 177, 179, 180, 192, 214, 216, 221, 222, 284, 320, 323, 347, 393, 400, 401
Ramón-Moliner, E., 145
Ranck, J. B., 378
Ratliff, F., 27
Reader, T. A., 423
Redburn, D., 194, 196
Redman, S. J., 91, 110, 112, 120, 415, 462
Reese, T. S., 145, 465
Regidor, J., 393, 400
Reh, T., 189
Renshaw, B., 98
Reubi, J. C., 339
Reymond, L., 172
Ribak, C. E., 150, 160, 353, 363, 415, 417
Richmond, B., 433
Richmond, F. J. R., 92
Riemann, H. J., 182
Riley, K. P., 249
Rinzel, J., 122, 463
Robbins, S. L., 347
Roberts, E., 358
Roberts, G. W., 339
Roberts, W. J., 102
Robson, J. G., 197
Rock, M. P., 398, 415, 417
Rockel, A. J., 390
Rodieck, R. W., 170, 172, 173, 176, 183, 191, 197, 252

Rodriguez, R., 335, 340, 342
Roffler-Tarlov, S., 237
Rogawski, 49
Rohde, B. H., 237
Rolls, E. T., 346, 387
Roman, F., 335
Romanes, G. J., 92
Romer, A. S., 214
Ronnevi, L. -O., 109
Rose, A., 200, 212
Rose, D., 29
Rose, P. K., 92
Rosenberg, M., 135
Rotter, A., 162
Rovira, C., 379
Rudomin, P., 94, 98, 106, 112, 114, 115, 116, 118
Rudy, B., 38, 44
Rumelhart, D. E., 384
Rustioni, A., 253
Ryall, R. W., 101
Rymer, W. A., 98

Saito, H., 175
Sakai, H., 177
Sakai, Y., 238
Sakanaka, M., 339
Salt, T. E., 252, 274, 424
Sandell, J., 195, 205
Sanderson, K. J., 254, 264
Sandoval, M. E., 238
Sanides, D., 323
Sanides, F., 323
Sarvey, J. M., 421, 422
Sasaki, K., 244
Satou, M., 330, 332, 338
Scharfman, H. E., 353, 363, 421, 422
Schein, S. J., 173
Schiller, P. H., 205
Schmechel, D. E., 423
Schmidt, M., 206
Schneider, S. P., 137, 139, 140, 141, 150
Schoenfeld, T. A., 140
Scholfield, C. N., 331, 332, 338
Schwartz, E., 195
Schwartz, S., 416
Schwartzkroin, P. A., 352, 353, 363, 368, 378, 381, 383, 408
Schwindt, P. C., 126, 127, 407, 408, 414, 423
Schwob, J. E., 320, 328
Scott, J. W., 137, 168
Scott, R. H., 113, 117
Segev, I., 118, 426, 429, 430, 465
Sejnowski, T. J., 73, 82, 333, 435, 465
Selemon, L. D., 285
Seress, L., 353, 363
Shapley, R., 197, 213, 252, 277
Shapovalov, A. I., 110, 116
Sharp, F. R., 149
Sharrard, W. J. W., 92

Shatz, C. J., 188
Shepherd, G. M., 3, 7, 16, 34, 38, 135, 139, 140, 141, 143, 145, 146, 148, 149, 154, 155, 157, 158, 163, 165, 167, 168, 327, 328, 340, 390, 429, 430, 437, 463, 465
Sherman, S. M., 175, 246, 248, 249, 252, 253, 254, 256, 259, 260, 261, 263, 264, 266, 270, 273, 274, 275, 276, 277
Sherrington, C. S., 124, 431
Shinoda, Y., 114
Shinozaki, H., 238
Shipley, M. T., 321
Shipp, S., 397
Shiraev, B. I., 110, 116
Siggins, G. R., 238, 239
Sillito, A. M., 27, 30, 252, 274, 416, 422
Sim, J. A., 332
Simon, H., 239
Sims, S. M., 60
Singer, W., 246, 277, 403, 411
Skeen, L. C., 150, 320
Skrede, K. K., 364
Slaughter, M. M., 194
Slawsky, M., 408
Sloper, J. J., 393, 397, 398, 400, 430
Smith, J. L., 129
Smith, R. G., 186, 189, 190, 200, 208, 211
Smith, S. J., 410
Smith, T. C., 110
Soltész, I., 430
Somogyi, P., 237, 293, 324, 393, 399, 400, 415, 418, 423, 430
Sonnhof, U., 109, 115, 116
Sorensen, K. E., 350
Sotelo, C., 224, 238, 244
Spear, P. D., 248, 249, 252, 277
Spreafico, R., 254–255
Squire, L. R., 346, 353, 358, 363, 384
Srinivasan, M. V., 200, 203
Stafstrom, C. E., 41, 407, 409, 424
Stanford, L. R., 175
Stein, P. S. G., 103
Steinberg, R. H., 173
Steinbusch, H. W. M., 340
Stelzer, A., 379
Stent, G. S., 373
Stephens, J. A., 127, 128, 129
Steriade, M., 43, 267, 269, 270, 271, 273, 275
Sterling, P., 170, 171, 177, 184, 185, 186, 187, 188, 189, 190–191, 195, 198, 199, 200, 205, 206, 207, 208, 211, 212
Stevens, C. F., 45, 328, 330, 332, 333, 382
Stevens, J. K., 465
Steward, O., 359
Stofer, W. D., 74
Stone, J., 252, 403
Stone, T. W., 421, 423
Storm-Mathisen, J., 379–380

Stripling, J. S., 335
Struble, R. G., 141, 364
Stryker, M., 188
Stuart, D. B., 92
Stuermer, C., 189
Suda, K., 253
Sugimori, M., 228, 232, 237, 242, 308, 407, 429
Swanson, L. W., 349, 352, 362, 364
Swindale, N., 426
Switzer, R. C., 317–318
Sypert, G. W., 97
Szentágothai, J., 221, 226, 393

Tachibana, M., 195
Takagi, H., 293, 298, 321
Takeuchi, Y., 239
Tanabe, Y., 344
Tank, D. S., 242
Tank, D. W., 345
Tauchi, M., 195, 196
Taylor, C. P., 49, 359
Taylor, D. A., 423
ten Bruggencate, G., 238
Thalman, R. H., 368, 378, 421, 422
Thier, P., 196
Thomson, A. M., 399, 415, 420
Torre, V., 467
Toyama, K., 399, 415
Tracey, D. J., 98
Traub, R. D., 118, 347, 354, 363, 366, 381, 383
Tsairis, P., 94
Tseng, G. -F., 320, 322, 324, 328, 330, 331, 332, 333, 334, 337, 338, 339, 340
Tsien, R. W., 38, 41, 43, 62
Ts'o, D. Y., 399, 405
Tsukamoto, Y., 202, 203
Tsumoto, T., 253, 416, 421
Turey, M., 237
Turner, D. A., 381

Ueki, A., 315
Ulfhake, B., 92
Usowicz, M. M., 381

Valcana, T., 237
Valentino, K. L., 339
Valverde, F., 320, 323, 324, 328
Vandermaelen, C. P., 41
Vanderwolf, C. H., 378
Van Essen, D., 397
Van Hoesen, G. W., 285
Van Keulen, L., 98
Vardi, N., 193, 206
Vertes, R. P., 378
Vidyasagar, T. R., 416
Vincent, S. R., 422, 423
Vinogradova, O. S., 378
Vogel, M., 188
Voight, T., 196

Volkind, R. A., 243
Voogd, J., 221

Walberg, F., 238
Walker, J. E., 336–337
Walls, G. L., 172
Walmsley, B., 91, 98, 110, 112, 120, 129, 130, 415
Walters, C. P., 141
Wässle, H., 175, 176, 182, 183, 184, 195, 196
Watanabe, K., 337, 340
Watkins, J. C., 54, 380
Weight, F. F., 69, 106, 107, 382
Weitsen, H. A., 69
Wellis, D. P., 168
Werblin, F. S., 192, 198
West, R. W., 180
Westbrook, G. L., 273, 337, 369, 381
Westenbroek, R. E., 338–339
Westgaard, R. H., 364
Westrum, L. E., 325
White, E. L., 143, 393, 398, 415, 417
Whittaker, V. P., 239
Whitteridge, D., 396, 418, 437
Wiesel, T. N., 29, 189, 401, 403, 405
Wigstrom, H., 373, 410, 411
Williams, D. R., 184
Williams, J. T., 41
Williams, R. W., 249

Wilson, C. J., 286, 290, 293, 302, 303, 307, 309, 310, 312, 314, 345
Wilson, J. R., 256, 260
Wilson, M. A., 24, 345
Winson, J., 362, 378
Wong, R. K. S., 43, 317, 347, 363, 379, 383, 408
Wong-Riley, M. T. T., 403
Woolf, N. J., 422
Woolf, T. B., 165
Wouterlood, F. G., 327
Wu, P. H., 239
Wu, S. M., 189, 208
Wurtz, R. H., 315, 316

Yamada, W., 75, 76
Yamamoto, C., 364
Yang, X. L., 189, 208
Yarom, Y., 43, 62, 243
Yeterian, E. H., 285
Yoshinda, M., 296
Yoshioka, K., 113
Young, A. B., 115, 237

Zaborszky, L., 330
Zador, A. M., 382
Zbicz, K. L., 382
Zeki, S., 397
Zengel, J. E., 124, 126
Zucker, C. L., 180

SUBJECT INDEX

Acetylcholine (ACh), 52
  in the hippocampus, 379
  in the neocortex, 422–423
  in the olfactory cortex, 340
Action potential threshold, 37–38
Adenosine, 52
Adenosine triphosphate (ATP), 38
Afferents
  cortical, in the basal ganglia, 282–286
  direct action of, on motoneurons, 96–98
  monoaminergic, in the cerebellum, 238–239
  in the neocortex, 395–397
  in the olfactory bulb, 135–139
  retinal, 249–252
  in the spinal cord, 88–89, 90–92
  in the thalamus, 248–254
Afterhyperpolarization (AHP), 271–273
Amacrine cells in the retina, 18–19, 170, 179–180
  as source of neurotransmitters, 195–196

Amino acids
  excitatory
    in the hippocampus, 380–381
    in the neocortex, 420–421
    in the olfactory cortex, 336–338
  molecular mechanisms of amino acid synapses, 51
Amygdala, 318
Anterior olfactory nucleus (AON), 137
Apical dendrites
  in the neocortex, 425–426
  in the olfactory cortex, 322–323
Ascending tracts in the spinal cord, synaptic organization of, 106–108
Associative learning, 18
Autoassociation in the hippocampus, 384–387
Autonomic ganglia, 70–71
  functions of, 87
Autonomic nervous system, three divisions of, 67–68

Axoaxonic (chandelier) cells in the neocortex, 399–400
Axons
  collateral synaptic connections onto neostri-
    ated neurons, 293
  comparison between dendrites and, in the
    retina, 192
  efferent, in the basal ganglia, 289–290
  of Purkinje cells, 221

Basal ganglia, 279–316
  basic circuit for, 294–296
  cell population in, 290–291
  dendritic properties of, 307–312
  functional operations in, 312–316
    complex integrative tasks, 315–316
    natural firing pattern, 312–315
    mosaic organization of the neostriatum,
      299–302
  neuronal elements in, 281–290
    efferent axons, 289–290
    inputs, 281–286
    interneurons, 288–289
    principal neurons, 286–288
  neurotransmitters in, 296–299
    input fibers, 298–299
    interneurons, 298
    spiny neurons, 296–298
  synaptic actions in the neostriatum, 302–
    307
    actions of the cortical and thalamic in-
      puts, 302–305
    actions of the interneurons, 307
    actions of substantia nigra inputs, 305–
      307
  synaptic connections in the neostriatum,
    291–294
    axon collateral connections, 293
    cortical and thalamic connections, 291
    interneuron connections, 293–294
    output connections, 294
    substantia nigra connections, 291–
      293
  three parts of, 280–281
Basket cells
  in the cerebellum, 220–221
  in the hippocampus, 358
  in the neocrotex, 400
Biogenic amines as neurotransmitters in the
  neocortex, 423
Bipolar cells in the retina, 194–195
Brain, basic circuit of a region of, 28–29
Brainstem afferents in the thalamus, 253
Brainstem inputs
  into the hippocampus, 350
  into thalamic relay cells, 275–276

Cable equations, 440
  deriving, 440–442
  solving, 442–443
Calcium (Ca) ionic currents, 39, 41–43

synaptic activity in the neocortex and,
  408–410
Caudate nucleus of the basal ganglia, 280,
  281
Centrifugal fibers
  in the olfactory bulb, 161–162
  in the retina, 180–182
Cerebellar nuclear (CN) neurons, 230
  intrinsic properties of, 231
Cerebellum, 214–245
  basic circuit organization of, 226–228
    cerebellar cortex-deep nuclei circuit,
      226–228
    climbing fiber circuit, 226
    mossy fiber circuit, 226
  dendritic properties of, 239–242
    microelectrode recordings, 239–241
    optical recording, 242
  functional circuits of, 242–245
  intrinsic membrane properties in, 228–230
    cerebellar nuclear cells, 230
    Purkinje cell, 228–230
  neuronal elements in, 216–221
    cerebellar nuclei, 221
    input elements, 216–218
    intrinsic elements, 220–221
    output elements, 218–219
  neurotransmitters in, 237–239
    GABA, 238
    glutamate, 237–238
    monoaminergic afferents, 238–239
    mossy fibers, 239
  synaptic actions in, 230–237
    climbing fiber action on Purkinje cells,
      230–234
    inhibitory synapses in the cortex, 234–
      236
    parallel fiber action on Purkinje cells,
      234
    Purkinje cell action on cerebellar nuclear
      cells, 236–237
  synaptic connections in, 221–226
    cerebellar nuclei, 225–226
    granular cell layer, 222
  molecular layer, 222–225
  Purkinje cell layer, 222
Cerebral cortex
  basic circuit for the synaptic organization
    of, 22, 24
  See also Hippocampus; Neocortex; Olfac-
    tory cortex
Chemical synapses, 48, 49–51
  factors that control synaptic potential ampli-
    tude at, 131
Cholinergic fibers
  of the olfactory bulb, 161–162
  of the thalamus, 275–276
Climbing fibers, 216, 226
  Purkinje cell connections and, 222–224,
    230–234
Commissural fibers, 349–350

Communication, *see* Neuronal communication
Computational properties of the hippocampus, 383–388
Cone bipolar neurons, 177
Cones of the retina, 173
  visual transduction and, 191–192
Contralateral hippocampus, commissural fibers from, 349–350
Cortical afferent fibers
  of the basal ganglia, 282–286
  of the thalamus, 252–253
Cytoarchitecture of the piriform cortex, 319–320

Daylight circuit of the retina
  on-alpha ganglion cell for, 205–207
  on-beta ganglion cell for, 199–205
Dendrites, 3, 4
Dendritic electrotonus, 117, 439–473
  steady-state, 440–456
    deriving the cable equation, 440–442
    effect of dendritic branching and termination on electrotonic spread, 449–453
    effect of dendritic diameter on electrotonic spread, 445–446
    effect of membrane resistance on electrotonic spread, 446–448
    significance of the characteristic length, 443–445
    significance of the input resistance of a neuron, 453–456
    solving the equation, 442–443
  transient, 456–473
    spread of transient potentials, 456–459
    synaptic integration, 465–473
    synaptic potentials in dendritic spines, 462–465
    synaptic potentials in dendritic trees, 459–462
Dendritic integration, 13–18
Dendritic properties
  of the basal ganglia, 307–312
  of the cerebellum, 239–242
  of the hippocampus, 381–383
    active properties, 382–383
    passive properties, 381–382
  of the neocortex, 424–430
    apical dendrites, 425–426
    excitatory input, 426–427
    inhibition of spines, 430
    nonsaturating spines, 428–430
    saturation, 427–428
    spine action potential, 428–429
    spines, 426
  of the olfactory bulb, 162–167
    granule cell, 165–167
    mitral cell, 162–165
    periglomerular cell, 167
  of the olfactory cortex, 340–344
    dendritic events evoked by odor, 342–343

    integration of synaptic inputs, 343–344
    sequence of synaptic events in pyramidal cells, 340–342
  of the retina, 192–193
    comparison between axons and dendrites, 192
    feedforward elements, 192–193
    lateral elements, 193
  of the spinal cord, 117–124
    complications, 118–120
    dendritic cable properties, 117–118
    effect of dendritic location on synaptic potentials, 120–124
  of the thalamus, 261–263
    interneurons, 261–263
    relay cells, 261
Dendritic spines
  fine structure of, 17
  synaptic potentials in, 462–465
Dendritic trees
  subunit organization within, 18–19
  synaptic potentials in, 459–462
Dendrodendritic inhibition by granule cells, 154–158
Dentate gyrus, 347–350
  principal neurons of, 352
Directional selectivity, 26, 27–28
Disynaptic reflex pathways in the spinal cord, 98–101
Dopamine, 52
Dorsal root ganglion cell (DRG), 4
Dorsal spinocerebellar tract (DSCT), 106–108
Dorsal thalamus, 246, 247
Double-bouquet cells in the neocortex, 400

Efferent axons in the basal ganglia, 289–290
Electrogenic sodium-potassium pump, 38
Electrotonic potentials, 439–440
Electrotonic properties of the spinal cord, 117–124
Electrotonus, *see* Dendritic electrotonus
Enkephalins, 52
Enteric nervous system, 67
Entorhinal cortex, 318, 349
Ephatic interactions, 48, 49, 50
Epithalamus, 246, 247
Equilibrium potential (E), 33, 34
Eulaminate (homotypical) cortex, 392
Excitatory amino acids
  in the hippocampus, 380–381
  in the neocortex, 420–421
  in the olfactory cortex, 336–338
Excitatory systems in the olfactory cortex, 327–329
External plexiform layer (EPL) of the olfactory bulb, 145–146

Fast postsynaptic potentials (FPSPs), 53–57
  excitatory, 53–56, 74–76
  inhibitory, 56–57
Feedback excitation, 23–24

Feedback (recurrent) inhibition, 12–13
Feedback interneurons, 330
Feedforward excitation, 22–23
Feedforward inhibition, 11–12

Gamma-aminobutyric acid (GABA), 19, 52
  in the cerebellum, 238
  in the hippocampus, 378–379
  -mediated IPSPs, 56–57
  in the neocortex, 421–422
  in the olfactory cortex, 338–339
GABAergic inputs to thalamic relay cells,
    274–275
Ganglion cells, 69–70
  in the retina, 170, 175–176, 196–199
    on-alpha (parallel circuit for daylight),
      205–207
    on-beta (circuit for daylight), 199–205
    on-beta (circuit for starlight), 208–212
    on-beta (circuit for twilight), 207–208
Gap junction, 48–49
Gating functions of the thalamus, 270–273
  relay versus burst response modes, 270
  RNT control of response modes, 270–271
  roles of other intrinsic conductances, 271–
    273
Giant aspiny interneuron, 288–289
Glial membranes in the olfactory bulb, 147
Glomerular layer in the olfactory bulb, 143–
    145
  synaptic actions of, 151–154
Glutamate, 52
  in the cerebellum, 237–238
  synaptic potentials mediated by release of,
    55
Glutamatergic inputs to the thalamic relay
    cells, 273–274
Glycine-mediated IPSPs, 57
Golgi cells, 220
Granular layer
  of the cerebellar cortex, 234
  of the olfactory bulb, 146–147
Granule cells (GC), 140, 141, 220
  dendritic properties of, 165–167
  dendrodendritic inhibition by, 154–158
  in the hippocampus, 358
  in the olfactory bulb, 160–161

Hebbian synapse, 374
Hebb's postulate for learning, 373–374
Hilar cells in the hippocampus, 358
Hippocampus, 346–388
  basic circuit for, 359–362
    closing the loop, 361
    extrinsic neuromodulation, 364
    local interactions, 362–364
    longitudinal circuitry, 364
    staggered projections, 361–362
    trisynaptic circuit, 359–361
  computational properties of, 383–388
  dendritic properties of, 381–383

active properties, 382–383
  passive properties, 381–382
  neuronal elements in, 347–353
    inputs, 349–350
    intrinsic neurons, 352–353
    principal neurons, 350–352
    quantitative estimates of cell population,
      353
  neurotransmitters in, 378–381
    acetylcholine, 379
    excitatory amino acids, 380–381
    GABA, 378–379
    norepinephrine, 379–380
  synaptic actions in, 364–378
    endogenously generated responses, 365–
      367
    experimentally elicited responses, 367–
      371
    long-term potentiation, 371–378
    theta rhythm, 378
  synaptic connections in, 353–359
    electrical interactions, 359
    mossy-fiber synapse, 354–358
    Schaffer collateral synapse, 353–354
    synapses from basket cells, 358
    synapses from hilar cells, 358
    synapses onto granule cells, 358
Histamine, 52
Horizontal cells in the retina, 170, 177–179,
    195

Inhibitory operations of synaptic circuits, 11–
    13, 24–28
  in the cerebellar cortex, 234–236
  directional selectivity of, 26, 27–28
  in the olfactory cortex, 329–330
  receptive field organization of, 26, 27
  rhythmic generation, 25, 26
Inhibitory postsynaptic potentials (IPSPs)
  GABA mediated, 56–57
  glycine mediated, 57
Inner plexiform layer (IPL) of the retina,
    186–188
Input fibers, 3, 4
  in the basal ganglia, 281–286, 298–299
  in the cerebellum, 216–218
Insular cortex, 318
Interglomerular microcircuits, 149
Interneurons
  in the basal ganglia, 288–289, 298
  feedback, 330
  in the spinal cord, 89, 94–95
  synaptic connections onto neostriated
    neurons, 293–294
  in the thalamus, 254–256, 261–263
Interplexiform cells in the retina, 170, 180
Intraglomerular microcircuits, 147–149
Intrinsic circuits, 20–28
Intrinsic conductances of thalamic neurons,
    267–270, 271–273
Intrinsic currents, 62–66

Intrinsic neurons, 3, 4
  in the hippocampus, 352–353
  in the olfactory bulb, 140–141
  in the olfactory cortex, 323–324
  in the peripheral ganglia, 70–71
  in the retina, 176–182
Ionic actions of neurotransmitters, 53–62
  fast postsynaptic potentials, 53–57
  slow postsynaptic potentials, 57–62
Ionic currents in neuronal membranes, 32–
    35, 38–46
  calcium (Ca) currents, 39, 41–43
  potassium (K) currents, 39, 44–46
  sodium (Na) currents, 39, 40–41
Ionic pumps, 38

Lateral geniculate nucleus (LGN), 247
  as the prototypical thalamic nucleus, 248
Learning, associative, 18
Lemniscal afferents, 249–252
Local synaptic circuits, 20–28
  basic circuit for cortical organization, 24
  excitatory operations, 20–24
  inhibitory operations, 24–28
Locus coeruleus (LC), 136, 137
Long-term potentiation (LTP), 347
  in the hippocampus, 371–378
  in the neocortex, 410–411
Lumbar sympathetic ganglia, 68

Medial geniculate nucleus (MGN), 247
Membranes, *see* Neuronal membranes
Microcircuits, 8–13
  inhibitory operations, 11–13
  interglomerular, 149
  intraglomerular, 147–149
  synaptic convergence, 9, 10–11
  synaptic divergence, 8–10
Microelectrode records of cerebellum's den-
    dritic properties, 239–241
Mitral cells, 19, 137–139
  in the olfactory bulb, 162–165
Molecular layer of the cerebellar cortex, in-
    hibitory synapses in, 234–236
Monoaminergic afferents, 238–239
Monoamines in the olfactory cortex, 340
Mossy fibers, 216–218
  in the cerebellum, 226, 239
  in the hippocampus, 354–358
    long-term potentiation in, 371–373
Motoneurons (motor neurons) in the spinal
    cord, 4, 89, 92–94
  direction action of afferents on, 96–98
  intrinsic properties of, 126–127
  mechanisms of synaptic actions in, 110–
    117
    excitatory synaptic mechanisms, 110–
    112
    inhibitory synaptic mechanisms, 115–
    116
    other excitatory systems, 114–115

presynaptic inhibition, 116–117
    quantization of synaptic action, 112–113
    transmitter release at Ia synapses, 113–
    114
  recruitment of, 124–131
Motor coordination, 215
Motor cortex, 392
Multisynaptic reflex arcs in the spinal cord,
    98

Natural firing patterns of neostriated neurons,
    312–315
Neocortex, 389–438
  basic circuit for, 401–405
    columnar organization, 401–403
    cortical output, 404–405
    serial and parallel circuits, 403
  dendritic properties of, 424–430
    apical dendrites, 425–426
    excitatory input, 426–427
    inhibition on spines, 430
    nonsaturating spines, 429–430
    saturation, 427–428
    spine action potentials, 428–429
    spines, 426
  functional properties of, 430–438
    neuronal basis for visual perception,
      431–433
    new designs for the visual circuits, 433–
      434
    parallel processing in neural networks,
      434–436
  neuronal elements in, 392–397
    afferents, 395–397
    smooth neurons, 393–395
    spiny neurons, 393
  neurotransmitters in, 420–424
    acetylcholine, 422–423
    biogenic amines, 423
    excitatory amino acids, 420–421
    GABA, 421–422
    neuropeptides, 423–424
  synaptic actions in, 405–420
    excitatory synapses, 414–415
    inhibitory synapses, 415–420
    intrinsic membrane properties, 405–414
  synaptic connections in, 397–400
    smooth neurons, 399–400
    spiny cells, 397–399
Neostriatum, 280, 281
  mosaic organization of, 299–302
  See also Basal ganglia
Neuromodulation, neurotransmission versus,
    52–53
Neuronal communication, 48–51
  chemical synapses, 48, 49–51
  ephaptic interactions, 48, 49, 50
  gap junctions, 48–49
Neuronal elements
  in the basal ganglia, 281–290
    efferent axons, 289–290

inputs, 281–286
  interneurons, 288–289
  principal neurons, 286–288
in the cerebellum, 216–221
  cerebellar nuclei, 221
  input elements, 216–218
  intrinsic elements, 220–221
  output elements, 218–219
in the hippocampus, 347–353
  inputs, 349–350
  intrinsic neurons, 352–353
  principal neurons, 350–352
  quantitative estimates of cell population, 353
in the neocortex, 392–397
  afferents, 395–397
  smooth neurons, 393–395
  spiny neurons, 393
in the olfactory bulb, 133–143
  cell populations, 142–143
  inputs, 135–137
  intrinsic neurons, 140–141
  principal neurons, 137–140
in the olfactory cortex, 319–325
  cell population, 324–325
  central inputs, 320
  connection between olfactory cortical areas, 321
  cytoarchitecture, 319–320
  inputs, 320
  intrinsic neurons, 323–324
  outputs, 321
  principal neurons, 321–323
in the peripheral ganglia, 69–71
  inputs, 69
  intrinsic neurons, 70–71
  principal neurons, 69–70
in the retina, 173–184
  cell populations, 182–183
  circuits, 183–184
  ganglion cells (output elements), 175–176
  intrinsic elements, 176–182
  receptors (input elements), 173–175
in the spinal cord, 88–95
  interneurons and tract cells, 89, 94–95
  motoneurons, 92–94
  primary afferents, 90–92
in the thalamus, 248–256
  inputs, 248–254
  interneurons, 254–256
  relay neurons, 254
Neuronal membranes
  action potential of, 35–38
  ionic currents present in, 32–35, 38–46
    calcium (Ca) currents, 39, 41–43
    potassium (K) currents, 39, 44–46
    sodium (Na) currents, 39, 40–41
  properties of, 46–48
  resting potential of, 35–36
Neuronal microcircuits, *see* Microcircuits

Neurons
  as an integrative unit, 19–20, 21
  neuronal operations and their underlying biophysical mechanisms, 22
  triad of neuronal elements, 3, 4
  See also types of neurons
Neuropeptides in the neocortex, 423–424
Neuropsychological studies of the hippocampus, 384
Neurotransmission versus neuromodulation, 52–53
Neurotransmitters
  in the basal ganglia, 296–299
    input fiber, 298–299
    interneurons, 298
    spiny neurons, 296–298
  in the cerebellum, 237–239
    GABA, 238
    glutamate, 237–238
    monoaminergic afferents, 238–239
    mossy fibers, 239
  common responses in the CNS, 52
  in the hippocampus, 378–381
    acetylcholine, 379
    excitatory amino acids, 380–381
    GABA, 378–379
    norepinephrine, 379–380
  ionic actions of, 52–62
    fast postsynaptic potentials, 53–57
    slow postsynaptic potentials, 57–62
  in the neocortex, 420–424
    acetylcholine, 422–423
    biogenic amines, 423
    excitatory amino acids, 420–421
    GABA, 421–422
    neuropeptides, 423–424
  in the olfactory bulb, 158–162
    centrifugal fibers, 161–162
    granule cells, 160–161
    mitral cells, 160
    olfactory sensory neurons, 158
    peptides, 161
    periglomerular cells, 158–160
    tufted cells, 160
  in the olfactory cortex, 336–340
    acetylcholine, 340
    excitatory amino acids, 336–339
    GABA, 338–339
    monoamines, 340
    peptides, 339
  in the peripheral ganglia, 73–74
  in the retina, 193–196
    amacrine cells, 195–196
    bipolar cells, 194–195
    horizontal cells, 195
    peptides, 196
    photoreceptors, 194
  in the thalamus, 273–276
    brainstem inputs, 275–276
    GABAergic inputs, 274–275
    glutamatergic inputs, 273–274

Nicotinic cholinergic responses, 54–56
Noradrenaline-containing fibers, 161
Noradrenergic inputs to the thalamus, 276
Norepinephrine (NE), 52
  effect on neuronal activity pattern generated
    by thalamic relay neurons, 62, 63
  in the hippocampus, 379–380
Novelty filtering, 387
Nucleus acumbens in the basal ganglia, 280,
  281

Odor stimulation
  processing of, in the olfactory bulb, 167–
    169
  responses to, in the olfactory cortex, 342–
    343, 344–345
Olfactory bulb, 133–169
  basic circuit for, 28, 147–151
    centrifugal modulation, 150–151
    input processing, 147–149
    output control, 149–150
  dendritic properties of, 162–167
    granule cell, 165–167
    mitral cell, 162–163
    periglomerular cell, 167
  functional circuits in, 167–169
  neuronal elements in, 133–143
    cell populations, 142–143
    inputs, 135–137
    intrinsic neurons, 140–141
    principal neurons, 137–140
  neurotransmitters in, 158–162
    centrifugal fibers, 161–162
    granule cells, 160–161
    mitral cells, 160
    olfactory sensory neurons, 158
    peptides, 161
    periglomerular cells, 158–160
    tufted cells, 160
  synaptic actions in, 151–158
    dendrodendritic inhibition by granule
      cells, 154–158
    glomerular synaptic actions, 151–154
  synaptic connections in, 143–147
    external plexiform layer (EPL), 145–146
    glia, 147
    glomerular layer, 143–145
    granular layer, 146–147
Olfactory cortex, 37, 134, 317–345
  areas and connections of, 317–318
  basic circuit for, 327–330
    excitatory systems, 327–329
    inhibitory systems, 329–330
  dendritic properties of, 340–344
    dendritic events evoked by odor. 342–
      343
    integration of synaptic inputs, 343–344
    sequence of synaptic events in pyramidal
      cells, 340–342
  functional operations of, 344–345
    clues from neuronal circuitry, 345

responses to odor stimulation, 344–345
  neuronal elements in, 319–325
    cell population, 324–325
    central inputs, 320
    connection between olfactory cortical
      areas, 321
    cytoarchitecture, 319–320
    inputs, 320
    intrinsic neurons, 323–324
    outputs, 321
    principal neurons, 321–323
  neurotransmitters in, 336–340
    acetylcholine, 340
    excitatory amino acids, 336–338
    GABA, 338–339
    monoamines, 340
    peptides, 339
  synaptic actions in, 330–335
    effects of voltage sensitive channels,
      331–332
    postsynaptic potentials, 332–334
    sequential activation patterns, 335
    synaptic plasticity, 335
  synaptic connections in, 325–326
Olfactory epithelium, 134
Olfactory peduncle, 318
Olfactory tubercle, 318
On-alpha ganglion cell in the retina (parallel
  circuit for daylight), 205–207
On-beta ganglion cell in the retina
  circuit for daylight, 199–205
  circuit for starlight, 208–212
  circuit for twilight, 207–208
Optical recording of cerebellum's dendritic
  properties, 242
Orthogonalizing function of the hippocampus,
  387–388
Outer plexiform layer (OPL) of the retina,
  184–186

Parasympathetic nervous system, 67
Peptides
  neuropeptides in the neocortex, 423–424
  in the olfactory bulb, 161
  in the olfactory cortex, 339
  in the retina, 196
Periglomerular cells (PG), 140
  dendritic properties of, 167
  in the olfactory bulb, 158–160
Peripheral ganglia, 67–87
  basic circuit for, 71–73
  functions of autonomic ganglia, 87
  mammalian superior cervical ganglia, 83–
    86
  neuronal elements in, 69–71
    inputs, 69
    intrinsic neurons, 70–71
    principal neurons, 69–70
  neurotransmitters in, 73–74
  synaptic and membrane properties of, 74–
    83

fast excitatory postsynaptic potential, 74–76
slow excitatory postsynaptic potential, 80–83
voltage-gated conductances, 76–80
Photoreceptors in the retina, 194
Piriform cortex of the olfactory cortex, 318
basic circuit of, 327–330
cytoarchitecture of, 319–320
dendritic properties of, 340–344
functional operations of, 344–345
neuronal elements in, 319–325
neurotransmitters in, 336–340
olfactory discrimination in, 345
synaptic actions of, 330–335
synaptic connections in, 325–326
Plexi, 67
Postganglionic fibers, 70
Postsynaptic inhibition, microcircuits that mediate different types of, 12
Postsynaptic potentials (PSPs), 53–62
fast excitatory, 53–56, 74–76
fast inhibitory, 56–57
slow, 57–62
decrease in potassium currents, 60–62
increase in potassium conductance, 57–60
possible gating action of neurotransmitters, 62
Potassium (K) ionic currents, 39, 44–46
synaptic activity in the neocortex and, 407–408
Preganglionic fibers (preganglionics), 69
Primary afferents in the spinal cord, 88–89, 90–92
Primary visual cortex, 392
Principal neurons, 3, 4
in the basal ganglia, 286–288
in the hippocampus, 350–352
in the olfactory bulb, 137–140
in the olfactory cortex, 321–323
in the peripheral ganglia, 69–70
Projection neurons, 3
Purkinje cells, 218
action on cerebellar nuclear cells, 236–237
axons of, 221
climbing fiber-Purkinje cell connections, 222–224, 230–234
dendritic properties of, 239–242
intrinsic membrane properties of, 228–230
parallel fiber action on, 234
parallel fiber-Purkinje cell connection, 224
plasticity of connectivity of, 224–225
synaptic connections in Purkinje cell layer, 222
Putamen of the basal ganglia, 280, 281
Pyramidal cells
intrinsic and synaptic currents in, 62–66
in the olfactory cortex, 321–322, 332–333
sequence of synaptic events in, 340–342

Raphe nucleus (Ra), 136, 137
Receptive field organization (spatial contrast), 26, 27
Receptors in the retina, 173–175
Recurrent collaterals, 139
Relay neurons, 3
in the thalamus, 254, 256–260
Renshaw cells (spinal neurons), 98–101
Retina, 170–213
basic circuit for, 28, 189–191
dendritic properties of, 192–193
comparison between axons and dendrites, 192
feedforward elements, 192–193
lateral elements, 193
development and plasticity of, 188–189
functional circuits in, 196–212
ganglion cell function, 196–199
on-alpha ganglion cell, 205–207
on-beta ganglion cell, 199–205, 207–212
neuronal elements in, 173–184
cell populations, 182–183
circuits, 183–184
ganglion cells (output elements), 175–176
intrinsic elements, 176–182
receptors (input elements), 173–175
neurotransmitters and synaptic actions in, 193–196
amacrine cells, 195–196
bipolar cells, 194–195
horizontal cells, 195
peptides, 196
photoreceptors, 194
species difference in retinal structure and visual function, 172
synaptic connections in, 184–188
inner plexiform layer, 186–188
outer plexiform layer, 184–186
visual transduction in, 191–192
Retinal (lemniscal) afferents, 249–252
Rhythmic generation, 25, 26
Rod bipolar neurons, 177
Rods in the retina, 173
visual transduction and, 191–192

Schaffer collateral synapses, 353–354
long-term potentiation in, 373–378
Sensory neurons in the olfactory bulb, 158
Serotonergic inputs to the thalamus, 276
Serotonin, 52
-containing fibers, 161
Slow postsynaptic potentials (SPSPs), 57–62
decrease in potassium currents, 60–62
excitatory, 80–83
increase in potassium conductance, 57–60
possible gating actions of neurotransmitters, 62
Smooth neurons in the neocortex, 393–395, 399–401